RECENT DEVELOPMENTS OF ELECTRICAL DRIVES

Recent Developments of Electrical Drives

Best papers from the International Conference on Electrical Machines ICEM'04

Edited by

S. Wiak, M. Dems, and K. Komęza

Technical University of Lodz, Poland

Springer

A C.I.P. Catalogue record for this book is available from the Library of Congress.

ISBN-13 978-90-481-7144-6
ISBN-10 1-4020-4535-2 (e-book)
ISBN-13 978-1-4020-4535-6 (e-book)

Published by Springer,
P.O. Box 17, 3300 AA Dordrecht, The Netherlands.

www.springer.com

Printed on acid-free paper

CONTENTS

II. Control, Measurements, and Monitoring

3.2. Wind Generators

3.3. Use of Advanced Materials and New Technologies

PREFACE

Selected papers for SPRINGER MONOGRAPH, after final reviewing process were presented at the XVI International Conference on Electrical Machines ICEM'2004 which was held in Cracow, Poland, on September 5–8, 2004.

The International Conference on Electrical Machines (ICEM) is the only major international conference devoted entirely to electrical machines. Started in London in 1974, ICEM is now established as a regular biennial event. Following the very successful conferences in Istanbul in 1998, Helsinki in 2000, and Belgium in 2002 ICEM'2004 was jointly organized by Institute of Mechatronics and Information Systems, Technical University of Łódź, Poland (main organizer) in cooperation with a few Polish Universities and Polish Society of Applied Electromagnetics.

The Conference venue was Cracow. Historically, Cracow is Poland's most distinguished city, on a par with the most famous places in Europe, comprising rich and varied cultural heritage. The ancient royal capital of Poland for centuries has been constituting crossroads where influences of many traditions meet, namely traditions of the Italian, French, German, Austrian, and Jewish cultures.

Cracow's uniqueness made it one of the first places to be entered on the UNESCO World Cultural Heritage List in 1978. Alongside hundreds of wonderful monuments of architecture, you will see here exquisite collections of Polish, Western European, Jewish, Persian, and Japanese art. Cracow is the city of Copernicus and Pope John Paul II, of Penderecki and Wajda. It is a city of churches and museums, theatres and festivals, scholarship and music, a city brimming with life, and as ever the focus of Poland's cultural and academic spirit.

The aim of the Conference is to discuss recent developments of modeling and simulation methodologies, control systems, testing, measurements, monitoring and diagnostics, advanced software methodology, etc., applied in electrical drives. The meeting is intended to be a forum for applied mathematicians, computer and software engineers, and electrical engineers to exchange ideas, experience on the new developments, trends, and applications from industrial and academic viewpoints on the topic. An important goal of ICEM is also stimulating personal contacts and cooperation, especially between industrial and academic institutions.

Almost 930 papers have been submitted as digests, and after reviewing process, in which two referees have reviewed each paper, 525 papers have been accepted for the presentation at the Conference. Over 430 papers in full versions, after final reviewing, have been published on CD as Conference Proceeding. The papers published in the Conference Proceeding have been refereed by the sessions' chairmen for further publication as post conference issue. It is the tradition of the ICEM meetings that they comprise quite a vast area of computational and applied problems. Moreover, the ICEM conferences aim at joining theory and practice, thus the majority of papers are deeply rooted in engineering problems, being simultaneously of high theoretical level. A knowledge of physical phenomena is necessary for understanding the operation of electromagnetic devices, electromechanical converters, and electronic systems in general, such as sensors, actuators, solid state devices, integrated circuits, and Micro Electro Mechanical Systems (MEMS). In general the coexistence of electric, magnetic, thermal, and mechanical effects characterizes the global behavior of any electrical drives or systems. The main topics of ICEM'2004 meeting are listed below, selected to either oral or poster session:

Oral Session

- Industrial Applications
- Permanent Magnet Machines
- Special Machines
- Control Drives and Generators
- Controlled Drives for Special Applications
- Controlled Drives for Special Applications, Actuators
- Finite Element Methodology
- Modeling and Simulation
- Modeling and Simulation of Induction Motors
- Wind Generators
- Thermal, Acoustic Noise, and Vibration Aspects
- Testing, Measurements, Monitoring, and Diagnostics
- Transformers, Special Machines
- Use of Advanced Materials
- Linear Drives

Poster Session

- Permanent Magnet Machines
- Special Machines
- Special Machines, Actuators
- Finite Element Methodology
- Modeling and Simulation
- Wind Generators
- Thermal, Acoustic Noise, and Vibration Aspects
- Testing, Measurements, Monitoring, and Diagnostics
- Transformers, Use of Advanced Materials and New Technology
- Control Drives and Generators
- Linear Machines

It makes some order in reading but also it somehow represents the main directions, which are penetrated by researchers dealing with modern electrical drives. Looking at the content of the book of digests one may also notice that the more and more researchers go into the investigation of new applications of computer engineering, especially these connected with software methodology, CAD techniques, system control, and material sciences. The computational techniques, which have been under development during the last three decades and are being still developed serve as good tools for discovering new phenomena.

We, the editors of SPRINGER MONOGRAPH, would like to express our thanks to our colleagues who have contributed to this issue.

The book is composed of the papers, which were presented at the XVI International Conference on Electrical Machines—ICEM'2004, which was held in Cracow, Poland, on September 5–8, 2004. It consists of three chapters:

1. Theory and Advanced Computational Methods in Electrical Drives
2. Control, Measurements, and Monitoring
3. Electrical Drives Applications
   3.1 New Motor Structures
   3.2 Wind Generators
   3.3 Use of Advanced Materials and New Technologies

The papers accepted for this issue concerns the following leading problems:

- Computer modeling of wide range of electrical drives while the models are validated by experimental measurements. Moreover the optimization methodologies, based on stochastic, gradient, and genetic algorithms are used to increase alternator efficiency and power-to-weight ratio by changing building parameters. Then, multiobjectives optimization strategy is applied in order to find the best compromise between high efficiency and high power-to-weight ratio.
- Development of the software methodology, dedicated for 3D structure modeling by means of solid models of electromechanical converters.
- Generalized circuital modeling of electromechanical devices by means of the lumped-parameters in terms of equivalent circuits. The generalized treatment suppressing several of such hypotheses, leveraging on matrix notation and lagrangian notation for mechanical aspects to provide a powerful conceptual tool for the analysis of a wide class of devices.
- Novel sensorless operation of different type of electrical machines, theory, and applications. The methodology behind the controller operation is presented together with test results taken from industry and prototype systems.
- The numerical modeling of small size electromechanical converters with extremely high forces with review of different possible topologies rotary actuators in precision engineering applications with a low mechanical stiffness and ironless.
- Design of different type of electrical machines for a vehicle application, while the criterion linked with automotive applications (torque density, efficiency, flux weakening capability) for machine with distributed windings stator, and concentrated windings is set up. A new configuration of a transverse-flux permanent-magnet machine (TFPM) for a wheel-motor suitable as wheel motor are proposed.
- Design, optimization, and manufacturing of new structures machines made from classical and new materials like iron-powder, met-glass, and microcrystal materials.
- Complex study of different types of wind generators introducing stabilization aspects, variable speed controller, new control methodologies, dynamic response, design and construction, fuzzy logic, and digital controllers.

- Application of artificial intelligence (fuzzy logic and neural networks) in modeling and control of drive systems.
- Monitoring in real time (on line) while dedicated expert systems are implemented.

S. Wiak
M. Dems
K. Komęza
Institute of Mechatronics and Information Systems,
Technical University of Lodz,
Stefanowskiego 18/22
90-924 lodz, Poland
wiakslaw@p.lodz.pl
mdems@p.lodz.pl
komęza@p.lodz.pl

SECTION I
THEORY AND ADVANCED COMPUTATIONAL METHODS IN ELECTRICAL DRIVES

Introductory Remarks

The papers selected to first chapter are mainly focused on recently developed theory and computational techniques applied to modeling, simulation, and design of electrical drives.

The papers accepted for this chapter concerns the following leading problems:

- An analysis of the core losses under no-load conditions in turbine generators by utilizing a three-dimensional magnetic field calculation based on a finite element method. The analysis consists of two steps. First, we calculate the loss in laminated steel sheets from experimental data obtained with an Epstein frame. In the calculation the differences between the actual core loss and cataloged data, and as well the additional losses in metal parts other than the steel sheets are taken into account. Basing on the analysis results the total calculated core losses with measured values for two turbine generators are compared.
- The optimization of the loss calculation in a design program for salient pole synchronous machines, as very important issue of the design of hydro-generators reliability. Statistical methods are used to calibrate the loss calculation with measurements made during commissioning. Special importance is attached to the optimization of the no-load electromagnetic losses.
- A coupled model for accurate representation of the characteristics of permanent magnet synchronous motors, and proposed the determination of the direct axis reactance, "Xd", and the quadrature axis reactance, "Xq", by calculus and texts with the permanent magnet synchronous machine under generator duty. The starting and synchronization processes of the PMSM and the influence that on transient behavior of the motor produce different values of the main motor parameters is analyzed.
- A dynamic model of the Vernier Hybrid Machine (VHM), for use as a linear generator in a wave energy converter, has been presented and verified using near sinusoidal displacement data. The model forms the basic building block to investigate the performance and control of direct drive wave energy converters. A dynamic model capable of predicting the machine's behavior for this kind of mechanical excitation. Simple equivalent circuit models have been also found.
- A comparative study for permanent magnet synchronous motor (PMSM) with buried V-shape magnets and for a motor with unusual designed with U-shape magnets in every second pole is performed.
- The choice of the magnet blocks number over a pole must be considered as an optimization parameter acting on local phenomena such as the cogging torque and higher torque

S. Wiak, M. Dems, K. Komęza (eds.), Recent Developments of Electrical Drives, 1–2.

harmonics in PM motors. The technique that consists on the choice of an appropriate number of magnet blocks over a magnet pole cannot be done without considering the main parameters, which impose the principal machine performances such as the average torque. In addition to the reduction of the cogging torque and high torque harmonics, and reducing eddy currents inside the magnets as well.

- The numerical model for three-dimensional field analysis of electrostatic micromotors with stator and rotor symmetry axes located in different points. The proposed algorithm based on the mesh overlapping allows avoiding the mesh to be generated for the whole model and decrease the time of computation. Reduction of this time can be obtained by using separated submeshes for both stator and rotor generated only once, and only recalculating the part describing the air gap.
- A compound drive simulator is invented, where a finite element method (FEM) model of the electric motor is coupled with a frequency converter model and a closed-loop control system. The method is implemented for SIMULINK and applied on a 2 MW asynchronous machine drive. The results are validated by measurements and the performance is compared with an analytical motor model. It is proved that simulation with the FEM model provides very good results and gives much better insight in the motor behavior than the analytical model.
- An intuitive approach to the analysis of parasitic effects with particular emphasis on torque ripple. The approach is based on the notion of space phasor modeling. A good approximation is achieved in predicting the nature and the magnitude of torque ripple by the use of a relatively simple time-domain model.
- An analytical multiphysical model—electromagnetic, mechanic, and acoustic—in order to predict the electromagnetic noise of a permanent magnet synchronous machine (P.M.S.M.). The experimental design method, with a particular design: "trellis design", is used to build response surfaces of the noise with respect to the main factors. These surfaces can be used to find the optimal design or more simply, to avoid unacceptable designs of the machine, in term of noise for a variable speed application.
- Numerical methods of calculating the electromagnetic forces and of simulating the oscillation behavior of the stator end winding are introduced. The end winding oscillations of different turbo generators under forced vibrations are computed in a combined simulation. Also eigen-frequencies and eigen-modes are determined. Numerical simulation of oscillation behavior is found a useful tool in end winding design although model parameter identification still offers improvement potential.
- The design procedure of a drive consisting of a voltage supplied brushless motor and a lead screw transformation system. In order to reduce the cost and the weight of this drive an optimization of the main dimensions of each component considered as an interacting part of the whole system is conducted. An analysis is developed to define the interactions between the elements in order to justify the methodology. A specific application in then presented and comparisons are made between different solutions depending on different cost functions (max power, weight, cost . . .). With this procedure, the optimization is no longer limited to the fitting between separated elements but is extended to the system whose parameters are issued from the primitive design parameters of the components.

I-1. CORE LOSS IN TURBINE GENERATORS: ANALYSIS OF NO-LOAD CORE LOSS BY 3D MAGNETIC FIELD CALCULATION

A. Nakahara[1], K. Takahashi[1], K. Ide[1], J. Kaneda[1], K. Hattori[2], T. Watanabe[2], H. Mogi[3], C. Kaido[3], E. Minematsu[4], and K. Hanzawa[5]

[1]Hitachi Research Laboratory, Hitachi, Ltd., 7-1-1, Omikacho, Hitachi, Ibaraki 319-1292, Japan
anakaha@gm.hrl.hitachi.co.jp, takahiko@gm.hrl.hitachi.co.jp
[2]Hitachi Works, Power Systems, Hitachi Ltd., 3-1-1, Saiwaicho, Hitachi, Ibaraki 317-8511, Japan
kenichi_hattori@pis.hitachi.co.jp, isao@keyaki.cc.u-tokai.ac.jp
[3]Steel Research Laboratories, Nippon Steel Corp., 20-1, Shintomi, Futtsu, Chiba 293-8511, Japan
mogi@re.nsc.co.jp
[4]Flat Products Division, Nippon Steel Corp., 6-3, Otemachi, 2-chome, Chiyoda-ku, Tokyo 100-8071, Japan
[5]Yawata Works, Nippon Steel Corp., 1-1, Tobihatacho, Tobata-ku, Kitakyusyu, Fukuoka 804-8501, Japan

Abstract. Magnetic field analysis of no-load core loss in turbine generators is described. The losses in laminated steel sheets are calculated from the results of finite element magnetic field analysis. The additional losses in metal portions other than the steel sheets are also calculated. The sums of these losses were compared with the measured values for two generators and found to be 88% and 96% of the measured values. The results revealed that the additional losses made up a considerable part of the core losses.

Introduction

Turbine generators have been developed by using various design technologies to meet the needs of customers. Reliable estimation of losses is essential in designing highly efficient turbine generators [1–3].

Among various losses, core loss is one of the most difficult to estimate for two reasons:

1. The cataloged data of electrical steel sheets are measured for a rectangular shape in a uniform magnetic field. Electrical steel sheets in an actual machine, however, are processed into complex shapes, and the induced field is not uniform.
2. The measured core loss of a turbine generator seems to include additional losses. One of them is eddy current loss in the electrical steel sheets due to the axial magnetic flux. Others include losses in metal parts other than the steel sheets.

S. Wiak, M. Dems, K. Komęza (eds.), Recent Developments of Electrical Drives, 3–12.

This paper presents an analysis of the core losses under no-load conditions in turbine generators by utilizing a three-dimensional magnetic field calculation based on a finite element method. The analysis consists of two steps. First, we calculate the loss in laminated steel sheets from experimental data obtained with an Epstein frame. In this calculation, we take into account differences between the actual core loss and cataloged data. Second, we calculate the additional losses in metal parts other than the steel sheets. Based on the analysis results, we also compare the total calculated core losses with measured values for two turbine generators.

Calculation method

As noted above, the core losses are calculated in a two-step procedure. First, we calculate the loss in the laminated steel sheets by using the experimental data obtained with an Epstein frame. In this calculation, we take into account the rotational magnetic field and the harmonics.

Second, we calculate the additional losses. For metal parts other than the laminated steel sheets, we calculate the losses by three-dimensional finite element analysis. We also use the finite element method to calculate the losses due to the axial flux in the laminated steel sheets, because the data obtained with the Epstein frame do not include these losses.

Loss in laminated steel sheets

The loss due to the alternating field in the laminated steel sheets can be calculated from the experimental data with the following equation:

$$W_i = W_h + W_e = K_h B_{\max}^{\alpha} f + K_e B_{\max}^2 f^2 \qquad (1)$$

where W_i is the loss per weight of the sheets, W_h and W_e are the hysteresis and eddy current losses per weight, respectively, K_h and K_e are coefficients obtained with the Epstein frame, f is the frequency of the alternating magnetic field, and $B_{\max}$ is the maximum magnetic flux density occurring in one cycle.

Although the magnetic field in an Epstein frame is a static alternating field, the magnetic field in an actual generator is a rotational field with harmonics. Thus, the rotational and harmonic effects must be taken into account, and to calculate these effects, we apply two methods. We utilize the method proposed by Yamazaki [4] to calculate the hysteresis loss, and the Fourier series expansion method to calculate the eddy current loss.

In equation (1), it is assumed that W_h and W_e are proportional to f and f^2 respectively for any level of the magnetic flux density, B. The core loss, however, actually includes the excess loss due to the microstructure of a steel sheet [5–7]. In addition, the B-dependency of the hysteresis loss varies according to the level of B [8].

To consider the excess loss and the B-dependency of the hysteresis loss, various methods have been proposed. Though the eddy current loss is expressed by one term in equation (1), it is expressed by two terms in the methods proposed to consider the excess loss [4–6]. One term expresses the classical eddy current loss and is proportional to $B^2 f^2$. The other term expresses the excess loss and is assumed proportional to $B^{1.5} f^{1.5}$. On the other hand, a method proposed to express the B-dependency of the hysteresis loss changes the values of the exponent α and of K_h for different levels of B in equation (1) [8]. Different levels defined in this method are from 0 to 1.4 T, from 1.4 to 1.6 T, and from 1.6 to 2.0 T.

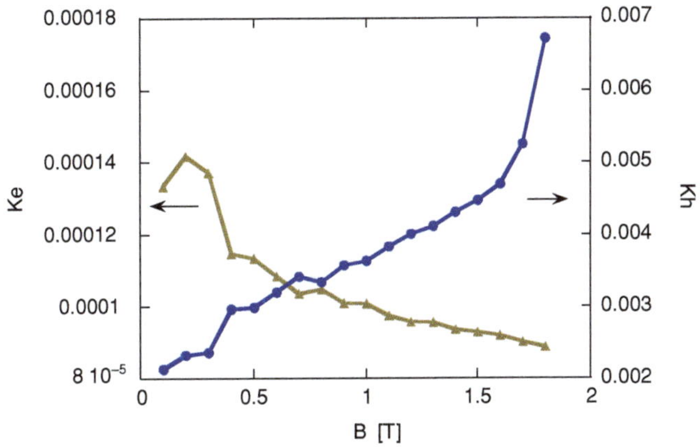

Figure 1. B-dependency of K_h and K_e.

These methods consider the B- or f-dependency of the core loss by changing the components of B or f. Nevertheless, it is difficult to completely express these complex dependencies. Additionally, the dependencies differ according to the kind of steel sheet.

Consequently, we propose a method to reflect the B- and f-dependencies of K_h and K_e. In equation (1), we assume that $\alpha = 1.6$, based on tests by Steinmetz [9]. Fig. 1 shows an example of the B-dependencies of K_h (circles) and K_e (triangles) obtained with an Epstein frame. In this case, the maximums of K_h and K_e are roughly twice and three times as large, respectively, as their minimums.

In Fig. 2, the dots represent the ratio, W_i/f, at different frequencies, where W_i is the loss in electrical steel sheets measured with an Epstein frame at 0.5 T for 50, 60, 100, 200, and 400 Hz. Dividing equation (1) by f gives the following equation:

$$W_i/f = K_h B_{\text{max}}^{1.6} + K_e B_{\text{max}}^2 f \qquad (2)$$

K_h and K_e can thus be derived from the slope and intercept of a line connecting two points, as shown in Fig. 2. For example, K_h (50–60 Hz) indicates the value of K_h derived from the

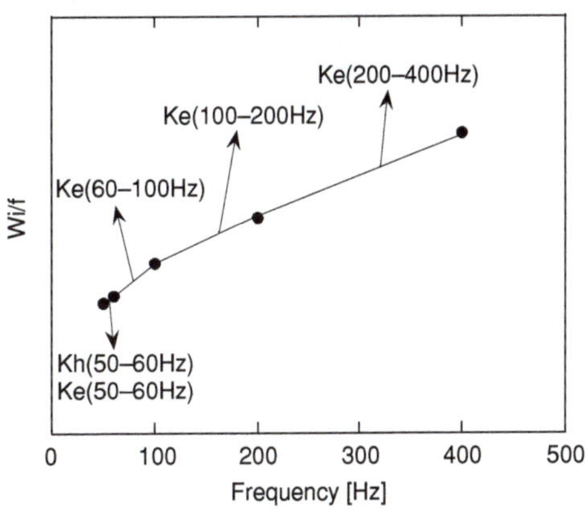

Figure 2. Derivation of K_h and K_e.

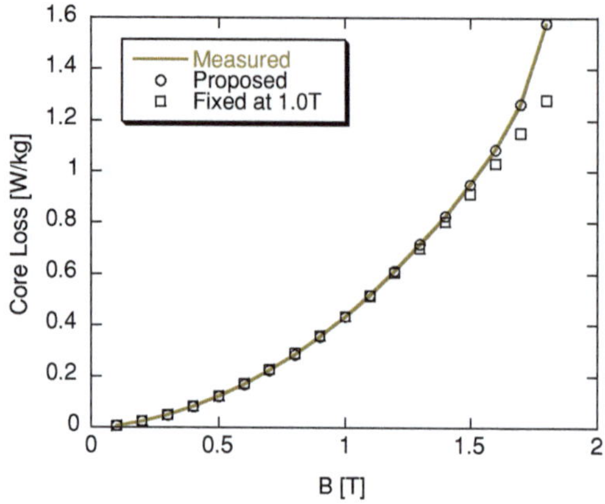

Figure 3. Core loss reproduced by proposed method.

points corresponding to 50 and 60 Hz, and it is applied over the range from 50 to 60 Hz in the calculation. By repeating this operation for each level of B, tables showing the values of K_h and K_e for various values of B and f can be constructed.

Fig. 3 shows the core loss data, with the line representing measured results. The circles represent values obtained by equation (1) in the proposed method, while the squares represent values obtained by equation (1) with K_h and K_e derived at 1.0 T and 50–60 Hz. As seen from the data, the approximation is not good enough. On the other hand, the measured values are accurately reproduced by the proposed method. Thus, the complex dependency can be expressed by generating sufficient quantities of data for B and f.

It is difficult to experimentally evaluate the genuine loss of the laminated steel sheets in an actual generator because the measured loss inevitably includes the additional losses in metal parts other than the steel sheets. For this reason, we compared the calculated values with the experimental results for a stator core model to verify the accuracy of the calculation. The results are plotted in Fig. 4. The difference between the calculated and measured values is within 10%.

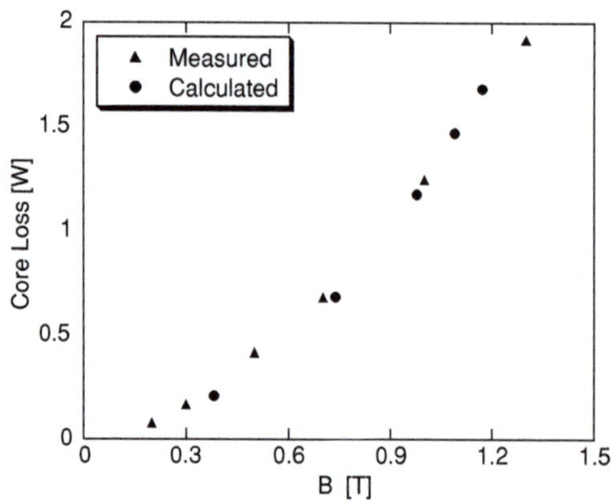

Figure 4. Core loss of the model core.

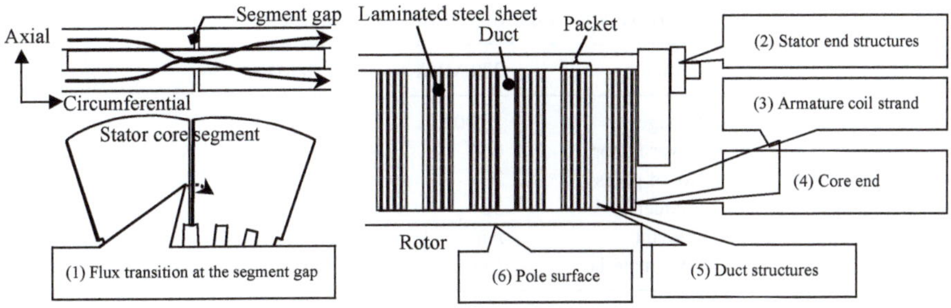

Figure 5. Causes of additional losses.

Additional losses

We can now calculate the additional losses, which are illustrated in Fig. 5. They are calculated with a local model for each portion, because calculating the additional losses with a whole generator model would take too long during the design phase. Fig. 6 depicts an example of a whole generator model for a two-pole machine, so the modeled region is half of the generator. The magnetic flux levels in the local models are coordinated to match the levels in the whole generator model. The local models separately account for the following portions of the generator:

1. *Flux transition at the segment gap.* There are gaps between two core segments in the laminated steel sheets, so the magnetic flux transfers from one layer to another at these gaps. As a result, eddy current losses due to the axial magnetic flux arise in the laminated steel sheets. These losses are calculated with a local model for several layers of steel sheets.
2. *Stator end structures.* The eddy current losses in the clamping flanges and the shields are calculated for each local model.
3. *Armature coil strand.* After calculating the magnetic flux density incoming to the armature end winding, the loss in the coil strand is calculated by a analytical formula.

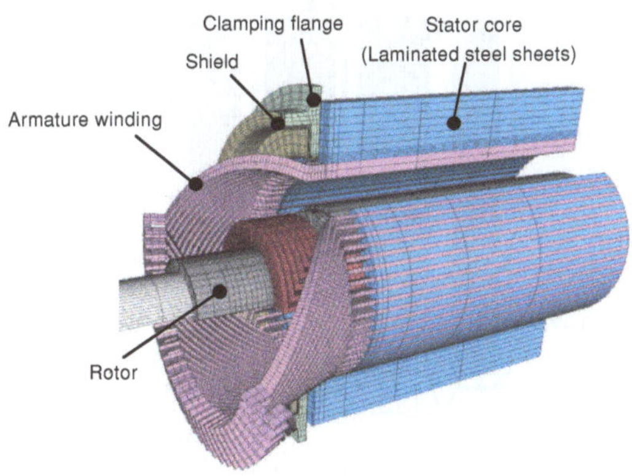

Figure 6. Whole generator model.

Table 1. Specifications of turbine generators

Rating	220 MVA	170 MVA
Voltage	18,000	13,200
Power factor	0.9	0.85
No. of poles	2	2
Frequency	50	50
Coolant	Air	H_2
Core material	NO	GO

4. *Core end.* The eddy current loss due to the axial magnetic flux is calculated for a local model of this portion.
5. *Duct structures.* The eddy current loss in the duct pieces is calculated.
6. *Pole surface.* The eddy current loss at the pole surface is calculated.

Results

Table 1 shows the specifications of the two turbine generators that we analyzed. These two generators have a typical difference in their core materials: one is made of non-grain-oriented steel sheets (NO), while the other's core is grain-oriented (GO).

Loss in laminated steel sheets

The stator core of a turbine generator has cooling ducts, as shown in Fig. 7. This causes the magnetic flux to concentrate at the corners of the steel sheets. To consider this concentration, we calculate the magnetic flux density of a one-packet model by using three-dimensional finite element analysis.

Fig. 8 shows the axial distributions of the radial magnetic flux. The triangles represent the magnetic flux density in the stator teeth, while the squares represent that in the stator

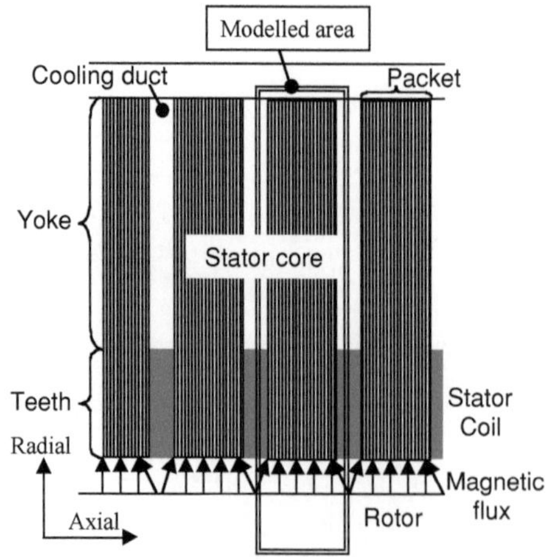

Figure 7. Cooling ducts.

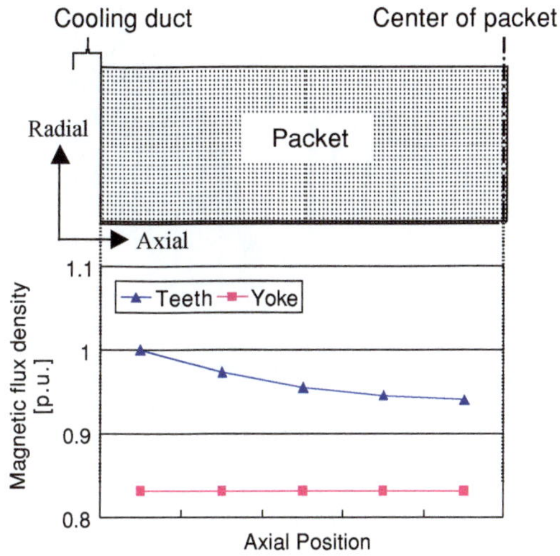

Figure 8. Concentration of magnetic flux at duct area.

yoke. The magnetic flux density in the yoke is constant in the region from the duct side to the center of the packet. On the other hand, the magnetic flux density in the teeth at the end is about 5% larger than that at the center. The eddy current loss due to the axial magnetic flux is calculated by using another model with finer elements.

The magnetic flux vectors and the distributions of the core loss density in the laminated steel sheets for the 220 MVA and 170 MVA machines are depicted in Figs. 9 and 10, respectively. The magnetic flux vectors are shown by the blue arrows in Figs. 9(a) and 10(a).

In Figs. 9(b) and 10(b), the red and blue areas represent regions of higher and lower loss density, respectively. The loss density is especially high at the tooth tips in both machines. It

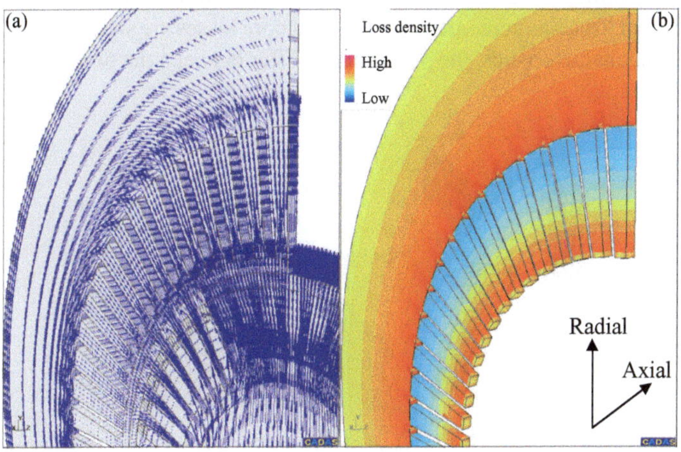

Figure 9. Loss density in laminated steel sheets (220MVA). (a) Magnetic flux vectors. (b) Distribution of loss density.

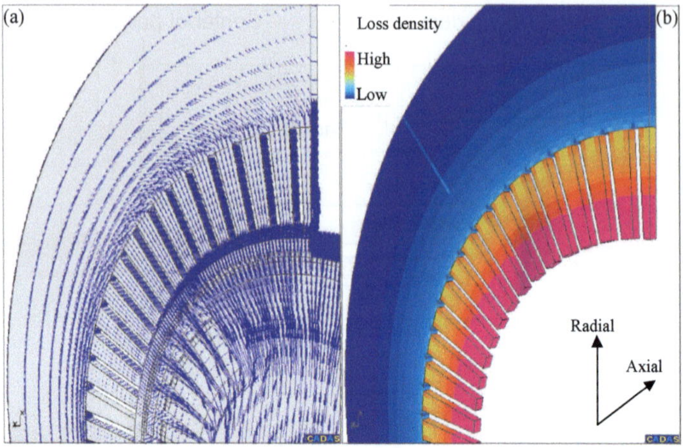

Figure 10. Loss density in laminated steel sheets (170 MVA). (a) Magnetic flux vectors. (b) Distribution of loss density.

is also high at the inner area of the stator yoke. The differences in loss distribution between the two machines are due to the different stator core materials.

The loss density in the stator yoke of the 170 MVA machine is lower than that of the 220 MVA machine because its stator core material is GO steel. In contrast, the loss density at the teeth of the 170 MVA machine is higher than that of the other machine due to the properties of the electrical steel sheets.

Additional losses

The eddy current loss densities in the clamping plate and shield are shown in Fig. 11. The red and blue areas represent high and low density, respectively. The loss is concentrated at the inner area in both parts because of the concentration of the magnetic flux there.

The additional losses as percentages of the total core losses are shown in Fig. 12. Reflecting the different characteristics, the percentages differ between the two generators. Several

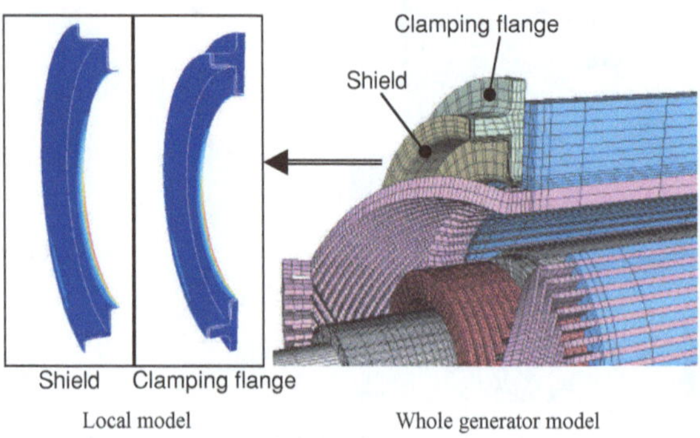

Figure 11. Eddy current loss of the shield.

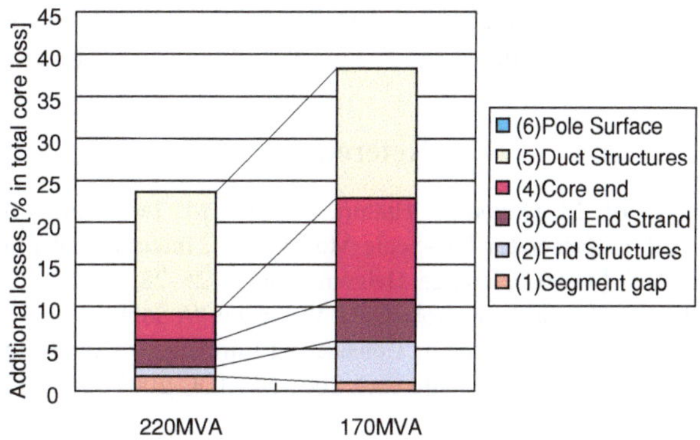

Figure 12. Calculation results of additional losses.

factors influence the additional losses, including the electrical design, the structure, and the materials.

Fig. 13 shows the calculation results for the total core losses. The calculated losses were 88% and 96% of the measured values for the 220 MVA and 170 MVA machines, respectively. In both cases, the additional losses make up a considerable part of the core losses. This confirms the necessity of calculating the additional losses when estimating the total core losses of turbine generators.

Conclusions

We have shown that the so-called core loss of a turbine generator includes various losses besides those produced in the laminated steel sheets of the core. We have also analyzed the causes of the losses in these sheets. Part of these losses can be calculated by considering the rotational field and the harmonics. Another part is due to the axial flux or field concentration. Additional losses result from the metal parts other than the steel sheets. By considering all of these losses, the total core losses of two different types of generators were calculated.

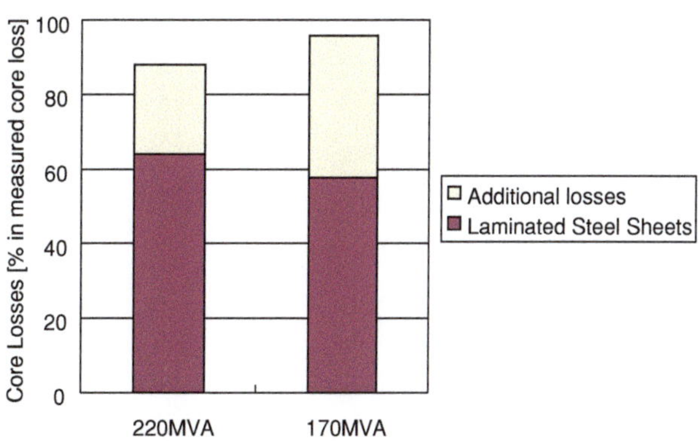

Figure 13. Calculated total core losses.

The differences between the calculated and measured total core losses were within 12%. This technique can thus contribute to the design of highly efficient turbine generators.

References

[1] K. Takahashi, K. Ide, M. Onoda, K. Hattori, M. Sato, M. Takahashi, "Strand Current Distributions of Turbine Generator Full-Scale Model Coil", International Conference Electrical Machines 2002 (ICEM 2002), Brugge, Belgium, August 25–28, 2002.

[2] K. Ide, K. Hattori, K. Takahashi, K. Kobashi, T. Watanabe, "A Sophisticated Maximum Capacity Analysis for Large Turbine Generators Considering Limitation of Temperature", International Electrical Machines and Drives Conference 2003 (IEMDC 2003), June 1–4, 2003, Madison, Wi.

[3] K. Hattori, K. Ide, K. Takahashi, K. Kobashi, H. Okabe, T. Watanabe, "Performance Assessment Study of a 250MVA Air-Cooled Turbo Generator", International Electrical Machines and Drives Conference 2003 (IEMDC 2003), June 1–4, 2003, Madison, Wi.

[4] K. Yamazaki, "Stray Load Loss Analysis of Induction Motors Due to Harmonic Electromagnetic Fields of Stator and Rotor", International Conference Electrical Machines 2002 (ICEM 2002), Brugge, Belgium, August 25–28, 2002.

[5] G. Bertotti, General properties of power losses in soft ferromagnetic materials, IEEE Trans. Magn., Vol. 24, pp. 621–630, 1988.

[6] P. Beckley, Modern steels for transformers and machines, Power Eng. J., Vol. 13, pp. 190–200, 1999.

[7] J. Anuszczyk, Z. Gmyrek, "The Calculation of Power Losses Under Rotational Magnetization Excess Losses Including", International Conference Electrical Machines 2002 (ICEM 2002), Brugge, Belgium, August 25–28, 2002.

[8] H. Domeki, Y. Ishihara, C. Kaido, Y. Kawase, S. Kitamura, T. Shimomura, N. Takahashi, T. Yamada, K. Yamazaki, Investigation of benchmark model for estimating iron loss in rotating machine, IEEE Trans. Magn., Vol. 40, pp. 794–797, 2004.

[9] C.P. Steinmetz, On the law of hysteresis, *AIEE Trans.*, Vol. 9, 1892, pp. 3–64. Reprinted under the title "A Steinmetz contribution to the AC power revolution" introduced by J.E. Brittain, Proc. IEEE, Vol. 72, pp. 196–221, 1984.

I-2. OPTIMIZED CALCULATION OF LOSSES IN LARGE HYDRO-GENERATORS USING STATISTICAL METHODS

Georg Traxler-Samek, Alexander Schwery, Richard Zickermann and Carlos Ramirez

ALSTOM (Switzerland) Ltd., Hydro Generator Technology Center, CH-5242 Birr, Switzerland
georg.traxler@power.alstom.com, alexander.schwery@power.alstom.com, richard.zickermann@power.alstom.com, carlos.ramirez@power.alstom.com

Abstract. A very important issue during the electrical design of hydro-generators is the reliability of the loss calculation in the manufacturer's design calculation program. The design engineer who has to guarantee the losses must be able to estimate the risk of liquidated damages when defining the guarantee values. This paper presents the optimization of the loss calculation in a design program for salient pole synchronous machines. Statistical methods are used to calibrate the loss calculation with measurements made during commissioning. Within this paper special importance is attached to the optimization of the no-load electromagnetic losses.

Introduction

In hydro power plants, the mechanical power of the water turbine is converted into electrical power mainly by three-phase synchronous generators with salient poles (see Fig. 1). These machines are built with an active power up to 800 MW. To reach the best efficiency of the turbine the speed of the generator is adapted to the hydraulic conditions resulting in typical speed ranges of generators from 67 to 1,500 rpm. The corresponding number of poles of the salient pole machine start from $2p = 4$ up to $2p = 90$ for a 50 Hz grid.

In the basic design phase of a hydro-generator the design engineer optimizes the electrical design regarding the electromagnetic load, the temperature rises, the losses, and the manufacturing costs without exceeding tolerable mechanical stresses in the machine at runaway speed. The main problem when calculating power losses in hydro-generators are deviations between the calculated and measured losses. These deviations can have several reasons:

1. Inaccuracies in the used loss calculation model,
2. Manufacturing tolerances,
3. Measuring tolerances during commissioning tests.

Especially when guaranteeing the losses, the design engineer must be able to estimate the risk of liquidated damages.

The analytical loss calculation is based on mathematical and physical calculation models. Due to the complexity of synchronous machines, inaccuracies in the loss calculation model

S. Wiak, M. Dems, K. Komęza (eds.), Recent Developments of Electrical Drives, 13–23.

Figure 1. Hydro-generator during installation of the rotor.

cannot be avoided. Furthermore material properties are only known with a limited precision. Especially in refurbishment projects such information is generally missing. In this case the design engineer is obliged to roughly estimate some important material parameters. The recalculation of the existing machine using the described loss calculation method can help to get an idea about the material properties.

Tolerances in the manufacturing process (as a worn out stator lamination punching tool for example) lead to non-predictable deviations between calculation and measurements. Finally measurements are affected by errors even though they are carried out according to international standards [1].

The uncertainties and the only limited accessibility for analytical algorithms make statistical methods a valuable help in order to improve the precision of the computed values and consequently be in-line with site measurements made during commissioning tests.

Method of loss calibration

To calibrate the analytical loss calculation using measurements, a series of reference machines and a series of test machines were defined. The *reference machines* are used to calibrate the losses by means of statistical methods, the *test machines* are used to validate the loss calibration results.

In the electrical design program the analytical loss calculation is subdivided into N parts assembled in an N dimensional loss vector $p_c^T = (P_1 \dots P_N)$. The components of this vector are the result of an analytical calculation algorithm. The loss vector is scaled with the measured sum of the loss components $p_{\Sigma m}$ to get the dimensionless expression

$$p = \frac{1}{p_{\Sigma m}} \cdot p_c \qquad (1)$$

By defining the N-dimensional vector $o^T = (1 \dots 1)$ we generally get

$$p_c^T \cdot o \neq p_{\Sigma m} \text{ resp. } p^T \cdot o \neq 1 \tag{2}$$

due to deviations between the calculated and measured losses. The aim is to get a good accordance between computation and measurement by defining a weighting factor kj for each of the N loss components

$$k = (k_1 \dots k_N) \tag{3}$$

such that $p^T \cdot k \approx 1$. The difference between the calculated and the measured value d is defined by

$$d = p^T \cdot k - 1 \tag{4}$$

The set of M *reference machines* is introduced with its scaled loss vectors $p_1 \dots p_M$. These vectors are assembled in a loss matrix

$$P = \begin{pmatrix} p_1^T \\ \vdots \\ p_M^T \end{pmatrix} = \begin{pmatrix} p_{11} & \cdots & p_{1N} \\ \vdots & \ddots & \vdots \\ p_{M1} & \cdots & p_{MN} \end{pmatrix} \tag{5}$$

The deviation vector $d(k)$ for a given set of weighting factors k including the set of reference machines is derived from equation (4)

$$d(k) = P \cdot k - o \tag{6}$$

An optimized set of weighting factors k can be found by minimization of the mean quadratic deviation δ

$$\delta = \sqrt{\frac{1}{M} d(k)^T \cdot d(k)} \tag{7}$$

This can be done with different optimization algorithms. In the following example the optimized factors are found by means of numerical methods described in [2].

No-load electromagnetic losses

The no-load electrical losses are the so-called *Iron Losses*. They are measured during commissioning when the machine is excited at the rated machine voltage. All the other losses, which exist at no-load operation with rated voltage (mechanical losses: air friction, fan losses, bearing friction, and rotor copper losses for the no-load excitation current) are subtracted and therefore not included in the *Iron Losses*.

The calculations and models shown in this document are based on research work from different sources (e.g. [3–13]). Some methods were used with the already existing calculation method. Other methods are new and mainly based on recent works.

Numerical simulations with the Finite-Element method [14–16] were used to confirm the analytical computations. For the integration in the calculation tool these methods would be too time consuming. The used electromagnetic no-load loss model contains $N = 10$ different partial losses:

1. Stator iron losses in teeth P_1 and yoke P_2

These losses are calculated with the well-known formula

$$P_{1,2} = k_{Fe} \cdot M \cdot c(B, f) \tag{8}$$

where f is the grid frequency, M is the mass of the stator teeth/yoke, and the function $c(B, f)$ defines the specific iron losses of the stator core lamination material in dependency of the magnetic flux density B and the frequency f. The factor k_{Fe} is based on experience and contains the influence of the air-gap field Fourier expansion harmonics.

2. Eddy current losses on the pole shoe surface due to tooth ripple pulsation P_3

These losses are calculated according to the two-dimensional analytical model described in [3]. In the air-gap region, the Laplace equation and in the pole shoe region, the Helmholtz equation are solved. As shown in Fig. 2(a) the tooth ripple pulsation of the magnetic flux density is replaced by a linear current density field wave

$$K(x, t) = K_0 \cdot \exp j(\omega t - kx) \tag{9}$$

where k is the wave number and ω the angular frequency of the tooth ripple pulsation. Saturation effects are taken into account with a surrogate relative permeability μ_r obtained

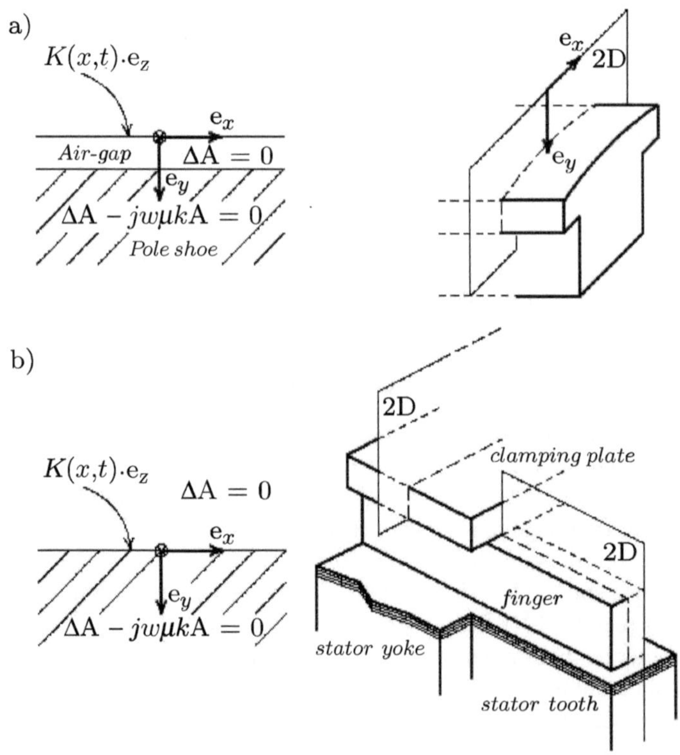

Figure 2. Analytical loss calculation models for the calculation of P3, P8, and P9.

iteratively in dependence on the tangential magnetic flux density on the pole shoe surface $B(H) = \mu_0 \mu_r H$. The 3D-effect of laminated poles is considered with a loss reduction coefficient [7].

3. Eddy current losses in the upper strands of the stator winding due to the radial magnetic field in the stator slot P_4

The radial magnetic field in the stator slot is composed of the magnetic field entering the slot computed with Conformal Mapping [12] and the additional magnetic field due to the tooth relief in case of saturated stator teeth. The eddy current losses in the strands are calculated with a simplified formula [13].

4. Circulating and eddy current losses in the Roebel bars due to the parasitic end region magnetic field P_5, P_6

The parasitic end region magnetic field is obtained by a two-dimensional end region Boundary Element model shown in Fig. 3. The obtained 2D magnetic field distribution is converted into cylindrical 3D-coordinates with

$$B_{3D}(r, \alpha, z) = B_{2D}(r, z) \cdot \exp(jp\alpha) \cdot \frac{D}{2r} \cdot f_p\left(\frac{d}{\tau_p}\right) \qquad (10)$$

where α is the tangential angle, D is the stator bore diameter, r the radial coordinate, and p the number of pole pairs. The function f_p takes into account the influence of adjacent poles with their negative orientation, d is the distance of the field calculation point from

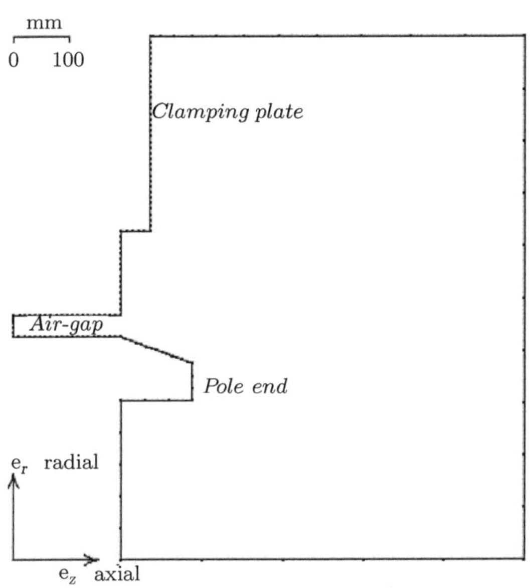

Figure 3. Simplified two-dimensional Boundary Element model of the end region.

the air-gap end and τ_p the pole pitch length. The circulating $P5$ and eddy current losses $P6$ are calculated by methods described in [13]. The Roebel bar is replaced by a network which contains the resistances, self- and mutual inductances of the strands. The parasitic end region magnetic field is introduced by means of voltage sources.

5. *Eddy current losses in the clamping fingers* P_7

The magnetic field in the clamping fingers is also calculated with the Boundary Element model shown in Fig. 3. The obtained magnetic flux density is corrected to take into account the effect of the stator slots. This is done with Conformal Mapping depending on the local slot geometry. The losses are calculated with a local eddy current model shown in [13].

6. *Eddy current losses in the stator core end laminations* P_8

The Boundary Element model shown in Fig. 3 is used to compute the magnetic field entering the stator core end tooth laminations. The magnetic flux density is corrected to take into account the effect of the stator slots (see item 5). The eddy current losses are computed by means of a local eddy current model shown in Fig. 2(b), where the core end laminations (solving the Helmholtz equation) and the surrounding air (Laplace equation) are modeled. The lamination effect cannot be taken into account. The pulsating magnetic field on the end laminations is introduced with a pulsating linear current density function

$$K(x,t) = \frac{K_0}{2} \cdot (\exp j(\omega t - kx) + \exp j(\omega t + kx)) \tag{11}$$

(angular frequency ω and wave number k), local saturation effects are taken into account iteratively (see item 3).

7. *Eddy current losses in the stator clamping plates* P_9

The calculation of the eddy current losses in the stator clamping plates is also based on the Boundary Element model (Fig. 3). The obtained magnetic flux density field wave on the clamping plates is applied to a local calculation model displayed in Fig. 2(b) [13]. The calculation method is similar to the calculation of eddy current losses in the stator core end laminations (item 6), the exciting magnetic field wave is introduced with a surrogate linear current density field wave

$$K(x,t) = K_0 \cdot \exp j(\omega t - kx) \tag{12}$$

(angular frequency ω and wave number k), local saturation effects are again considered by iteration (see item 3).

8. *Losses in the damper bars due to tooth ripple pulsations in the air-gap magnetic field* P_{10}

For the calculation of losses in the damper bars a simplified asynchronous squirrel cage model is applied. Neither the effect of the d- and q-axes nor the effect of a damper displacement are taken into account.

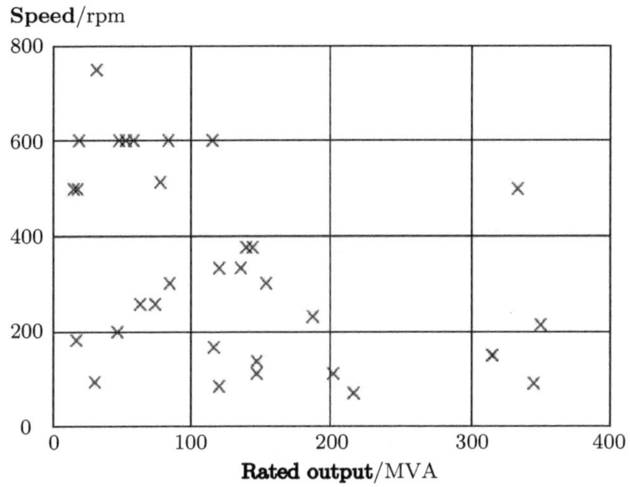

Figure 4. Range of reference and test machines used for the loss calibration tests.

Recalculation of existing machines

It is necessary to have a good and possibly large set of reference and sample machines. For all of these machines, a new electrical recalculation is performed using not only the original electrical calculation, but also a set of drawings with detailed information regarding

- *The main dimensions:* This is necessary, to be sure to get the electrical calculation of the machine which was actually built.
- *Material parameters:* It is obvious, that an exact knowledge of the used materials (for example the stator core lamination quality) is necessary.
- *Additional dimensions:* For the new loss calculation, some parameters, which were not taken into account in the old calculation must be available.
- *Measurements:* The measurement of the no-load test with rated voltage excitation and if possible also the air-gap measurement (stator roundness) must be available.

The loss evaluation method presented above is used to calibrate the no-load losses of large synchronous machines with salient poles. As shown in Fig. 4, a set of various machines is taken into account.

Optimization of theno-load electromagnetic losses

The aim of the statistical evaluation as described above is to find an optimum set of loss calibration weighting factors $k = (k_1 \dots k_N)$ where $N = 1, \dots, 10$. The range of these factors can be limited in order to allow the optimization process to take only physically meaningful factors into consideration:

$$k_{\min} \leq k \leq k_{\max} \tag{13}$$

The limits are set very carefully taking into account experience, certain detailed measurements and the results of special investigations.

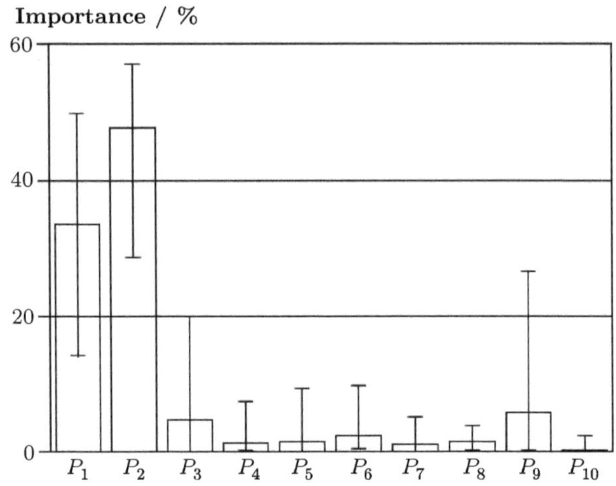

Figure 5. Importance of different loss types. Average, minimum, and maximum values.

To compare the optimum set of weighting factors with the classical calculation method the weighting factor $\mathbf{k}_{\text{old}}$ is defined, where the factors for the new developed partial losses are set to zero. Furthermore a not-calibrated set of weighting factors $\mathbf{k}_0$ is defined where all factors are set to one.

The relative importance of the partial losses $P_1, \ldots, P_{10}$ is shown in Fig. 5. The stator core losses in teeth P_1 and yoke P_2 are the most important items followed by eddy current losses in the clamping plates P_9 and eddy current losses in the pole shoe surface due to tooth ripple pulsation P_3. The high variation shows, that the order of importance can change significantly depending on the machine type.

The final evaluations are made with different sets of reference and test machines: Two runs were made with fixed evaluation factors ($\mathbf{k}_{\text{old}}$ and $\mathbf{k}_0$) and different evaluation runs were made with different distributed groups of $M = 21$ *reference*- and 12 *test machines*.

The loss calibration weighting factors were evaluated by minimizing the mean quadratic deviation.

The deviation histogram showing the frequency distribution of deviations $\mathbf{d}$ for the old calculation method using the weighting factor $\mathbf{k}_{\text{old}}$ is shown in Fig. 6, whereas the deviation histograms for the new calculation method are displayed in Figs. 7 and 8. Fig. 7 shows the frequency distribution with all evaluation weighting factors set to one, Fig. 8 shows the best evaluation run.

In all deviation histograms, a negative deviation means a more pessimistic calculation (higher losses calculated than measured) and consequently a positive deviation a too optimistic calculation (lower losses calculated than measured). The vertical lines show the $\pm 20\%$ and $\pm 10\%$ deviation band. The hatched bars (left bars) in Fig. 8 represent the test machines taken into account to test the loss calibration results while the white bars (right bars) represent the *reference machines* taken into account for the evaluation.

The new calculation method shows significant better results than the old calculation method. Even in the not-calibrated run, where all weighting factors are set to one, the frequency distribution of the deviations shows a smaller standard deviation. The loss

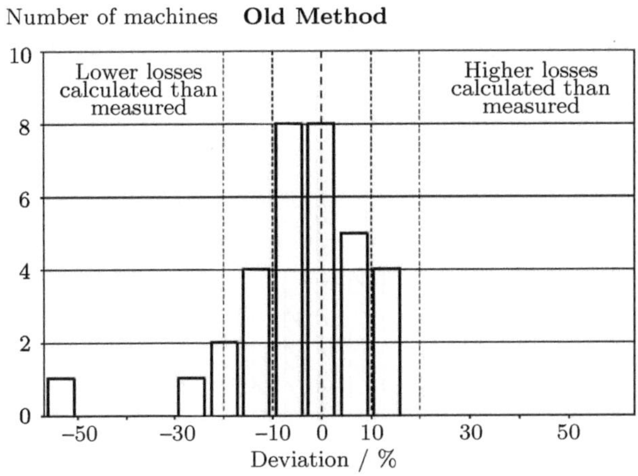

Figure 6. Frequency distribution of deviations d between calculation and measurement for the old loss calculation method.

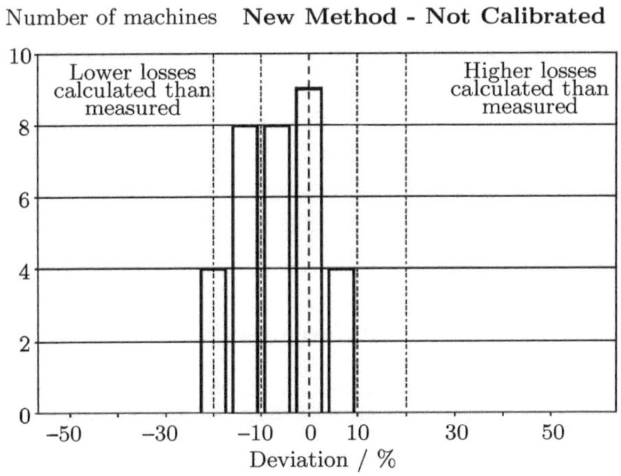

Figure 7. Frequency distribution of deviations d between calculation and measurement for the new loss calculation method, all evaluation factors set to one.

calibration with the *best* set of weighting factors does not improve the standard deviation but centers the deviations (mean value close to zero).

Conclusion

The presented calculation method shows that the loss calculation can be improved significantly with the help of statistical methods. The standard deviation of the frequency plot allows for an estimation of the risk when defining the guaranteed losses during a tender. As it is very time consuming to collect all the necessary machine data, the given calculation example uses only 33 *reference-* and *test machines*. For a good statistical statement this is

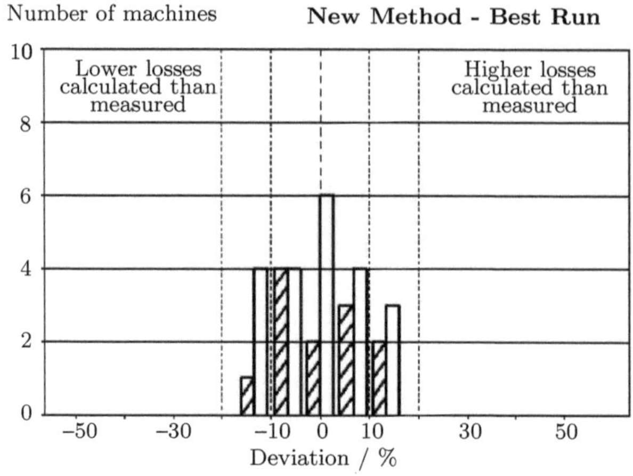

Figure 8. Frequency distribution of deviations d between calculation and measurement for the new loss calculation method after the loss calibration using a set of 21 *reference-* (white bars) and 12 test *machines* (hatched bars).

not enough. As the calibration process is an ongoing work it will be improved in the future with more and more measured machines.

The new method provides much more detailed results allowing the electrical design engineer to have a good idea of critical parts in the machine like the pole end design, the stator core end design and the winding overhang. This simplifies the decision process for special and cost-intensive design improvements like stepping or slitting of the stator core end laminations.

References

[1] IEC 60034-2, Rotating Electrical Machines. Part 2: Methods for Determining Losses and Efficiency of Rotating Electrical Machinery from Tests (excluding machines for traction vehicles), International Electrotechnical Commission, Switzerland, 1972.

[2] T. Coleman, M.A. Branch, A. Grace, Optimization Toolbox, For Use with MATLAB® : User's Guide, The Math Works, Inc., United States, 1990–1999.

[3] M.G. Barello, Courants de Foucault engendrés dans les pièces polaires massives des alternateurs par les champs tournants parasites de la réaction d'induit. Rev. Gen. Electr., Vol. 64, Issue 11, pp. 557–576, 1955.

[4] H. Bondi, K.C. Mukherji, An analysis of tooth-ripple phenomena in smooth laminated pole-shoes, Proc. IEE, Vol. 104 C, pp. 349–356, 1957.

[5] F. Fiorillo, A. Novikov, An improved approach to power losses in magnetic laminations under nonsinusoidal induction waveform. IEEE Trans. Magn., Vol. 26, No. 5, pp. 2904–2910, 1990.

[6] J. Greig, K. Sathirakul, Pole-face losses in alternators, Proc. IEE, Vol. 108 C, pp. 130–138, 1961.

[7] J. Greig, E.M. Freeman, Simplified presentation of the eddy-current-loss equation for laminated pole-shoes, Proc. IEE, Vol. 110, pp. 1255–1259, 1963.

[8] St. Kunckel, G. Klaus, M. Liese, "Calculation of Eddy Current Losses and Temperature Rises at the Stator End Portion of Hydro Generators", Proceedings on the 15th International Conference on Electrical Machines, ICEM, Brugge, Belgium, August, 2002.

[9] M.S. Lancarotte, A. Penteado, Estimation of core losses under sinusoidal or non-sinusoidal induction by analysis of magnetization rate. IEEE Trans. Energy Convers., Vol. 16, No. 2, pp. 174–179, 2001.

[10] D.C. Macdonald, Losses in Roebel bars: effect of slot portion on circulating currents, Proc. IEE, Vol. 117, No. 1, pp. 111–118, 1970.

[11] D.C. Macdonald, Circulating-current loss within Roebel bar stator windings in hydroelectric alternators, Proc. IEE, Vol. 118, No. 5, pp. 689–697, 1971.

[12] W. Schuisky, Berechnung elektrischer Maschinen, Wien: Verlag Springer, 1960.

[13] G. Traxler-Samek, Zusatzverluste im Stirnraum von Hydrogeneratoren mit Roebelstabwicklung, Dissertation, TU-Wien, 2003.

[14] M.T. Holmberg, "Three-dimensional Finite Element Computation of Eddy Currents in Synchronous Machines", Technical Report No. 350, Department of Electric Power Engineering, Chalmers University of Technology, Goteborg, Sweden, 1998.

[15] E. Schlemmer, F. Klammler, F. Mueller, "Comparison of Different Numerical Approaches for the Calculation of Eddy Current Losses in Large Synchronous Generators", Proceedings of the Seventh International Conference on Modeling and Simulation of Electrical Machines, Converters and Systems, ELECTRIMACS, Montreal, Canada, 2002.

[16] E. Schmidt, G. Traxler-Samek, A. Schwery, "3D Nonlinear Transient Finite Element Analysis of Eddy Currents in the Stator Clamping System of Large Hydro Generators", Proceedings of the 16th International Conference on Electrical Machines, ICEM, Cracow, Poland, 2004.

I-3. COUPLED MODEL FOR THE INTERIOR TYPE PERMANENT MAGNET SYNCHRONOUS MOTORS AT DIFFERENT SPEEDS

M. Pérez-Donsión

Electrical Engineering Department, Vigo University, Campus of Lagoas-Marcosende, 36200 Vigo, Spain

donsion@uvigo.es

Abstract. A coupled model for accurate representation of the characteristics of permanent magnet synchronous motors has been presented in this paper. The starting and synchronization processes of the PMSM, and the influence that on transient behavior of the motor produces the different values of the main motor parameters have been analyzed.

Introduction

Permanent Magnet Synchronous Motors (PMSM) are widely applied in industrial and robotic applications due to their high efficiency, low inertia, and high torque-to-volume ratio. Concerning with the design one of the greatest advantages of PMSM is that it can be designed directly for low speeds without any weakening in efficiency or power factor. An induction motor with a mechanical gearbox can often be replaced with a direct PMSM drive. Both space and cost will be saved, because the efficiency increases and the cost of maintenance decreases. A PMSM and a frequency converter form together a simple and effective choice in variable speed drives, because the total efficiency remains high even at lower speeds and the control of the whole system is very accurate. Since a low speed motor requires often a large amount of poles the number of stator slots per pole and phase is typically low. Thus the stator magneto motive force contains a lot of large harmonic components. Especially the fifth and the seventh stator harmonics are very harmful and tend to produce torque ripple at a frequency six times the supply frequency. At the lowest speed this might be extremely harmful.

The classical *d*-*q* model, uncoupled, linear and with constant parameter, applied to salient pole synchronous machines may be inadequate for accurate modeling and characteristics prediction of permanent magnet synchronous motors of interior type. It leads to important errors when evaluating machine performance or calculating the control circuits.

The lack of excitation control is one of the most important features of permanent magnet motors, as a consequence, the internal voltage of the motor rises proportionally to the rotor

S. Wiak, M. Dems, K. Komęza (eds.), Recent Developments of Electrical Drives, 25–37.

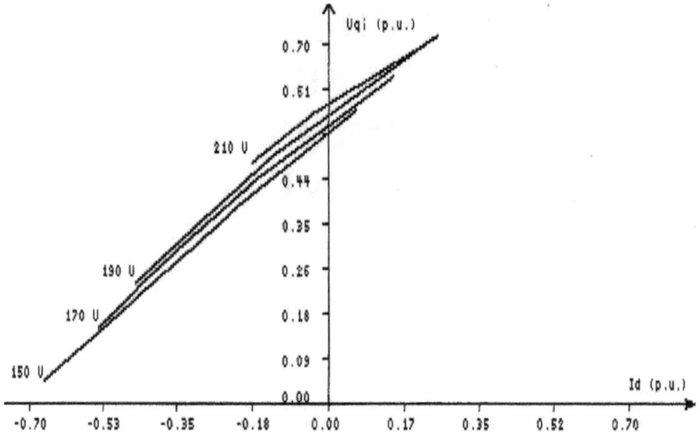

Figure 1. Graphic representation of *Vqi* vs. *Id*.

speed, and when the motor is working at constant horsepower mode its power factor becomes leading.

The behavior of permanent magnet machines of the interior type can be rather different than expected form the conventional two axis theory. For this reason, it is necessary to establish new models to take into account the magnetic flux redistribution phenomena along the rotor iron placed between the magnets and the air-gap.

On the other side due to the presence of permanent magnet excitation, the conventional methods of testing for determination of synchronous machine parameters cannot be applied in the case of permanent magnet machines, then it is necessary use tests procedures that differ from the classical methods applicable to wound field synchronous machines.

In order to observe the cross coupling phenomenon, we can measure and plot the curves of the interior voltage of the motor, "*Vqi*" vs. "*Id*," for the machine under study, Fig. 1.

The voltage steady state equations will be:

$$\begin{aligned} Vqi &= \alpha \cdot Vq - R_1 \cdot Iq = \alpha \cdot Eo + \alpha \cdot Xd \cdot Id \\ Vdi &= \alpha \cdot Vd - R_1 \cdot Id = \alpha \cdot Xq \cdot Iq \end{aligned} \tag{1}$$

Where "R_1" is the stator resistance, "E" is the induced voltage by the magnets, and α is a coefficient for take into account the operation at different speeds.

If the cross coupling effect didn't exist and considering constant excitation all curves $Vqi = f(Id)$ should cross at the same point for $Id = 0$. However they intersect in different points. We can see, Fig. 1, that for $Id = 0$ the distance between two curves *Vqi* is proportional to *Iq*, then we can think it is due to the magnetic coupling between d-q axis circuits, or in other words, the magnetic effects on the d-axis flux caused by q-axis current, of course we can consider the influence on the q-axis flux motivated by d-axis current. A possible solution for take into account this effect consist in the addition of a coupling term between the direct and the quadrature axis, then the model becomes:

$$\begin{aligned} Vqi &= \alpha \cdot Eo + \alpha \cdot Xd \cdot Id + \alpha \cdot Xqd \cdot Iq \\ Vdi &= \alpha \cdot Xq \cdot Iq + \alpha \cdot Xdq \cdot Id \end{aligned} \tag{2}$$

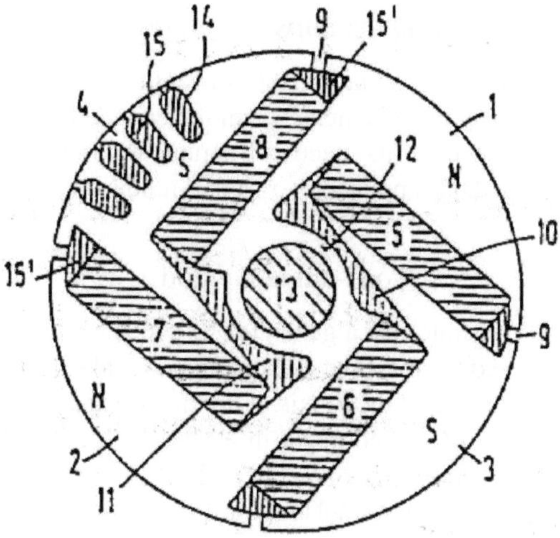

Figure 2. Rotor configuration of a SIEMOSYN interior type PMSM.

The effect of the term $\alpha \cdot Xdq \cdot Id$ depends of the configuration and dimensions of the PMSM and for the SIEMOSYN motors, Fig. 2, we have observed that it is practically negligible.

Then we can consider that the definition equations, *Vqi* and *Vdi*, for a SIEMOSYN PMSM, are:

$$\begin{aligned} Vqi &= \alpha \cdot Eo + \alpha \cdot Xd \cdot Id + \alpha \cdot Xqd \cdot Iq \\ Vdi &= \alpha \cdot Xq \cdot Iq \end{aligned} \tag{3}$$

and in Fig. 3 we can see the phasor diagram.

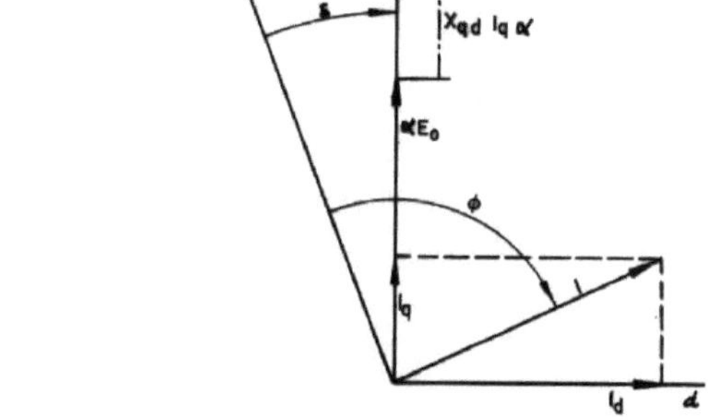

Figure 3. Phasor diagram for the SIEMOSYN interior type PMSM.

Synchronous reactances

Due to the presence of permanent excitation, the conventional methods of testing for determination of synchronous machine parameters cannot be applied in the case of a permanent magnet machine. Measurement of its electrical parameters requires test procedures that differ from the classical methods applicable to wound field synchronous machines.

Load-angle method

In this method, the MSIP operate like a generator, at synchronous speed, over a balanced three phase load. First we text the machine without load, we take the measurement of the *Eo* voltage and establish the position of the q-axis. After that we apply at synchronous machine different loads and we obtain the load angle in each case. In Fig. 4 we can see the text scheme for this method.

Taking into account the classical model and for different speeds (different frequencies), the phasor diagram is represented in Fig. 5.

And then the equations of the voltages over the d and q axis, are:

$$\begin{aligned} V \cdot Sin(-\delta) &= \alpha \cdot Xq \cdot Iq - R_1 \cdot Id \\ V \cdot Cos(-\delta) &= \alpha \cdot Eo - \alpha \cdot Xd \cdot Id - R_1 \cdot Iq \end{aligned} \tag{4}$$

For currents:

$$\begin{aligned} Id &= I_1 \cdot Sin(\phi - \delta) \\ Iq &= I_1 \cdot Cos(\phi - \delta) \end{aligned} \tag{5}$$

Replacing the d-q currents, into voltage equations, allows solution to direct and quadrature axis reactances, for $\alpha = 1$

$$\begin{aligned} Xd &= [Eo - V \cdot Cos(-\delta) - R_1 \cdot I_1 \cdot Cos(\phi - \delta)] / I_1 \cdot Sin(\phi - \delta) \\ Xq &= [V \cdot Sin(-\delta) - R_1 \cdot I_1 \cdot Sin(\phi - \delta)] / I_1 \cdot Cos(\phi - \delta) \end{aligned} \tag{6}$$

Where: α = actual frequency/base frequency, δ = load angle, and Φ = power factor angle.

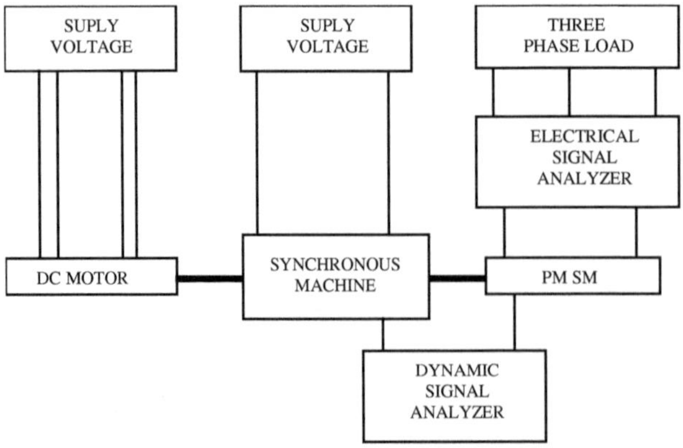

Figure 4. Text scheme load-angle method.

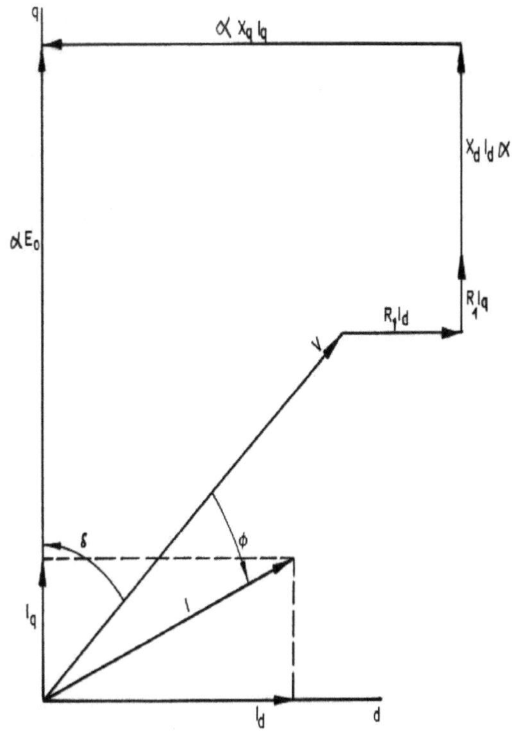

Figure 5. Phasor diagram model for a synchronous generator of salient poles at different speeds.

Using the expressions (6) we can calculate the reactances taken measurements for obtain the values of V, I_1, P, $Cos\,\phi$, and also the load angle (δ). Without load this angle is δ_0, Fig. 6. The load angle along the successive load test is calculated comparing the waveforms of the voltage supply and the reference signal.

In Fig. 7 we have represented the results obtained for the quadrature reactance *Xq*. Like we can see that the results are not constant if the *Iq* current change. We also have obtained this values by other procedure (current method) and we can conclude that both procedures are in a good agreement. This results are also in concordance with the obtained by other authors for PMSM of the interior type but with different geometries. Then we can say that this phenomena is common for all the interior type PMSM.

The values of the direct axis reactance, *Xd*, calculated by the equation (6) are not in agreement with the expected values of this reactance. We think this is because the d-axis flux consist of the combine action of magnets, d-axis current and q-axis current. The effect of *Iq* can be magnetizing or demagnetizing depending of the rotor geometry and it is not possible to separate by test the individual contributions of the magnet and the *Id* current to the total d-axis flux.

In Fig. 8 we can see *Xd* values vs. *Id* applied the classical model and calculated by the following equation (coupled model):

$$Xda = [Vd - Eo - R_1 \cdot Iq - Xdq \cdot Iq]/Id \tag{7}$$

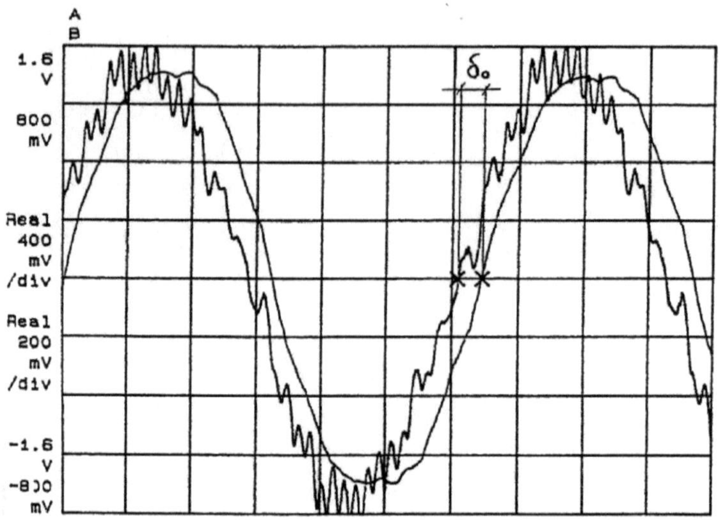

Figure 6. Charts for determination of the reference angle δ_0.

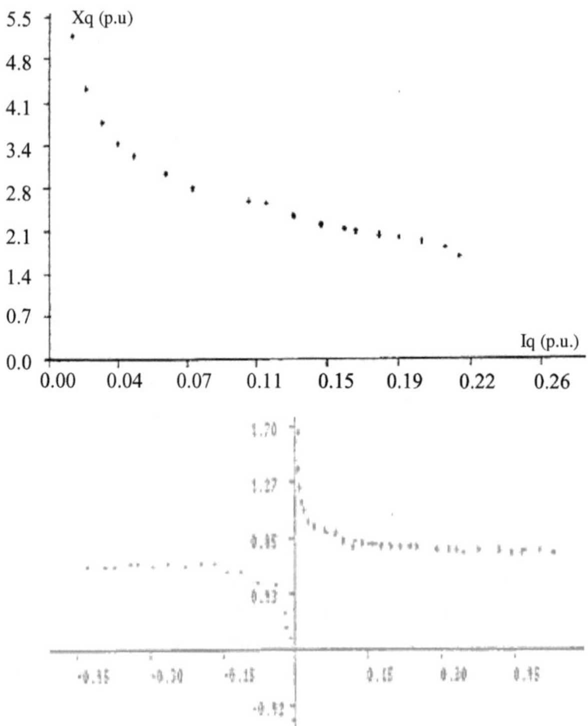

Figure 7. Graphic representation of Xq vs. Iq.

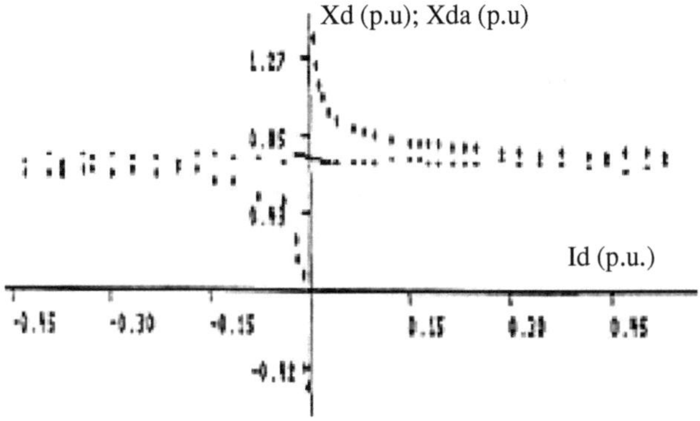

Figure 8. Graphic representation of *Xd* vs. *Id*. +, Values of *Xd* according with the classical model. –, Values of *Xd* take into account the cross coupling.

In Fig. 8 we can observe that the values of *Xd* with cross coupling are practically constant, which implies that, in this case, the most of the flux path on the *d*-axis is produced by the magnets.

In reference [4] we have developed the *Xqd* reactance determination and we have compared, in different cases, the simulation results using the classical model and the coupled model with the real measurements and we concluded that the values calculated using the coupled model are in better agreement with those obtained by text.

PMSM behavior

Now we have developed new texts and simulations for analyze other cases of the real operation of the PMSM. Then Fig. 9 show the good concordance between the curves speed-time obtained by simulation and by text. In this case we have used an acceleration ramp of 0 to 50 Hz during 0.45 s take into account a friction and ventilation torque of 0.011 pu and without load. It is curious observe the initial negative interval of the speed which depend on the initial angle between one of the motor phases and the direct axis. The effect of the saturation on the *q*-axis is take into account using the variation of the *q*-reactance with the *q*-axis current obtained by text.

In Fig. 10 we can observe the incidence that over the speed has a 0.25 pu sudden increase of the load and in Fig. 11 the influence that produce a sudden decrease of load, when previously the machine has obtained the permanent regimen.

The Fig. 12 represent the temporal evolution of the speed just after has take place a overload Sc, for different values of the permanent load torque before the disturbance.

The sudden application of the load produce an instantaneous decrease of the speed and then appear an positive asynchronous torque (Fig. 15) that helps to the rotor obtain one time more the synchronism. This asynchronous torque disappear just in the moment that the rotor obtain the synchronization. Like one can observe in Fig. 12 with the same value of the overload, the maximum slip obtained is lower for the higher level of the stationary initial

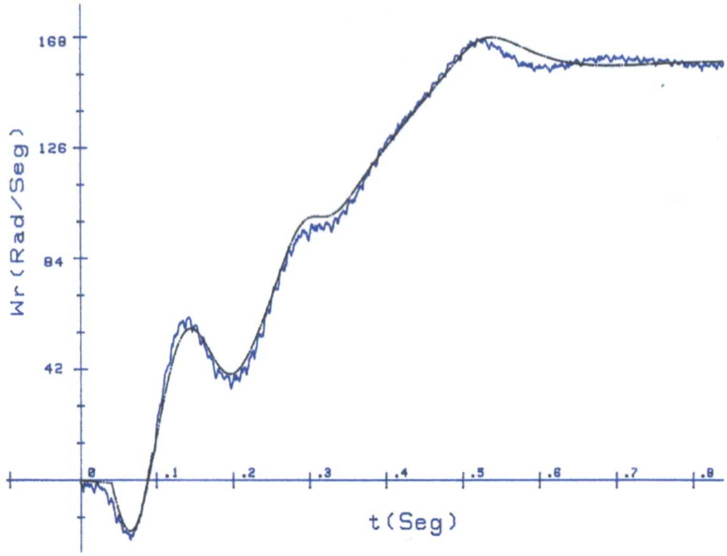

Figure 9. Graphic representation of speed vs. time during the started process, obtained by: -.-, applied the model (simulation) and taken measurements (continuous line).

load torque. At the same time this slip is so higher as so higher is the overload value and in consequence, for the same final load, so higher is the overload as higher is the maximum slip obtained. At the same time we can also observe that the time for which the maximum slip is obtained is practically the same in all cases.

It is interesting take notice in Fig. 12 that, one time that the motor obtain the synchronization, it can permit the application of sudden loads higher than it can synchronize when it start for the same inertia.

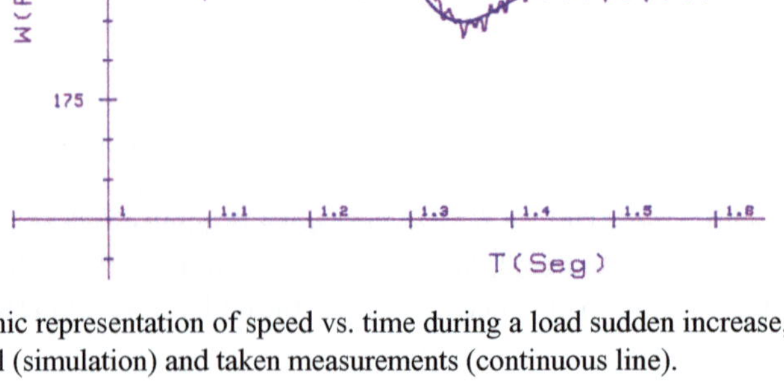

Figure 10. Graphic representation of speed vs. time during a load sudden increase, obtained by: -.-, applied the model (simulation) and taken measurements (continuous line).

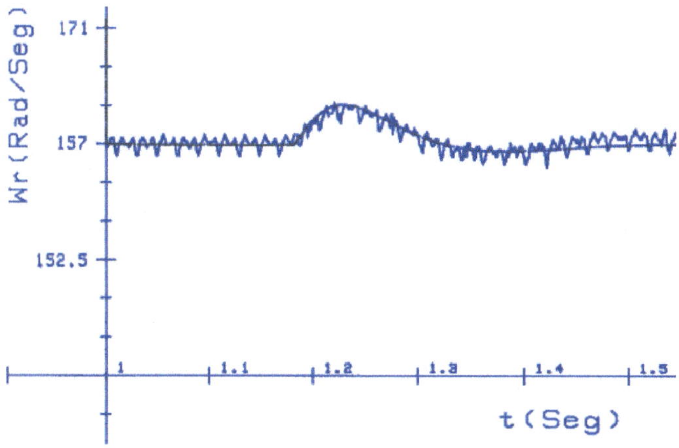

Figure 11. Graphic representation of speed vs. time during a load sudden decrease, obtained by: -.-, applied the model (simulation) and taken measurements (continuous line).

Then one of the most important factors that has influence about the transient behavior of the PMSM in front of a sudden increase/decrease of the load is the rotor inertia. A high value of the rotor inertia produce a large number of oscillations and if the value of the inertia is lower the response is more quicker, because the ratio torque/inertia is higher, but with the maximum slip more higher, Fig. 13.

In Fig. 14 we have represented the squirrel cage torque when take place a sudden decrease of the load and in Fig. 15 when the load increase. In both cases for the same values of the load torque (*Tl*) and overload (*Sl*).

In Fig. 16 we have represented the torque of the magnets and reluctance when take place a sudden increase of the load and in Fig. 17 when the load decrease. In both cases for the same values of the load torque (*Tl*) and overload (*Sl*). Logically the synchronous torques of permanent magnets and reluctance permit maintain the rotor in synchronism.

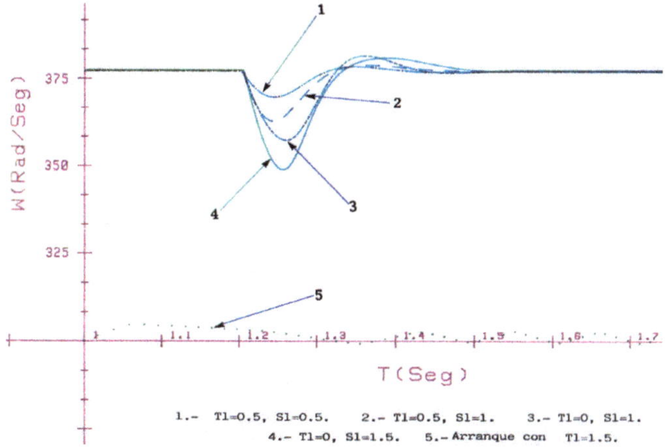

Figure 12. Graphic representation of speed vs. time during a load sudden increase.

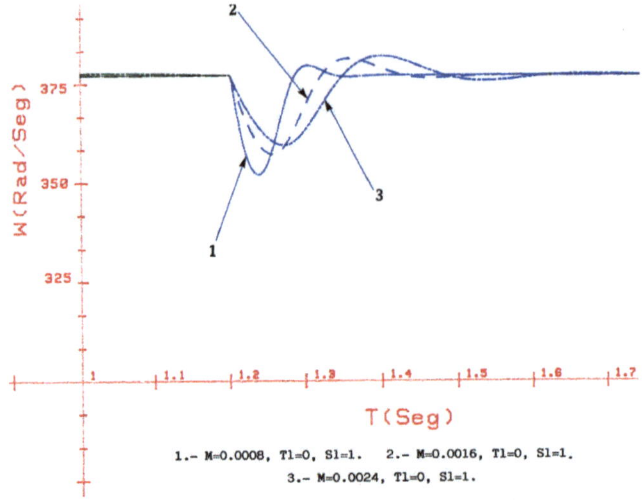

Figure 13. Graphic representation of speed vs. time for different inertia torque (M) with $Tl = 0$ and $Sl = 1$.

The permanent magnets influence about the transient behavior of the PMSM is very important. As higher are the equivalent currents of the magnets as lower are the slips of the transient response. In Fig. 18 we can observe that if the current of the magnet decrease below a certain value the motor is not capable of take up the overload and the motor lost the synchronism. The optimum value of this current depends, amount other factors, of the magnets braking torque at the synchronous speed proximity. If this value is overcome they will appear higher oscillations during the transient operation.

The rotor geometry and in consequence the relationship between the d-axis and q-axis reactances, also modify the PMSM behavior, as in the same way that for the equivalent current we must obtain an optimum value for the relation Xd/Xq and also it is important to obtain the most appropriate squirrel cage resistance value.

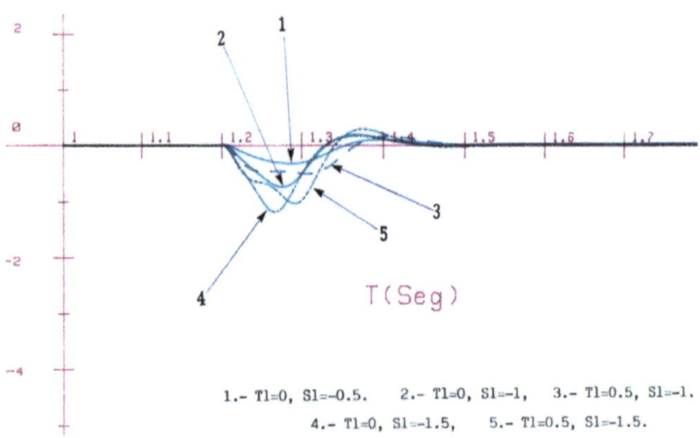

Figure 14. Graphic representation of the torque of the squirrel cage vs. time for a sudden decrease of the load.

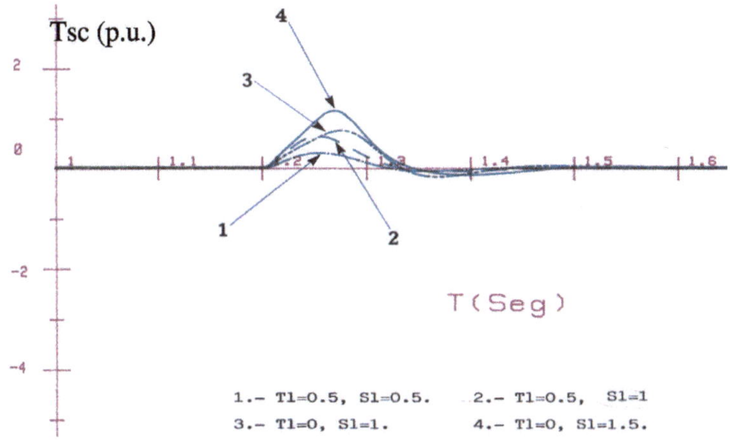

Figure 15. Graphic representation of the torque of the squirrel cage vs. time for a sudden increase of the load.

Really the number of variables that have influence about the starting and synchronization processes of a PMSM, take into account the motor and also the load, is very higher and then it is very difficult know in advance a set of necessary conditions for the correct synchronization of the PMSM. Then for develop an analyze of this type it is necessary take into account the parametric variation of the main magnitudes that have influence about the synchronization process.

In this particular case we have analyzed this process in term of his synchronization energy, specially we have considered the property "capacity of synchronization" of the motor, that

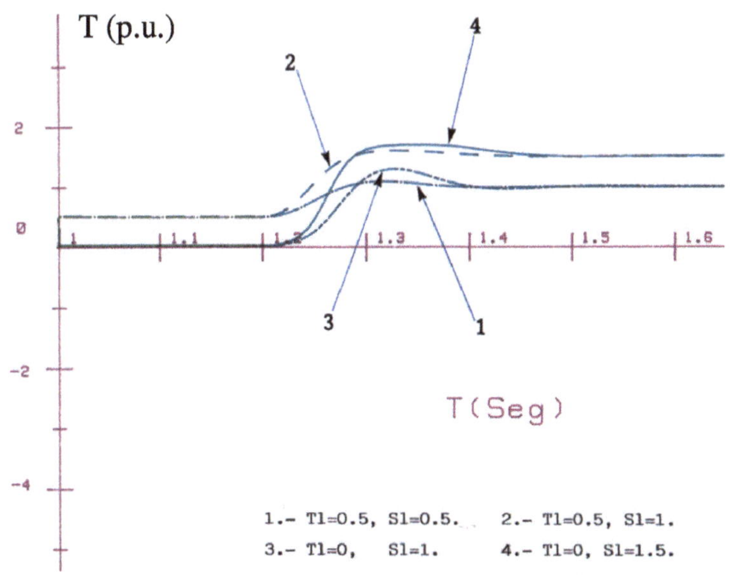

Figure 16. Graphic representation of the magnets and reluctance torques vs. time during a sudden increase of the load.

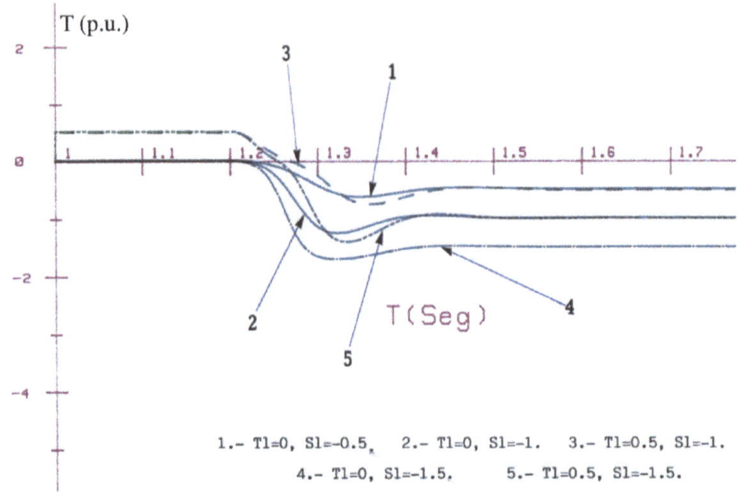

Figure 17. Graphic representation of the magnets and reluctance torques vs. time during a sudden decrease of the load.

we can defined it like a set of critical combinations of inertia and load torque in which the PMSM is capable to obtain the synchronization.

For obtain the synchronization energy we use a set of simple expressions that permit determine this magnitude in the last stage of the synchronous operation of the motor just when the machine describe a limit circle.

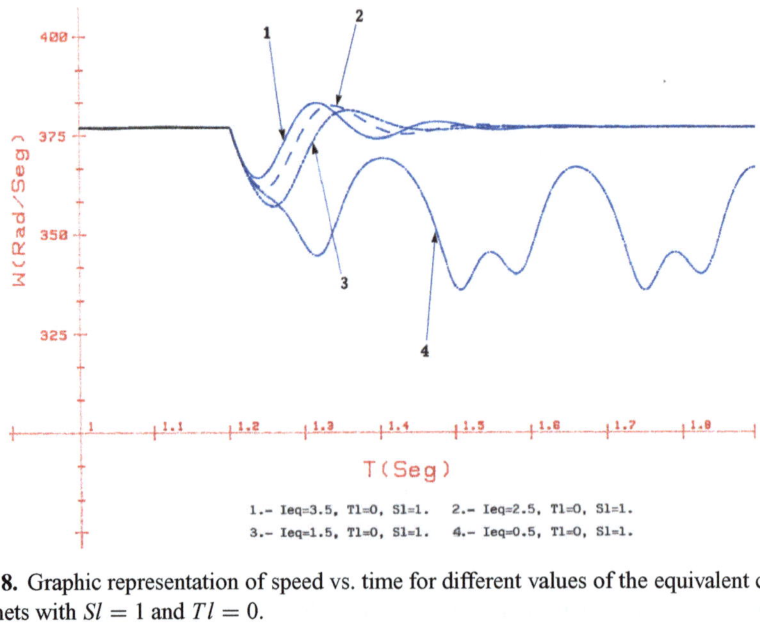

Figure 18. Graphic representation of speed vs. time for different values of the equivalent current of the magnets with $Sl = 1$ and $Tl = 0$.

The dynamic equation expressed in the torque-slip plane is, (8):

$$-\frac{1}{p}J \cdot w_0^2 \cdot s\frac{ds}{d\delta} = Ts(\delta) + Ta(s) - Tc(s) \tag{8}$$

Where: J is the combination inertia of the motor and the load, *Ts* is the sum of all the synchronization torques, *Ta* include all the asynchronous average torques, and *Tc* is the sum of the load, slip, and ventilation torques.

The equation (8) describe the critical trajectories of the polar slips on the load angle-slip plane.

Conclusions

We have developed along this paper a coupled model for accurate representation of the characteristics of permanent magnet synchronous motors and we have proposed the determination of the direct axis reactance, "*Xd*," and the quadrature axis reactance, "*Xq*," by calculus and texts with the permanent magnet synchronous machine under generator duty.

We also have analyzed the starting and synchronization processes of the PMSM and the influence that on transient behavior of the motor produce different values of the main motor parameters.

References

[1] J. Salo, T. Heikkilä, H.T. Pyrhönen, "New Low-Speed High-Torque Permanent Magnet Synchronous Machine With Buried Magnets", International Conference on Electrical Machines (ICEM 2000), Espoo, Finland, 2000, pp. 1246–1250.

[2] F. Parasiliti, P. Poffet, A model for saturation effects in high field permanent magnet synchronous motors, IEEE Trans. Energy Convers., Vol. 4, No. 3, pp. 487–494, 1989.

[3] M.P. Donsion, M.F. Ferro, Motores sincronos de imanes permanentes, Research book published by the University of Santiago de Compostela, Spain.

[4] M.P. Donsión, J.F. Manzanedo, C. Iglesias, "Coupled Model of the Interior Type Permanent Magnet Synchronous Motor. Application to a Siemosyn Motor", International Conference on Electrical Machines (ICEM'94), París, France, 1994, pp. 144–147.

[5] M.F. Ferro, M.P. Donsion, "Torques Analysis in Permanent Magnet Synchronous Motors", IASTED Power High Tech'89, Valencia, Spain, 1989, pp. 271–275.

[6] M.F. Ferro, M.P. Donsión "Transient Behavior of Permanent Magnet Synchronous Motors Under Sudden Change in Load", IASTED Ninth International Symposium, Modelling, Identification and Control, Innsbruck, Austria, pp. 406–410.

[7] M.F. Ferro, M.P. Donsión, "Specific Characteristics of the Interior Type Permanent Magnet Synchronous Motors. Aplication so a Siemosyn 1FU3134", International AEGEAN Conference on Electrical Machines and Power Electronics, Turkey, Vol. 2, pp. 378–382.

I-4. DYNAMIC MODELING OF A LINEAR VERNIER HYBRID PERMANENT MAGNET MACHINE COUPLED TO A WAVE ENERGY EMULATOR TEST RIG

M.A. Mueller[1], J. Xiang[2], N.J. Baker[2], and P.R.M. Brooking[2]

[1]Institute for Energy Systems, School of Engineering and Electronics, University of Edinburgh, Edinburgh, EH9 3JL, UK

markus.mueller@ed.ac.uk

[2]School of Engineering, University of Durham, Science Site, South Road, Durham, DH1 3LE, UK

J.Xiang@lboro.ac.uk, n.baker@lancaster.ac.uk, p.brooking1@btinternet.com

Abstract. A vernier hybrid machine has been developed for use as a linear generator in a wave energy converter. Accurate predictions for power capture require testing this machine in a nonsinusoidal manner. A dynamic model capable of predicting the machine's behavior for this kind of mechanical excitation is presented. Simple equivalent circuit models have been found to be unsuitable for these machines and a flux-linkage map approach is instead used. Experimental results are used to verify this approach and the functioning of a unity power factor controller.

Introduction

The vernier hybrid permanent magnet machine (VHM) is a member of the family known as variable reluctance permanent magnet machines. These machines are known to produce air gap shear stresses significantly higher than conventional machines. Weh et al. [1] measured a shear stress in the transverse flux machine (TFM) of the order of 200 kN/m^2. However, the TFM is a very complex machine to construct. Mecrow and Jack [2] investigated the use of VRPM topology in a more conventional machine structure and demonstrated that improvements over conventional machines could be made. Spooner and Haydock [3] developed the VHM, which is easier to construct than the TFM, but also benefits from high shear stress. A shear stress of 106 kN/m^2 has been measured for a prototype linear VHM [4]. Combined with the high shear stress capability and the effect of magnetic gearing the VHM is a suitable machine for low speed high torque (or force) applications. One such application is in direct drive wave energy converters as proposed by Mueller and Baker [5].

In this application a linear generator is directly coupled to the wave energy device, such as a heaving buoy or the Archimedes Wave Swing [6] so that the generator experiences the same displacement as the device. Ideal waves are monochromatic resulting in sinusoidal motion. Under this condition the displacement of the generator is well known, but the induced voltage is variable in both frequency and magnitude. In real sea conditions waves from different frequencies combine to give a very random motion. Wave data has been collected

S. Wiak, M. Dems, K. Komęza (eds.), Recent Developments of Electrical Drives, 39–49.

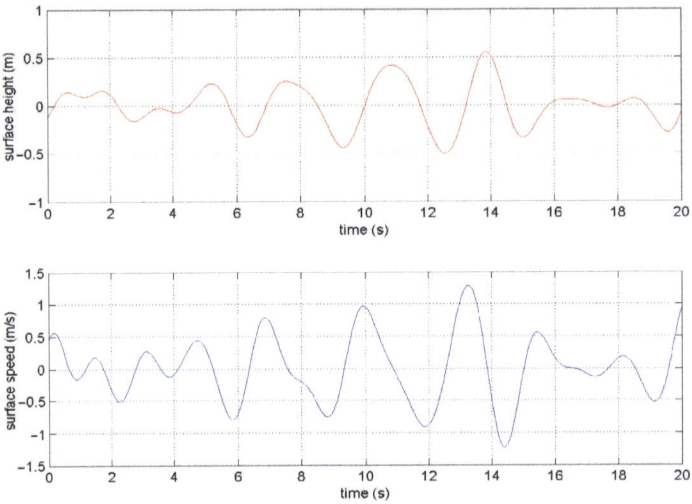

Figure 1. Open circuit voltage in a linear VHM.

over many years using instrumentation buoys, and data from one particular buoy has been collated to demonstrate the nature of the actual displacement a direct drive generator will experience. The data is stored as one dimensional wave data, a compact form summarizing the displacement characteristics of the sea surface [7]. Fig. 1 shows the recreation of typical displacement and velocity waveform obtained from the data of one such buoy. The amplitude and period of the signal have been scaled down to preserve the shape of the signal whilst allowing for a manageable sized test rig [8].

It is quite clear that the machine is expected operate in a very dynamic environment. Hence a dynamic model of the VHM has been developed to investigate its performance and how it interacts with the wave energy device. The model can be used to investigate the use of the electrical generator itself to control and tune the actual characteristics of the device. An indication of how this can be achieved is given in the paper. In addition the model has been developed to enable the designer to investigate the conditions under which the machine has to operate to overcome the inherent high inductance and hence to extract maximum power. Based upon the displacements in Fig. 1 the power output is likely to be random and pulsating. Output from the dynamic model can be used to investigate control strategies for energy storage to smooth the power.

The model is described in the paper and results are presented to verify it and to illustrate its use with this application in mind. It has been developed using MATLAB and incorporated into SIMULINK such that it can be represented as a block to enable it to be connected to models of different marine energy prime movers.

Experimental machine and test rig

A prototype linear VHM has been designed and built, details of which are given in [9], and a photo is shown in Fig. 2. Table 1 summarizes the main geometrical data. The translator is driven via a crank mechanism connected to a variable speed drive via a 14:1 step down gearbox. In this way the translator can be driven at frequencies typical of those expected in

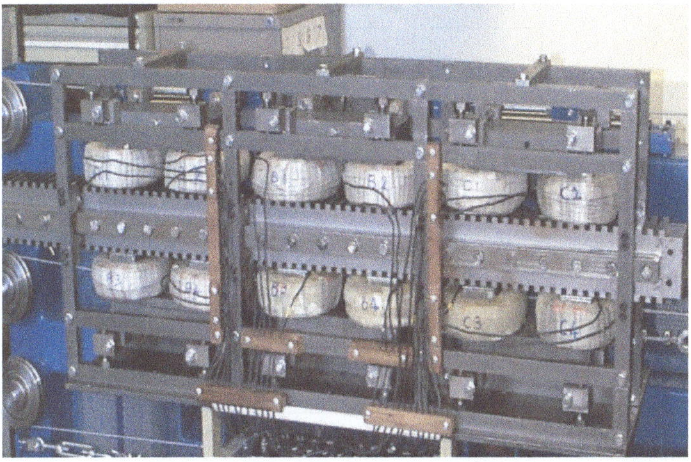

Figure 2. Linear VHM prototype.

a direct drive wave energy converter. The output of the generator is fed into a three-phase ac/ac converter, the details of which can be found in reference [4].

Model description

Fig. 3 shows a block diagram of the complete generator model including a block to maximize the power output of the electrical generator and the prime mover. The complete electromechanical model consists of three blocks, A, B, and C, corresponding to the prime mover, generator, and power conversion equipment, respectively. The contents of each block will be described in this section.

Block A: Prime mover model

In this application the prime mover is a marine energy device, which could be a wave or tidal stream energy converter. The inputs to the model are the force imparted by the waves

Table 1. Main dimensions of the prototype

Magnet pitch (mm)	12
Magnet thickness (mm)	4
Air gap (mm)	1
Core length (mm)	100
Magnets per pole	6
Rotor pole pitch (mm)	24
Rotor slot depth (mm)	10
C-core slot depth (mm)	100
C-core slot width (mm)	144
Core back (mm)	50
Turns per coil	240
Coil resistance (Ω)	0.3
Coil inductance (H)	0.14

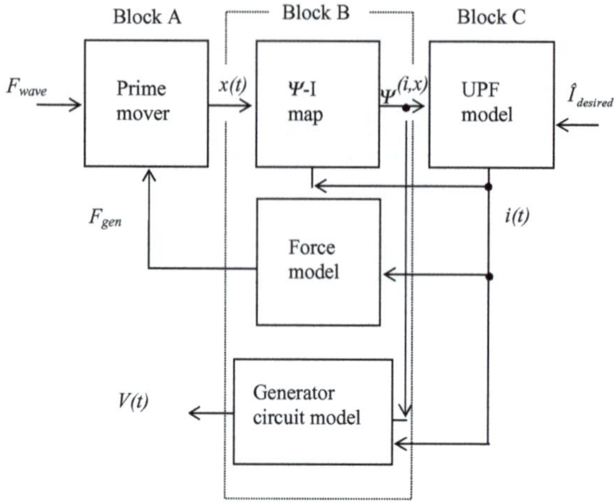

Figure 3. Model block diagram.

say and the generator reaction force, with displacement being the output. A wave energy device is simply modeled as a mass-spring damper system according to equation (1).

$$F_{wave} - F_{gen} = Kx + B\dot{x} + M\ddot{x} \tag{1}$$

where F_{wave} is the force on the device from the incident wave, F_{gen} is the generator reaction force, M is the device mass, K is the buoyancy force, and B is the mechanical resistance. The buoyancy force is essentially a spring force. The mechanical resistance is the sum of the radiation resistance due to waves being created by motion of the device and any mechanical and viscous losses.

A test rig has been developed to emulate the mass-spring damper system in a wave energy device. The structure of the rig is shown in Fig. 4 and has a similar response to equation (1).

An induction motor stepped down through a gearbox drives the crank which drives the rotor (or translator) of the linear generator via the springs and steel cables. Without the springs the amplitude of the rotor is 0.2 m, but with the additional energy from the springs the rotor can achieve much greater amplitudes, depending upon the choice of spring

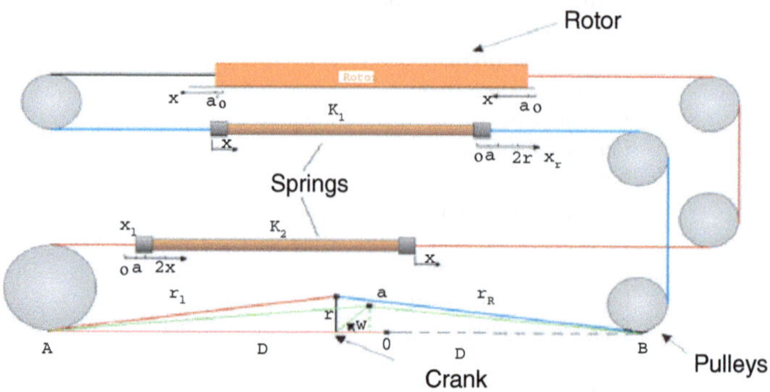

Figure 4. Wave energy emulator.

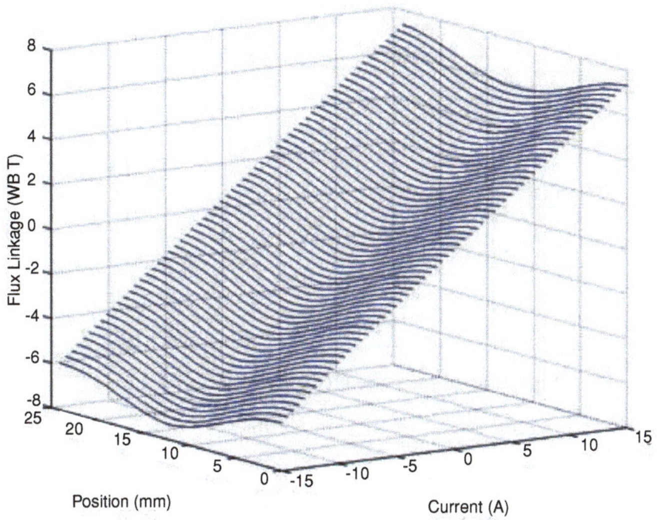

Figure 5. Flux-linkage map for the VHM prototype.

stiffness and the friction. The displacement, $x(t)$, is nearly sinusoidal, thus representing monochromatic waves. In this case F_{wave} is now the force imparted to the crank by the drive motor, K is the spring stiffness, M is the mass of the translator, and B represents the mechanical losses in the rig, which are principally in the pulleys.

Block B: Generator flux-linkage model

The flux-linkage-position map provides a complete electromagnetic description of the machine. Because of the small magnet pitch in the VHM it was found that a simple equivalent circuit approach used for PM synchronous machines was not accurate enough. 2D finite element analysis was used to generate the flux-linkage map, which is essentially used as a look-up table in the model. Knowing the position and current the flux-linkage is then determined, which is used in the force and circuit models. Fig. 5 shows the flux-linkage map generated from a 2D finite element model [9].

Block B: Generator force model

The flux-linkage map generated from 2D finite element analysis is used to determine the co-energy at a particular position and coil current. Force is then calculated from the rate of change of co-energy with displacement according to equation (2):

$$F = \frac{\partial W}{\partial x} \qquad (2)$$

Block B: Generator circuit model

The generator circuit model is shown in Fig. 6, in which the no-load induced emf and inductance are lumped into one represented by the rate of change in flux linkage. The

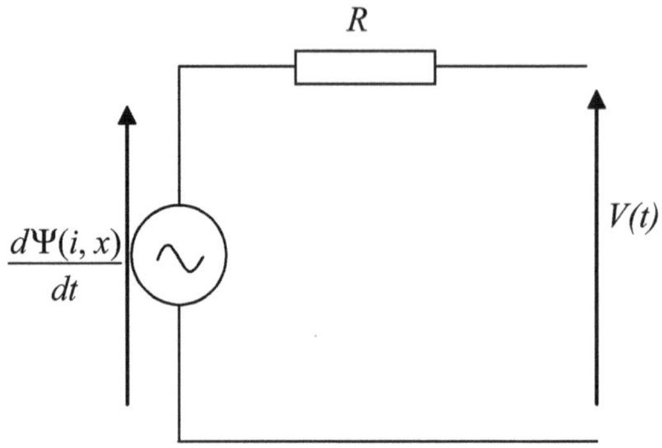

Figure 6. Phase equivalent circuit for the VHM.

terminal voltage is described in equation (3).

$$V(t) = \frac{d\Psi(x, i)}{dt} - i(t)R \tag{3}$$

Block C: Unity power factor correction model

A three-phase active rectifier connected to the terminals of the VHM is controlled to ensure that the generator current is in phase with the induced emf in order to maximize the power generated at the terminals. By controlling the machine in this way reactive power flows from the active rectifier to compensate for the high inductance in the machine. This has been implemented and demonstrated on the prototype machine [4]. A measure of the no-load induced emf is required so that PWM signals can be generated to control the switches in the active rectifier and thus ensuring the current tracks the emf exactly. Search coils are used on the rig to achieve this, but in the simulation the induced emf is obtained from the flux-linkage map at zero armature current (equation (4)). The desired armature current is simply scaled from the no-load induced emf according to the ratio of the peak desired current and the peak no-load induced emf as shown in equation (5).

$$E(t) = \frac{d\Psi(x, i = 0)}{dt}. \tag{4}$$

$$i(t) = \frac{\hat{I}_{desired}}{\hat{E}_{no\ load}} E(t). \tag{5}$$

The generator current then feeds the force model and the flux-linkage model and hence used to calculate the new position, the next value of flux linkage, and so on.

Verification of the generator models in block B

Experimental results were used to verify the model and algorithm developed. Displacement measurements taken from the test rig were fed into the model to generate these results. Fig. 7 shows the no-load emf.

Fig. 8(a,b) shows experimental and computed results for the prototype machine, when operating in unity power factor mode.

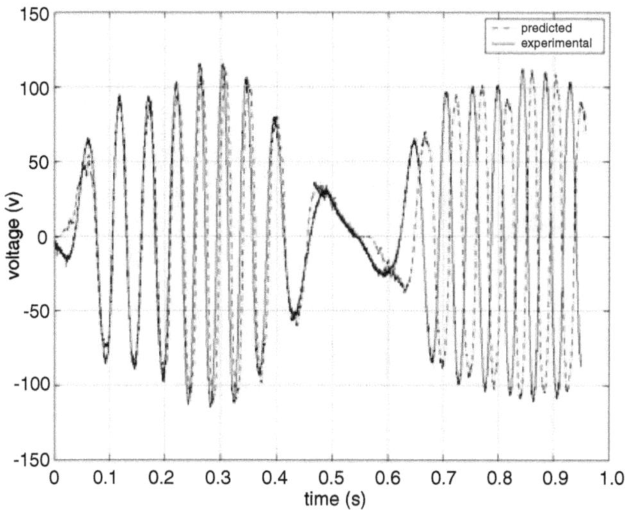

Figure 7. Calculated and measured no-load emf.

The correlation between experimental and simulated results is very good, giving confidence in the model, which can then be used to investigate the machine performance under various loading conditions.

Fig. 9 shows the variation in the total three-phase power at the output obtained by subtracting the I^2R loss from the product of no-load emf and generator current. There will be additional iron losses and eddy current losses in the permanent magnets, which should be included to obtain total efficiency.

Frequency control of a wave energy converter

The results presented in "Verfication of the Generator Models in Block B" have been generated for a known displacement measured from the test rig and simply fed into the model by-passing the force model. This was done simply to verify the electrical parts of the model and hence served its purpose. However, in any electromechanical system the interaction between the electrical and mechanical system are of interest. Fig. 10 shows the force data generated from the generator force model. Also included in the graph are results of force using an analytical model. The force model based upon the flux-linkage map has been verified using experimental results in [10].

The wave emulator test rig is a mass-spring damper system in which the amplitude is given by equation (6) and the resonant frequency is equal to the root of the ratio of stiffness to mass. On the test rig the mass is 190 kg, the

$$X = \frac{F_{wave}}{\sqrt{M^2\left(\omega^2 - \frac{K}{M}\right)^2 + B^2\omega^2}} \tag{6}$$

spring stiffness is 8,000 Nm and the friction (B) is equal to 148 N/m, which was estimated by parameter identification methods from experimental test results. Fig. 11 shows the frequency characteristics of the test rig for different values of B.

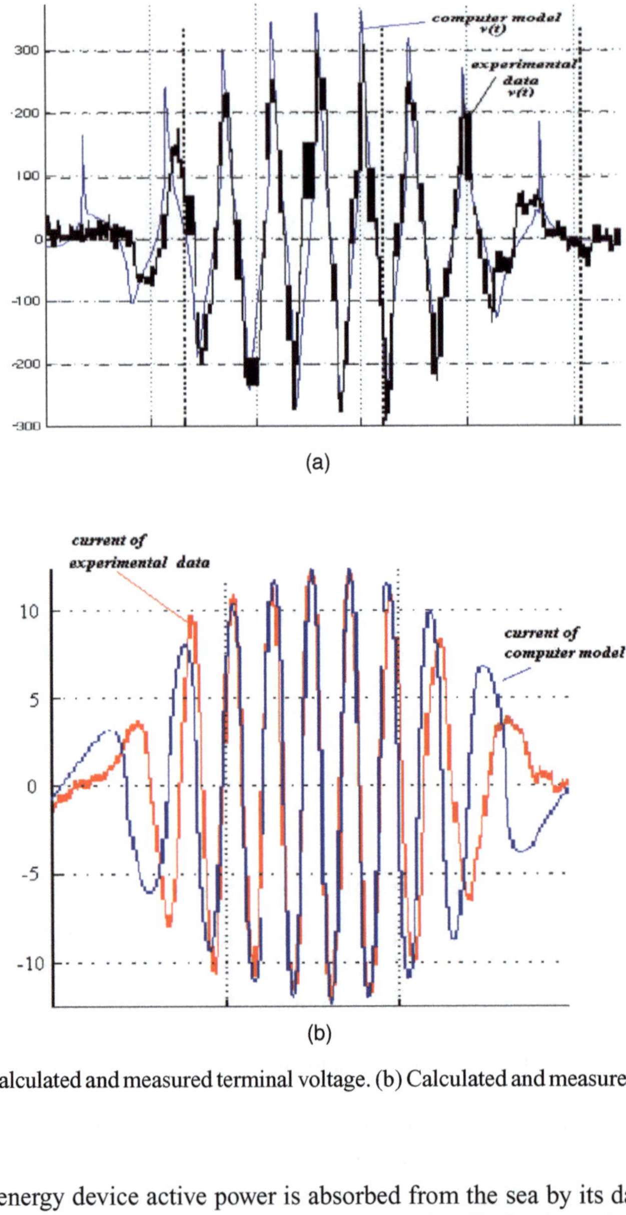

Figure 8. (a) Calculated and measured terminal voltage. (b) Calculated and measured generator phase current.

In a wave energy device active power is absorbed from the sea by its damping components. These are divided into mechanical, viscous, and radiation loses in addition to the electrical damping force providing the electrical power conversion. In addition some of the incident wave energy is used to supply the energy stored in the device mass and device spring stiffness. The electrical analogy of this would be reactive power. At resonance no reactive power is supplied to the device from the sea. When the device is operating at off-resonance points a method of supplying the reactive power externally is required to optimize the energy captured. Externally applied forces that modify the stiffness of the system have been proposed as a means of frequency control. In order to investigate how the generator

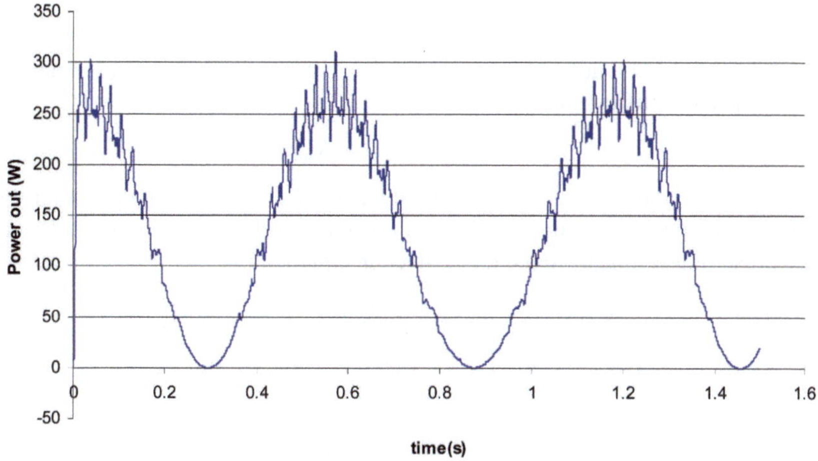

Figure 9. Total three-phase output power.

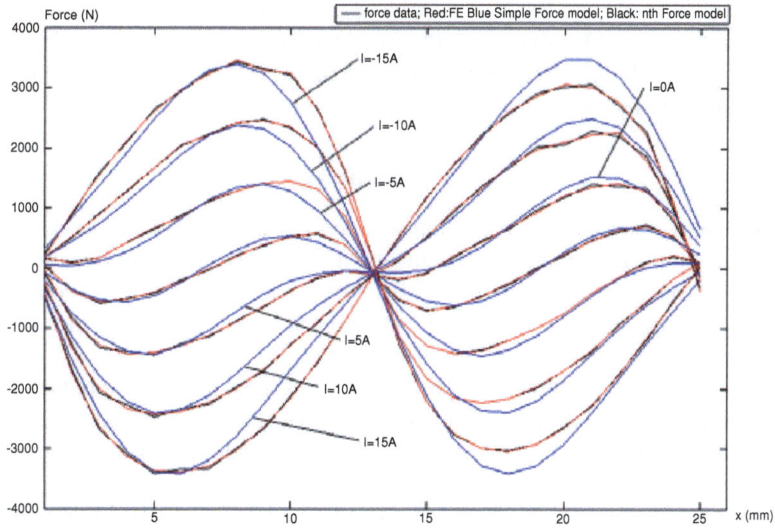

Figure 10. Generator force data for one phase.

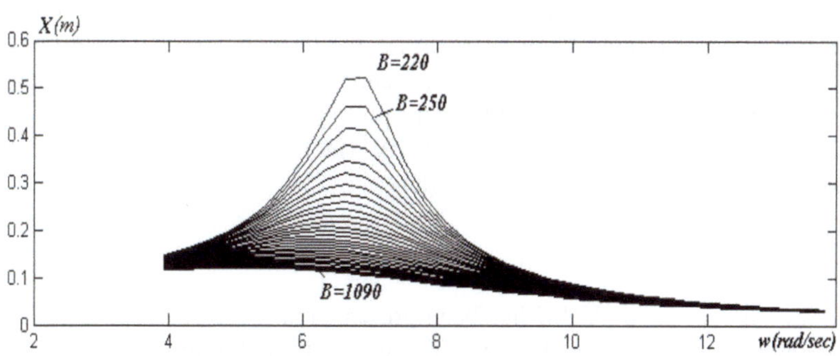

Figure 11. Frequency characteristics of the test rig.

reactive force could be utilized it is represented as the sum of two forces: a damping force and an equivalent stiffness force.

$$F_g = B_g \dot{x} - K_g x \tag{7}$$

where B_g is the equivalent electrical damping and K_g is the equivalent generator stiffness. The frequency response is modified to include the generator reaction force (equation (8)) and the resulting resonant frequency is given in equation (9).

$$X = \frac{F_{wave}}{\sqrt{M^2\left(\omega^2 - \frac{K_w - K_g}{M}\right)^2 + \left(B_w + B_g\right)^2 \omega^2}} \tag{8}$$

$$\omega_0 = \sqrt{\frac{K_w - K_g}{M}} \tag{9}$$

By controlling the stiffness component of force in equation (7) it is possible to modify the frequency characteristics of the device. The two components of forces in equation (7) are perpendicular to one another. Resolving the currents into components 90 degrees to one another will enable control of the generator damping and stiffness force. Control of the latter will enable the frequency characteristics of the device to be modified and hence optimize the energy captured.

Discussion

Marine energy converters, in particular wave energy devices, are highly dynamic devices. Directly coupling a linear electrical generator to the device requires a dynamic model of the generator in question in order to investigate performance under realistic conditions. Such a modeling tool enables the designer to compare and assess electrical generator technology before going to the next stage of production.

As expected the output power from the device shown in Fig. 8 is pulsating due to the reciprocating nature of the motion. Energy storage is required to ensure smooth power flow from a single device which could be investigated by including an energy storage and control block in the overall model.

This paper has described in detail a generator model for the VHM represented by block B in Fig. 3. It forms the basis of a system model including any prime mover model or control models to optimize performance of the whole systems. An indication of how the model might be used to control the frequency characteristics has been given in "Frequency Control of a Wave Energy Converter." Since the generator model is the basic building block in the system, the designer must have confidence in it. The model has been verified using experimental results obtained from the prototype in Fig. 2. A sample of experimental and calculated results is shown in Fig. 8(a,b), which shows very good correlation giving confidence in the electrical generator model.

Conclusion

A dynamic model of the VHM has been presented and verified in this paper using near sinusoidal displacement data. The model forms the basic building block to investigate the performance and control of direct drive wave energy converters.

Acknowledgements

The authors would like to thank Durham University for providing facilities to do this work and the Engineering and Physical Sciences Research Council for funding (Grant no. 38299).

References

[1] H. Weh, H. Hoffman, J. Landrath, "New Permanent Magnet Excited Synchronous Machine with High Efficiency at Low Speeds", Proceedings of the International Electrical Machines Conference, Pisa, Italy, September 1988, pp. 35–40.

[2] B.C. Mecrow, A.G. Jack, "A New High Torque Density Permanent Magnet Machine Configuration", Proceedings of the International Electrical Machines Conference, Cambridge, MA, USA, September 1990.

[3] E. Spooner, L. Haydock, Vernier hybrid machines, IEE Proc. Part B Electr. Power Appl., Vol. 150, No. 6, pp. 655–662.

[4] M.A. Mueller, N.J. Baker, P.R.M. Brooking, J. Xiang, "Low Speed Linear Electrical Generators for Renewable Energy Applications", Proceedings of the Linear Drives in Industrial Applications Conference, Birmingham, UK, September 2003.

[5] M.A. Mueller, N.J. Baker, "A Low Speed Reciprocating Electrical Generator", IEE Power Electronics, Machines and Drives Conference, Bath, April 2002.

[6] H. Polinder, B.C. Mecrow, A.G. Jack, P. Dickinson, M.A. Mueller, "Linear Generators for Direct Drive Wave Energy Converters", Proceedings of the International Electrical Machines and Drives Conference, Madison, WI, 2003.

[7] M.J. Tucker, Waves in Ocean Engineering: Measurement, Analysis, Interpretation, Ellis Horwood Series in Marine Science, 1991, ISBN 0-13-932955-2.

[8] N.J. Baker, M.A. Mueller, P.J. Tavner, "Development of Reciprocating Test-Rig for Wave and Tidal Power at the New and Renewable Energy Centre", Proceedings of Marine Renewable Energy Conference, Newcastle, July 2004.

[9] M.A. Mueller, N.J. Baker, Modelling the performance of the vernier hybrid machine, IEE Proc. Part B Electr. Power Appl., Vol. 150, No. 6, pp. 649–654, 2003.

[10] J. Falnes, Ocean Wave and Oscillating Systems: Linear Interactions Including Wave-Energy Extraction, Cambridge University Press, London, 2002, ISBN 0521782112.

I-5. FINITE ELEMENT ANALYSIS OF TWO PM MOTORS WITH BURIED MAGNETS

J. Kolehmainen
ABB Oy, Electrical Machines, FI-65101 Vaasa, Finland
jere.kolehmainen@fi.abb.com

Abstract. In this paper, a permanent magnet synchronous motor (PMSM) with buried V-shape magnets is compared to a motor with unusual design with buried U-shape magnets in every second pole. It is shown that the motor design with U-shape magnets has same electrical properties than the design with V-shape magnets.

Introduction

Permanent magnet synchronous motors (PMSM) with buried magnets have been considered in a wide range of variable speed drives. A buried magnet design has many advantages compared to designs with surface mounted and inset magnets. Flux concentration can be achieved which induces higher air gap flux density. Higher air gap flux density give a possibility to raise torque of a machine. The buried magnets construction also gives a possibility to form air gap and get smoother torque [1]. The rotor can also be produced easier. Some of the different rotor with buried magnets types are presented in Fig. 1.

Buried magnet designs give the possibility to reduce reluctance by narrowing and lengthening the magnets but keeping the amount of the magnets the same. By using buried magnets in V-shape or radial magnets, there are limits to reducing reluctance. Designs with U-shape magnets in every pole have good properties of both designs with V-shape and radial magnets [2]. However, with a design with U-shape magnets in every second pole it is possible to reduce reluctance further.

In this paper two buried magnet machines are compared, one with V-shape magnets and another with U-shape magnets in every second pole. The analysis is done by using time stepping and static calculations with Finite Element Method (FEM) [3]. Also these machines with different magnetic width and length are considered.

Motor designs

Both designs with buried magnets inside the rotor make the assembly of the rotor easier compared to the other designs. Rotor disks keep the magnets in place and no extra reinforcing bandage is needed. The magnets are inserted into punched slots in the laminated rotor iron. The example of design with buried magnets in V-shape is shown in Fig. 2 and with buried U-shape magnets in every second pole in Fig. 3.

S. Wiak, M. Dems, K. Komęza (eds.), Recent Developments of Electrical Drives, 51–58.

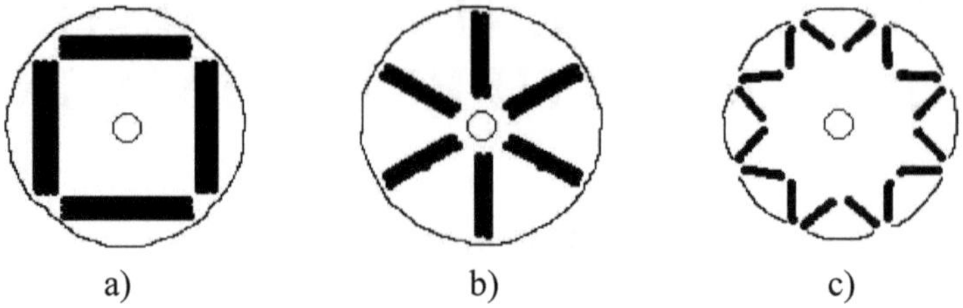

Figure 1. Rotor constructions of buried permanent magnet motors with (a) tangential magnets, (b) radial magnets, and (c) V-shape magnets.

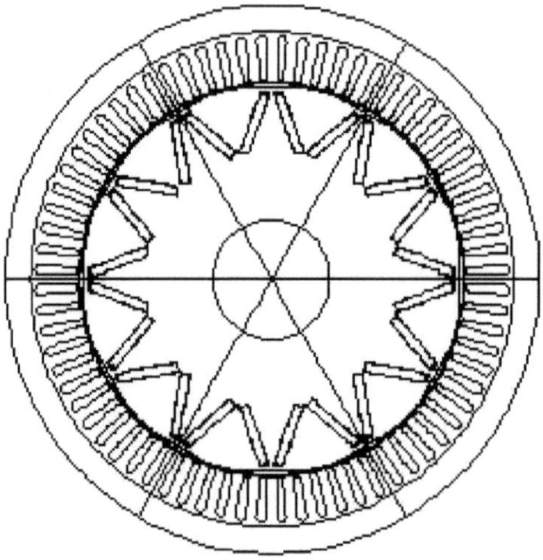

Figure 2. 12-pole PM motor design with magnets in V-shape.

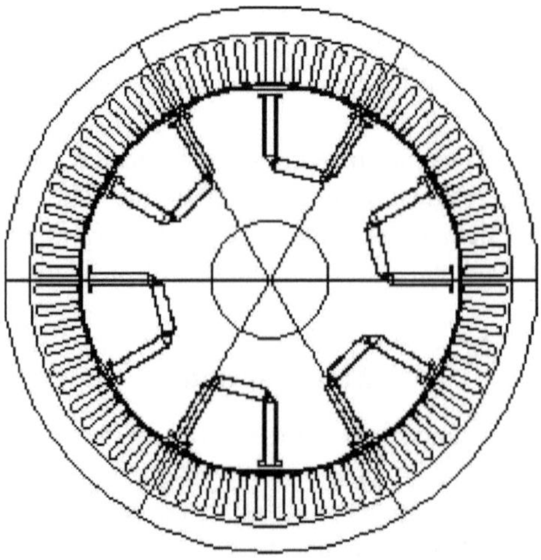

Figure 3. 12-pole PM motor design with magnets in U-shape in every second pole.

The only difference of these two motors is in their internal rotor structure. Areas of the magnets are same and in the structure of Fig. 3 magnets per one pole are thinner and longer. Also, a structure where magnets per one pole have same width and length is considered. In addition, all sizes of the iron bridges between the magnets and air gap are the same.

The number of magnet pieces in U-shape design is also reduced to 3/4 of number in V-shape design. This saves time for inserting magnets to rotor.

Calculation results

The electrical properties of the motors with V-shape and the U-shape designs are studied. Studied motor data is shown in Table 1. Calculations are done with the time stepping method with FEM [3]. Properties are studied with different loads.

In calculations voltage source and delta connection is used. Because of the different structure of rotors, two poles of each construction are modeled. Circuit of calculations is shown in Fig. 4. In the circuit there are three voltage sources, six winding connection and three end winding resistances and three end winding inductances. In all time stepping calculations, voltage angle of the stator and amplitude are same. Calculations are started with different rotor angles and stopped when transient phenomena is over. Constant rotor speed is used.

The flux lines of three example designs with nominal load are shown in Figs. 5–7. The packing of the flux can also be seen. Every second pole in the U-shape designs is different which means that the structure between two poles is not symmetric. In Figs. 5 and 6, total length, width, and area of magnets per one rotor pole are same.

Fig. 8 shows flux densities in the stator teeth as a function of time with nominal load calculations of V-shape and U-shapeA designs. The effect of difference of designs can be

Table 1. Motor data

Shaft height	280 mm
Power	27.5 kW
Voltage	370 V
Current	45 A
Pole number	12
Speed	300 rpm

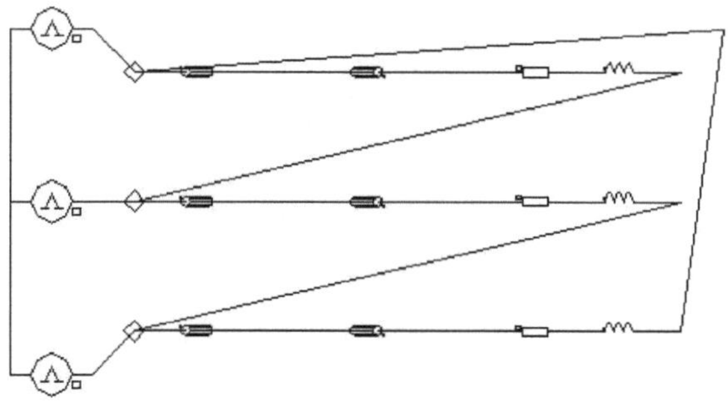

Figure 4. Circuit used in calculations.

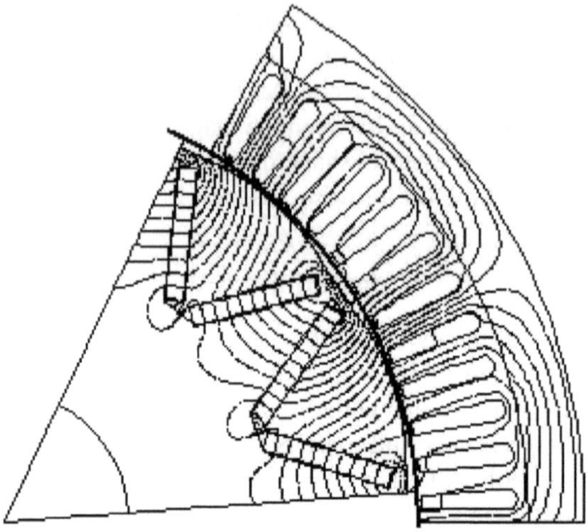

Figure 5. Packing of flux with nominal load and original design (V-shape).

seen. It is relatively small. It can also be seen that absolute value of flux is periodically symmetric between two poles in our U-shapeA design. Flux is also symmetric with U-shapeB design. No deviation of symmetry can be seen.

Fig. 9 shows flux densities of V-shape and U-shapeA designs produced only by magnets in the stator teeth with different rotor angles. Length and width of magnets per one rotor pole are same. Maximum and average flux densities of V-shape and U-shapeA designs are 1.463 T, 1.420 T and 0.932 T, 0.926 T. Flux densities with U-shapeA design is slightly smaller because of small effect of gaps between the magnets.

Nominal and maximum loads of our three example designs are calculated with time stepping calculations. In Table 2, the calculation results are compared to experimental results of V-shape design. It can be seen that calculation of V-shape design gives a correct

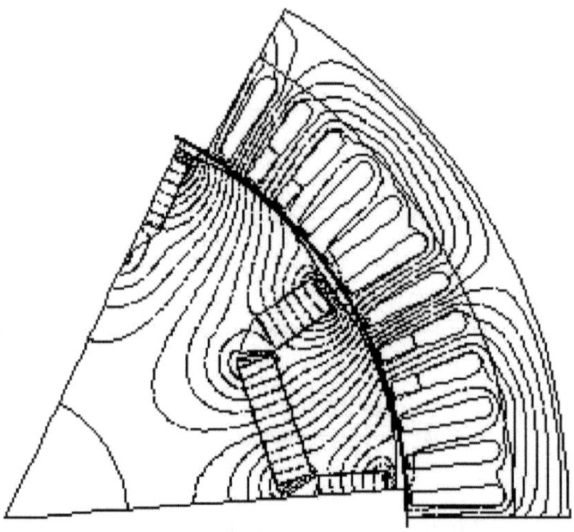

Figure 6. Packing of flux with nominal load and new design A (U-shapeA).

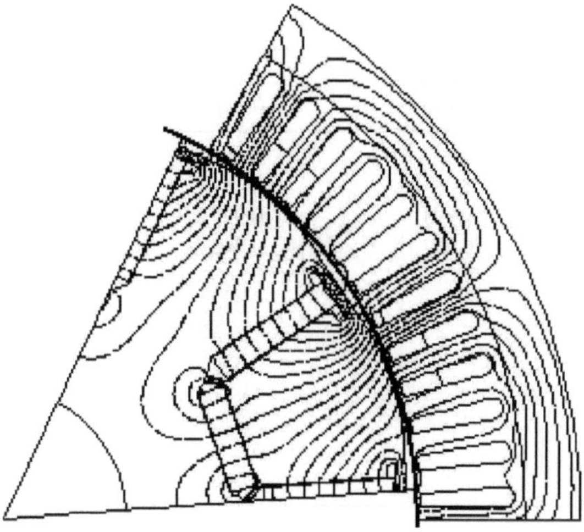

Figure 7. Packing of flux with nominal load and new design B (U-shapeB).

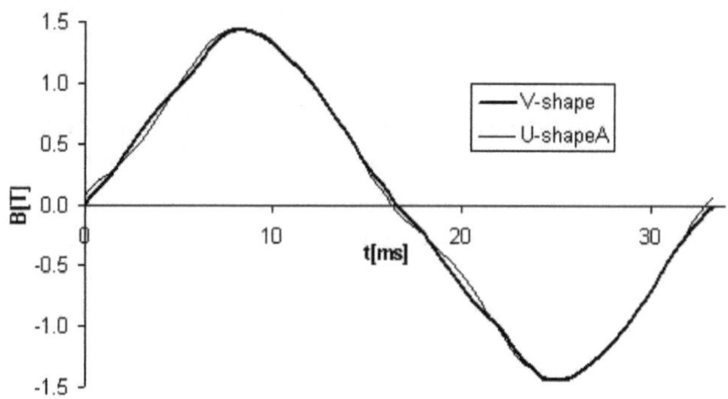

Figure 8. Flux densities of V-shape and U-shapeA designs in the stator teeth on one period with nominal loads.

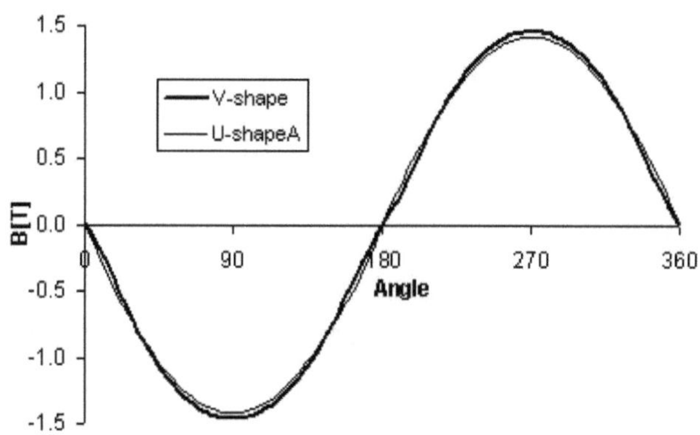

Figure 9. Flux densities of V-shape and U-shapeA designs produced only by magnets in the stator teeth with different rotor angles.

Table 2. Comparison of nominal load results

	Measured V-shape	Calculated V-shape	Calculated U-shapeA	Calculated U-shapeB
T_n[Nm]	875	875	875	875
I_n[A]	46.5	45.5	45.8	45.4
Cos ω	0.993	0.997	0.997	1.000
η	0.929	0.946	0.939	0.938
Angle		32.6	29.7	25.4
T_k [Nm]		1,688	1,557	1,418
I_k [A]		122.5	113.8	98.5
Cos ω		0.856	0.841	0.804
Angle		114.1	111.5	99.2
Magnets thickness		7.3	7.3	5.15
Magnets width		52	52	72.2
Magnets area		379.6	379.6	371.83

current. Only copper losses in stator winding are taken account in efficiency η calculations. Other losses are relatively small.

In the table first calculated U-shapeA results are calculated with the design with same total magnet length and thickness per pole than with V-shape design. The second calculated results U-shapeB are calculated with design with longer and thinner magnets per pole. Magnets thickness, width, and area are also shown in the table with unit of mm. Dimensions of V-shape design are real dimensions of one magnet and for U-shape designs dimension are values which corresponds values of V-shape design.

The maximum output torque with the first U-shapeA design is smaller than with the V-shape design and it has also smaller load angle difference. This is due to smaller reluctance torque and effect of iron bridges between the magnets. Torque and reluctance torque curves are shown in Fig. 10.

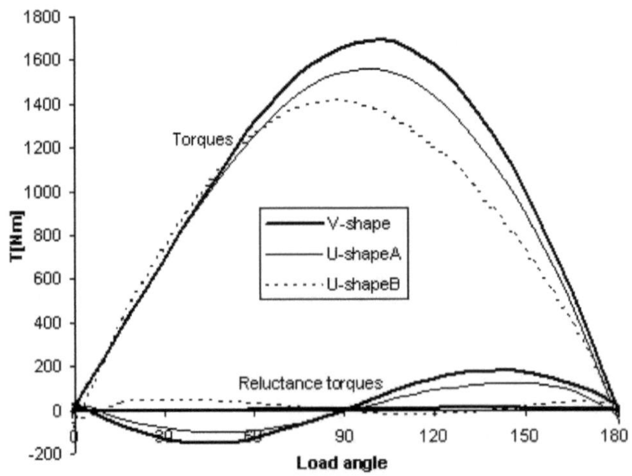

Figure 10. Torque and reluctance torque of motors with V- and U-shape designs as a function of load angle.

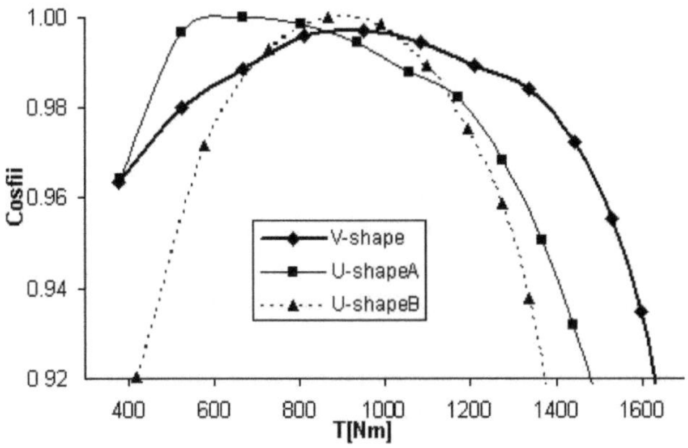

Figure 11. Power factor as a function of torque.

Reluctance torque is larger with V-shape than with U-shapeA design, because the magnetic structure of rotor. By comparing torques of U-shapeA and U-shapeB designs can also see the effect of decreasing of magnet thickness. Reluctance and maximum torque is smaller with thicker magnets.

Power factors of V-shape and U-shape designs are shown as a function of torque in Fig. 11. Power factor of the motor with the U-shapeA design is larger up to the nominal point and with the higher torque it is smaller. Nominal torque of the motors is 875 Nm and usually the motors are used with partial loads with different speeds. Hence, the motor with the U-shape magnets is usually in the torque range with better power factor. Also the maximum torque decreases because of the smaller reluctance. With the longer and thinner magnets in the V-shapeB design there is smaller maximum torque and higher power factor with nominal load as can be expected.

There is significant difference of torques between V- and U-shapeA designs. This is shown in the Fig. 12. With the U-shapeA design, the oscillation of torque is with the frequency of magnets going over stator phase. With the V-shape design oscillation frequency is two times of frequency with U-shapes, because two magnets go over one stator phase with V-shape

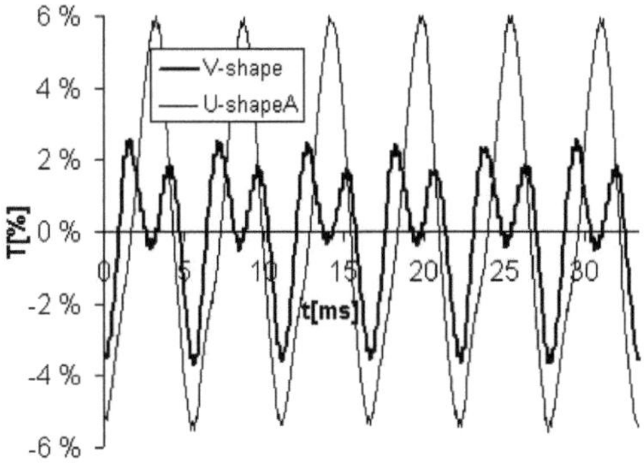

Figure 12. Torque oscillations of V-shape and U-shapeA designs.

design while one magnet going over one stator phase with U-shapeA design. In addition the amplitude is smaller with V-shapes.

Conclusion

It is shown that the PM motor with the U-shape magnets in every second pole works as well as the conventional PM motor with the V-shape magnets in every pole. Asymmetrical structure of pole pairs in this design cause no asymmetry to the magnetic field of air gap. This design gives a possibility to get higher flux densities with the same amount of magnets. The number of magnet pieces is also reduced.

Torque oscillation with U-shapeA design is too high compared to V-shape design. This could be avoided with using different stator slots or iron structure near the magnets and air gap.

In conclusion, this new solution gives more possibilities to produce buried permanent motors with better power factor and efficiency.

References

[1] J. Salo, T. Heikkilä, J. Pyrhönen, T. Haring, "New Low-Speed High-Torque Permanent Magnet Synchronous Machine With Buried Magnets", International Conference Electrical Machines (ICEM 00), Vol. 3/3, Espoo, Finland, August 28–30, 2000, pp. 1246–1250.

[2] F. Libert, J. Soulard, J. Engström, "Design of a 4-pole Line Start Permanent Magnet Synchronous Motor", International Conference Electrical Machines (ICEM 02), Brugge, Belgium, August 25–28, 1998, p. 173.

[3] Flux2D Software, www.cedrat.com.

I-6. DESIGN TECHNIQUE FOR REDUCING THE COGGING TORQUE IN LARGE SURFACE-MOUNTED MAGNET MOTORS

R. Lateb[1], N. Takorabet[1], F. Meibody-Tabar[1], J. Enon[2] and A. Sarribouette[2]

[1]INPL–GREEN, 2 avenue de la forêt de Haye, 54516 Vandoeuvre-lès-Nancy, France
ramdane.lateb@ensem.inpl-nancy.fr, Nouredinne.Takorabet@ensem.inpl.nancy.fr, Farid.Meibody-Tabar@ensem.inpl-nancy.fr
[2]Converteam, 4 rue de la Rompure, 54250 Champigneulles, France
jacques.enon@converteam.com, alain.saribouette@converteam.com

Abstract. An approach based on magnets segmentation is introduced for minimizing the cogging torque of surface-mounted permanent magnet motors. The authors show that the magnet segmentation has also an effect on the ripple torque especially on its high order harmonics. However, this technique has a small effect on the main performances of the motor such as the average torque. So, the segmentation number is chosen according to the choice of the magnet span and the stator winding. An approach based on Fourier analysis is used to justify the numerical results obtained by finite elements method.

Introduction

The use of high power surface-mounted permanent magnet (PM) motors in different industrial applications (windmill generator, marine propulsion, traction . . .) is very attractive thanks to their high torque density [1]. The design of PM motors must take into account the requirements of such applications. One of the most important constraints is the mechanical shaft vibrations, especially at low speeds, that can be avoided by reducing the amplitude of torque harmonics. This can be achieved by using a wide range of techniques proposed by several authors [2]. Some of these techniques are based on modifying the current waveforms to cancel torque pulsations for any PM motor with known electromotive force (EMF) waveform [3]. If the EMF is not sinusoidal, high dynamic current waveform is required which is difficult to apply by high power inverters. The other techniques are based on structural solutions.

The first structural solution consists in an adapted choice of the stator winding for a given number of stator slots [4]. Even if this solution allows to minimize the ripple torque rate due the interaction between stator currents and rotor magnets, it has no effect on the cogging torque since the stator slots design is chosen. As well known in PM motors,

S. Wiak, M. Dems, K. Komęza (eds.), Recent Developments of Electrical Drives, 59–72.

cogging torque arises from the interaction between magnets and slotted iron stator. Another solution consists to optimize the magnets pole angular width (span). Even if this geometrical parameter allows to reduce the cogging torque, it is often used to maximize the average torque and to minimize the ripple torque rate for an imposed current waveform. It has shown that a classical requirement to eliminate the cogging torque first harmonic is to choice a magnet pole span almost a multiple of slot pitch [5–7]. It's obvious that motors with closed slots or slotless stator have no cogging torque. These solutions, which lead to mechanical difficulties, can be approached by introducing magnetic wedges in slots that reduces significantly the cogging torque. It has been shown that choosing equal tooth and slot width may diminish the fundamental component of the cogging torque [5]. In fact, for a given slot pitch, there is several values of tooth width which minimize the cogging torque. A teeth-pairing design with different tooth widths as well as the teeth notching design is other techniques that allow to reduce the cogging torque [8]. One of the most popular techniques to reduce cogging torque is to skew the stator lamination stack or rotor magnets [9]. Ideally, the cogging torque is vanished with a skewing angle of integer multiple of the cogging torque period. Stator skewing is less interesting because of a more complex construction. To make easier the rotor manufacturing, the rotor skewing may be done by placing the PM axially skewed by several discrete steps [9]. All the techniques cannot be related in this paper but one can find a large bibliography in [2].

In high power PM motors, the magnet pole span is too large for being realized in only one block. For technical and cost considerations, each rotor pole is often realized with several elementary magnet blocks with the same polarity (magnet segmentation). In this paper, the authors present the effect of the magnets segmentation on the machine performances. One can expect that the segmentation of the magnets does not modify significantly the main performances since the magnet span conserves the same value. However, the magnets segmentation modifies locally the air-gap magnetic field distribution which leads not only to a significant modification of the cogging torque but also the back EMF harmonics.

In the second section, using analytical approaches, we present different techniques reducing the cogging torque. For the sake of the analyze of the cogging torque minimization, the analytical method is useful but not enough precise to evaluate its exact shape and its amplitude. In the third section we present the used numerical finite elements method. In the last part of this paper, for different values of the magnet span, we analyze the influence of magnets segmentation on the performances (Average torque, Harmonics of back EMF, cogging torque, Pulsating torque) of a surface-mounted PM motor.

The obtained results and the carried out analysis highlight that for a given value of a magnet pole span, it exists an optimal number of magnet blocks per pole for which the best compromise between average torque and pulsating torque rate may be achieved.

Used techniques for cogging torque minimization

The cogging torque being caused by the interaction between the rotor magnets and the stator teeth. The main parameters which affect significantly its shape and amplitude are: the

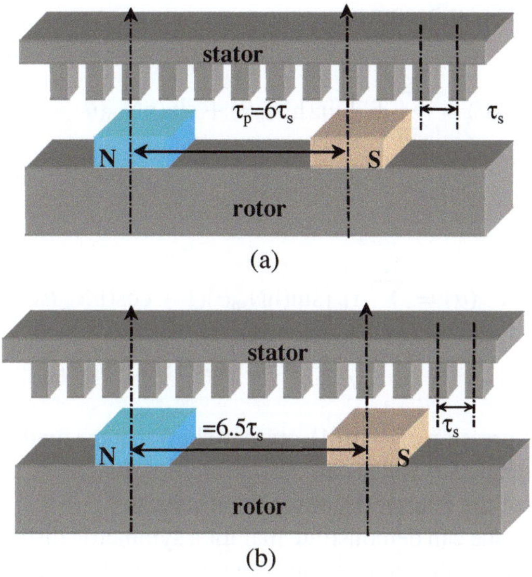

Figure 1. Two cases of symmetric distribution: (a) two poles, 12 slots; (b) two poles, 13 slots.

number of stator teeth, the magnet pole span, and the magnet segmentation. In the following we detail how to choose these parameters to minimize the cogging torque.

Number of slot

In a PM motor with N_{sp} stator slots per pole pair, the contribution T_{cm} of one magnet to the cogging torque is of the form:

$$T_{cm}(\theta) = \sum_{h=1}^{\infty} T_h \sin(hN_{sp}\theta) \tag{1}$$

Where T_h, is the Fourier coefficient of the hth harmonic. θ is the electrical angular position.

To illustrate the interest of an odd number of stator slots per pole pair, we consider a pole pair of a PM motor as shown in Fig. 1. The cases of $N_{sp} = 12$ (Fig. 1a) and $N_{sp} = 13$ (Fig. 1b) are considered.

In Fig. 1(a), we can observe that each magnet has the same relative position with respect to the stator teeth. The cogging torque per pole pair is twice the contribution of one magnet.

$$T_{cp}(\theta) = 2\sum_{h=1}^{\infty} T_h \sin(hN_{sp}\theta) \tag{2}$$

In the second case, the magnets have not the same relative position with respect to the stator teeth. Since one magnet has a half slot pitch electrical shift angle $\theta_0 = \dfrac{\pi}{N_{sp}}$, the cogging torque per pole pair becomes:

$$T_{cp}(\theta) = \sum_{h=1}^{\infty} T_h \left[\sin(hN_{sp}\theta) + \sin(hN_{sp}(\theta - \theta_0))\right] \tag{3}$$

Hence:

$$T_{cp}(\theta) = \sum_{h=1}^{\infty} T_h \left[\sin(hN_{sp}\theta)(1 + \cos(hN_{sp}\theta_0)) - \cos(hN_{sp}\theta)\sin(hN_{sp}\theta_0)\right] \tag{4}$$

By replacing the expression of θ_0 in (4), one can show that the fundamental of the cogging torque ($h = 1$) is eliminated as well as all odd harmonics. Indeed, in this case only the even harmonics ($2 \times N_{sp}$) of the Fourier decomposition subsist.

In the general case, one can demonstrate that for a symmetric distribution of N_p magnets and N_s slots in the motor, the fundamental frequency of the cogging torque is the least common multiple (LCM) of N_p and N_s.

$$N_L = LCM(N_p, N_s) \tag{5}$$

High power low speed PM motors offer the possibility to use a large number of slots per pole that allows to increase the frequency of the cogging torque first harmonic. Moreover, as illustrated above, using an odd number of slots per pole pair doubles the frequency of the cogging torque.

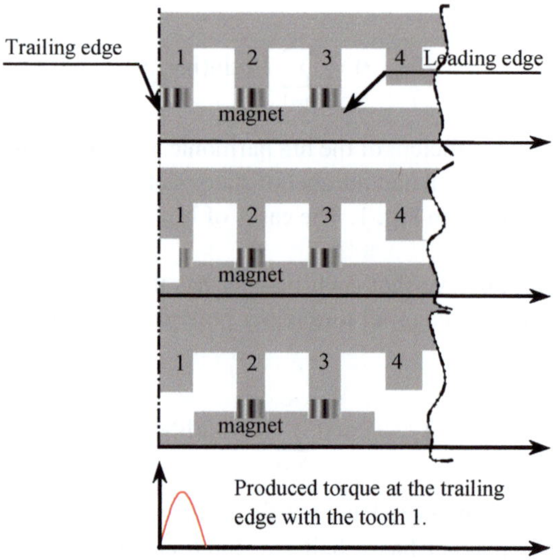

Figure 2. Simple model of cogging torque mechanism.

Magnet span

The motor cogging torque is mainly caused by the interactions of magnets edges (trailing and leading edges) and stator teeth. Thus, the study of the cogging torque can be reduced to the analysis of these interactions.

By ignoring the effect of rotor curvature, magnet leakage flux, and fringing flux and by bringing back to a rectangular field problem, one can represent the mechanism of cogging torque production by the simple model illustrated in Fig. 2, where the dashed lines represent the area in which the main part of magnetic energy is stored.

For the three magnet positions shown in Fig. 2, the variation of the magnetic energy under the tooth 1 during the passage of the trailing edge is illustrated, while the energy under the other teeth (2, 3, 4) doesn't vary. The produced torque due to the passage of the trailing edge is proportional to this energy variation. It is a periodic function that can be expressed as:

$$T_{trailing}(\theta) = T_0 + \sum_{h=1}^{\infty} T'_h \cos(hN_{sp}\theta) \tag{6}$$

Let's define α_m as the electrical shift angle between the leading and trailing edges. The torque produced by the leading edge is the opposite of the one produced by the trailing edge shifted by α_m.

$$T_{leading}(\theta) = -T_0 - \sum_{h=1}^{\infty} T'_h \cos(hN_{sp}(\theta - \alpha_m)) \tag{7}$$

Then the expression of the cogging torque, produced by this magnet, becomes:

$$T_{cm}(\theta) = \sum_{h=1}^{\infty} T'_h \left(\cos(hN_{sp}\theta) - \cos(hN_{sp}(\theta - \alpha_m))\right) \tag{8}$$

Under the assumptions mentioned above, from (8) it can be easily shown that the cogging torque can be eliminated by choosing:

$$\alpha_m = \frac{2\pi k}{N_{sp}} = \tau_s k \tag{9}$$

with k an integer and τ_s the electrical slot pitch angle.

This condition, indicating that the angle α_m must be a multiple of a slot pitch, is obtained under the mentioned assumptions, which don't take into account the effect of rotor curvature, magnet leakage flux and fringing flux. Considering these phenomena, a zero cogging torque cannot be achieved. However, according to different authors, using finite element analysis, the cogging torque may be minimized for a magnet span of $\alpha_m = (n + 0.14)\tau_s$ by ignoring the effect of rotor curvature [5] or $(n + 0.17)\tau_s$ by considering the effect of rotor curvature [6] or $(n + 0.25)\tau_s$ for linear motors [7].

Magnets segmentation

As previously said, for manufacturing and cost reasons, in large permanent magnet motor each rotor pole is often realized with several elementary magnet blocks with the same polarity (magnets segmentation). In the following, a curved shape magnet is used for each

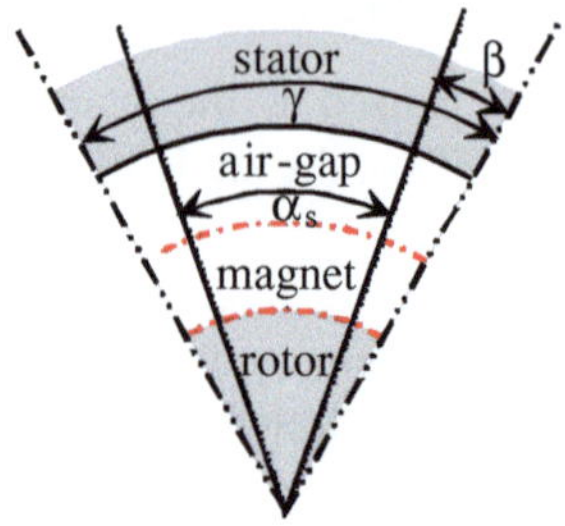

Figure 3. Representation of an elementary RSMM.

nonsegmented rotor pole and rectangular cross-section magnet blocks are used for realizing each pole of a segmented permanent magnet machine. For the sake of simplicity, we will use the next abbreviations:

- RSMM: Rectangular Surface-Mounted Magnets with parallel magnetization.
- CSMM: Curved Surface-Mounted Magnet per pole.

For a CSMM motor, the mechanical air-gap is constant (CSMM is delimited by the red dashed lines in Fig. 3), while in the RSMM motor the air-gap varies.

Fig. 3 shows the geometrical representation of an elementary RSMM. The relation between the whole magnet pole span α_m and the elementary block magnets span γ is:

$$\begin{cases} \alpha_m = N\gamma \\ \gamma = \alpha_s + 2\beta \end{cases} \tag{10}$$

where N is the number of elementary magnet blocks forming each pole, α_s the opening angle of the elementary magnet facing the air-gap, and β half of the opening angle of the slit between two elementary magnet blocks. The span of each elementary magnet block is expressed in term of the slot pitch in the same way adopted above:

$$\begin{cases} \gamma = (n \pm \varepsilon)\tau_s \\ 0 \le \varepsilon < 1 \end{cases} \tag{11}$$

where n is an integer.

Using the results given in [6], the optimal span of each elementary magnet block should be such as $\varepsilon \approx 0.17$. However this result has been obtained for CSMM motor. For rectangular magnets (RSMM) motors the optimum value of ε will be probably modified.

Numerical analysis

Finite elements method is used for the computation of the machine characteristics. In order to increase the precision of the results and especially the computation of the cogging torque, a finer mesh is applied all around the air-gap [10], at each rotor position the meshing is renewed. Thanks to geometrical and electrical symmetries, only one pole pair of the machine is considered, it allows a minimum time consuming. In addition, for each design we made the calculations for 60 different rotor positions over a slot pitch. The number of nodes is more than 40,000 nodes. The computations take into account the saturation of iron core.

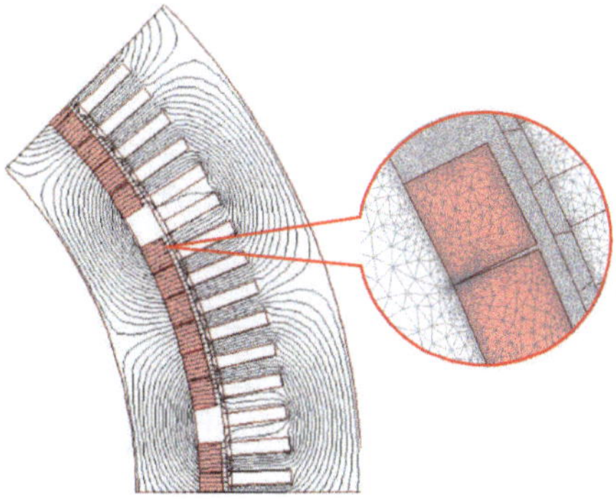

Figure 4. PM motor with six rectangular magnet blocks.

The cogging torque is computed through the Maxwell weighted stress tensor method [11] and the used software for the computation is FEMM [12].

Fig. 5 shows a cross-section view of a PM motor in three configurations where each magnet pole is divided in two, three, or four blocks. In the following, the performances of the segmented PM motor having N elementary magnet blocks (N = 1, ..., 6) per pole are computed (Fig. 6). These performances will be presented vs. the total magnet span $\alpha_m = N\gamma$.

Results

The computations are performed for a 6 MW, 16-poles, 170 rpm, surface-mounted PM motor supplied by sinusoidal waveform currents. An adapted stator winding with a fractional slot number per pole and per phase (15 slots per pole pair) allows to reduce considerably the space harmonics of stator magnetomotive force (MMF), especially the fifth one which is totally cancelled. Another advantage of using an odd slot number over a pole pair is to increase the pulsation of the cogging torque. So, for the studied topology the cogging torque period is $2\pi/30$ electrical degrees. Magnetic wedges used in the slot openings (isthmus),

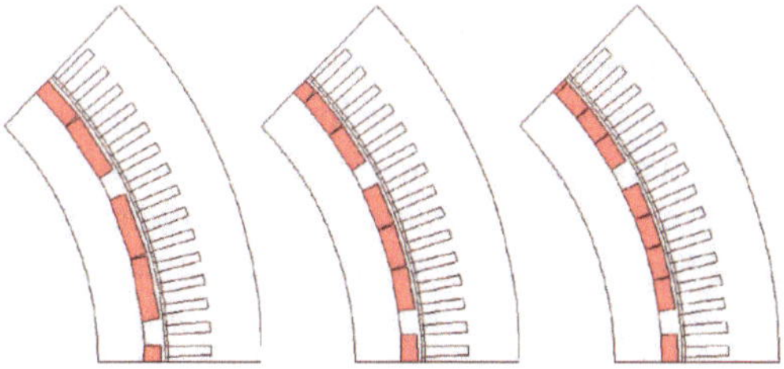

Figure 5. Cross-section view of one pair pole of the PM motors with: (a) two magnet blocks per pole; (b) three magnet blocks per pole; (c) four magnet blocks per pole.

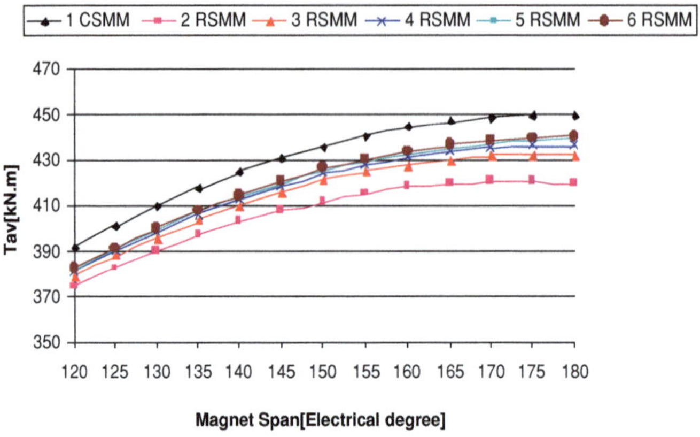

Figure 6. Average torque vs. magnet span for each structure.

allows to reduce the amplitude of the cogging torque. The slot-opening angle is less than half a slot pitch τ_s.

Assuming that the phase currents are function of the electrical rotor position θ (self-controlled PM motor), the electromagnetic torque T_{em}, which is the sum of the cogging torque, T_c and the current-magnets interaction torque T_{e-i}, can be expressed by:

$$\begin{aligned} T_{em}(\theta) &= T_c(\theta) + T_{e-i}(\theta) \\ &= T_c(\theta) + \frac{1}{\Omega}\sum_{j=1}^{q} i_j(\theta) \times e_j(\theta) \end{aligned} \tag{12}$$

where:

Ω is rotor angular speed.
i_j and e_j are the current and back EMF of the jth phase.
q is the phase number (q = 3).

The phase currents are assumed to be sinusoidal while the back EMF contains harmonics. These harmonics are at the origin of the pulsating component of T_{e-i}, called ripple torque T_r. So the total pulsating torque T_{cr} is the mean value of T_{e-i} is the average torque T_{av}.

Fig. 6 shows the average torque vs. magnet pole span for the PM motors with N elementary magnet blocks ($N = 1, \ldots, 6$) per pole. It is obvious that the nonsegmented PM motor (CSMM) in which a curved magnet per pole is used, has the highest average torque since the magnet volume is more important and the air-gap is constant. The PM motor with two magnets blocks per pole (two RSMM) has the lowest average torque because the average air-gap is more important which affects the air-gap flux density. The four, five, and six RSMM configurations present almost the same average torque for a given value of magnet pole span. Within sight of Fig. 6, one finds a classical result, which shows that on one hand, beyond $\alpha_m \approx 165°$ the profit in average torque is weak compared to the cost generated by the increase of the magnet volume. On the other hand, for $\alpha_m < 145°$ the average torque is relatively weak.

Varying the magnet span influences not only the average torque but also the cogging torque and the ripple torque. As these two pulsating torque components depend differently on the magnet span, a compromise should be made in order to minimize the total pulsating

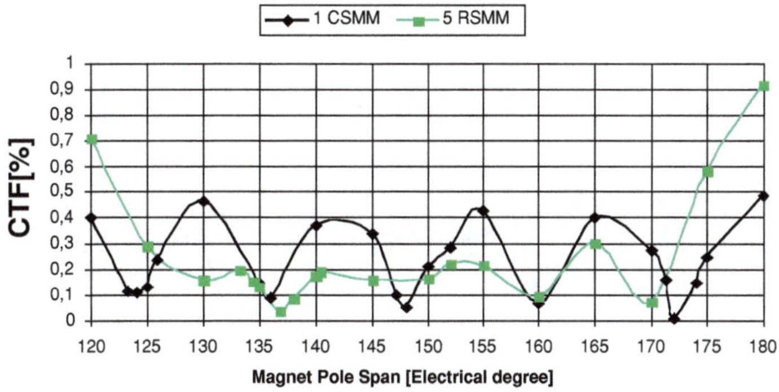

(a)

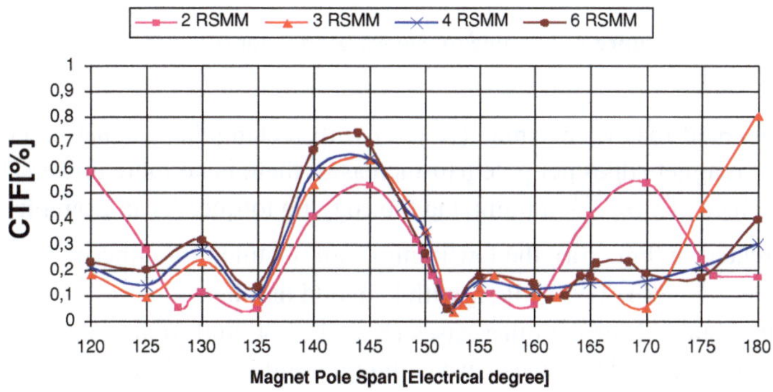

(b)

Figure 7. Cogging torque factor vs. magnet pole span.

torque T_{cr}. In order to carry out a comparative study of various configurations, we use the next criteria:

$$CTF = \frac{T_{cpp}}{T_{av}}; \quad PTF = \frac{T_{crpp}}{T_{av}} \tag{13}$$

where CTF and PTF are respectively the Cogging Torque Factor and the Total Pulsating Torque Factor. T_{cpp} and T_{crpp} are respectively the peak-to-peak cogging torque and the peak-to-peak total pulsating torque.

In order to illustrate clearly the effect on the magnets subdivision on the cogging torque, the results are presented on Fig. 7(a,b). The CTF obtained for the CSMM configuration ($N = 1$) presents several minima (Fig. 11a) achieved for the following values of the magnet pole span: $\gamma = \frac{\alpha_m}{1} = (n/2 + 0.17)\tau_s$ with $n = 10$ to 14. In [5] the authors have found $\alpha_m = (n + 0.17)\tau_s$. These two results are not in contradiction, because the studied stator has an odd slot number per pole pair which doubles the number of minima of the cogging torque.

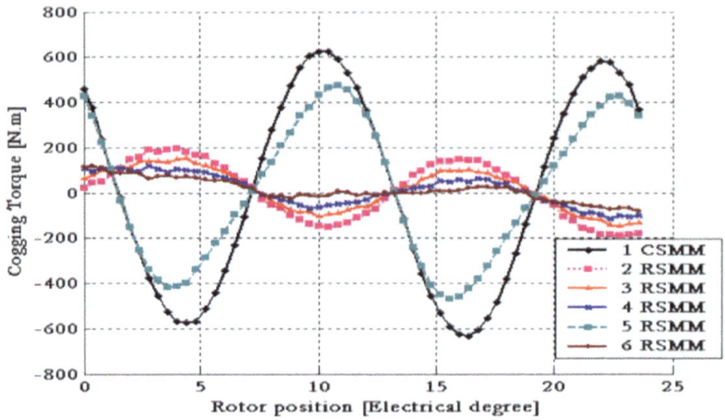

Figure 8. Cogging torque waveforms for $\alpha_m = 152°$.

For a segmented magnet machine ($N \neq 1$) with rectangular magnet blocks (RSMM), the number of magnet edges per pole pair increases, there is no more universal rule giving the optimal values of $\gamma = \frac{\alpha_m}{N}$ minimizing the cogging torque. As an example, for the case ($N = 5$) presented on Fig. 7(a), the few minima are obtained for either $\gamma = \frac{\alpha_m}{5} = (n/2 - 0.17)\tau_s$ or $\gamma = \frac{\alpha_m}{5} = (n/2 + 0.14k)\tau_s$ with k an odd number.

The results obtained for the other cases (N = 2, 3, 4, and 6) are gathered in Fig. 7(b) because they present similar shapes. Two common minima are clearly distinguished for the magnet pole span $\alpha_m \approx 135°$ and $\alpha_m \approx 152°$.

From the investigations presented above, it is clear that one cannot extract a general rule that reduces the cogging torque rate. However, one can affirm that there is some configurations offering the possibility to reduce considerably the cogging torque as shown in Fig. 8. Indeed, for $\alpha_m = 152°$, the weakest cogging torque is achieved with six RSMM per pole. However the most important criterion is to maximize the average torque and reduce as possible the total pulsating torque. So we present in Fig. 9 the total pulsating torque factor (PTF) evolution vs. the magnets span for different configurations ($N = 1, \ldots, 6$). The six curves have the same shape but the amplitude of the PTF varies slightly according to segmentation number N. Referring to Fig. 9, the best choice (PTF $\approx$ 1.2%) should be $N = 3$ ($\alpha_m \approx 165°$) or $N = 4$ ($\alpha_m \approx 160°$). According to Fig. 6 these two configurations lead nearly to the same average torque but N = 4 corresponds to lower magnet volume.

Even if the most important criterion is to maximize the average torque and to reduce as possible the total pulsating torque, a special care must be taken to the reduction of lower torque harmonics (6 and 12). As these harmonics are due the low harmonics (5, 7, 11, and 13) of back EMF, we study in the following the simultaneous effects of the magnet span and the segmentation in N blocks on their amplitudes.

Fig. 10 presents the magnitude evolution of the main back EMF harmonics (7th, 11th, and 13th). Note that the fifth harmonic of the back EMF is null thanks to the adopted fractional winding.

For the seventh harmonic, all the curves are almost identical (except for three RSMM structure), which shows that the segmentation has not a real influence on the seventh harmonics. For this latter, the minimum is obtained for a magnet pole span of 155°. The magnet segmentation seems to have a significant effect on the 11th and 13th harmonics as shown in Fig. 10(b,c). However, all the curves shown on Fig. 10(b,c) present two minima which do not coincide with the seventh harmonic one (Fig. 10a). Then to reduce the amplitude of the sixth torque harmonic, one has to choice a magnet pole span such as the seventh harmonic of the back EMF is weak compared to 11th and 13th.

For the studied machine with an odd number of slots per pole pair, the choice of an adapted winding allows to suppress the fifth EMF harmonic. The choice of an appropriate span ($\alpha_m \approx 155°$) allows to make a good compromise between the increase of the average torque and the reduction of the sixth torque harmonic. This can be achieved with a segmentation number N equal to 4, 5, or 6 magnet blocks.

Among these values (N = 4, 5, 6), for the studied machine, the choice of N = 4 leads not only to the weakest value of the total pulsating torque (Fig. 9) but also to the weakest value of the cogging torque (Fig. 11).

Taking into account the obtained results in the case of the studied machine, we showed that the winding type, the stator slots number and the magnet pole span remain the main parameters acting on the principal performances (Average Torque, sixth torque harmonic) of the machine. For the appropriate choice of these main parameters, a well-adapted choice of the segmentation number of blocks allows to reduce the cogging torque and the total pulsating torque as well.

Conclusion

PM motors are finding expanded use in high power directly driven applications where torque smoothness is essential. Cogging torque in PM motors is among the undesired effects contributing to the motor's output ripple, vibration, and noise. It can be substantially reduced by the combination of several well-known techniques.

For manufacturing and cost reasons, in large permanent magnet motor, each rotor pole is often realized with several elementary magnet blocks with the same polarity (magnets segmentation). In this paper we have shown that the choice of the magnet blocks number over a pole must be considered as an optimization parameter acting on local phenomena such as the cogging torque and higher torque harmonics.

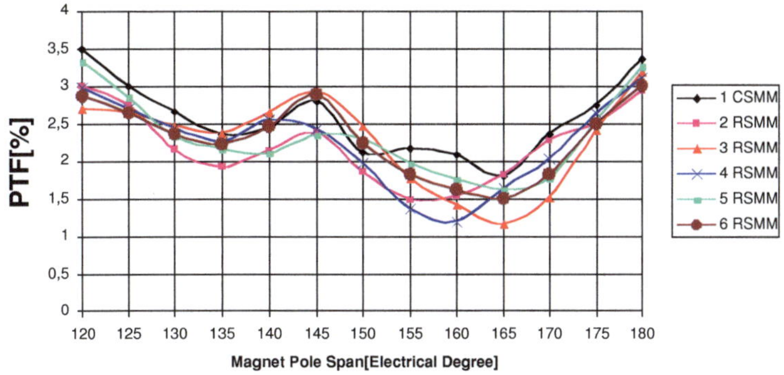

Figure 9. Pulsating torque factor vs. magnet pole span.

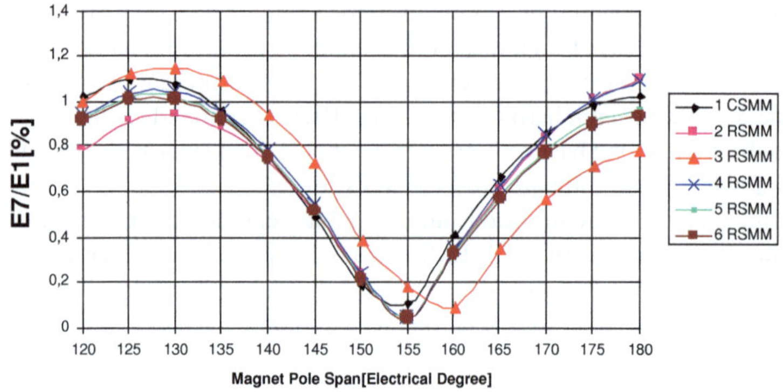

(a)

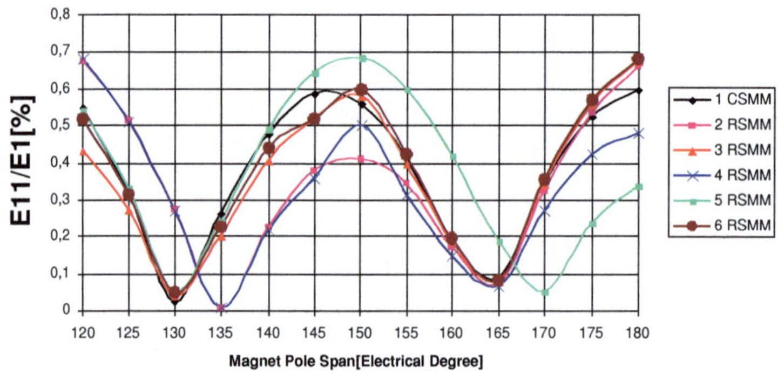

(b)

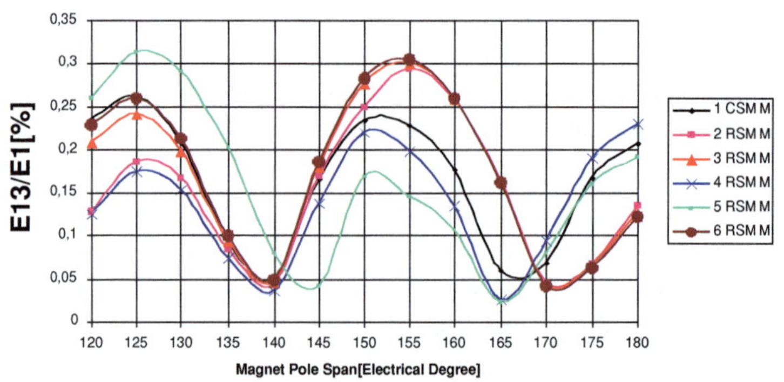

(c)

Figure 10. 7th, 11th, and 13th harmonics back EMF vs. magnet span.

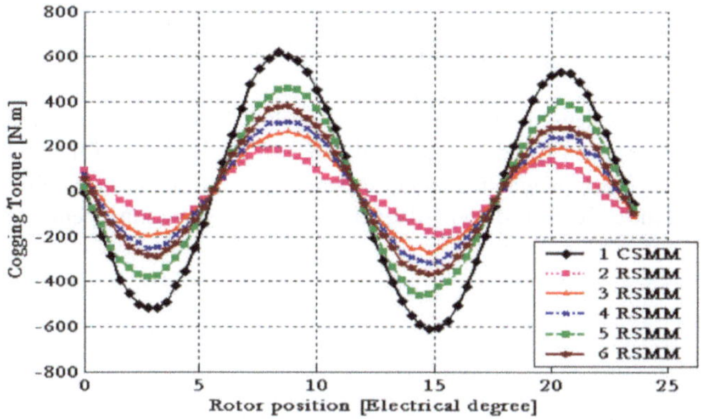

Figure 11. Cogging torque waveforms for a magnet span corresponding to 155° (electrical degree).

The technique that consists on the choice of an appropriate number of magnet blocks over a magnet pole cannot be done without considering the main parameters, which impose the principal machine performances such as the average torque.

In addition to the reduction of the cogging torque and high torque harmonics. Another important effect of magnet subdivision is to reduce eddy currents inside the magnets. This may be achieved by the choice of a segmentation number around 6.

References

[1] A. Arkkio, N. Bianchi, S. Bolognani, T. Jokinen, F. Luise, M. Rosu, "Design of Synchronous PM Motor for Submersed Marine Propulsion Systems", International Conference on Electrical Machines (ICEM 2002), Paper No. 523, Brugge, Belgium, August 25–28, 2002.

[2] T.M. Jahns, W.L. Soong, Pulsating torque minimization techniques for permanent magnet AC motors drives—a review, IEEE Trans. Ind. Electron., Vol. 43, No. 2, pp. 321–330, 1996.

[3] J.-P. Martin, F. Meibody-Tabar, B. Davat, "Multiple-phase Permanent Magnet Synchronous Machine Supplied By VSIs Working Under Fault Conditions", IEEE Industry Applications Conference, 2000, 35th IAS Annual Meeting, Roma, Italy, October 2000.

[4] L. Parsa, L. Hao, H.A. Toliyat, "Optimization of Average and Cogging Torque in 3-Phase IPM Motor Drives", IEEE Industry Applications Conference, 2002, 37th IAS Annual Meeting, Vol. 1, October 13–18, 2002, pp. 417–424.

[5] T. Li, G.R. Slemon, Reduction of cogging torque in permanent magnet motors, IEEE Trans. Magn., Vol. 24, No. 6, pp. 2901–2903, 1988.

[6] T. Ishikawa, G.R. Slemon, A method of reducing ripple torque in permanent magnet motors without skewing, IEEE Trans. Magn., Vol. 29, No. 2, pp. 2028–2031, 1993.

[7] K.-C. Lim, J.-K. Woo, G.-H. Kang, J.-P. Hong, G.-T. Kim, Detent force minimization techniques in permanent magnet linear synchronous motors, IEEE Trans. Magn., Vol. 38, No. 2, pp. 1157–1160, 2002.

[8] S.-M. Hwang, J.-B. Eom, Y.-H, Jung, D.-W. Lee, B.-S. Kang, Various design techniques to reduce cogging torque by controlling energy variation in permanent magnet motors, IEEE Trans. Magn., Vol. 37, No. 4, pp. 2806–2809, 2001.

[9] D.C. Hanselman, Effect of skew, pole count and slot count on brushless motor radial force, cogging torque and back EMF, IEE Proc. Electron. Power Appl., Vol. 144, No. 5, pp. 325–330, 1997.

[10] D. Howe, Z.Q. Zhu, The influence of finite element discretization on the prediction of cogging torque in permanent magnet excited motors, IEEE Trans. Magn., Vol. 28, No. 2, pp. 1080–1083, 1992.

[11] F. Henrotte, G. Deliége, K. Hameyer, "The Eggshell Method for the Computation of Electromagnetic Forces on Rigid Bodies in 2d and 3d", Proceedings of the 10th Biennial IEEE Conference on Electromagnetic Field Computation, CEFC'2002, June 2002, p. 30.

[12] D. Meeker, Finite Element Method Magnetics Software, www.http://femm.foster-miller.com.

I-7. OVERLAPPING MESH MODEL FOR THE ANALYSIS OF ELECTROSTATIC MICROACTUATORS WITH ECCENTRIC ROTOR

Piotr Rembowski and Adam Pelikant

Institute of Mechatronics and Information Systems, Technical University of Lodz, Poland, ul. Stefanowskiego 18/22, 90-924 Lodz, Poland

prembowski@poczta.onet.pl, apelikan@mail.p.lodz.pl

Abstract. The numerical model for three-dimensional field analysis of electrostatic micromotors with stator and rotor symmetry axes located in the different points has been presented. The results of the numerical tests confirm the thesis about the correctness of the model. Short CPU time is obtained even with quite big number of mesh elements.

Introduction

The paper presents numerical model for three-dimensional field analysis of electrostatic micromotors with stator and rotor symmetry axes located in different points. Due to a very small size of micromachines it is impossible to place the rotor in such a position that would provide ideal symmetrical air gap between electrodes. There is no algorithm which fully covers this kind of asymmetry. Solving this problem through commercial applications leads to mesh generating for each single analyzed position, which means increased time of the analysis. The application of the mesh overlapping lets one avoid repeated mesh generating for the whole model and decrease the time of computation. Reduction of this time can be obtained by using separated submeshes for both stator and rotor generated only once, and only recalculating the part describing the air gap.

Possible applications

There are many technical solutions with purposely designed nonsymmetrical air gap. One of the most important applications is the possibility of using ferroelectrics in the construction of the wobble motor (Fig. 1). This solution results in a considerable increase of torque. In this kind of motors the air gap asymmetry is so important that it must not be omitted in computation.

With the exception of a few approximated analytical modeling cases, there are no known algorithms for an efficient solution that deals with big asymmetries of the air gap in the electrostatic microactuators, especially with ferroelectrics.

S. Wiak, M. Dems, K. Komęza (eds.), Recent Developments of Electrical Drives, 73–83.

Figure 1. SEM photo of a 100 μm-diameter, 2.5 μm-gap, wobble micromotor with a free bearing.

A new solution was developed to compute models with nonsymmetrical air gap. This was possible due to extending mesh overlapping model.

What is more, it is possible to apply the algorithm for models with leant rotor rotation axis. The sloping can result from technical inaccuracy as well as constructor's intention. So far the problem of mesh overlapping in all three dimensions has not been so far considered in literature.

Model

A numerical, finite element method based algorithm has been constructed to solve problems mentioned above. The integral form of the second Maxwell equation with application of the Gauss law (1) is a base to formulate mathematical equation describing the analyzed object.

$$\oiint_S (\varepsilon \cdot gradV)\, dS = 0 \tag{1}$$

The equation (1) applied in finite element method with the approximation on each mesh element with weigh functions λ_i leads to formula (2), where V_i means the values of the potentials of the nodes.

$$\sum_j \iint_{S_j} \varepsilon_j\, grad\left(\sum_{\Omega_e} \lambda_i V_i\right) dS_j = 0 \tag{2}$$

Assuming cylindrical coordinate system, the second degree polynomial as approximating function in a single element was in form (3):

$$a_0 r(\vartheta r)z + a_1 r(\vartheta r) + a_2(\vartheta r)z + a_3 rz + a_4 r + a_5(\vartheta r) + a_6 z + a_7 = 0 \tag{3}$$

In the consequence of the above one gets a system of linear equations (4) with symmetric, well conditioned, positive definite matrix with 27 nonzero elements in each row.

$$[M]\{V\} = \{Q\} \tag{4}$$

Mesh overlapping

In the overlapping mesh model the stator and the rotor are represented by two separate meshes. An air gap is included in both of them. There is a common region consisting of at least one common layer along all the height of the model. Values of potentials of peripheral nodes are determined by boundary conditions and linear approximation based on values in neighboring nodes of the other mesh. There are two cases of solving the problem: first when the centers of the stator and the rotor are shifted by a distance which is smaller than one third of the air gap width (Fig. 2) and the second when the shift is larger (Fig. 3).

In both cases the rotor and stator meshes cover the air gap on the area whose width is equal to the smallest distance between the rotor and the stator electrodes. In connection with the above, in the former case both meshes cover the whole area of the air gap and one single layer of elements (the last one) can be used for mesh overlapping.

In the latter there is a need to extend one of the meshes (by adding additional layers of nodes) to cover the whole area of the air gap. In this case it is necessary to use more layers of elements in mesh overlapping computation.

In the symmetrical model both meshes have a common surface, in the air gap area, along all the height of the model. In this case nodes of the stator and the rotor meshes for overlapping bounds have only different angle θ (Fig. 4). In this figure nods belonging to the rotor mesh have numbers starting from the letter i, and nods belonging to the stator mesh have numbers starting from the letter j. The letters k and n signify the numbers of nodes in rotor and stator meshes for the constant radius.

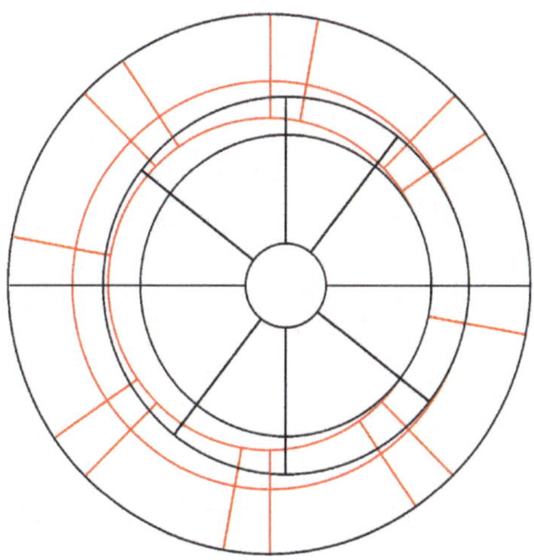

Figure 2. Generated meshes for the shit less than one third of the air gap width.

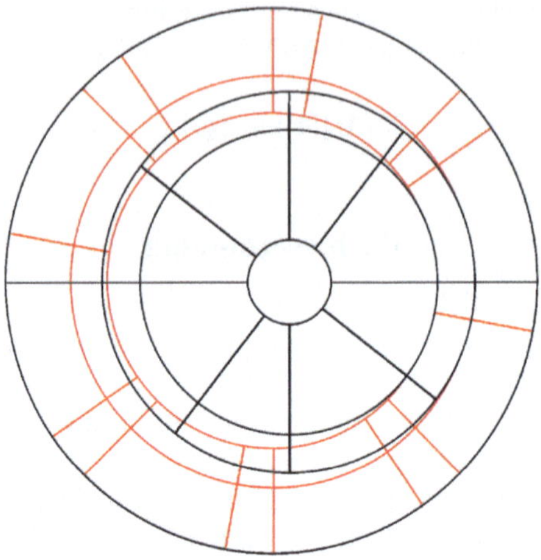

Figure 3. Generated meshes for the shit larger than one third of the air gap width.

Taking into account the stator node with number j neighboring with nodes i and $i+1$ of the rotor mesh, f the linear interpolation the equation for this node can be described as follows (5):

$$V_J \Delta\theta_i - V_i(\gamma_i) - V_{i+1}(\Delta\theta_i - \gamma_i) = 0 \qquad (5)$$

Where $\Delta\theta_i$ is the angle between nodes with the numbers i and $i+1$, and γ_i is the angle between the stator mesh node j and the rotor mesh node i.

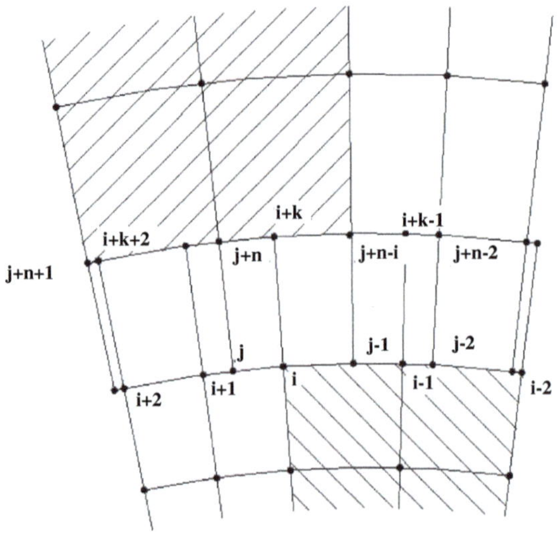

Figure 4. Part of the one level ($z = $ const.) of the mesh with the overlapping region (parts of electrodes dashed, dimensions enlarge)—symmetrical air gap.

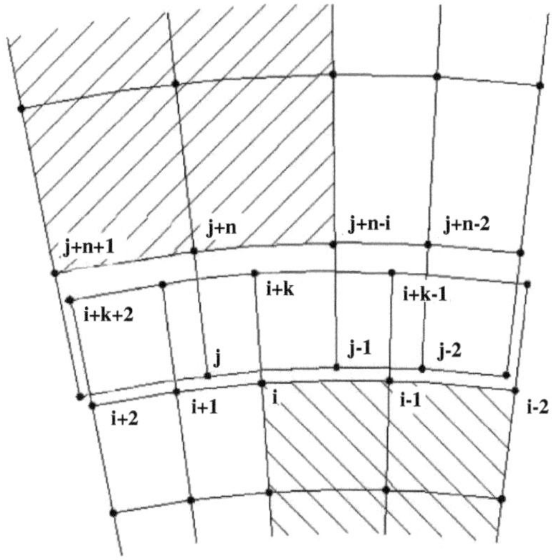

Figure 5. Part of the one level (z = const.) of the mesh with the overlapping region (parts of electrodes dashed, dimensions enlarge)—nonsymmetrical air gap.

In the nonsymmetrical model nodes of the stator and the rotor meshes for overlapping bounds have different angle θ and radius r (Fig. 5). The stator mesh node with number j neighbors with four nodes i, $i+1$, $i+k$, and $i+k+1$ of the rotor mesh.

The equation describing the value of the potential in the j node can be written down using bilinear interpolation function in the following form (6):

$$V_j(\gamma_i, r_i) = a_0\gamma_i r_i + a_1\gamma_i + a_2 r_i + a_3 \tag{6}$$

Using equation (6) for each node at both boundaries (outer for the rotor mesh and inner for the stator mesh) one obtains sub matrix of main matrix $[M]$ containing five nonzero elements for each row. As a result one gets nonsymmetrical system of linear equations, which is solved using LDU decomposition method with permutation matrix (7).

$$[M] = [P][L][D][U][\tilde{P}] \tag{7}$$

Presented algorithm was implemented in a numerical program, which allows determining a field distribution for every angular position of the rotor and every possible movement of its rotation axis.

Integral parameters

The application allows calculating integral parameters for every position of the rotor—in particular the system energy that can be written down in general in form (8):

$$W = \iiint_\Omega w\, d\Omega = \iiint_\Omega \left(\int_0^E \mathbf{D}\, d\mathbf{E} \right) d\Omega \tag{8}$$

According to Maxwell stress tensor, formula defining force components can be described as follows (9):

$$\mathbf{F} = \iint_S \frac{1}{\varepsilon}\left[(\mathbf{D}\times\mathbf{n})\mathbf{D} - \frac{1}{2}|\mathbf{D}|^2\,\mathbf{n}\right] d\mathbf{S} = \left(\varepsilon \iint_S \frac{E_n^2 - \mathbf{E}_t^2}{2}\, d\mathbf{S}\right)\mathbf{n} + \left(\varepsilon \iint_S E_n \mathbf{E}_t\, d\mathbf{S}\right)\mathbf{t} \tag{9}$$

Using explicit choose shape functions λ_i in formula (8), one can calculate the total system co-energy as the sum of the energy accumulated in each of the mesh elements (10).

$$W = \sum_e \sum_{\Omega_e} \left(\varepsilon \iiint_{\Omega_e} grad^2 V_i\, \lambda_i\, d\Omega_e\right) \tag{10}$$

Proceeding in the same way with the general expression (9) leads to formulas describing force components in relation to surface S, which consists of the sum of elementary surfaces S_i in the single mesh elements. As the result one obtains the normal force component in form (11):

$$F_n = \frac{\varepsilon_0}{2} \sum_i \iint_{S_i} (grad_n^2 V_i\, \lambda_i - grad_t^2 V_i\, \lambda_i)\, dS_i \tag{11}$$

By analogy the tangent force components can be written down as follows (12):

$$\mathbf{F}_t = \sum_i \iint_{S_i} \left[\varepsilon_0 (grad_n V_i \lambda_i \cdot grad_t V_i \lambda_i)\right] dS_i \tag{12}$$

Numerical verification

The basis of the verification of the presented model was the numerical experiment. Air gap energy was calculated in the part common for both the rotor and the stator meshes and obtained results were compared. The quality of energy calculation was determined on the basis of numerical testing of the convergence of the solutions from both meshes (Figs. 6 and 8). The influence of mesh density on the value of energy accumulated in the air gap was analyzed for different positions of rotor (rotation and shift). It allows determining minimal mesh density for given accuracy of computations. A clear tendency of both curves to reach the same value was observed. It means a convergence of energy value and exact value. The convergence was observed irrespective of the rotor's location. However, the slope of the curve changes, which results from different energy values for different locations of the rotor.

At the same time the difference between the energy value calculated from stator mesh and the energy value calculated from rotor mesh was computed (Fig. 7). Convergence to zero of the above difference was observed. Like before the tendency appears irrespective of the rotor's position.

Convergence of solutions determined on the basis of the values of potentials in the nodes of both rotor and stator meshes confirm the thesis that the implemented model is correct.

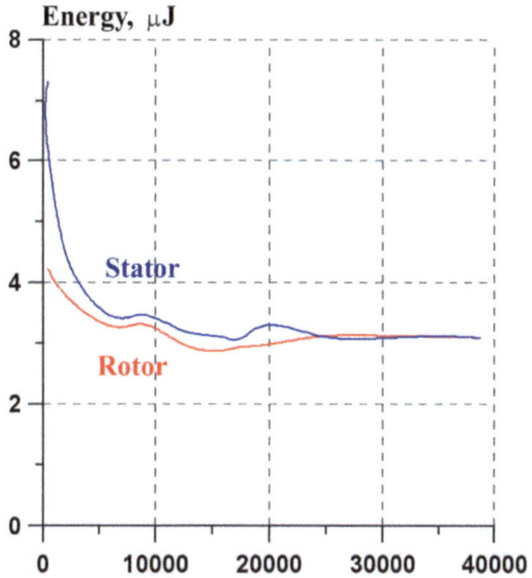

Figure 6. Influence of the mesh density on the value of air gap energy for rotor position 0°.

Calculating the changes of energy value for different rotor angular position (Fig. 9) allows determining static torque as follows (13).

$$M = \frac{\partial W}{\partial \gamma} \tag{13}$$

As a matter of fact, the approach based on Maxwell's tensor is used (11, 12), whereas the above formula (13) is only a method of confirming the correctness of the results.

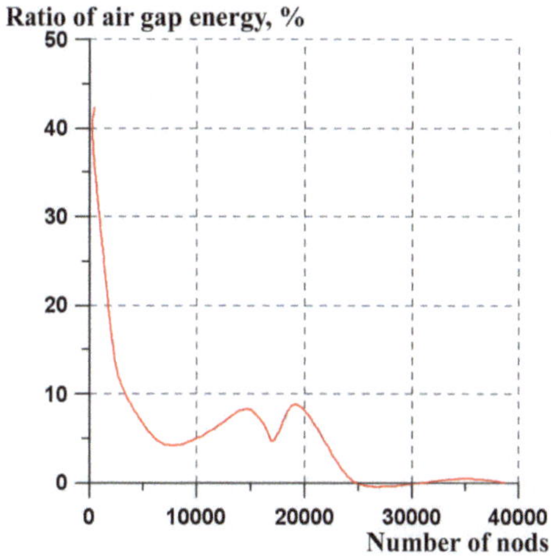

Figure 7. Influence of the mesh density on the value of air gap energy for rotor position 30°.

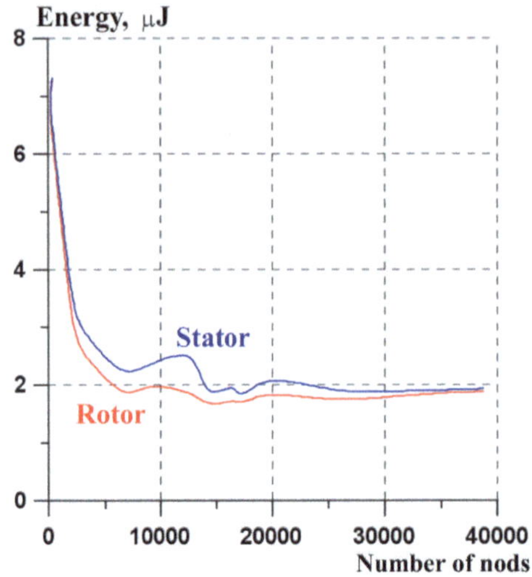

Figure 8. Ratio of air gap energy calculated from stator and rotor meshes.

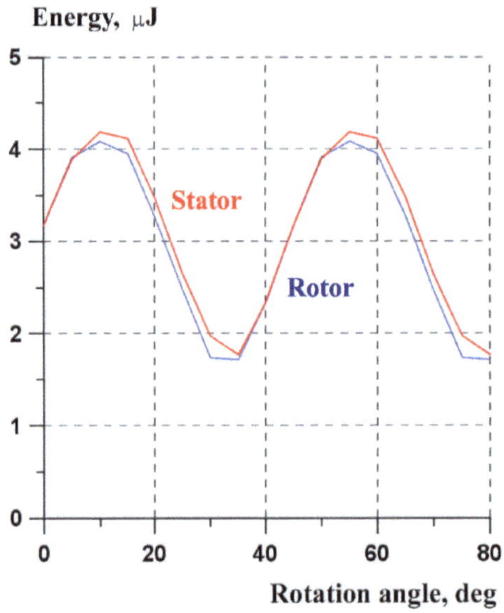

Figure 9. Dependence of air gap energy on the rotation angle of the rotor (about 25,000 mesh elements).

Conclusions

Another step in developing the model will be extending it to the analysis of microacutators with leant rotation axis. It will require interpolation by three variable function and not one variable function (symmetrical model) or two variable function (model with shifted rotation axis) as so far.

The most important conclusion resulting from the carried out studies on the three-dimensional model for the analysis of the electrostatic micromotors is that it allows effective analysis for any position of the rotor—both rotation and rotation axis shift.

Presented algorithm allows correct and exact representation of the changing width air gap in the model. Since significant part of the main matrix rows is calculated only once and it's only recalculated fragment is the one representing the air gap, it is possible to reduce computation time.

The results of the numerical tests confirm the thesis about the correctness of the model. Short computation times are obtained even with quite big number of mesh elements.

References

[1] G. Hainsworth, D. Rogger, P. Leonard, 3*D* finite element modeling of conduction supports in coilguns, IEEE Trans. Magn., Vol. 31, No. 3, pp. 2052–2055, 1995.

[2] M. Mehergany, S.F. Bart, L.S. Tavrow, J.H. Lang, S.D. Senturia, M.F. Schelecht, A study of three microfabricated variable capacitance motors, Sens. Actuators, A21–A23, pp. 173–179, 1990.

[3] A. Pelikant, Analiza polowo-obwodowa silników elektrostatycznych i elektromagnetycznych zasilanych impulsowo, Wydawnic two Politcchniki Łodikicy, Zesryty Nau Kowe nr 908, Rozprowy Naukowe, Z. 111 2002.

[4] R. Perrin-Bit, J. Coulomb, A three dimensional fine elements mesh connection for problem with movement, IEEE Trans. Magn., Vol. 31, No. 3, pp. 1920–1923, 1995.

[5] I. Tsukerman, Overlapping finite elements for problems with movement, IEEE Trans. Magn., Vol. 28, No. 5, pp. 2247–2249, 1992.

I-8. COUPLED FEM AND SYSTEM SIMULATOR IN THE SIMULATION OF ASYNCHRONOUS MACHINE DRIVE WITH DIRECT TORQUE CONTROL

S. Kanerva[1], C. Stulz[2], B. Gerhard[3], H. Burzanowska[2], J. Järvinen[3] and S. Seman[1]

[1]Laboratory of Electromechanics, Helsinki University of Technology, P.O. Box 3000, FI-02015 HUT, Finland

sami.kanerva@hut.fi, slavomir.seman@hut.fi

[2]ABB Switzerland Ltd, Large Drives, Austrasse, CH-5300 Turgi, Switzerland

christian.stulz@ch.abb.com, halina.burzanowska@ch.abb.com

[3]ABB Oy, Electrical Machines, P.O. Box 186, FI-00381 Helsinki, Finland

bilal.gerhard@ch.abb.com, jukka.jarvinen@fi.abb.com

Abstract. A compound drive simulator is presented, where a finite element method (FEM) model of the electric motor is coupled with a frequency converter model and a closed-loop control system. The method is implemented for SIMULINK and applied on a 2-MW asynchronous machine drive. The results are validated by measurements and the performance is compared with an analytical motor model. It is shown that simulation with the FEM model provides very good results and gives much better insight in the motor behavior than the analytical model.

Introduction

As the demands for performance of electric drive systems increase, also the simulation software must follow the requirements. Designers of frequency converters and electric motors rarely work in the same location, but they must be able to model both parts of the drive as accurately as possible. Naturally, different expertise is required to model electrical machines or power electronics, but the key issue is to couple these models together in a way that experts in both fields can profit from each other by using the most advanced simulation models in their design.

Accurate modeling of digital control systems requires simulation in multiple timescales, because different sampling times are used for measurement, filtering, estimation, and modulation. By including all detailed functions and sample times, it is possible to create very accurate simulation models of the converter control. In such a case, however, also a detailed electrical machine model is needed in order to get the maximum advantage of the drive model.

Finite element method (FEM) is a widely known method to model electrical machines with high accuracy. For standard-type machines, two-dimensional field solution coupled

S. Wiak, M. Dems, K. Komęza (eds.), Recent Developments of Electrical Drives, 83–92.

with simple circuit equations of the windings is usually accurate enough, when the cross-section geometry and material properties are known [1].

The most problematic in the drive simulation is to couple the FEM computation with the converter model. Most obvious method would be to couple the converter model in the FEM code and solve all the equations simultaneously with uniform time steps [2,3]. However, such an approach is hardly applicable to a detailed converter model with digital closed-loop control because of the amount of programming, and the demand for common time step length would make the simulation too heavy with respect to existing computing facilities.

Reference [4] presents an indirect method for coupling time-stepping FEM simulation with SIMULINK using multiple sample times for different parts of the system model. The method was applied to a cage induction motor and a frequency converter with direct torque control (DTC). The model of the control system was developed in order to investigate control-related topics and verified for steady-state and transient operation of the drive. In its original state, it was using a motor model that was based on analytical equations.

In this paper, the same method is applied to an asynchronous motor drive with DTC. The frequency converter model is based on a real application, comprising a detailed model of the digital control system. The frequency converter model is implemented in SIMULINK and it is coupled with a two-dimensional FEM model of the asynchronous motor. The system is simulated in steady-state and transient operation, and the simulation results are validated by a comparison with the measured results.

Compound model of inverter-fed electrical drive

The general structure of the compound drive simulator is shown in Fig. 1. The model is implemented in SIMULINK but the execution of the model is controlled from within

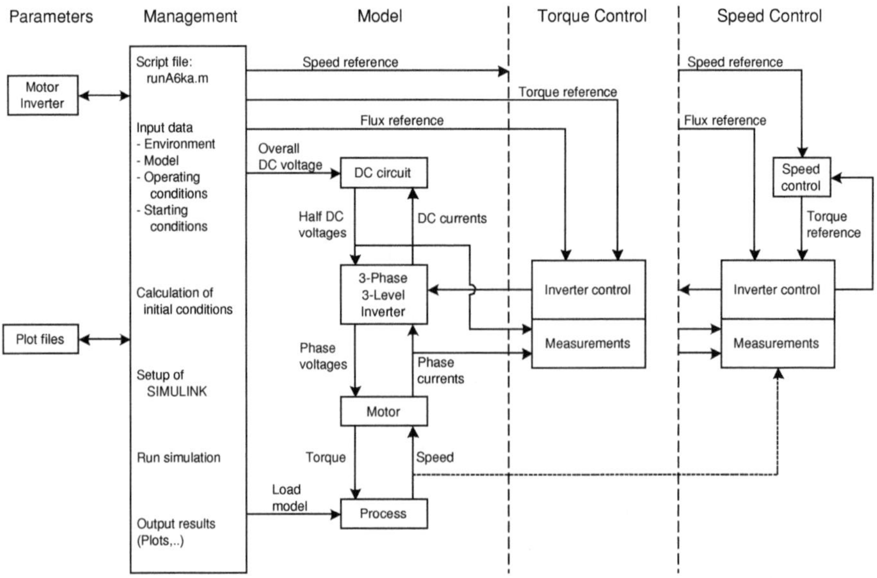

Figure 1. General structure of the drive simulator.

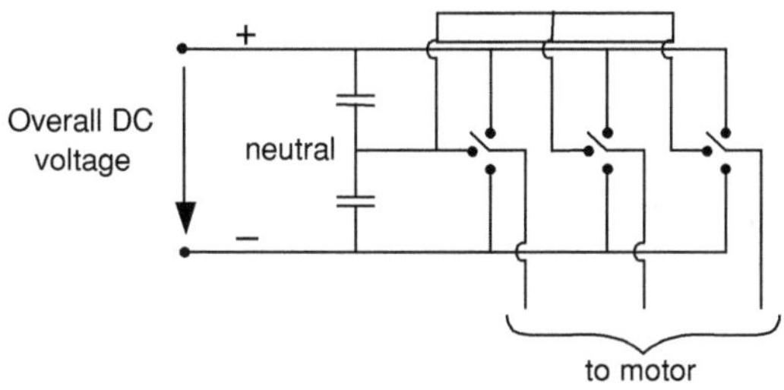

Figure 2. Simulation model of the three-phase three-level inverter.

MATLAB. This allows to specify the plant parameters, operating and starting conditions very easily. Based on the selected operating conditions, the initial conditions for continuous and discrete time states are determined. This allows to start the simulation in a reasonably stable operating point. The machine model in this drive simulator can be selected to be the simple analytical or the precise FEM-based model.

The main components of the simulated plant are DC circuit, inverter, motor, process, and control. Two basic control schemes can be selected: torque or speed control.

Inverter and DC link

The three-level inverter is modeled as a set of ideal switches, which can connect the phase voltages to either plus, neutral, or minus potential of the DC links. Fig. 2 gives a rough overview. The switching pattern is given by the drive control. The status of the switches together with the phase currents determines the currents in the DC bus bars of the DC link. The current in the neutral bus bar is used to calculate the potential of the neutral point of the DC link. The phase voltages transferred to the motor terminals are defined by DC link voltages and switching pattern.

Analytical motor model and load

The analytical motor model is used for simulations that will be compared to the FEM-based motor model. It is based on the well-known space vector representation of the asynchronous machine. It uses both the stator and the rotor fluxes as state variables. The following features are present in the model:

- constant air-gap and sinusoidal flux distribution along the air-gap
- no iron losses
- resistances and inductances are independent of frequency and temperature
- the magnetizing inductance can saturate with increasing main flux

The model needs phase voltages and speed as inputs and produces phase currents and air-gap torque as outputs.

The driven process is described by the differential equation of motion. A single inertia is used. The load torque may follow several functions of the speed (constant, linear, quadratic, or mixed). The mechanical mass is driven by the electromagnetic torque of the motor and gives the speed as output.

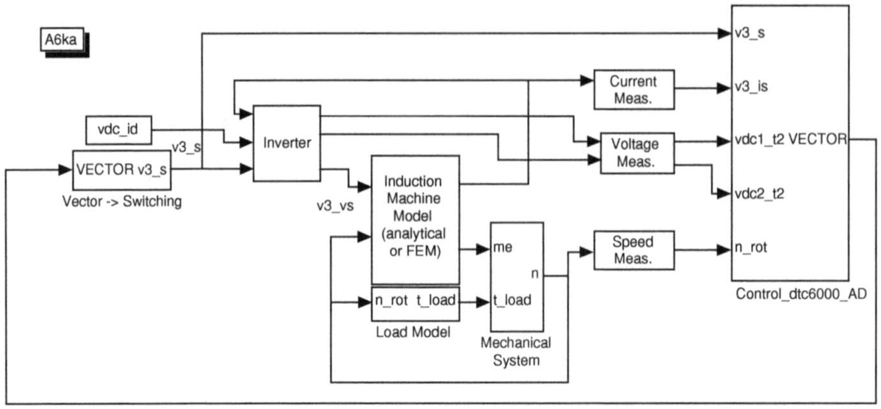

Figure 3. Model of the drive system implemented in SIMULINK.

Control

The control model describes speed/torque control using a DTC algorithm. The main functions of the ACS6000 drive are implemented as discrete functions on different time levels to appropriately represent the behavior of the real drive. The detailed description of the DTC control cannot be in the scope of this paper.

The top level of the SIMULINK environment is shown in Fig. 3.

Model of the asynchronous motor

Modeling by finite element method (FEM)

The FEM model of the motor is based on two-dimensional finite element method and circuit equations of the windings [1]. The magnetic field in the core region is calculated using magnetic vector potential formulation, in which the vector potential and current density have only z-axis components.

The phase windings in the stator or rotor are modeled as filamentary conductors with uniformly distributed current flowing through all the coils that belong to the same phase. The rotor bars are modeled as solid conductors, in which the current density varies according to eddy currents. The sources of the magnetic field are the phase currents, the voltage drop in the rotor bars and the magnetic force of the permanent magnets, depending on the type and construction of the machine.

The relations between voltage and current are determined in the circuit equations of the stator and rotor windings, which also include the end-winding impedances and the short-circuit rings. As a result, only phase voltages are needed as an electrical input for the FEM model. The electromagnetic torque is calculated by virtual work principle, and the movement of the rotor is determined from the equation of motion. At each time step, new position is calculated for the rotor and the air-gap mesh is refined.

FEM block for SIMULINK

The FEM computation is implemented as a functional block in SIMULINK using dynamically linked program code (S-function), as illustrated in Fig. 4. The stator voltage

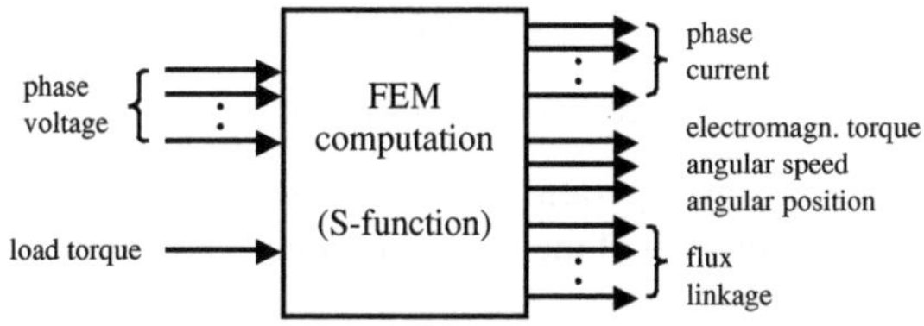

Figure 4. Functional block of the FEM computation.

Table 1. Characteristics of the asynchronous motor drive

	Asynchronous motor		Frequency converter
P_N	2 MW	P_N	9 MW
U_N	3150 V	U_{max}	3300 V
I_N	436 A	I_{max}	1645 A
f_N	40 Hz	f_N	0–75 Hz
n_N	792 rpm		

and the load torque on the shaft are given as input variables and the phase currents, electromagnetic torque, rotor position, and the stator flux linkage are obtained as output variables.

The mathematical coupling between the FEM model and SIMULINK is weak, which means that the internal variables of the subsystems are solved separately and updated to each other with one-step delay. Accordingly, there is no need to use uniform step size in the whole model, which provides flexibility and computation-effective simulation due to the different timescales in the system model.

Characteristics of the motor model

Table 1 presents the ratings and characteristics of the drive, including the asynchronous motor and the frequency converter. Because of symmetry, the finite element mesh of the motor covers half of the cross section, comprising 13,143 nodes and 6,518 quadratic triangular elements. The geometry of the modeled region is presented in Fig. 5.

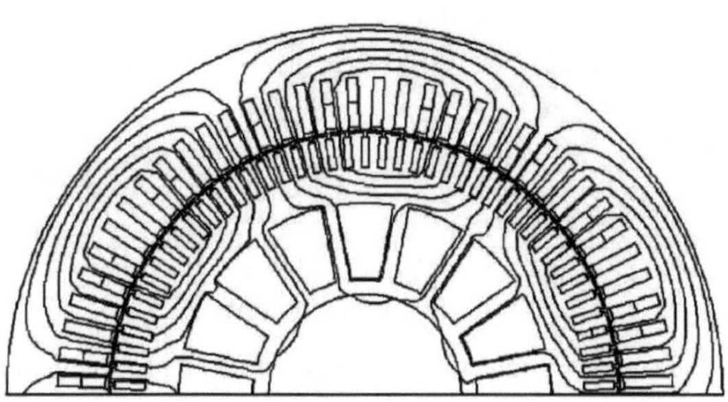

Figure 5. Geometry of the asynchronous motor model.

Results

Steady-state operation

In order to study the steady-state operation of the drive, the time-stepping simulation was run at 600 rpm, which is about 75% of the nominal speed. The nominal torque 24 kNm was applied, resulting in the nominal stator current.

The time step was 12.5 μs for the drive model with analytical motor model. When the measurement and control are modeled in different time levels, it takes 66 s to run 1 s simulation on a 900 MHz Pentium 4 PC (Matlab release 13SP1). The same case was also simulated with FEM motor model, when 12.5 μs time step was used for the drive model and the FEM computation was executed at 100 μs steps. Here the computation time is remarkably longer, it would take about 14 h to run 1 s. Naturally, the simulation time can be lowered to about one third by using linear elements. The results were validated by comparing them with measurements.

Due to the stochastic nature of the DTC control strategy, direct comparison of the waveforms doesn't give much information. Instead, the results are gathered from several cycles of the fundamental frequency and Fourier analysis is performed to find out the harmonic content of the waveforms. Fig. 6 presents the spectrum of the line-to-line supply voltage obtained by FEM and analytical motor models in comparison with the measured spectrum, and Fig. 7 presents the corresponding results for the phase current. The fundamental components are scaled out from the figures in order to see the differences in higher harmonics. In the voltage spectrum, distinctive difference is seen between the FEM model and analytical model in certain frequencies, but otherwise they follow each other closely and also correspond very well with the measured results. In the current spectrum, the difference between FEM model and analytical model is significant in all harmonic components. It is also seen that the current spectrum obtained by the FEM model agrees very well with the measured spectrum.

Good agreement between the simulated and measured results shows that the control model behaves correctly in the simulations and the weak coupling between frequency

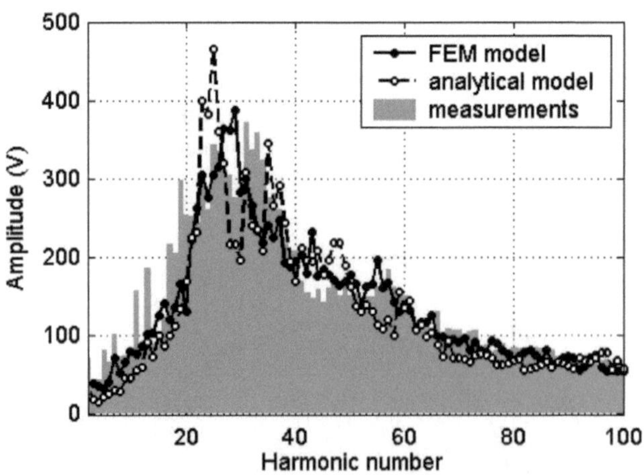

Figure 6. Spectrum of the supply voltage obtained by the FEM and analytical models and compared with the measured spectrum.

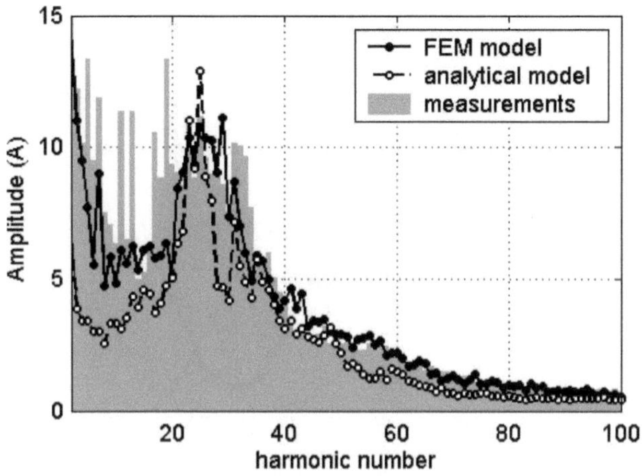

Figure 7. Spectrum of the phase current obtained by the FEM and analytical models and compared with the measured spectrum.

converter and FEM models gives correct and reliable results. Furthermore, the voltage spectrum reveals that the analytical model is adequate for modeling the control system in steady state, but the differences in the current spectrum clearly proves the better accuracy of the FEM model over the analytical model in the harmonic analysis of the phase current. This is also illustrated in Fig. 8, which presents the impedance of the motor for the measured frequency range. The impedance obtained by the FEM model follows closely the measurements until 4 kHz, whereas the analytical model shows two times higher impedance at the same frequency.

The measured losses of the motor in steady-state operation were 58.8 kW, and the losses estimated by the FEM-based motor model were 60.6 kW, which shows excellent capability

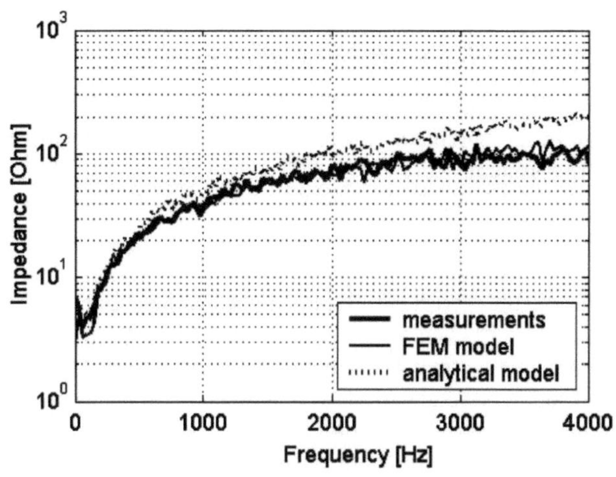

Figure 8. Impedance of the motor obtained by the FEM model, analytical model, and measurements.

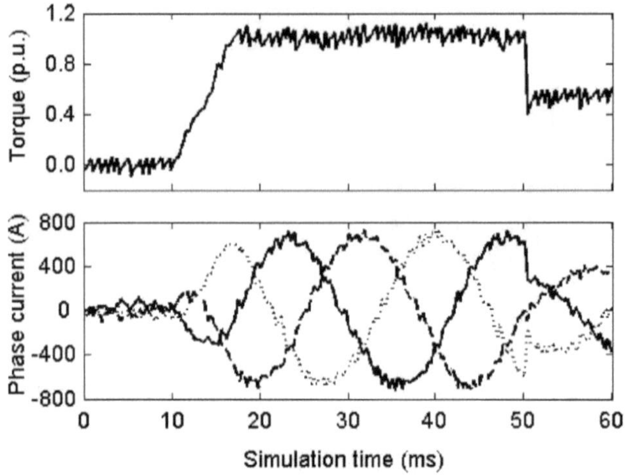

Figure 9. Electromagnetic torque and phase currents, when the torque is changed from zero to nominal and from nominal to 0.5 pu.

for loss prediction. In FEM, the copper losses in the coils were determined from the resistance and current density, and the iron losses were determined from the harmonic components of the supply voltage and the loss factors provided by the iron sheet manufacturer.

Transient operation using FEM model

After validating the drive simulator with steady-state measurements, the drive was simulated in transient operation. A torque step from 0 to 1.0 pu was applied, when the motor was running at nominal speed. After a while, the torque was changed to 0.5 pu. The electromagnetic torque and the phase currents of the motor are presented in Fig. 9.

In another transient simulation, rotational speed was changed from nominal to 0.3 pu, while operating at no load conditions. The inertia of the motor was reduced in order to have a faster speed change. The electromagnetic torque and rotational speed are presented in Fig. 10.

In both transient simulations, the control system responds well to reference changes. Although not validated by measurements, the results show the capability to simulate transients of the controlled drive system.

Determination of the initial state

Traditionally, finding out the correct initial state for the FEM computation has been problematic. Especially with static frequency converter models, simulation of the startup transient may take hours of computation time. Even if the simulation is started from an initial field obtained by sinusoidal supply, several periods of fundamental frequency must be time stepped, until the transient has stabilized in the motor.

In the presented simulation environment, the initial transient converged remarkably faster than in the previous studies. This is due to the calculated initial states for all state variables

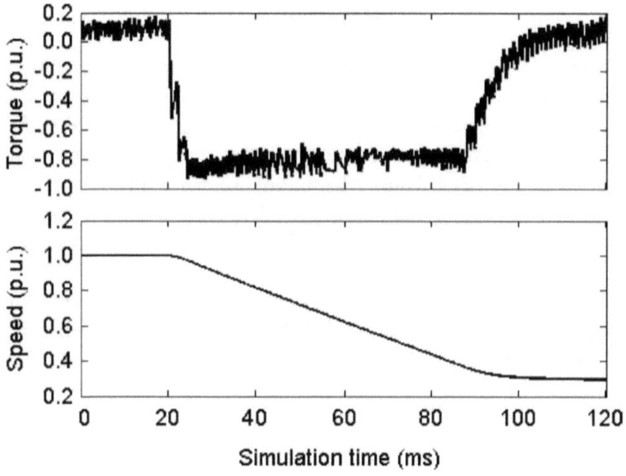

Figure 10. Electromagnetic torque and rotational speed, when the speed reference was changed from nominal to 0.3 pu.

and accurate closed-loop model of the control system, which estimates the magnetic state of the motor and controls the supply voltage to set the motor in the required operating point as quickly as possible. In other words, the simulation model operates exactly as the real drive system.

Conclusion

This paper presents a drive simulator system comprising a three-level inverter, speed/torque control by DTC algorithm and an analytical or FEM-based motor model. The FEM model of the motor is coupled with SIMULINK using indirect approach. This means that different parts of the drive system can be simulated simultaneously, but using different time steps.

An asynchronous machine drive was simulated using analytical and FEM-based motor models and the results were compared with measurements. In the supply voltage spectrum, agreement with measured results was excellent for both motor models. In the current spectrum, agreement with the measurements was clearly better with the FEM-based model. In transient simulation, the control system responds very well to the changes in reference values.

Using the proposed methodology, the FEM model of the motor and the frequency converter model can be designed separately and easily combined for coupled simulation.

With the developed simulation environment, the initial states for the analytical motor model and the FEM computation are achieved very rapidly. Based on the results, analytical motor model is suitable for control design, but FEM model is needed for detailed analysis of the saturation and frequency dependence of the motor parameters. As well, the motor losses obtained by the FEM computation agree very well with the measurements. In general, the simulation results with the FEM model are very accurate and reliable, which leads to benefits in the design and development of advanced control algorithms.

References

[1] A. Arkkio, "Analysis of Induction Motors Based on the Numerical Solution of the Magnetic Field and Circuit Equations", Acta Polytechnica Scandinavica, Electrical Engineering Series, No. 59, 1987, p. 97. Available at http://lib.hut.fi/Diss/198X/isbn951226076X/.

[2] J. Väänänen, Circuit theoretical approach to couple two-dimensional finite element models with external circuit equations, IEEE Trans. Magn., Vol. 32, No. 2, pp. 400–410, 1996.

[3] A.M. Oliveira, P. Kuo-Peng, N. Sadowski, M.S. de Andrade, J.P.A Bastos, A non-a priori approach to analyze electrical machines modeled by FEM connected to static converters, IEEE Trans. Magn., Vol. 38, No. 2, pp. 933–936, 2002.

[4] S. Kanerva, S. Seman, A. Arkkio, "Simulation of Electric Drive Systems with Coupled Finite Element Analysis and System Simulator", 10th European Conference on Power Electronics and Applications (EPE 2003), Toulouse, France, September 2–4, 2003.

I-9. AN INTUITIVE APPROACH TO THE ANALYSIS OF TORQUE RIPPLE IN INVERTER DRIVEN INDUCTION MOTORS

Ö. Göl[1], G.-A. Capolino[2] and M. Poloujadoff[3]

[1]Electrical Machines and Drives Research Group, University of South Australia, Australia GPO Box 2471, Adelaide SA-5001, Australia

ozdemir.gol@unisa.edu.au

[2]Energy Conversion and Intelligent Systems Laboratory, Université de Picardie Jules Verne 33, rue Saint Leu, 80039 Amiens Cedex 1, France

gerard.capolino@u-picardie.fr

[3]Université de Pierre et Marie Currie—Case 252, 4 place jussieu, 75252 Paris, France

mpo@ccr.jussieu.fr

Abstract. An intuitive approach of parasitic effects with particular emphasis on torque ripple has been proposed successfully. It is shown that a good approximation can be achieved in predicting the nature and the magnitude of torque ripple by the use of a relatively simple time-domain model.

Introduction

It is well known that, when an induction motor is driven from a non-sinusoidal supply, problems may arise due to the presence of supply harmonics. For instance it is well known that the use of a six-step inverter may lead to the creation of parasitic effects such as torque pulsations accompanied by noise and vibration. It is less well known that torque ripple along with associated disturbances can also be present in the case of drive systems which emulate a sine wave, such as field orientation control schemes if and when they are driven into overmodulation.

Various methods of analysis have been proposed to assess the extent of the effect of supplying a motor from a non-sinusoidal source [1–3]. Of these, methods which are based on frequency domain analysis yield results which provide no interpretation of time-domain results, thus not allowing the significance of supply harmonics in terms of parasitic behavior to be appreciated when a harmonic-riddled source is used.

This paper proposes an intuitive approach to the analysis of parasitic effects with particular emphasis on torque ripple. The approach is based on the notion of space phasor modeling [4]. It is shown that a good approximation can be achieved in predicting the nature and the magnitude of torque ripple by the use of a relatively simple time-domain model.

S. Wiak, M. Dems, K. Komęza (eds.), Recent Developments of Electrical Drives, 93–100.

Basic considerations

Both direct phase models [5] and orthogonal models (generally referred to as d-q models—based on Park's two reaction theory [6]) have been used in analyzing the time-domain performance of asynchronous motors. The former have been considered to be more relevant to the modeling of polyphase machines since directly measurable physical quantities are present in the model and effects of winding asymmetry and supply unbalance can be assessed with relative ease. But the use of the latter has been far more pervasive.

On the other hand, it seems to have gone unnoticed that space phasor models offer a valid and interesting alternative. They intrinsically contain the elements of both direct phase models and orthogonal models, making the progressive or the retrogressive transition between space phasor models and others possible. Furthermore they correctly model the rotating field within the machine space. Thus their adoption for modeling may arguably constitute an "intuitive" approach.

The space phasor concept

The transition from a direct phase model to a space phasor model can be effected by bestowing "vector" attributes upon the time-variant electromagnetic quantities of the machine. Thus the sum of stator and rotor phase currents for a three-phase machine in space phasor notation become

$$\tilde{I}_S = \frac{2}{3}\left(I_{SA} + \tilde{a} I_{SB} + \tilde{a}^2 I_{SC}\right) \tag{1}$$

$$\tilde{I}_S = \frac{2}{3}\left(I_{SA} + \tilde{a} I_{SB} + \tilde{a}^2 I_{SC}\right) \tag{2}$$

Stator phase voltages can also be expressed as a single space phasor quantity as

$$\tilde{U}_S = \frac{2}{3}\left(U_{SA} + \tilde{a} U_{SB} + \tilde{a}^2 U_{SC}\right) \tag{3}$$

Similar considerations apply to flux linkages, namely

$$\tilde{\lambda}_S = \frac{2}{3}\left(\lambda_{SA} + \tilde{a} \lambda_{SC} + \tilde{a}^2 \lambda_{SC}\right) \tag{4}$$

where

$$\tilde{a} = e^{j\frac{2\pi}{3}} \tag{5}$$

$$\tilde{a}^2 = e^{j\frac{4\pi}{3}} \tag{6}$$

It must be emphasized that the complex j-operator used in the definition of the unit space phasors $\tilde{a}$ and $\tilde{a}^2$ has a completely different connotation from the one used in electrical circuit analysis: it designates a spatial shift of the quantity with which it is associated.

Equations (1) to (4) imply that a single space phasor can be constructed on the basis of individual phase windings of the polyphase motor. Alternatively, especially if the transition is from a transformed model as in the case of orthogonal models, the aggregate stator and rotor currents in space phasor notation can also be expressed as

$$\tilde{I}_S = I_\alpha + jI_\beta \tag{7}$$
$$\tilde{I}_R = I_d + jI_q \tag{8}$$

Similar considerations apply to both the stator and rotor phase voltages and flux linkages, namely

$$\tilde{U}_S = U_\alpha + jU_\beta \tag{9}$$
$$\tilde{U}_R = U_d + jU_q \tag{10}$$
$$\tilde{\lambda}_S = \lambda_\alpha + j\lambda_\beta \tag{11}$$
$$\tilde{\lambda}_R = \lambda_d + j\lambda_q \tag{12}$$

The machine model

With the foregoing considerations, a space phasor model describing the electromagnetic behavior of the entire machine can be devised; remarkably, consisting of a single model equation for stator and rotor phase windings respectively, that is

$$\tilde{U}_S = R_S\tilde{I}_S + p\tilde{\lambda}_S \tag{13}$$
$$\tilde{U}_R = R_R\tilde{I}_R + p\tilde{\lambda}_R \tag{14}$$

where $\tilde{U}_R = 0$ for the singly excited induction motor.

In terms of electrical circuit model parameters the equations can also be written as

$$\tilde{U}_S = R_S\tilde{I}_S + L_S p\tilde{I}_S + \frac{3m}{2} pR_S\tilde{I}_R \tag{15}$$
$$0 = R_R\tilde{I}_R + L_{!R}\left(p\tilde{I}_R + jp\vartheta\tilde{I}_R\right) + \frac{3m}{2}\left(p\tilde{I}_S + jp\vartheta\tilde{I}_S\right) \tag{16}$$

Together with the equation of motion given below, this deceptively simple model can be deployed to analyze the behavior of a polyphase induction motor in the time domain.

$$T_{\text{elec}} = Jp\omega + D\omega + T_{\text{load}} \tag{17}$$

In the above equations, p denotes the time derivative of the variable it precedes.

The electromagnetically developed torque can be obtained as:

$$T_{\text{elec}} = \frac{3m}{2}\Im\left(\tilde{I}_R\tilde{I}_S^*\right) \tag{18}$$

The supply model

If the induction motor is to be operated in a variable speed drive, then the non-sinusoidal nature of the supply voltage must be taken into account in modeling the drive to reflect the effect on machine performance of the harmonic content of the supply voltage. In the case of a voltage source inverter configured in six-step mode, illustrated in Fig. 1, the terminal voltages V_A, V_B, and V_C are as depicted in Fig. 2. For the purposes of this discussion, the inverter model shown here assumes ideal switches. Fig. 2 depicts the resultant voltages at

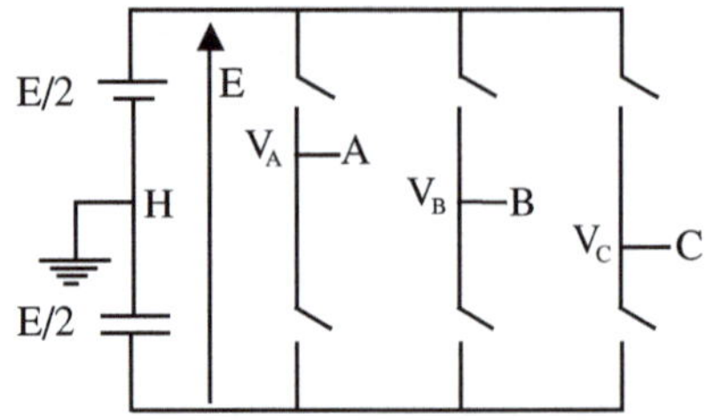

Figure 1. Voltage source inverter.

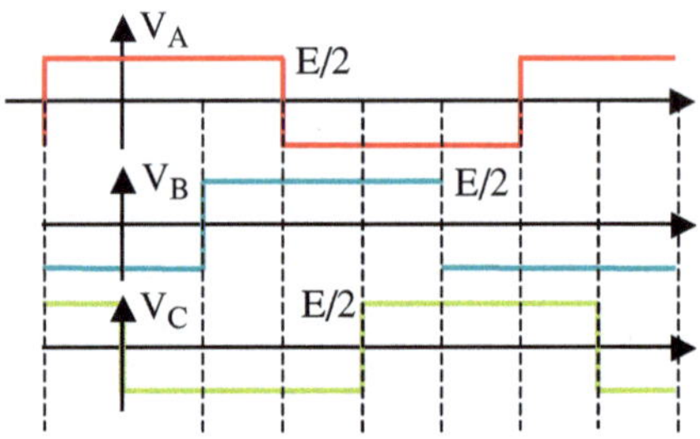

Figure 2. Terminal inverter voltages.

the inverter terminals, leading to "six-step" voltages across the stator phase windings of the motor.

The space phasor form of the resultant "six-step" voltage applied to the motor terminal can be conveniently obtained in terms of orthogonal components as

$$\tilde{U}_S = U_\alpha + jU_\beta \tag{19}$$

$$U_\alpha(t) = \sqrt{\frac{2}{3}}\left(V_A - \frac{V_B + V_C}{2}\right) \tag{20}$$

$$U_\beta(t) = \frac{1}{\sqrt{2}}(V_B - V_C) \tag{21}$$

The Fourier expansions of U_α and U_β give

$$U_\alpha = \sqrt{\frac{2}{3}}\frac{2E}{\pi}\sum_{k=1}^{\infty}\frac{\sin(2k-1)\frac{\pi}{6} + \sin(2k-1)\frac{\pi}{2}}{2k-1}\cos(2k-1)\omega t \tag{22}$$

$$U_\beta = \frac{1}{\sqrt{2}}\frac{4E}{\pi}\sum_{k=1}^{\infty}\frac{\cos(2k-1)\frac{\pi}{6}}{2k-1}\sin(2k-1)\omega t \tag{23}$$

Obviously, not all harmonics in (22) and (23) are significant in terms of causing parasitic behavior. Only those harmonics which are significant and can profoundly affect performance need be considered in the supply model.

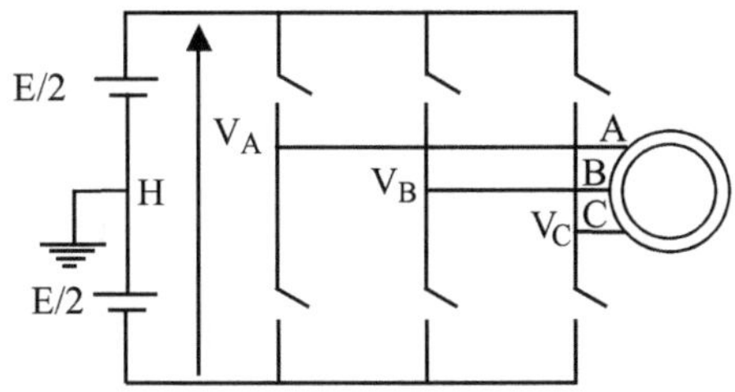

Figure 3. A basic inverter-induction motor drive.

Drive system model

An inverter driven induction motor can be modeled by combining the space phasor model of the machine with the supply model representing the non-sinusoidal voltage source inverter. Fig. 3 illustrates the ensuing model.

All electromagnetic terms in (15) and (16) are expressed as space phasors by advancing from the actual three-phase machine model to an orthogonal model with *alpha-beta* and *d-q* windings, the voltages and currents of which are combined to give a deceptively simple representation of the drive system. Furthermore, it becomes possible to assess the effects of supply harmonics by simply including (or injecting) the significant supply harmonics into $\tilde{U}_S$ of (15).

Simulation results

The simulated *alpha-beta* terminal voltages containing the significant harmonics have been drawn into the simulation. With these voltages the ripple torque for no-load and load conditions shown in Figs. 4 and 5 are predicted, obtained by solving the system equations of (15) to (17).

The simulation results show that the torque ripple is already of a considerable magnitude when the motor is not loaded. Under load, the ripple band is seen to widen. Electromagnetic quantities not shown here provide supportive evidence for the deterioration. The results are significant in that they indicate that the simulation method used is capable of estimating parasitic torque behavior in advance.

Experimental verification

The validity of the approach was tested experimentally for an inverter driven three-phase cage induction motor of 1.5 kW rating. Field orientation control with strong overmodulation was employed, resulting in supply harmonics both the order and magnitude of which resembled that of the six-step inverter at the motor terminals. Table 1 gives the relevant data for the test motor with which the foregoing simulations were conducted.

Fig. 6 depicts the measured torque ripple band under steady state operating conditions with full load. As can be seen, both the behavior and the relative magnitude of the torque reflect strongly those of the simulation. Fig. 7 gives the torque ripple at no load. The ripple magnitude is seen to have increased with load when compared with the no-load regime: an observation which is also supported by simulation.

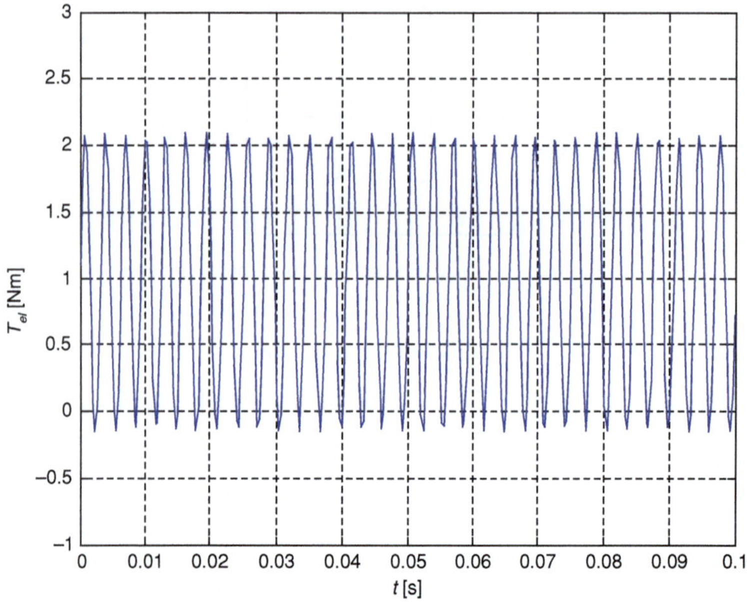

Figure 4. Torque ripple at no load (simulated).

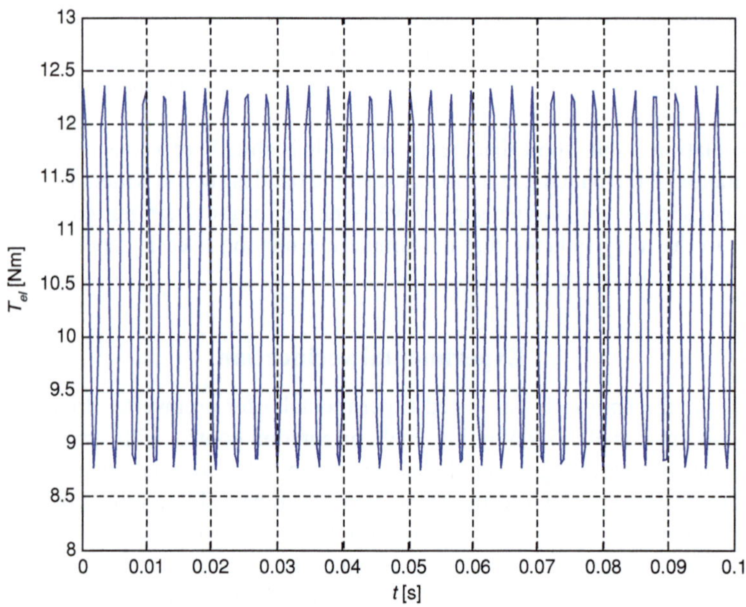

Figure 5. Torque ripple at full load (simulated).

Table 1. Data for the test motor

Number of pole pairs (P)	2
Rated voltage (U_N)	380/220 Y/Δ V
Rated frequency (F)	50 Hz
Rated power (P_N)	1.5 kW
Rated speed (n_N)	1,405 rpm
Rated current (I_N)	3.7 A
Main inductance (L_{1h})	382 mH
Stator inductance (L_s)	396 mH
Rotor inductance (L_r)	393 mH
Stator resistance (22°C) (R_s)	5.0 Ω
Rotor resistance (22°C) (R_r)	4.1 Ω
Rotor inertia (J_r)	0.008 kgm^2

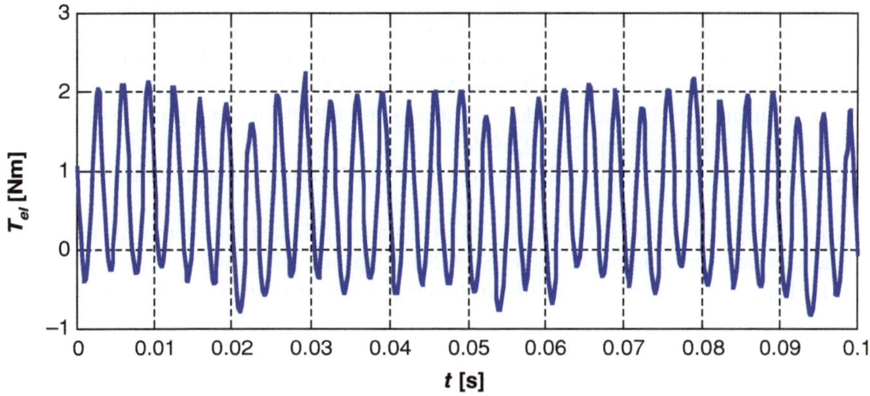

Figure 6. Torque ripple at no load (measured).

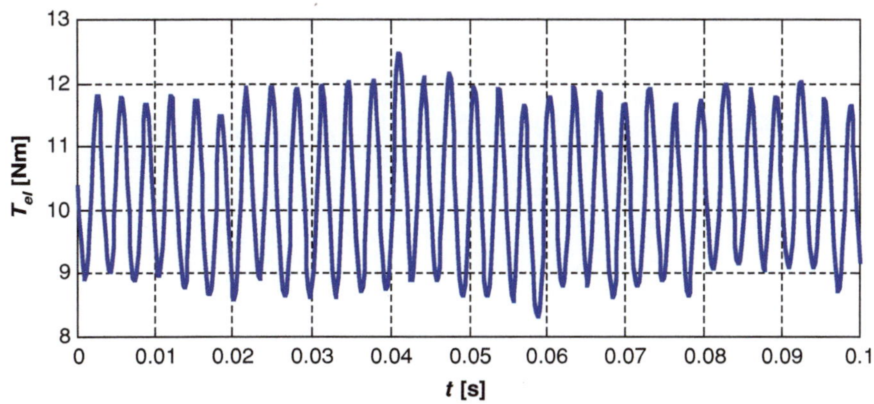

Figure 7. Torque ripple at full load (measured).

Evidently, the simple model representation does not allow the fine detail in the ripple band to be predicted in detail including the measured swings in the torque fluctuations. However the approximation achieved is satisfying.

Conclusion

The intuitive method of analysis based on the space phasor concept yields adequately accurate information about the nature of possible ripple torque generation in inverter driven induction motors. It is easy to assimilate and produces credible results with minimal computational effort. Although the approach has been demonstrated for a six-step inverter drive, it is equally applicable to more sophisticated drive systems.

References

[1] T. Lipo, P.C. Krause, H.E. Jordan, Harmonic torque and speed pulsations in a rectifier-inverter induction motor drive, IEEE Trans., Vol. PAS-88, No. 5, pp. 579–587, 1969.

[2] S.T.D. Robertson, K.M. Hebbar, Torque pulsations in induction motors with inverter drives, IEEE Trans. Ind. Gen. Appl., Vol. IGA-7, No. 2, pp. 318–323, 1971.

[3] G.B. Klimann, A.B. Plunkett, Modulation strategy for a PWM inverter drive, IEEE Trans. Ind. Appl., Vol. IA-15, No.1, pp. 72–79, 1979.

[4] K.P. Kovacs, J. Racz, Transiente Vorgänge in Wechselstrommaschinen, Budapest: Academiai Kiado, 1959.

[5] R.J.W. Koopman, Direct simulation of AC machinery including third-harmonic effects, IEEE Trans. Power Apparatus Syst., Vol. PAS 88, No. 4, pp. 465–470, 1969.

[6] R.H. Park, Two-reaction theory of synchronous machinery—Part I, AIEE Trans., Vol. 48, pp. 716–730, 1929.

I-10. VIBRO-ACOUSTIC OPTIMIZATION OF A PERMANENT MAGNET SYNCHRONOUS MACHINE USING THE EXPERIMENTAL DESIGN METHOD

S. Vivier[1], A. Ait-Hammouda[1], M. Hecquet[1], B. Napame[1], P. Brochet[1] and A. Randria[2]

[1]L2EP—Ecole Centrale de LILLE Ecole Centrale de Lille, Cité scientifique, B.P. 48, 59651 Villeneuve D'Ascq Cedex, France

stephane.vivier@ec-lille.fr, michel.hecquet@ec-lille.fr, ait-hammouda.amine@ec-lille.fr, pascal.brochet@ec-lille.fr, Bellemain.Napame@ec-lille.fr

[2]Alstom—2 Av de Lattre de Tassigny, 25290 Ornans, France

andry.randria@transport.alstom.com

Abstract. The aim of this paper is to use an analytical multi-physical model—electromagnetic, mechanic, and acoustic—in order to predict the electromagnetic noise of a permanent magnet synchronous machine (PMSM). Afterward, the experimental design method, with a particular design: "trellis design," is used to build response surfaces of the noise with respect to the main factors. These surfaces can be used to find the optimal design or more simply, to avoid unacceptable designs of the machine, in term of noise for a variable speed application.

Introduction

The majority of the electric machines operate at variable speed. In most of cases, it involves a generation of noise and vibrations, for a given speed and frequency.

For industries of manufacture, but also with the increasingly rigorous European standards, it is necessary to take into account the noise and the vibrations from the design stage.

A classical method used to study electromagnetic phenomena is the finite element method (FEM) in magneto-dynamics including the coupling with electrical circuits. However, in the case of strong coupling, taking into account the electromagnetic, vibro-acoustic, and thermic models in the same time would need a considerable computing effort. This would make the structure optimization practically impossible. In order to solve this problem, an analytical approach is considered instead.

The aim of this work is to develop and use an analytical multi-physical model—electromagnetic, mechanic, and acoustic—of a synchronous machine with permanent magnets. The complete model is coded using the data-processing tool MATLAB®, making possible the determination of fast and simple prediction models of the acoustic noise.

S. Wiak, M. Dems, K. Komęza (eds.), Recent Developments of Electrical Drives, 101–114.

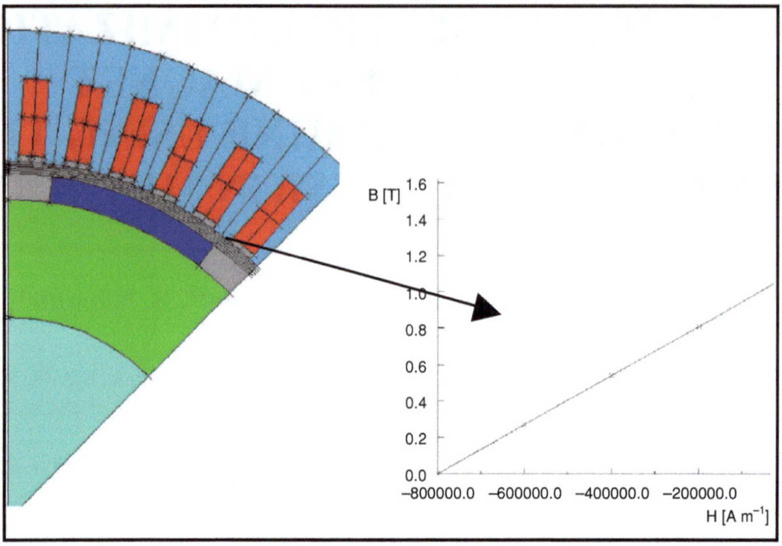

Figure 1. 1/8 of synchronous machine with magnet characteristic.

In order to reduce noises and vibrations, two main ways can be considered: by the control of the machine excitation [1], or by modifying the system structure. In this work, only the second solution is explored.

Three models are presented: electromagnetic, mechanical of vibration, and acoustic. For each of them, comparisons with FEM and experiments have been made.

Lastly, a study of sensitivity is presented in order to deduce the influential—or significant—factors on the noise. For that, the technique of the experimental designs is used. More particularly, the modeling of the noise will be achieved thanks to the new "trellis" designs. Several response surfaces are given; they represent the noise according to influential factors, with respect to different speeds of the machine.

These surfaces are useful to deduce the parts of the design space to avoid.

Presentation of the synchronous machine

This machine is composed of eight rotor poles and 48 stator slots. The power of this machine is about 250 kW (Fig. 1).

Analytical models

The vibration analysis of electrical machines is a rather old problem. During the 40s and 50s, it was deeply studied by various researchers [2 to 5]. Vibrations of electromechanical systems are due to excitation forces. Some of them have a magnetic origin. Other sources of vibrations, such as aerodynamic conditions, bearings, etc., will not be considered in this paper.

An analytical model, considering electromagnetic phenomena, mechanical vibrations, and acoustic noises, was developed to take into account the overall noise produced by a variable induction in the air-gap [6 to 9] and by forces applied to the various structures.

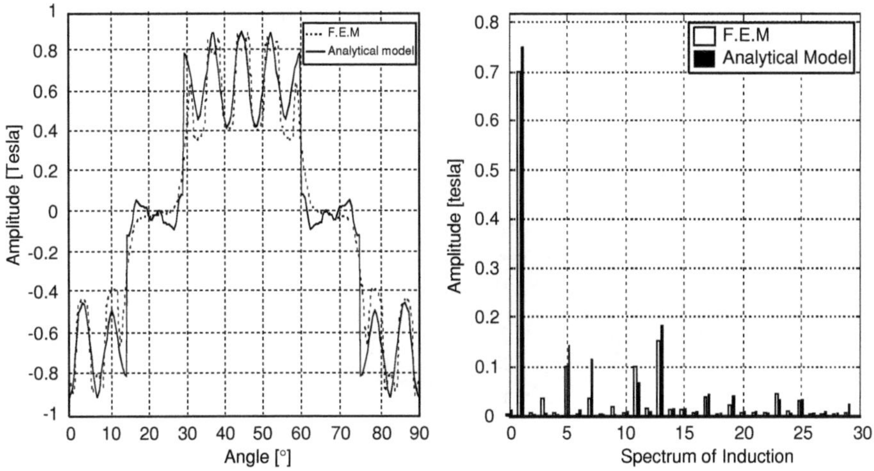

Figure 2. Comparison of the form induction and FFT.

Electromagnetic model

It is assumed that forces in the air-gap of the machine are the main mechanical excitation. To characterize induction in the air-gap, the proposed method is based on the calculation of the air-gap permeance (P_e) and the magnetomotive force (*mmf*) [6 to 8]. To establish the analytical expression of the permeance, some assumptions are made:

- the magnetic circuit has a high permeability and a linear characteristic,
- the tangential component of the air-gap flux density is negligible relative to the radial component.

Results are given in Ref. [10] and just a comparison is recalled by Fig. 2.

Using the finite element software OPERA-2D [11], the air-gap induction created by the magnet rotor as a function of space and time has been also calculated. In Fig. 2, a comparison on induction wave shapes vs. the angle is presented. The comparison results are very satisfactory, the induction distribution and the harmonic values determined analytically are validated numerically, as shown in Refs. [6,12].

The FFT of the radial forces vs. time (t) and angle (θ) is presented below in Fig. 3.

Vibratory model

Vibrations are the consequence of the excitation of the mechanical system by electromagnetic forces. Once the forces applied to the stator have been determined, the study of vibrations is possible. They correspond to the deformations whose amplitudes have to be calculated. For that purpose, some parameters have to be determined:

- the damping,
- the mode shapes and resonance frequencies for each mode.

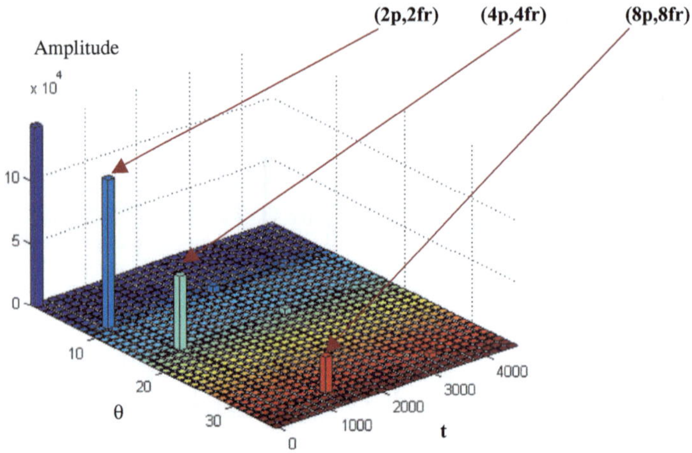

Figure 3. FFT 2D of radial force ($f_r = p'N$).

For the damping coefficient, we have used the experimental measurements and the software PULSE [13] to determine the resonance frequencies, the mode shapes, and the damping. For example, some results are detailed in Fig. 4.

The studied analytical model takes into account the yoke, the frame, the teeth, and the winding. The self vibration modes of the stator structure are determined, in various configurations: yoke only, yoke + teeth, yoke + teeth + carcass, and yoke + teeth + winding + carcass [14].

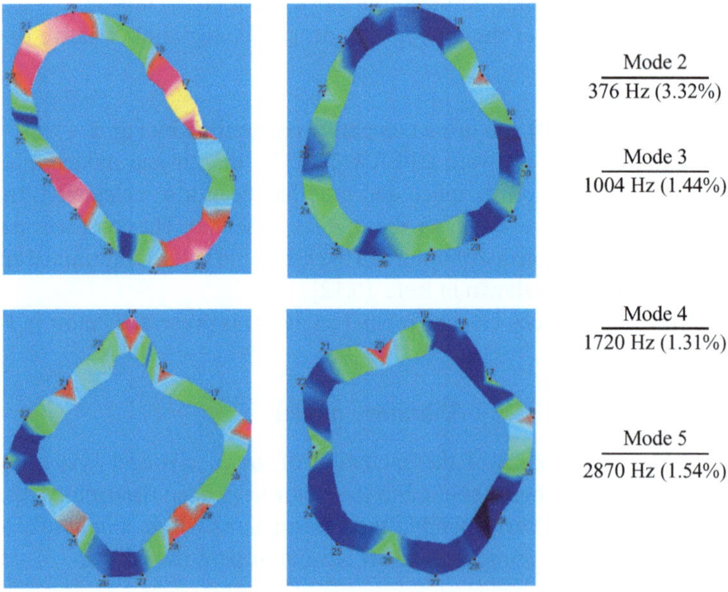

Figure 4. Mode shape, resonance frequencies with (damping coefficient) obtained by measurements.

Table 1. Resonance frequencies (Hz) for each mode

No. mode	Analytical model	FEM	No. mode	Analytical model	Experimental
0	3,063	3,151	0	2,736	2,855 (1.46)
2	243	268	2	308 416 (3.22)	376 (3.32)
3	688	732	3	871 1,140 (1.74)	1,004 (1.44)
4	1,319	1,349	4	1,670 1,968 (2.44)	1,720 (1.31)
5	2,134	2,078	5	2,702 2,944 (1.46)	2,870 (1.54)

(yoke and teeth only) (Complete stator: yoke + teeth + winding + carcass)

Some results are presented in Table 1, with an experimental comparison.

Resonance frequencies of the stator have been obtained by impact testing measurements, realized thanks to the impact test method. The comparison of the results with the analytical model are very satisfactory.

Let us note that the damping coefficient ξ_a cannot be given theoretically. However, Jordan [4] considers that for a synchronous machine, it stands between 0.01 and 0.04 (Table 2). The total vibratory spectrum obtained by our analytical model is presented in Fig. 5.

The simulation results agree well with the theory. In addition, the proximity of the frequency of excitation mode 0 with the frequency of the resonant mode 0 (at 2,844 Hz) explains the vibration peak located around 2,900 Hz. However let us point out that precautions must be taken when analyzing the results.

The model giving the induction values is not perfect (the saturation phenomenon is neglected) and the formulae of Timar [3] giving the vibrations are also approximated. What is of interest is to determine the frequency of the main peaks and to be able to range their amplitudes.

In order to study the vibrations generated by the operating conditions, an accelerometer is positioned on the frame of the machine. It measures the deformations of the frame. The vibratory spectrum gives lines identical to those obtained by the noise measurement; it displays a dominant line situated at 2,900 Hz, that corresponds to the theoretical excitation mode 0 predicted at 2,844 Hz (Fig. 6).

Table 2. Main characteristics of the machine

Speed	3,555 rpm
Frequency of the supply f_r	237
Rotational frequency f_{rot}	237/p
Frequencies of components of forces (multiple of $2f$)	$h * 237 (h = 2, 4, 6, \dots)$

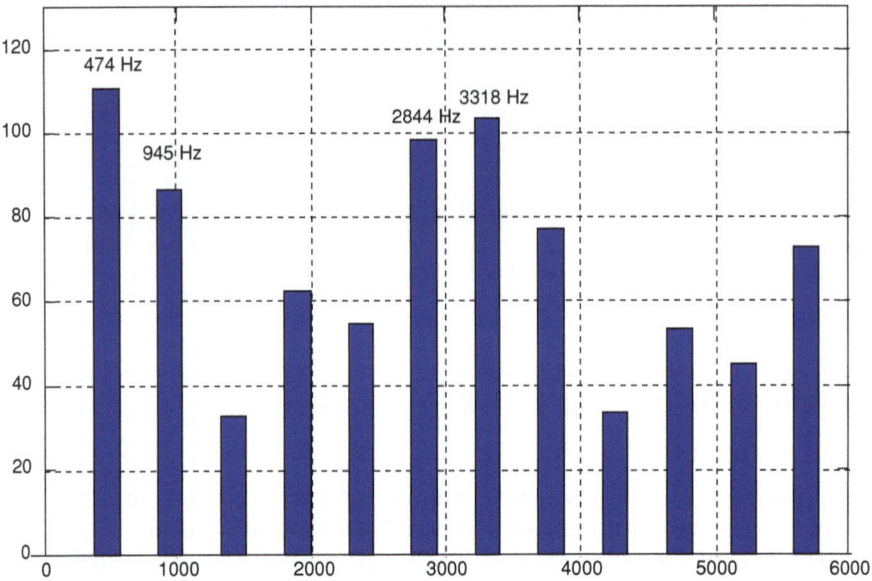

Figure 5. Total analytical vibratory spectrum with 3,555 rpm.

In order to study the vibrations generated by the operating conditions, an accelerometer is positioned on the frame of the machine. It measures the deformations of the frame. The vibratory spectrum gives lines identical to those obtained by the noise measurement; it displays a dominant line situated at 2,900 Hz, that corresponds to the theoretical excitation mode 0 predicted at 2,844 Hz (Fig. 6).

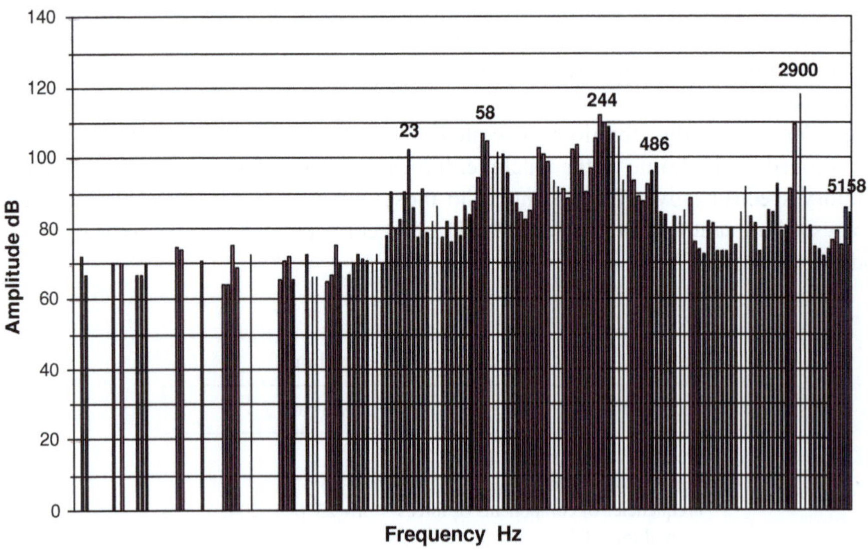

Figure 6. Vibratory spectrum measured with 3,555 rpm with 1/12 of octave.

Acoustic model

Acoustic intensity $I(x)$ can be written as a function of the frequency, the amplitude of vibrations, the mode order and the stator surface [5,9].

$$I(x) = \frac{\sigma 8200 f_r^2 Y_{md}^2 S_e}{4\pi x^2 (2m+1)}$$

The coefficient σ is called factor of radiation. It represents the capacity of a machine to be a good sound generator and can be calculated through two different ways according to whether one assumes the machine to be a sphere or a cylinder. σ is a factor which varies with λ (wavelength) and the diameter of the machine. It also depends on the mode shape [3]:

$$\sigma = f\left(\pi \frac{D}{\lambda}\right), \quad \lambda = \frac{c}{f_r}$$

c: Traveling speed of sound (344 m/s); f_r:Vibration frequency.

It appears that $I(x)$ is inversely proportional to the order of the mode, in addition the acoustic intensity is proportional to the square of the vibration amplitude. In general, we define I and W in decibels. We thus define the levels of acoustic pressure, acoustic intensity, and sound power as follows:

$$L_p = 20\log\left(\frac{P}{P_0}\right), \quad L_i = 10\log\left(\frac{I}{I_0}\right), \quad L_w = 10\log\left(\frac{W}{W_0}\right)$$

with:

$$P_0 = 20\,\mu\text{Pa}, \quad I_0 = 10^{-12}\,\text{W/m}^2, \quad W_0 = 10^{-12}\,\text{W}$$

The spectrum of the total noise obtained by our analytical model is presented below (Fig. 7).

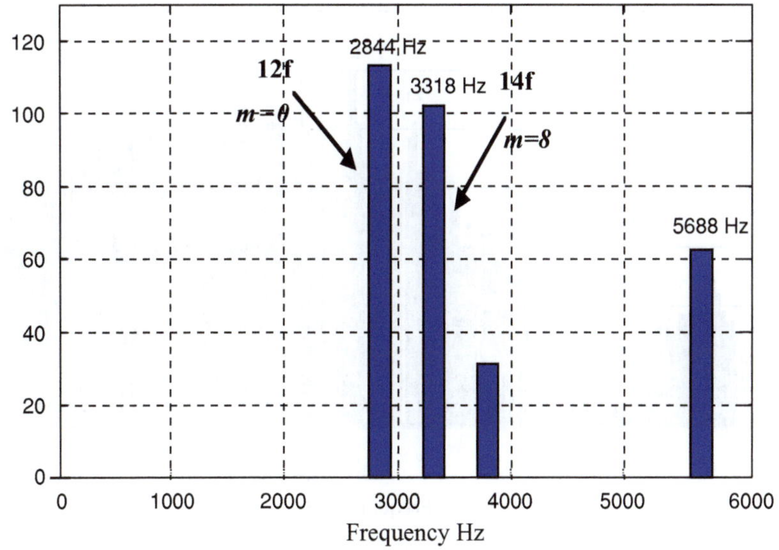

Figure 7. Spectrum of the noise of the simulated PMSM.

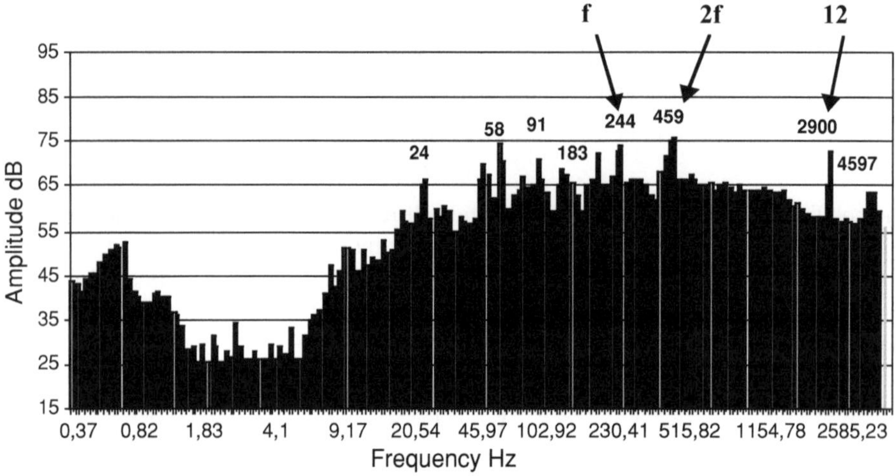

Figure 8. Spectrum of the noise of PMSM measured with 3,555 rpm (1/12 octave).

Fig. 8 presents the measured acoustic noise spectrum at the same speed (3,555 rpm). Lines are located at the same frequencies as in the vibration spectrum.

The first line determined by measurements is located at 2,900 Hz (12*f*). In theory, the harmonic of teeth (12*f*) is located at 2,844 Hz.

The lines at low frequencies (between 24 and 459 Hz) are not found in theory, because they are mainly related to the background noise. They are not generated by the PMSM, but by the driving motor and the ventilator (Fig. 9).

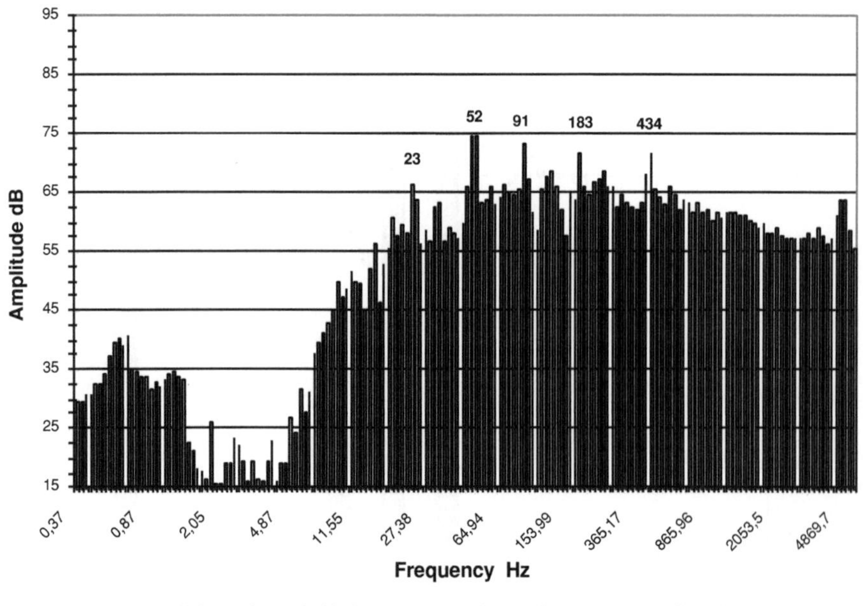

Figure 9. Spectrum of the noise of driving motor and ventilator measured with 3,555 rpm (1/12 octave).

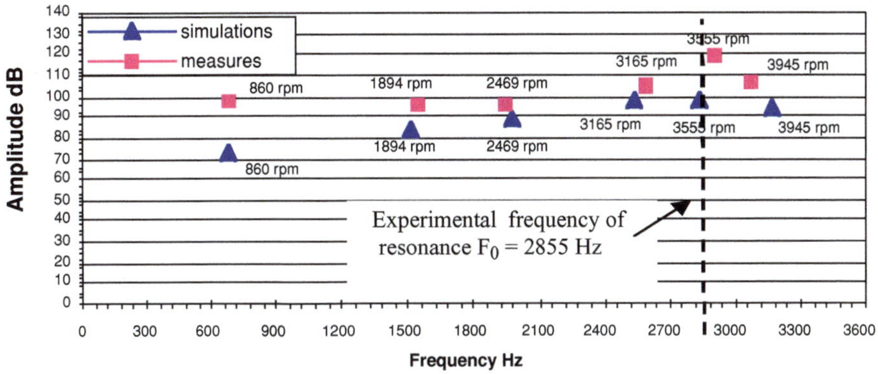

Figure 10. Level of the 12th harmonic vs. rotation speed. Vibratory comparisons.

After this comparison, the permanent magnet synchronous machine was tested at various speeds, that allowed us to highlight a particularly dangerous speed. Moreover, some results are overestimated but the quality of those is respected (Fig. 10). In spite of the inaccuracies, major lines appear, which is of primary importance in view of noise reduction.

To know which lines to reduce does not require to know its amplitude precisely. Its frequency, on the other hand, must be well given. Lastly, taking into account the complexity of the studied phenomena and the many steps of calculations making it possible to lead to the results, the latter seem very satisfactory.

Screening analysis

Once the different models finalized and assembled into a single "coupled model," it becomes possible to study the variations of the main variables representing the vibration sources. This is achieved by the building of response surfaces, and by the launching of optimizations. The privileged tool employed is the Experimental Design Method [15,16].

First of all, a sensibility analysis using the global coupled model is described.

The overall audible noise produced by the synchronous machine stands as the studied variable (the response). An analytical relation linking the noise amplitude with five variation sources (the factors) has been established:

- the stator slot opening (l_{se});
- the height of the yoke (h_{yoke});
- the opening of permanent magnet (alp);
- the width of the air-gap (e);
- the height of the permanent magnet (h_{mag}).

A screening design [17] is calculated. It gives the ability to determine the influent factors, with respect to the response, inside the design space. This domain is implicitly defined by the intervals of variation, for the five factors (Table 3).

The Fig. 11 gives a representation of the influence of each factor on the noise.

Firstly, it shows that the opening of the permanent magnets (alp) is a very influential factor, since its variation from its middle value (31°) to its upper limit (32°) makes the noise increase by about 15 dB.

Table 3. Intervals of variation—screening analysis

Factors	Lower bound	Upper bound
l_{se}	l_{se} min	l_{se} min + 20%
h_{yoke}	h_{yoke} min	h_{yoke} min + 20%
Alp	30°	32°
E	e_{min}	e_{min} + 20%
h_{mag}	10 mm	12 mm

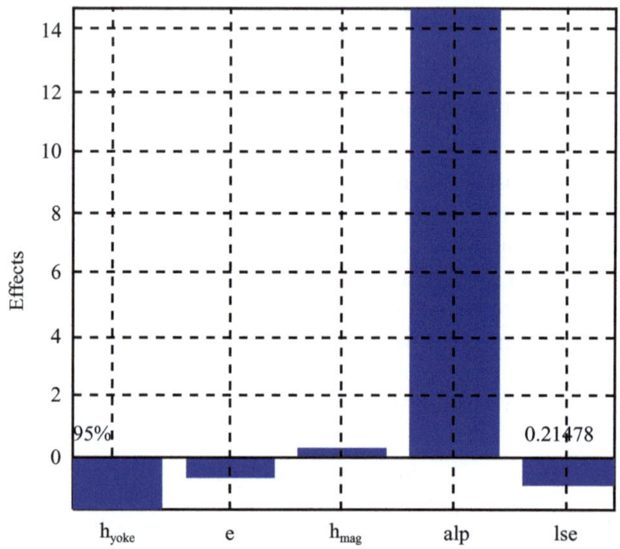

Figure 11. Factor influences on the noise amplitude.

The height of the yoke (h_{yoke}), the width of the air-gap (e), and the stator slot opening (l_{se}) are also significant factors according to this figure, since they all exceed the two 95% significance levels. It means that the probability to declare these factor influential although they are not, is equal to 5%. They all have a negative influence on the noise variations: their values have to be increased to reduce the noise amplitude.

The height of the magnets (h_{mag}) is not considered as an influent factor, if the same significance level is used.

It is important to keep in mind that these conclusions only hold inside the design domain.

The previous results have been obtained for a fixed rotor speed (3,000 rpm). Imposing different speeds do not change the relative influence of the factors. However, one can say that effect values increase for speeds around 3,000 rpm. This aspect will be confirmed by the following study.

Research of optimal conditions

In a second stage, our purpose was to "model" the part of the conception domain in which the global audible noise produced by the PMSM was smaller than a predefined limit: 80

Table 4. Intervals of variation—modeling stage

Factors	Lower bound	Upper bound
l_{se}	l_{se} min	l_{se} min + 20%
h_{yoke}	h_{yoke} min	h_{yoke} min + 50%
alp	26°	34°
e	e_{min}	e_{min}+ 50%
N	3,500 rpm	4,500 rpm

dB was considered as the maximal admissible noise intensity. This frontier for the noise—the response—has been computed with respect the same factors except the height of the permanent magnet (h_{mag}) and in addition, the motor speed (N).

These factors have been selected thanks to screening analyses realized with the complete coupled model. Their intervals of variation are given by Table 4.

Since we want to have a good description of the variations of the noise with respect to the five factors, it is necessary to increase the number of the different levels taken by each factor. Considering five levels is in general enough. Such a configuration leads to $5^5 =$ 3,125 experiments with the use of a grid design—that is a multi-level full factorial design. This number of evaluations of the coupled model is relatively large, and it can be interesting to take advantage of the new "trellis" deigns [17].

Trellis designs can be described as multi-level fractional grid design. They are build from fractional two-level factorial designs judiciously superposed (Fig. 12). For this reason, under important hypotheses, they keep their interesting mathematical characteristics, such as for instance the orthogonality property.

When five factors are present, it is possible to use the two-level fractional factorial design defined as 2^{5-2}, that is the quarter ($2^2 = 4$) of the corresponding full factorial design 2^5. When this design is used to build the trellis design, it leads to definition of a five-level incomplete grid, with only 795 experiments—instead of 3,125 with a complete grid. It takes approximately 16 h to compute this trellis design on a PC.

Instead of using the 795 values of noise directly, we have exploited the interesting relative location of the experiments inside the design domain: an iterative procedure has been applied to estimate the noise values for each of the $3,125 - 795 =$ 2,330 initially nonevaluated experiments. It has been shown that the overall error made for these 2,330 interpolations, realized thanks to the 795 initial experiments, is lower than 0.8%.

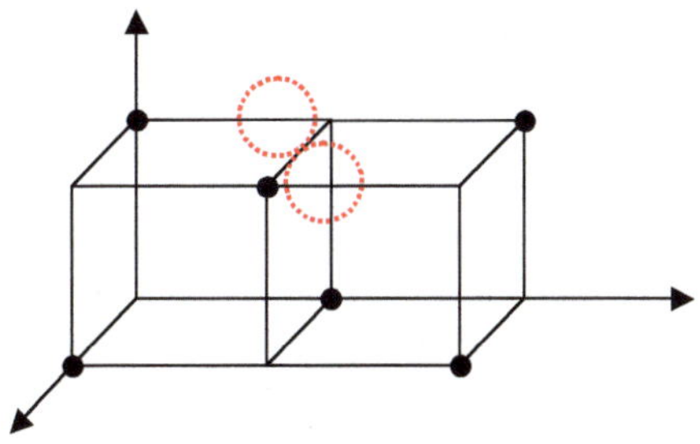

Figure 12. Example of experiment sharing between two fractional designs (2^{3-1}).

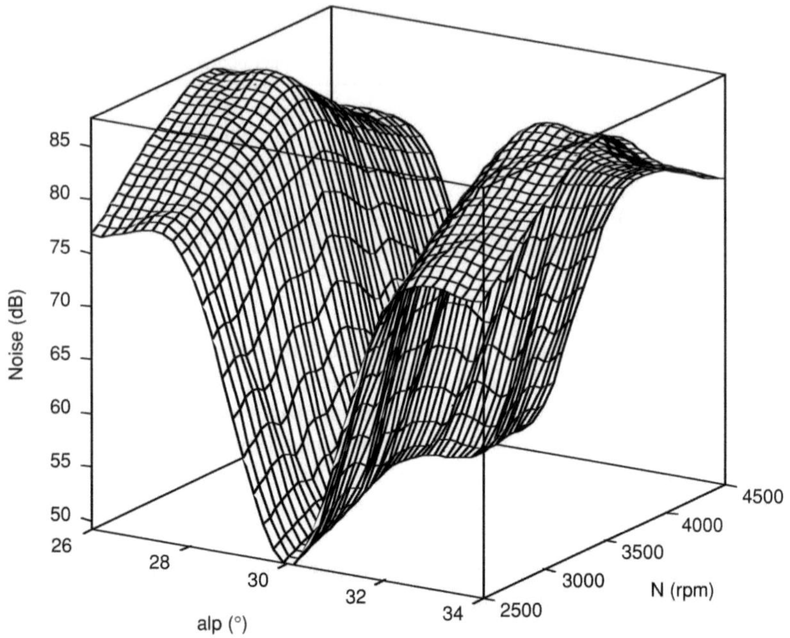

Figure 13. Noise variations vs. "alp" and "N."

The different results that follow are deduced from the 795 first experimental points mixed with the estimated ones.

It quickly appeared that the opening of the permanent magnets (alp) and the motor speed (N) were the two most influential variables over the noise production. The following response surface shows the corresponding variations, as shown in Fig. 13.

It is very clear that the noise is strongly reduced when the permanent magnet area—in fact the corresponding angular opening—is equal to 30°. This result is confirmed by practical considerations.

The rotor speed has also a neat influence over the noise production. A resonance phenomenon is visible near the speed value 3,500 rpm, whatever the factor alp values.

The influence of the three other factors are relatively small in comparison. However, we can notice that the decrease of h_{yoke} leads to move the resonance point toward lower rotation speed values.

The 80 dB limit can be graphically represented with respect to alp, N, and h_{yoke}, thanks to iso-value surfaces (Fig. 14).

Two iso-value surfaces are represented: one showing the noise equal to 80 dB, and the other to 84 dB. The graphic is nearly symmetrical: 30° standing as the central value for the magnet opening (alp). Then, the admissible subspace of the conception domain is modeled by the zone delimited by the two central 80 dB surfaces. This indicates that it is always possible to conceive a PMSM. generating a noise lower than 80 dB, provided that the magnet

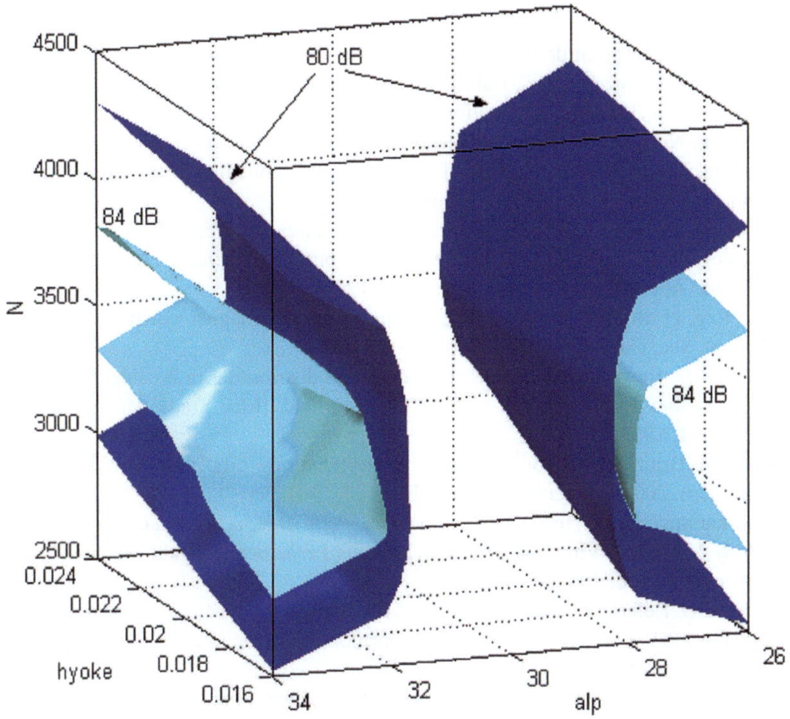

Figure 14. Noise iso-value surfaces (80 and 84 dB) vs. "*N*," "h_{yoke}", and "alp."

opening is chosen between 28.5° and 31.5°. This interval can be extended for particular rotor speeds greater than 4,000 rpm or lower than 2,700 rpm.

Conclusion

The purpose of this work is to present some results obtained from the exploitation of a complete coupled model of a permanent magnet synchronous machine. Different multi-physical aspects are considered: electromagnetic, mechanic, and acoustic phenomena are taken into account thanks to a single analytical model.

The Experimental Design Method is the privileged tool used to make the complex relationships between the main variables appear.

The first study—a screening analysis—shows that, whatever the rotor speed considered, the angular opening is a very influential factor: the particular value 30° is certainly the best choice. It is more difficult to set the other factors, since the rotor speed interacts with them. However, the height of the permanent magnets is declared nonsignificant in term of acoustic noise.

The second study is designed to work on more precise data. For that purpose, a trellis design with five levels per factor and only 795 experiments, is computed. The advantageous properties of this type of design allow the subsequent evaluations of 2,330 other points, with an excellent accuracy, leading to practical design choices for lowering the limited noise.

References

[1] M. Gabsi, "Conception de machines spéciales et de leurs alimentations. Réduction du bruit d'origine électromagnétique", Habilitation à diriger des recherches, Juillet 1999.

[2] S. Timochenko, Théorie des vibrations, Librairie Polytechnique CH Beranger, 1939.

[3] P.L. Timar, Noise and Vibration of Electrical Machines, Elsevier, 1989.

[4] H. Jordan, Electric Motor Silencer—Formation and Elimination of the Noises in the Electric Motors, W. Giradet-Essen Editor, 1950.

[5] S.J. Jang, Low-Noise Electrical Motors, Oxford: Clarendon Press, 1981.

[6] N. Boules, Prediction of no-load flux density distribution in permanent magnet machines, IEEE Trans. Ind. Appl., Vol. IA 21, No. 4, pp. 121–124, 1985.

[7] J.D.L. Ree, N. Boules, Torque production in permanent magnet synchronous motors, IEEE Ind. Appl. Soc. Conf. Rec., Vol. 87, pp. 15–20, 1987.

[8] Z.Q. Zhu, D. Howe, Instantaneous magnetic field distribution in brushless permanent magnet DC motors. Part III: Effect of stator slotting field, IEEE Trans. Magn., Vol. 29, No. 1, pp. 143–151, 1993.

[9] R. Corton, Bruit magnétique des machines asynchrones, procédure de réduction passive et active, thèse, 2000, Université d'Artois, France.

[10] A. Ait-hammouda, M. Hecquet, M. Goueygou, P. Brochet, A. Randria, "Analytical Approach to Study Noise and Vibration of a Synchronous Permanent Magnet Machine", ISEF'2003, Maribor, September 18, 2003, CD.

[11] OPERA_2D, Reference Manual, VECTOR FIELDS, http://www.vector-field.co.uk.

[12] R. Breahna, P. Viarouge, "Space and Time Harmonics Interactions in Synchronous Machines", Proceedings of Electrimacs, 1999, pp. 45–50.

[13] Brüel and Kjaer, PULSE System: Modal Test Consultant, http://www.bksv.com.

[14] S.P. Verma, L. Wen, "Experimental Procedures for Measurement of Vibration and Radiated Acoustic Noise of Electrical Machines", Power System Research Group 2002, ICEM 2002, p. 432.

[15] J.J. Droesbeke, J. Fine, G. Saporta, Plans d'expériences—Applications à l'entreprise, Ed. TECHNIP, 1997.

[16] J. Goupy, La Méthode des plans d'Expériences, Paris: Dunod, 1988.

[17] S. Vivier, "Stratégies d'optimisation par plans d'expériences et Application aux dispositifs électrotechniques modélisés par éléments finis", Thèse de doctorat, Université des Sciences et Techniques de Lille, July 2002.

I-11. ELECTROMAGNETIC FORCES AND MECHANICAL OSCILLATIONS OF THE STATOR END WINDING OF TURBO GENERATORS

A. Grüning and S. Kulig

Institute of Electrical Drives and Mechatronics, University of Dortmund, D-44227 Dortmund, Germany,

arne.gruening@uni-dortmund.de, stefan.kulig@uni-dortmund.de

Abstract. Numerical methods of calculating the electromagnetic forces and of simulating the oscillation behavior of the stator end winding are introduced. The end winding oscillations of different turbo generators under forced vibrations are computed in a combined simulation. Also eigenfrequencies and eigenmodes are determined. The obtained results are surveyed by measurements. Numerical simulation of oscillation behavior is found a useful tool in end winding design although model parameter identification still offers improvement potential.

Introduction

Due to the complex structure, the design of the stator end winding of large turbo generators and especially of the appendant support fixture still offers a huge potential for optimization. Primarily the capability of the stator end winding to perform oscillations owing to the operant electromagnetic forces and the resultant eventuality of damages like fatigue or even cracks of the insulation gives reason to accomplish improvements [1].

When optimizing the end winding support fixture in order to reduce the occurring oscillations and therewith the risk of damages, detailed knowledge of the vibration behavior under steady-state as well as under transient conditions is very beneficial.

The most viable method to obtain this knowledge is the accomplishment of numerical computer simulations, whereas detailed measurements are difficult to perform and therefore only sometimes used to verify the simulation results. As can be seen in Fig. 1, the stator end winding of a turbo generator is a very complex entity with a huge number of components of different mechanical properties. But due to the support fixture it is also acting as a complex composite structure. Therefore the appliance of simulation methods using a three-dimensional model of the complete end winding is precondition for obtaining reliable and useful results.

In this connection numerical simulation methods based on a three-dimensional modeling of the end region and facilitating investigations of the vibration behavior under various operation modes become increasingly the aim of development.

S. Wiak, M. Dems, K. Komęza (eds.), Recent Developments of Electrical Drives, 115–126.

Figure 1. Stator end winding.

In terms of the impact on the stator end winding a certain mode of operation can be characterized by the currents in the windings and, if the influence of the rotor winding is taken into account, the rotor movement, which determines the position of the rotor winding. On this account the procedure of investigating the vibration behavior of the stator end winding comprises the electromagnetic computation of the three-dimensional distribution of the forces generated by the currents and afterward the three-dimensional simulation of the mechanical oscillation due to these forces. The implementation of such a simulation method represents an interdisciplinary task combining electromagnetic and mechanical problems [2].

The present paper summarizes the results of an investigation, which was performed by a team of mechanical and electrical engineers over a period of about five years, concerned with the development and coupling of three-dimensional numerical electromagnetic and mechanical simulation methods.

Electromagnetic simulation

A numerical simulation method based on the application of Biot-Savart's law to line circuit segments was used to compute the three-dimensional distribution of electromagnetic forces acting on the stator end winding.

Computing the magnetic flux density B by Biot-Savart's law, materials with non-linear magnetization characteristics like iron cannot be considered directly. A practicable way to use this method nevertheless is to replace all iron parts by additional fictive current distributions emulating the influence of the iron parts on the magnetic field, according to the method of images. The consideration of both the original circuits and the additional fictive current distributions as line circuits enables the application of Biot-Savart's law in the following form:

$$\vec{B} = \frac{\mu_0 I}{4\pi} \cdot \oint \frac{\vec{dl} \times \vec{r}}{r^3} \tag{1}$$

In a numerical computation of the flux density B generated by a three-dimensional assembly of line circuits, these can be modeled by discrete line circuit segments. B is obtained using the principle of superposition after applying a discrete form of (1) to each segment.

The three-dimensional distribution of electromagnetic forces can afterward be obtained by computing the vector of force acting on each discrete line circuit segment of the length l and carrying the current I using Lorentz' law [3]:

$$\vec{F} = I \cdot (\vec{l} \times \vec{B}) \tag{2}$$

By using line circuits instead of conductors with a finite cross section and a given current density it may be expected that the forces will be computed up to 25% to high [3]. But due to the fact that both the modeling effort and the computing time could be reduced substantially this simplification was found admissible.

The described method was implemented in a numerical computation program and applied to calculate the three-dimensional distribution of electromagnetic forces acting on the stator end winding of different turbo generators.

A typical three-dimensional model of the end region used in the computation is shown in Fig. 2. It consists predominantly of the line circuit segments modeling the bars of the stator end winding.

The influence of the rotor end winding is not directly taken into account. Instead of that the magnetic flux generated by the rotor is emulated by line circuits in the area of the retaining ring. The assumption underlying this simplification is, that the stator flux during transients is approximately mirrored by the rotor. Therefore the currents in these additional line circuits are equal to the currents in the stator windings but have opposite direction.

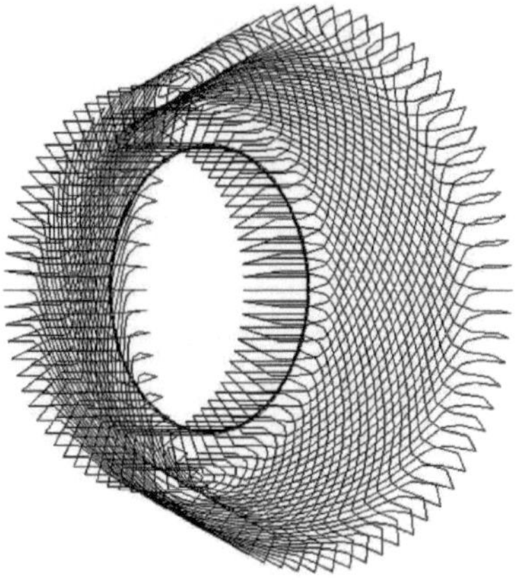

Figure 2. End winding model for electromagnetic force computation.

The influence of the stator core is considered according to the method of images by mirroring the complete model of line circuits at the end core plane. The impact of other magnetizable parts in the end region, like the rotor shaft, is neglected [3].

Additional investigations performed in [4] and also by the authors showed that both the influence of the rotor shaft and the influence of the end core plane are of minor importance.

By repeated computations the applied method is able to calculate discrete time functions of the electromagnetic forces using discrete time functions of currents as input data. Thus it was possible to investigate the transient behavior of the electromagnetic forces, like for example during a three-phase terminal short circuit, which represents the standard check of large electrical machines previous to initial operation.

Before accomplishing the force computation the discrete time functions of the currents in the stator windings have to be determined. Due to the fact that measurements with a sufficient time resolution are rare, the currents were computed using the numerical network simulation program NETOMAC [5], based on the Park transformation. As a likewise practicable method to obtain discrete time functions of currents, also a two-dimensional finite-difference time-stepping method as described in [6] was applied. Using the finite-difference method more accurate results with differences to measurements of less than 5% can be obtained, but at the price of an extensively longer computation time [1].

In order to be impressed on the nodes of the mechanical finite-element model, the results of the electromagnetic force computation are given by force vectors instead of uniform loads. The magnitude of the force vector acting on a certain node is therefore dependent on the length of the respective line circuit segment, which has to be considered when discussing the force distribution. Depending on the treated generator the number of nodes differs from about 500 to about 2,000, with three time functions of the force components at each node.

The electromagnetic forces acting on the end winding of a 90 MVA two-pole turbo generator during a three-phase short circuit are given in Fig. 3, Fig. 4 shows the forces on the end winding of a 1,500 MVA four-pole generator.

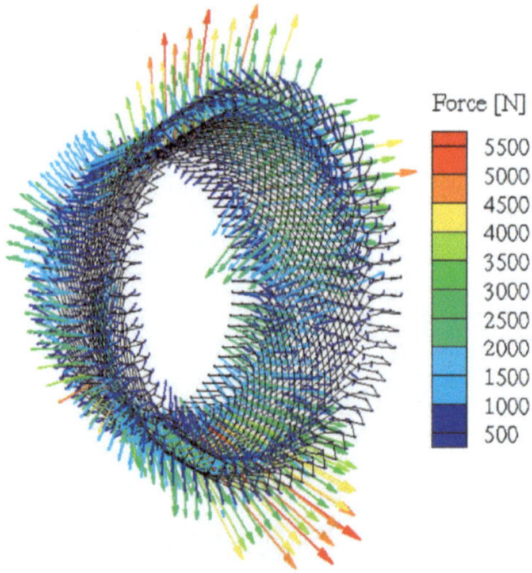

Figure 3. Forces on the end winding of a two-pole generator.

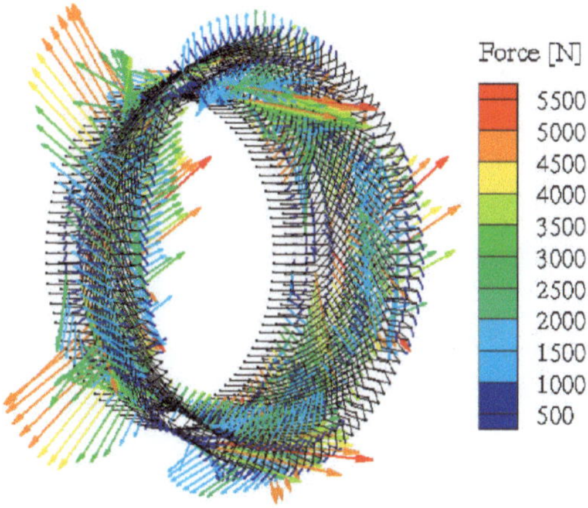

Figure 4. Forces on the end winding of a four-pole generator.

Both figures show the force distribution at a moment 10 ms after short circuit occurrence, when the currents in the stator windings approximately reach their maximum values. Whereas the maximum forces on the end winding of the two-pole generator occur approximately at the same time, the maximum forces on the four-pole generator end winding occur earlier, approximately 8 ms after short circuit occurrence.

The highest magnitudes in the force distribution occur in the area of the involute parts of the winding coils nearer to the end core plane. In both force distributions the number of poles of the corresponding generator is visible. Beside radial and tangential force components also strong axial force components occur.

The maximum uniform loads acting on the end winding of the two-pole generator amount to approximately 40,000 N/m, in the four-pole generator the uniform loads reach values around 70,000 N/m. This corresponds to the higher power density of the four-pole generator.

Figs. 5 and 6 show the force distribution of both the two-pole and the four-pole generator in a plane parallel to the end core plane at different moments around the occurrence of the force maximum. In both figures the force distribution enables the classification of the winding coils corresponding to the respective phase. Indeed the force distribution is non-sinusoidal [1].

As can be seen in Fig. 6, the tangential components in the force distribution of the four-pole generator are much more pronounced than in the force distribution of the two-pole generator.

Fig. 7 shows the time functions of the force components acting on a location in the middle of the involute part of the upper layer bar of the first coil of phase a of the two-pole generator, which has to sustain the largest electromagnetic forces during the three-phase terminal short circuit.

All components consist of a fraction oscillating with the system frequency and a constant fraction, both decaying exponentially. With the constant fraction decreasing the time functions are increasingly dominated by a fraction of twice the system frequency [1].

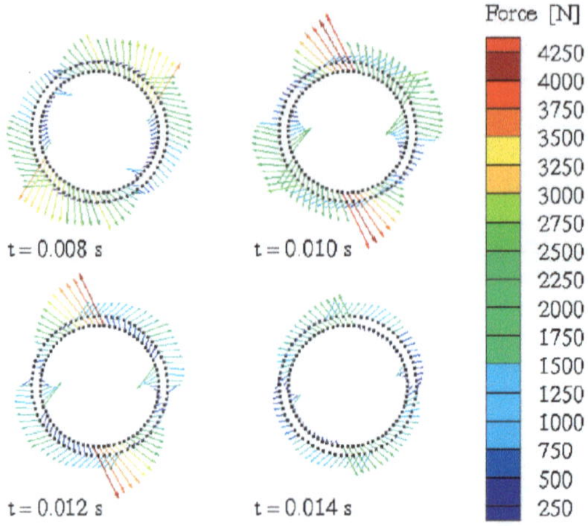

Figure 5. Plane force distribution in the two-pole generator end region at different moments.

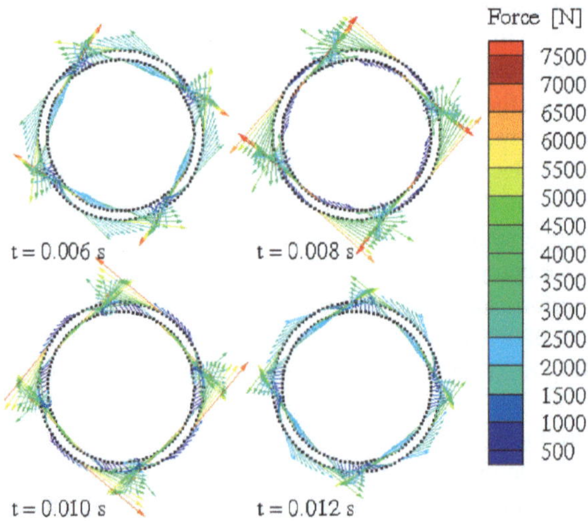

Figure 6. Plane force distribution in the four-pole generator end region at different moments.

Alternative computation

In the case of the two-pole generator the determination of the electromagnetic forces during a three-phase terminal short circuit was also accomplished using an alternative numerical computation method described in [4].

This method is characterized by a modeling of the stator end winding coils with a finite cross section. Furthermore this method considers the magnetizable rotor shaft using the integral equation method based on the separation of the magnetic field H into a zero-divergence fraction and an irrotational fraction [4]. Compared to the computation method introduced above, the rotor end winding is explicitly modeled and the rotor movement is

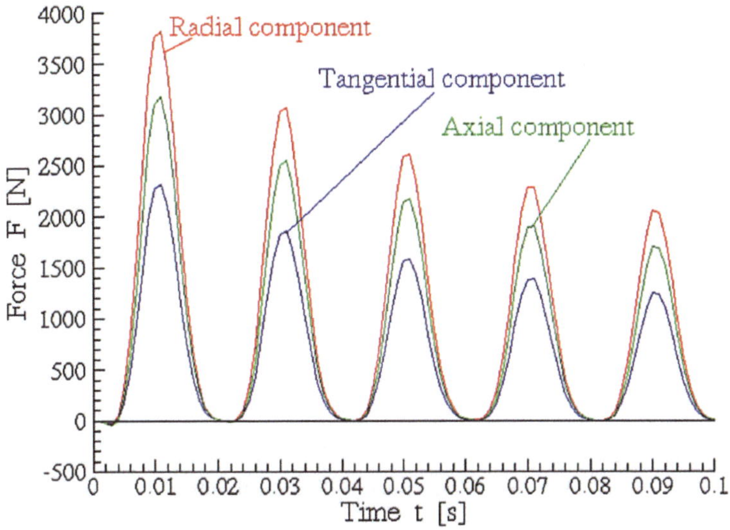

Figure 7. Time function of force.

taken into account. The influence of the stator core is considered similarly by the method of images. The model underlying this computation method is shown in Fig. 8.

Due to the fact that the computation time is extensively longer when applying this method, the results of both methods were compared at a single moment 10 ms after short circuit occurrence. In Fig. 9 the force distribution along the upper layer bar of the first coil of phase a calculated by both methods is compared.

Regarding the characteristic distribution of electromagnetic forces both methods provide comparable results. The differences correspond widely to the differences between the two end region models like for example the way of modeling the bars of the stator end winding.

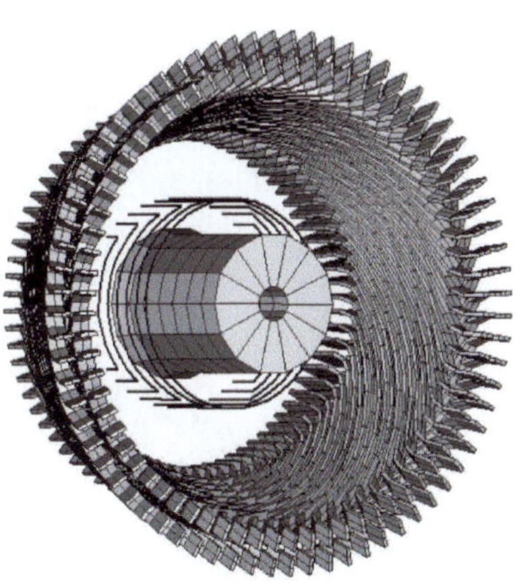

Figure 8. End winding model for alternative electromagnetic force computation.

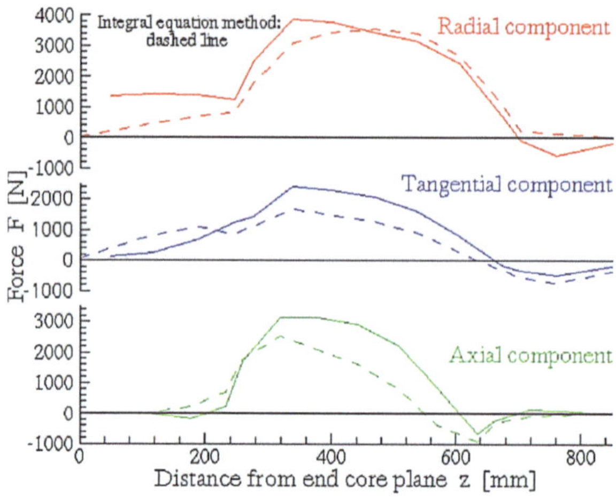

Figure 9. Comparison of force distribution.

Mechanical simulation

The simulation method used to compute the vibration behavior of the stator end winding is based on the application of mechanical finite-elements using the results of the electromagnetic simulation as input data [1].

Depending on the manufacturer and the cooling principle the construction of the stator end winding varies. The subsequently introduced mechanical finite-element model of the stator end winding of a 170 MVA air-cooled turbo generator, which represents a characteristic construction, is shown in Fig. 10.

The finite-element model comprises the bars of the upper layer and the lower layer which are connected in pairs at the coil nose, the pressure ring which borders the stator core and represents the boundary of the simulation area and the coil support brackets mounted on the pressure ring. The model is completed by the support rings surrounding the end winding and the bandages, which fix the bars to each other, the bars to the support rings and also the

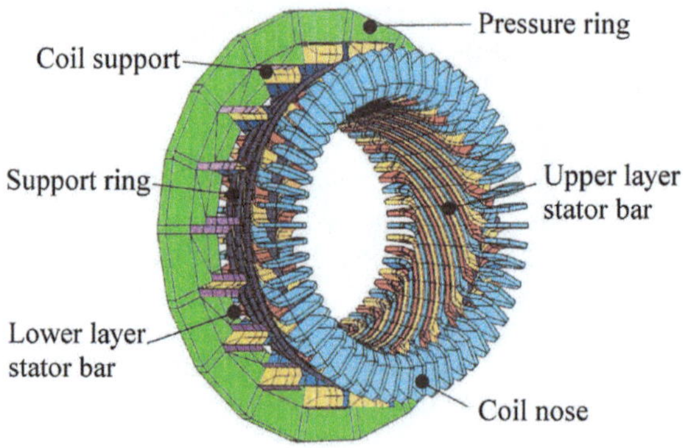

Figure 10. Mechanical finite-element model of the stator end winding [2].

support rings to the support brackets [2]. The components of the model are determined by their masses, stiffness, and damping characteristics. The complete model consists of about 20,000 bricks [1].

Both the bars of the stator end winding and the bandages have a very complex structure, the modeling of which would exceed the calculating capacity. Therefore these components were replaced by substitutes of approximately the same mechanical properties. To determine the parameters of the original bars and bandages, which show a non-linear behavior and also temperature dependence, extensive measurements have been accomplished.

The bars of the stator winding consist of multiple twisted conductors, each covered with insulating varnish. These are enclosed by epoxy resin impregnated glass silk tape. To obviate the modeling of such a complex structure the bars of the stator end winding were modeled as massive bodies of comparable mass distribution, stiffness and damping characteristics. To determine the stiffness parameters, bending tests were accomplished. Dynamic parameters were estimated by additional oscillation tests. Beside the measurements a detailed mechanical finite-element model of a stator bar was implemented and used to compute the parameters of different stator bars numerically.

The stiffness parameters of the bandages were determined by conducting static and dynamic tests in a servohydraulic test facility. Tests at different temperatures showed a significant dependency of the stiffness of the bandages on the temperature.

The performance of numerical simulations while varying certain model parameters showed that the accuracy of the simulation results is decisively determined by certain model parameters, especially by the stiffness of the bandages [2]. In this connection the precise determination of suchlike parameters still offers a potential for further developments.

Results and verification

A combined electromagnetic and mechanical simulation of the end winding oscillation behavior under forced vibrations was accomplished for three representative turbo generators differing in cooling principle, power class, and therefore in the stator end winding construction.

The deformation due to the electromagnetic forces acting on the stator end winding of a 170 MVA air-cooled turbo generator computed at one moment during a three-phase terminal short circuit is shown in Fig. 11. As can be seen the oscillation behavior owns characteristics of the one of a composite structure.

The time functions of the displacement computed at a location in the middle of the involute part of an upper layer bar are shown in Fig. 12, using the Cartesian coordinate system underlying the mechanical simulation method.

Reaching values around 1,000 μm in the area of the coil nose, the deformations are around 30 times higher than under steady-state conditions. But due to the damping characteristics of the end winding structure and according to the time behavior of the electromagnetic forces during the three-phase terminal short circuit the amplitudes of the displacement decrease relatively fast [2].

In the case of a 90 MVA air-cooled turbo generator the computation results of the oscillation behavior during a three-phase terminal short circuit were compared to measurements conducted by the manufacturer. At a location on the coil nose of different stator end winding coils acceleration sensors were used to measure the displacement by integrating the signal

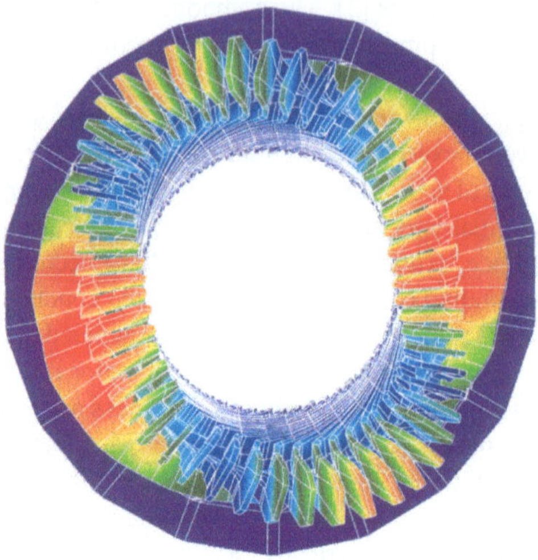

Figure 11. Deformation of stator end winding during a three-phase terminal short circuit.

of the sensors. The comparison of the maximum values of the displacement functions is given in Table 1.

As can be seen, except for one location the results of the numerical simulation and the measurements show a good congruence with deviations of mostly noticeable beneath 40% [1].

Beside the combined simulation the mechanical simulation method was applied to determine eigenfrequencies and eigenmodes of the treated end winding constructions. Therefore a rotating sinusoidal force distribution with solely a radial force component was impressed on the mechanical finite-element model. By varying the rotational speed of

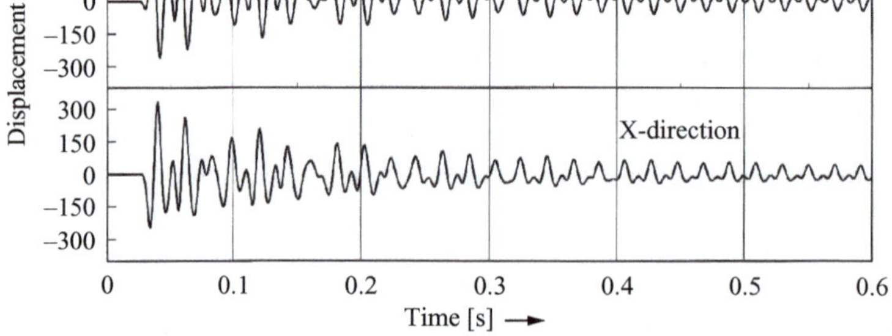

Figure 12. Time functions of displacement during a three-phase terminal short circuit [2].

Table 1. Comparison and displacement [1]

Location of coil nose	Displacement			
	Computed		Measured	
	Tangential (μM)	Radial (μM)	Tangential (μM)	Radial (μM)
3 h	934	1,216	—	1,174
6.75 h	1,059	1,055	558	1,215
7.5 h	1,344	1,220	—	877
9 h	931	1,159	879	934
9.75 h	925	1,103	—	1,502
10.5 h	874	1,021	1,068	1,644
11.25 h	1,003	989	—	993

the force distribution different eigenmodes were excited like for example the four-node oscillation of the stator end winding of a 500 MVA hydrogen-cooled generator shown in Fig. 13. The eigenfrequency corresponding to the shown oscillation amounts to 72 Hz. The four-node oscillation represents an eigenform, which can indeed easily be excited in the end winding of turbo generators [1].

A number of eigenmodes with the appendant eigenfrequencies of the 90 MVA air-cooled turbo generator is given in Table 2.

Table 2. Eigenmodes and eigenfrequencies

Eigenmode	Eigenfrequency (Hz)
Torsional oscillation	35.2
Two-node oscillation	41.4
Four-node oscillation	73.6

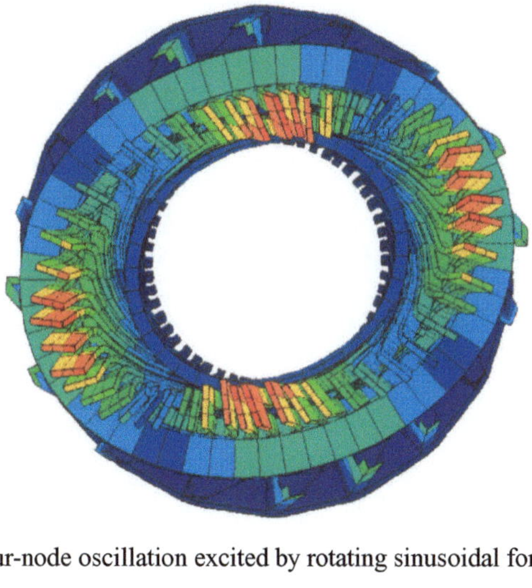

Figure 13. Four-node oscillation excited by rotating sinusoidal force distribution.

A resonance measurement accomplished for the 90 MVA generator by using the acceleration sensors mentioned above showed a clear resonance at 70.5 Hz, which corresponds good to the computed frequency of the four-node oscillation. But repeated measurements at different temperatures showed, that the measured frequency varied slightly with increasing temperature.

Conclusions

The electromagnetic and mechanical modeling of the turbo generator end region and especially the determination of certain model parameters still contain a great potential of improvement. Anyhow, three-dimensional numerical simulation of the end winding oscillation behavior of turbo generators may emerge as a useful tool in design and development of large electrical machines.

References

[1] O. Drubel, S. Kulig, K. Senske, End winding deformations in different turbo generators during 3-phase short circuit and full load operation, Electr. Eng., Vol. 82, pp. 145–152, 2000.
[2] K. Senske, S. Kulig, J. Hauhoff, D. Wünsch, "Oscillation Behaviour of the End Winding Region of a Turbo Generator During Electrical Failures", Conference Proceedings CIGRE, Yokohama, October 29, 1997.
[3] C.-G. Richter, Berechnung elektromagnetischer Kräfte auf die Spulenseiten im Wickelkopf von Turbogeneratoren, Institute of Electrical Machines and Drives: University of Hannover, 1996.
[4] B. Frei-Spreiter, Ein Beitrag zur Berechnung der Kräfte im Wickelkopfbereich großer Synchronmaschinen, Institute of Electrical Machines: Swiss Federal Institute of Technology Zürich, 1998.
[5] B. Kulicke, Digitalprogramm NETOMAC zur Simulation elektromechanischer undmagnetischer Ausgleichs-vorgänge in Drehstromnetzen, Elektrizitätswirtschaft, VDEW, Vol. 78, No. 1, pp. 18–23, 1979.
[6] R. Ummelmann, Erweiterung eines Finite-Differenzen-Zeitschritt-Programmsystems auf Synchronmaschinen, Institute for Electrical Machines, Drives and Power Electronics: University of Dortmund, 1997.

I-12. OPTIMIZATION OF A LINEAR BRUSHLESS DC MOTOR DRIVE

Ph. Dessante[1], J.C. Vannier[1] and Ch. Ripoll[2]

[1]Service EEI Ecole Supérieure d'électricité (Supelec), Plateau de Moulon, 91192 Gif sur Yvette, France,
philippe.dessante@supelec.fr jean-claude, vannier@supelec.fr
[2]Renault Research Center—Guyancourt,
christophe.ripoll@renault.com

Abstract. The paper describes the design of a drive consisting of a voltage supplied brushless motor and a lead-screw transformation system. In order to reduce the cost and the weight of this drive an optimization of the main dimensions of each component considered as an interacting part of the whole system is conducted. An analysis is developed to define the interactions between the elements in order to justify the methodology. A specific application in then presented and comparisons are made between different solutions depending on different cost functions (max power, weight, cost, . . .). With this procedure, the optimization is no longer limited to the fitting between separated elements but is extended to the system whose parameters are issued from the primitive design parameters of the components.

Introduction

The system studied in this paper is a linear electrical drive system realized with a voltage supplied brushless motor whose shaft is mechanically connected to a lead-screw drive device. The aim of this system is to drive a load along a linear displacement.

The specifications concerning the load consist mainly in two parts. Firstly, it has to apply a rather high static force at standstill as for instance to overcome some static friction force. Secondly, it has to be driven from one point to another point in a given time. This second part implies a dynamic force and a maximum speed depending on the kind of displacement function is chosen.

A discussion to chose the displacement function is important because as the motor will have a limited torque capacity, it may be necessary not to accelerate neither too early nor too late when it is entering the constant power region. Consequently this definition can have consequences on the system size. At the very beginning it may be considered sinusoidal or corresponding to a bang-bang acceleration.

System modeling

A general presentation of the system is given in Fig. 1 where the power source is supposed to be a battery bank.

S. Wiak, M. Dems, K. Komęza (eds.), Recent Developments of Electrical Drives, 127–136.

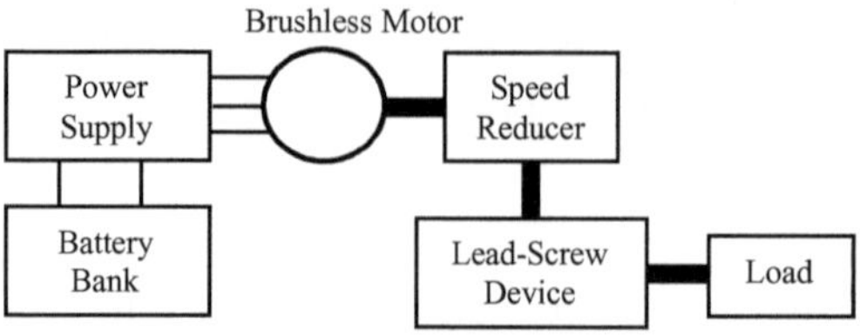

Figure 1. System main components.

Concerning the kinematic model, the lead-screw is represented by its transformation ratio deduced from the screw pitch while the speed reduction system introduces a speed transformation ratio. These two components can be represented by the global transformation ratio, ρ, between the motor shaft angle and the linear load displacement:

$$\theta = \rho x \tag{1}$$

This association between a speed reducer of a given ratio, N, and screw of a given pitch, τ, gives the resulting value for the transformation ratio:

$$\rho = \frac{2\pi N}{\tau} \tag{2}$$

This ratio is used to convert the load specifications in motor specifications. The load displacement is directly changed in angle variation and the forces are converted in torques taking into account the efficiency of each component. During acceleration the motor inertia leads to a difference between the output torque and the electromagnetic torque.

In this application two sorts of torques are to be generated by the drive system. A static torque (at zero speed) can be necessary to reach the breakaway force on the load just before it starts to move. It can either represent the torque needed to maintain the load in a position when an external force is applied. With a given force, F'_{sta}, the static torque is given by:

$$C_{sta} = \frac{F'_{sta}}{\rho} \tag{3}$$

When the load speed is increased, generally the motor has to generate a torque with two components. This second sort of torque is called the dynamic torque. It contains a part corresponding to the force required to accelerate the load and a second part to accelerate the rotor and the transmission system. This part is represented by the inertia, J_{mot}, of the motoring part. With a given dynamic resistive force, F'_{dyn}, an equivalent mass of the load m'', a friction coefficient f', the dynamic torque for an acceleration γ at a speed v on the load can be expressed as follows (4).

$$C_{dyn} = \left[J_{mot}\rho + \frac{m''}{\rho} \right] \gamma + \frac{f'v + F'_{dyn}}{\rho} \tag{4}$$

The two types of torque are dependent on the transformation ratio level.

For the static torque it is obviously interesting to use a high value of the transformation ratio because the corresponding torque value will decrease and this will reduce the motor constraint (Fig. 2).

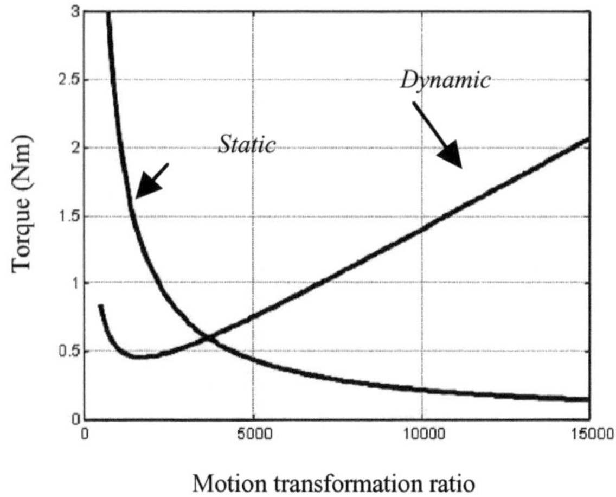

Figure 2. Torques vs. transformation ratio (5).

For the dynamic torque, the increase of the transformation ratio will reduce the component of the torque needed to drive the load but it will increase the torque required to accelerate the motoring parts mainly constituted of the rotor of the electrical motor.

Consequently a first limitation appears when choosing the value of the transformation ratio. It is not possible to retain a high value without having to generate a high dynamic torque.

If we first consider the situation illustrated in Fig. 2 for a fixed rotor inertia, it corresponds to the case of the total force required by the load in dynamic mode, F'_{dyn_tot}, whose value is referenced to the static force as:

$$F'_{dyn\_tot} < F'_{sta}/2 \tag{5}$$

In this case, a good value for ρ could the one observed at the intersection between the two curves [1–3]:

$$\rho_i = \sqrt{\frac{F'_{sta} - F'_{dyn\_tot}}{J_{mot}\gamma}} \tag{6}$$

With that value the torque to be generated by the motor is minimal.

Secondly we consider the case of a greater relative value for the total dynamic force needed by the load as:

$$F'_{sta}/2 < F'_{dyn\_tot} < F'_{sta} \tag{7}$$

As it can be observed in Fig. 3, the dynamic torque will be minimal after the intersection between the two curves. For this reason, a good value for the transformation ratio could be in that case the one corresponding to the minimization of the dynamic torque:

$$\rho_0 = \sqrt{\frac{F'_{dyn\_tot}}{J_{mot}\gamma}} \tag{8}$$

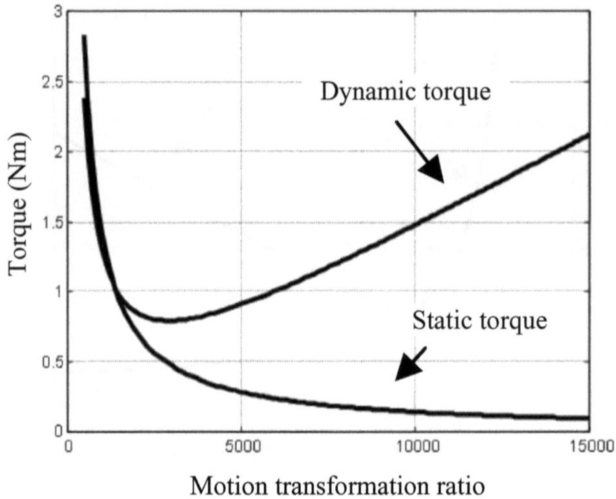

Figure 3. Torques vs. transformation ratio (7).

In that case, the torque to be generated by the motor is minimal with this choice.

As the dynamic torque also depends on the value of the rotor inertia which will be defined during the motor design the situation is more complex and will be discussed.

Other constraints [4] are also to be considered. The load duty cycle is generally defined and leads via the *rms* and the average values of the load dynamic to the definition of the corresponding *rms* torque:

$$C_{rms}^2 = \left[\frac{F_{rms}}{\rho}\right]^2 + \left[\frac{f' v_{ave}}{\rho}\right]^2 + \frac{2F'_{dyn} f' v_{ave}}{\rho^2} + \gamma_{rms}^2 \left[\rho J_{mot} + \frac{m''}{\rho}\right]^2 \tag{9}$$

Among the limits concerning the motor, there could be a maximal rotor speed and the actual speed has to be considered:

$$\Omega = \rho v \tag{10}$$

This expression clearly indicates that the augmentation of the mechanical transformation ratio will need higher rotor speed for the motor.

The motor supply and the battery tank characteristics introduce a limitation of the power consumption. This finally depends on the efficiency reached by the motor and on the power consumed by the load. The efficiency of a motor can be estimated from its main characteristics and the peak consumed power can then be defined:

$$P_{dyn} = C_{dyn}\hat{\Omega} = (J_{mot}\rho^2 + m'')\gamma v + (f'v + F'_{dyn})v \tag{11}$$

For the motor design, different levels of complexity in modeling are available. To simplify, it is possible to define the main dimensions by using the peak torque, the *rms* torque, and the rotor inertia as follows [5]:

$$J_{mot} = \frac{1}{2}\pi \mu_v R^4 L \tag{12}$$

$$C_p = 4pH_0\, BRLE \tag{13}$$

$$C_n = 2\pi AB\gamma_p R^2 L \tag{14}$$

Consequently, these three relationships introduce three main dimensions parameters for the design: the rotor radius R, the rotor length L, and the permanent magnet thickness E.

The remaining parameters are more or less constant or weakly dependent on the motor size.

They are defined as:

p = pole's number.
H_0 = magnet's peak magnetic field.
B = airgap flux density.
A = stator excitation level.
γ_p = pole's overlapping factor.
μ_v = mass density.

Concerning the converter, the volume of silicon can be linked to the maximum power value needed by the motor to drive the load.

Optimization

The established relationships are used to define the constraints in the optimization procedure.

The motor peak torque has to be greater than the static and the dynamic torques. The nominal torque is also greater than the required *rms* torque.

$$C_p > C_{sta} \tag{15}$$

$$C_p > C_{dyn} \tag{16}$$

$$C_n > C_{rms} \tag{17}$$

The maximum power consumption is to be kept below the maximal value supplied by the battery bank. The mechanical transformation device introduces inertia in the system equations. Furthermore it needs a volume that will be a part of the total volume allowed to the system.

$$P_{max} > P_{dyn} \tag{18}$$

Some technical constraints have to be added in order to be able to define a feasible motor. It concerns the maximal rotor speed and the ratio between the rotor length and the diameter.

$$\Omega_{max} > \hat{\Omega} \tag{19}$$

$$aR > L > bR \tag{20}$$

A minimum relative value is needed for L to be kept in the domain of validity of the previous expressions (12–14). A maximum value is settled to avoid the definition of a too thin rotor with a high length to diameter ratio as it could be required to reduce the rotor inertia (12). A quasi fixed ratio can also be imposed by the choice of the coefficients a and b.

Depending on the application, different cost functions can be minimized. For instance, if the weight is the principal criteria, the motor size will be reduced. If the volume is to be kept as low as possible, the mechanical transformation system size will be an issue.

Results

We present here the results concerning the definition of the motor and the motion transformation ratio whose dimensions are optimized for a given load.

In this example, the load characteristics are the followings:

$$\begin{aligned} F'_{sta} &= 900\ \text{N} & \hat{\gamma} &= 1\ \text{m/s/s} \\ F'_{dyn} &= 450\ \text{N} & \hat{v} &= 35\ \text{mm/s} \\ F_{rms} &= 90\ \text{N} & \gamma_{rms} &= 0.1\ \text{m/s/s} \\ f' &= 0\ \text{N/s} & v_{ave} &= 28\ \text{mm/s} \\ m'' &= 1\ \text{kg} \end{aligned}$$

The optimization procedure uses the constraints (15)–(20) and searches a set of values for R, L, E, and ρ which minimizes the motor peak torque. It appears that the mass is minimized as a consequence.

As boundaries are used to limit the variation of these parameters to feasible values it appears that the result is always for the upper boundary value for the transformation ratio.

In Fig. 4, the evolution of the main rotor dimensions with the maximum authorized transformation ratio value are presented.

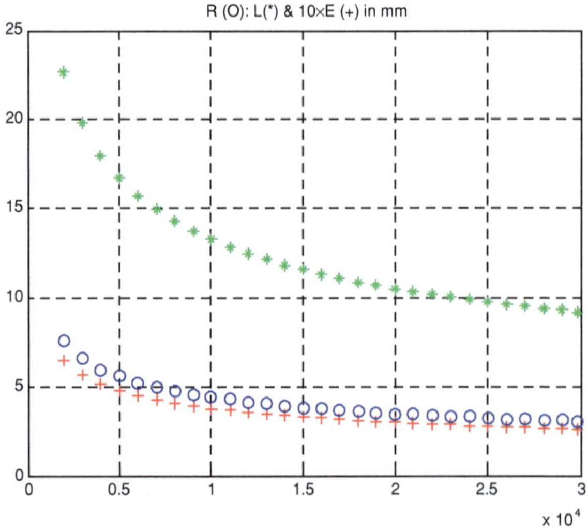

Figure 4. Rotor dimension R, L, E vs. ρ_{max} per meter.

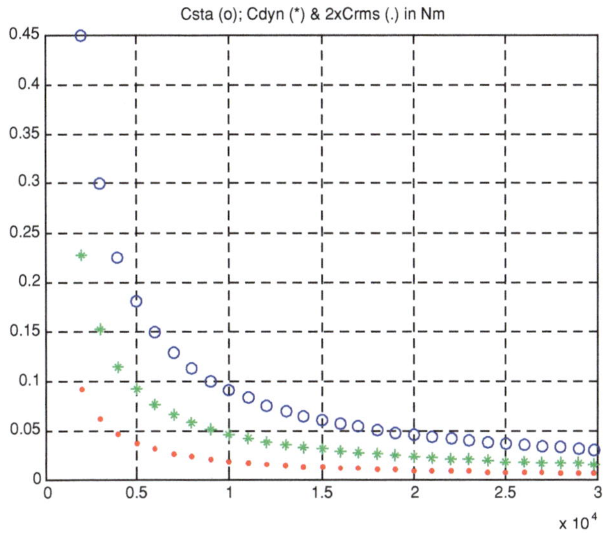

Figure 5. Motor torques vs. ρ_{max} per meter.

The rotor mass as its inertia are decreasing as long as the maximum value for ρ is increased.

In Fig. 5 the evolution of the torques is presented too.

As it was observed before, the static torque diminishes when the ratio increases. But in that procedure, it is observed for the dynamic torque as well and the good value for ρ is the maximum permitted value.

This main difference is due to the fact that the rotor inertia changes its value when the ratio does so. This could be a very important constraint for the motor design. In the presented design procedure, some constraints (20) have been introduced to avoid such design difficulties.

For every value of the maximum ratio, the rotor inertia can be evaluated and the previous good values for ρ (6) and (8) can be calculated too. They give the corresponding torques presented in Fig. 6.

In that particular case the values are almost the same because the total dynamic force is near half the static force. We can notice that the "good" ratio value is much more important than the permitted ratio value. Consequently, the torques values are lower than the values obtained at the boundary of the domain.

Finally, among the different values proposed by the design procedure, it is necessary to retain one of them to design the motor. A criterion can be the maximum rotor speed.

In Fig. 7, the evolution of the maximum rotor speed with the maximum transformation ratio is presented.

These speed values are rather common values for electrical motors. For small motors the choice of a maximum speed of 6,000 or 9,000 rpm is reasonable.

When the optimization procedure succeeds in defining a feasible motor, a more complex model is used to calculate all the dimensions.

In Fig. 8 is presented a view of one of these motors.

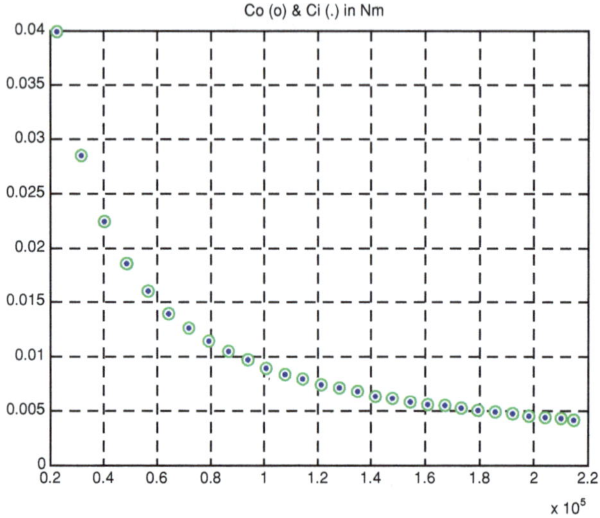

Figure 6. Former minimum torques vs. ρ_{opt} per meter.

The airgap diameter is 8 mm and the outer diameter is close to 19 mm. The rotor length is 12 mm and the inertia is 0.022 kgmm^2. NdFeB magnets are used to magnetize the airgap with gives a flux density equals to 0.8 T. The resulting active mass is 20 g. With the housing the resulting mass will be slightly higher. The original commercial motor used to drive this application had a mass equal to 100 g.

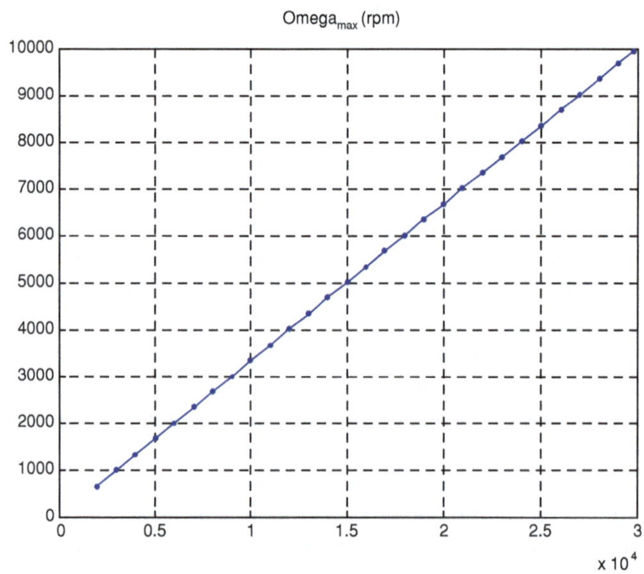

Figure 7. Maximum rotor speed vs. ρ_{max} per meter.

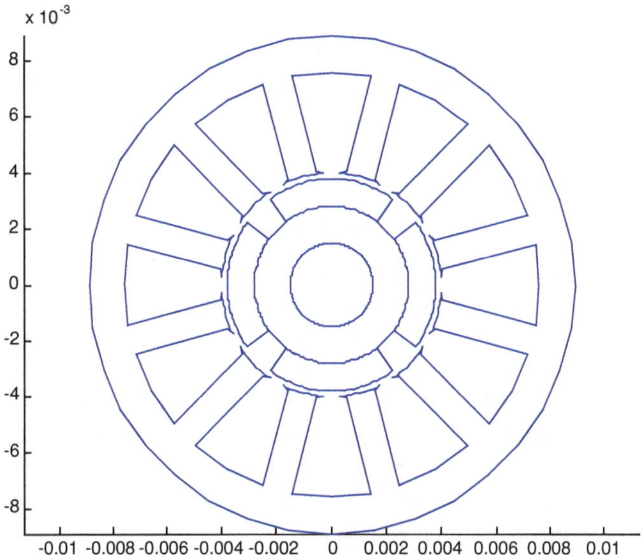

Figure 8. Resulting motor dimensions.

In Fig. 9, a simulation of the flux lines distribution is obtained with FEM analysis. This permits to verify the values expected from the design procedure.

The nominal torque is 10 mNm and the peak torque is at least 50 mNm. The maximum speed should be 6,600 rpm to drive the load at its maximum speed. At maximum power, the motor efficiency is about 50% if it is assumed that the joule losses are predominant.

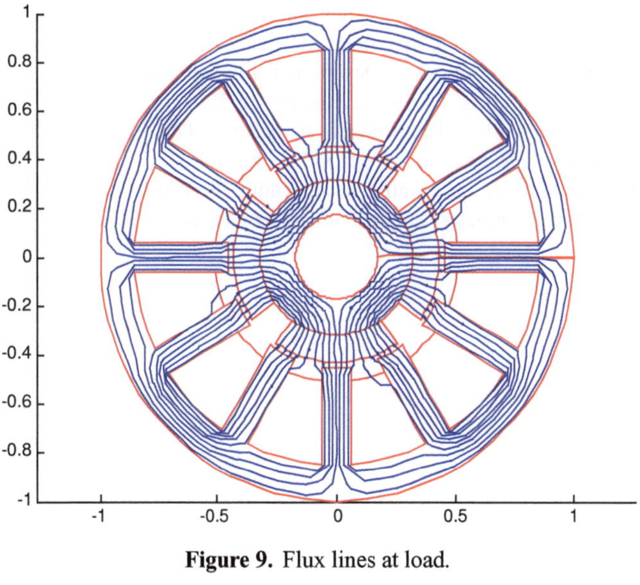

Figure 9. Flux lines at load.

A value of 20,000/m for the motion transformation ratio can be obtained with a lead-screw pitch of 3 mm and a speed reduction gearbox with a 9.5 ratio.

It can be observed that when the motor size decreases, the rotor speed increases which leads to the definition of a larger mechanical transformation system. This is another constraint which can be considered.

Conclusion

In this paper, an electromechanical conversion system is analyzed resulting in a modeling of the components. The model has to be inversed to link the dimensions to the performances for each component involved in the power conversion system. Consequently the whole system dimensions are available for the aggregate optimization of the system. This procedure permits a correct association between the components and can lead to a smaller volume or a smaller weight than it could be defined with a separated element optimization. The results presented have shown the interest to optimize simultaneously the rotor main dimensions and the transformation. Actually, this procedure avoids the risk of having to design a nonfeasible motor with a too low inertia for a given torque.

As it needs the complete specific design of a dedicated motor, it is reserved for rather expensive application (aircraft, space, . . .) with severe criteria or for very large scale application (automotive, . . .).

As for this type of application, the total mass of the system is to be considered, a complete modeling of the transformation system is needed as for the electronic converter. This could be presented in a further work.

References

[1] E. Macua, C. Ripoll, J.-C. Vannier, "Optimization of a Brushless DC Motor Load Association", EPE2003, Toulouse, France, September 2–4, 2003.

[2] E. Macua, C. Ripoll, J.-C. Vannier, "Design, Simulation and Testing of a PM Linear Actuator for a Variable Load", PCIM2002, Nürnberg, Germany, May 14–16, 2002, pp. 55–60.

[3] E. Macua, C. Ripoll, J.-C. Vannier, "Design and Simulation of a Linear Actuator for Direct Drive", PCIM2001, Nürnberg, Germany, June 19–21, 2001, pp. 317–322.

[4] M. Nurdin, M. Poloujadoff, A. Faure, Synthesis of squirrel cage motor: A key to optimization, IEEE Trans. Energy Convers., Vol C6, pp. 327–335, 1991.

[5] C. Rioux, Théorie générale comparative des machines électriques établie à partir des équations du champ électromagnétique, *Revue générale de l'Electricité* (*RGE*), Vol. t79, No. 5, pp. 415–421, mai 1970.

SECTION II
CONTROL, MEASUREMENTS, AND MONITORING

Introductory Remarks

The papers selected to first chapter are mainly focused on recently developed techniques of control, measurements, and monitoring of electrical drives.

The papers accepted for this chapter concerns the following leading problems:

- The discuss of fundamental equations used in high-frequency signal-based interior permanent-magnet synchronous motor (IPMSM) position estimators. For this purpose, an IPMSM model is presented that takes into account the nonlinear magnetic condition, the magnetic interaction between the two orthogonal magnetic axes, and the multiple saliencies. Using the novel equations, some recently proposed motion-state estimators are described. Simulation results reveal the position estimation error caused by estimators that neglect the presence of multiple saliencies or that consider the magnetizing current in the d-axis only.
- A novel sensorless control method for Syn.RM. The sensorless control is based on the modified flux linkage observer, which is proposed by authors for permanent magnet synchronous motors (PMSM). The observer is able to estimate the modified flux linkage and the electromotive force (EMF) simultaneously, and the motor speed and the rotor position are calculated from these estimated values. But as same as the other method, the precision of the observer-based estimation is affected by the parameter fluctuations. The new estimation method for Syn.RM using the modified flux linkage observer with an estimation error correct function. A proportional-integral (PI) type controller is added to the system to compensate the estimation error. It operates that the estimated magnitude of the flux corresponds to the nominal value.
- A novel scheme for vector control is presented that aims to improve some of the weaknesses of the sensorless vector control. Among indirect rotor flux-oriented control (IR-FOC), some of the aspects that can be improved are the low speed behavior, current control, and parameter detuning. The present scheme includes temperature estimation to correct the deviation in steady state, a new control scheme with skin effect estimation to improve the transient accuracy, and an over-current protection to be able to have a control on the stator current while allowing a good performance. The proposed scheme is designed form the Matlab/Simulink environment and is experimentally tested using a 1-kW induction motor and a TMS320C31 DSP proving its validity and usefulness.
- An integrated design and direct torque control (DTC) of inverter-fed interior permanent magnet synchronous motors (IPMSMs) for wide-speed operation with high torque capability is developed. The double-layer IPM rotor design is accounted for IPMSMs requiring

S. Wiak, M. Dems, K. Komęza (eds.), Recent Developments of Electrical Drives, 137–139.

a wide torque-speed envelope. A novel approach for the generation of the reference stator flux-linkage magnitude is developed in the proposed IPMSM DTC scheme to insure extended torque-speed envelope with maximum-torque-perstator-current operation range below the base speed as well as constant-power flux-weakening and maximum-torque-perstator-flux operation regions above the base speed.

- The technique to obtain optimal torque control parameters of switched reluctance motor (SRM). A relationship between dc-link voltage and rotor speed is used, reducing the number of control parameters. Using a nonlinear motor model, surfaces are created describing torque, torque ripple, and efficiency as function of rotor speed and the main control parameters. The advantage of this technique is an off-line optimization platform and the simplicity to create additional surfaces.
- A model for the magnetoelastic coupling is presented and used in the simulations of an induction machine. The goal of these simulations is to establish the effect of the magnetostriction on the vibrations of rotating electrical machines. For this purpose, an original method for the calculation of magnetostrictive forces is presented. It is shown that the magnetostriction affects the vibrations of rotating electrical machines by increasing or decreasing the amplitudes of velocities measured at the outer surface of the stator core of the machine. These velocity are the ones responsible for acoustic noise. Furthermore, the stress-dependency of the magnetostriction adds to the increase of the above amplitudes. The modeling of vibrations and noise of electrical machines should take into account the effect of magnetostriction and its stress-dependency.
- The structure and air-borne noise of an induction machine with squirrel-cage rotor are estimated. For this, different types of surface-force excitations and rotational directions are regarded for the first time. In general the calculated structure-borne sound-levels suit the acceleration measurements of the industrial partner very well. The acoustic noise levels differ from those. The comparison of the different excitations show that it is necessary to take the rotor excitation into account. In case of pure stator excitation, e.g., the first stator-slot harmonic at 720 Hz does not reach as significantly high levels as expected although it is one of the strongest orders measured.
- A theoretical approach which permits to study the evolution of each flux density air-gap component trough the stator. The aim of this method is to find, by computation, the magnitude of measured spectral lines. The study is made on the couples of toothing spectral lines and justifies why these couples do not have the same magnitude, what is not obvious in a first approach where the practical spectrum is directly compared with this one of the air-gap flux.
- Four methods to determine the mechanical characteristics (natural frequencies, mode numbers) of electrical machine stators are developed. Result comparison concerns analytical laws, a Finite Element software, a modal experimental procedure, and a method based on analogies between mechanic and electric domains. Simple structures are studied in order to analyze the validity of each method with accuracy. The fixation of a stator yoke allows to observe the modifications of the mechanical behavior.
- The tracking control design of an induction motor, based on input-output linearization with magnetic saturation included. Magnetic saturation is accounted for by the nonlinear magnetizing curve of the iron core and is used in the control design, the observer of state variables, and in the load torque estimator. Experimental results show that the proposed input-output linearizing tracking control with the included saturation behaves better than

the one without saturation. It also introduces smaller position and speed errors, and better motor stiffness.

- A novel control scheme for PWM rectifier-inverter system is proposed. Fast control strategies such as line voltage sensorless virtual flux (VF) based direct power control with space vector modulator (DPC-SVM) for rectifier and direct torque control with space vector modulator (DTC-SVM) for inverter side are used. These strategies lead to good dynamic and static behavior of the proposed control system—direct power and torque control-space vector modulated (DPTSVM). Simulations and experiment results obtained show good performance of the proposed system. Additional power feedforward loop from motor to rectifier control side improved dynamic behaviors of the power flow control. As a result, better input-output energy matching allows decreasing the size of the dc-link capacitor.

II-1. A GENERAL DESCRIPTION OF HIGH-FREQUENCY POSITION ESTIMATORS FOR INTERIOR PERMANENT-MAGNET SYNCHRONOUS MOTORS

Frederik M.L.L. De Belie, Jan A.A. Melkebeek, Kristof R. Geldhof, Lieven Vandevelde and René K. Boel

Electrical Energy Laboratory (EELAB), Department of Electrical Energy, Systems and Automation (EESA), Ghent University (UGent), Sint-Pietersnieuwstraat 41, B-9000 Gent, Belgium

frederik.debelie@ugent.be, Jan.Melkebeek@ugent.be, Kristof.Geldhof@ugent.be Lieven.Vandevelde@ugent.be, Rene.boel@ugent.be

Abstract. This paper discusses fundamental equations used in high-frequency signal based interior permanent-magnet synchronous motor (IPMSM) position estimators. For this purpose, an IPMSM model is presented that takes into account the nonlinear magnetic condition, the magnetic interaction between the two orthogonal magnetic axes and the multiple saliencies. Using the novel equations, some recently proposed motion-state estimators are described. Simulation results reveal the position estimation error caused by estimators that neglect the presence of multiple saliencies or that consider the magnetizing current in the *d*-axis only.

Introduction

Vector control of a high-dynamical, high-performance interior permanent-magnet synchronous motor (IPMSM) requires the stator flux linkage vector. For small stator currents, this flux is mainly generated by the high-grade permanent magnets, buried within the rotor. In a lot of drives, using field-oriented control, the rotor flux vector is considered instead of the stator flux linkage vector. Moreover, the rotor flux direction can be approximated by the rotor position, measured with a mechanical sensor.

During the last 15 years, motion-state estimation methods have been developed with the intention to remove the expensive mechanical transducer, which, due to temperature variations and mechanical vibrations, produces measurements of low reliability. Modern sensorless drives try to estimate the motion states from measurements of electrical variables. Filtering techniques and observing strategies are used to estimate the back-EMF vector and from that the rotor speed and angle. However, for a slow rotor motion, small signals have to be measured or calculated that are disturbed strongly by noise produced by normal operation of the PWM and motor. As a result, the precision of such estimators in the low speed region is insufficient to control the motor in a stable and efficient way.

S. Wiak, M. Dems, K. Komęza (eds.), Recent Developments of Electrical Drives, 141–153.

To obtain accurate position estimations at low speed, in recently proposed estimation methods, a high-frequency voltage are supplied which generates a high-frequency variation of small amplitude in the stator flux linkage [1–8]. In an IPMSM, without damper effects e.g. due to short-circuited windings or eddy currents, this flux variation mainly occurs in the main flux instead of the leakage flux path. If saturation occurs or an important reluctance variation along the air gap exists, it will be shown that the high-frequency current response will be modulated with the air-gap flux position additionally to the rotor angle.

For an IPMSM, considerable reluctance variations, called magnetic saliencies, can be detected due to the buried placement of the magnets within the rotor. Furthermore, several stator teeth are saturated due to the presence of an important permanent magnetic flux.

This paper discusses fundamental equations used in high-frequency signal based IPMSM position estimators. For this purpose, the small signal dynamic flux model, presented in [9], is used which takes into account the nonlinear magnetic condition and the magnetic interaction between the direct and the quadrature magnetic axis. An addition to the model is given to tackle the presence of multiple saliencies. By using the novel equations some recently proposed motion-state estimators are described. It is shown that the higher the inductance difference between the two orthogonal magnetic axes, the higher the position estimation resolution. Furthermore, simulation results reveal the position estimation error caused by estimators that disregard the existence of multiple saliencies or that consider the magnetizing current vector in the d-axis only.

General description of a PMSM

Small signal dynamic flux model

To obtain accurate position estimations at low speed, in recently proposed estimation methods a high-frequency voltage is supplied, which generates a high-frequency variation of small amplitude in the stator, flux linkage [1–8]. This implies that, to describe position estimators, a small signal dynamic flux model can be used. In an IPMSM, without the damper effects e.g. due to short-circuited windings or eddy currents, the flux variation generated by the high-frequency voltage mainly occurs in the main flux ϕ_m instead of the leakage flux path. As a result, the small signal dynamic model of a saturated synchronous machine, presented in [9], can be used. This model is given by the flux equation

$$\Delta\Phi_m^{qd}(t) = \begin{pmatrix} L_{qmt}\cos 2\mu + L_{qmo}\sin^2\mu & \frac{1}{2}(L_{qmt} - L_{qmo})\sin(2\mu) \\ \frac{1}{2}(L_{dmt} - L_{dmo})\sin(2\mu) & L_{dmt}\sin 2\mu + L_{dmo}\cos^2\mu \end{pmatrix} \Delta\boldsymbol{i}_m^{qd}(t) \tag{1}$$

written in a reference frame (qd) fixed to the physical quadrature and direct axis and with $\boldsymbol{i}_m$ the magnetizing current, with, see Fig. 1, L_{qmo}, L_{dmo} the chord-slope magnetizing inductances and L_{qmt}, L_{dmt} the tangent-slope magnetizing inductances in quadrature and direct magnetic axis respectively, μ the angle between the q-axis and the vector $\boldsymbol{i}_m$ and Δ denoting the small variation of a vector. In a current controlled drive, the vector $\boldsymbol{i}_m$ and the angle μ are regulated to a constant during steady state. By using the flux equation (1) the saturation level in both magnetic axes is assumed to be determined by $\boldsymbol{i}_m$ and as a result the proposed small signal dynamic flux model includes cross saturation or magnetic

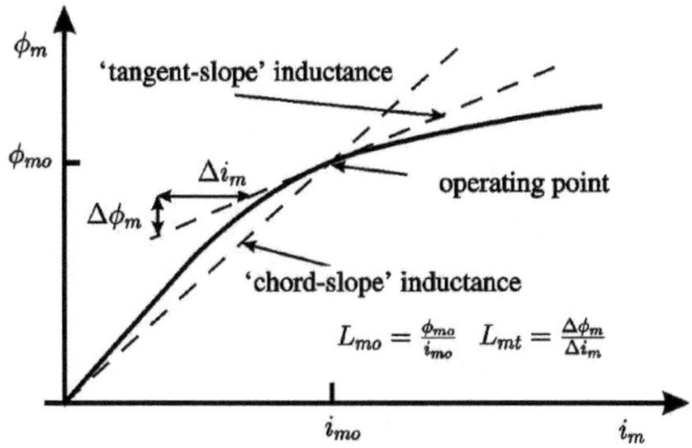

Figure 1. Magnetizing characteristic with i_{mo} and ϕ_{mo} the average modulus of i_m and ϕ_m respectively.

interaction between d- and q-axis. However, this model neglects possible stator leakage flux saturation.

In some high-frequency signal based sensorless drives, a small high-frequency stator current is supplied instead of a voltage. Therefore, fundamental equations used in position estimators are given for current as well as voltage sources. Nevertheless, it will follow from the discussion that both methods can be described in a similar way.

High-frequency current source

An estimation algorithm using a high-frequency current source measures the high-frequency flux response. For these estimators, the flux equation (1) is written in a complex notation, with the real axis parallel to the q-axis, as

$$\Delta\underline{\phi}_m^{qd}(t) = \underline{l} \cdot \Delta\underline{i}_m^{qd}(t) \tag{2}$$

with the complex inductance $\underline{l}$ given by

$$\underline{l} = L + \underline{L}_{rel} - \underline{L}_{sat} \tag{3}$$

where

$$L = \frac{L_q + L_d}{2} \tag{4}$$

$$\underline{L}_{rel}(\alpha) = \frac{L_q - L_d}{2} e^{-j2\alpha(t)} \tag{5}$$

$$\underline{L}_{sat}(\alpha, \mu) = \frac{\Delta L_q - \Delta L_d}{2} e^{-j2\mu} + \frac{\Delta L_q + \Delta L_d}{2} e^{j2(\mu-\alpha(t))} \tag{6}$$

$$L_q = \frac{L_{qmo} + L_{qmt}}{2} \qquad L_d = \frac{L_{dmo} + L_{dmt}}{2} \tag{7}$$

$$\Delta L_q = \frac{L_{qmo} - L_{qmt}}{2} \qquad \Delta L_d = \frac{L_{dmo} - L_{dmt}}{2} \tag{8}$$

with $\alpha(t)$ the angle between $\Delta \underline{i}_m$ and the q-axis. Indicating the complex conjugate with the operator *, the relationship in (2) can alternatively be written as

$$\Delta \underline{\phi}_m^{qd}(t) = \left\{ \frac{L_q + L_d}{2} - \frac{\Delta L_q - \Delta L_d}{2} e^{-j2\mu} \right\} \Delta \underline{i}_m^{qd}(t) + \left\{ \frac{L_q - L_d}{2} - \frac{\Delta L_q + \Delta L_d}{2} e^{j2\mu} \right\} \Delta \underline{i} *_m^{qd}(t) \quad (9)$$

High-frequency voltage source

A lot of position estimators use a high-frequency flux generated with a voltage source, and measure the current response. For these estimation methods, the flux equation (1) is written as

$$\Delta \underline{i}_m^{qd}(t) = \underline{r} \cdot \Delta \underline{\phi}_m^{qd}(t) \quad (10)$$

with the complex reluctance $\underline{r}$ given by

$$\underline{r} = M(L + \underline{L}_{rel} + \underline{L}_{sat}) \quad (11)$$

where

$$M = \frac{1}{L_q L_d + (L_q \Delta L_d - L_d \Delta L_q)\cos(2\mu) - \Delta L_q \Delta L_d} \quad (12)$$

$$L = \frac{L_q + L_d}{2} \quad (13)$$

$$\underline{L}_{rel}(\beta) = -\frac{L_q - L_d}{2} e^{-j2\beta(t)} \quad (14)$$

$$\underline{L}_{sat}(\beta, \mu) = \left(-\frac{\Delta L_q - \Delta L_d}{2} + \frac{\Delta L_q + \Delta L_d}{2} e^{-j2\beta(t)} \right) e^{j2\mu} \quad (15)$$

with β the angle between $\Delta \underline{\phi}_m$ and the q-axis. The relationship in (10) can alternatively be written as

$$\Delta i_m^{qd}(t) = \frac{1}{(L_q L_d + (L_q \Delta L_d - L_d \Delta L_q)\cos(2\mu) - \Delta L_q \Delta L_d)} \times \left\{ \left\{ \frac{L_q + L_d}{2} - \frac{\Delta L_q - \Delta L_d}{2} e^{j2\varpi} \right\} \Delta \underline{\phi}_m^{qd}(t) - \left\{ \frac{L_q - L_d}{2} - \frac{\Delta L_q + \Delta L_d}{2} e^{j2\mu} \right\} \Delta \underline{\phi}_m^{*qd}(t) \right\} \quad (16)$$

Discussion

As most estimators are based on a current response to a high-frequency voltage variation, the following discussion will be restricted to such strategies. However, as the equations for an estimator, using a high-frequency current source, are similar to those in (10)–(16), the following discussion applies to both cases.

From the reluctance $\underline{r}$ in (11) it can be seen that, in addition to a current change in phase with $\Delta \underline{\phi}_m$ due to L in (13), two important components in the current variation can be distinguished. As follows from (14), a part of the current change is proportional to the inductance difference between q- and d-axis and is phase shifted from $\Delta\underline{\phi}_m$ over -2β. Another current variation, according to (15), is linked with the differences between the chord-slope inductance and the tangent-slope inductance in both q- and d-axis.

If the saturation level is low, the chord-slope inductance equals almost the tangent-slope inductance. Consequently, the inductance in (15) becomes small and a phase shift between $\Delta\underline{i}_m$ and $\Delta\underline{\phi}_m$ is the result of the inductance difference in (14) only. Clearly, the component in (14) reflects the reluctance variation along the air gap with extrema in both orthogonal magnetic axes.

The reluctance $\underline{r}$ in pu, in the case of an unsaturated salient-pole synchronous machine, is shown in Fig. 2(a). It is given for various values of the inductance difference between the two magnetic axes. The trajectories of $\Delta\underline{i}_m$ for a circular trajectory of $\Delta\underline{\phi}_m$, shown in Fig. 2(b), are elliptical with axes of symmetry in q- and d-directions, corresponding to the point of minimum and maximum modulus of $\underline{r}$ respectively. Furthermore, for a given value of β, the higher the difference between the q- and d-inductance, the higher the angle between $\Delta\underline{i}_m$ and $\Delta\underline{\phi}_m$.

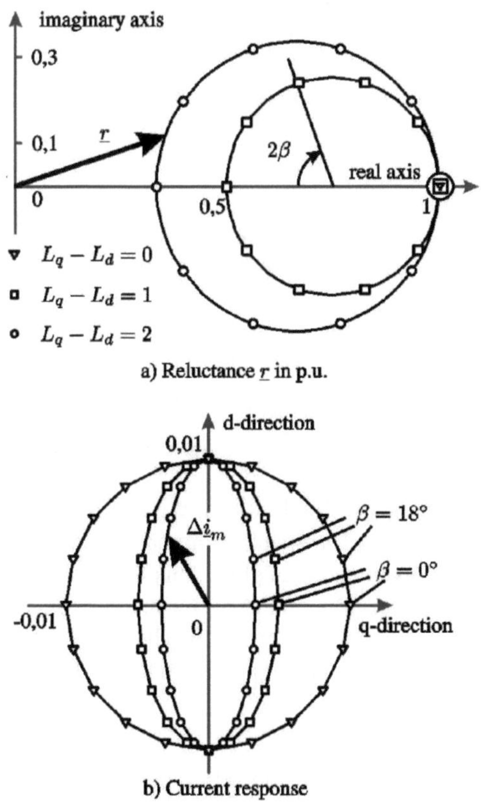

Figure 2. Reluctance $\underline{r}$ in pu and current response to small flux variations in the case of magnetic saliency with β as parameter.

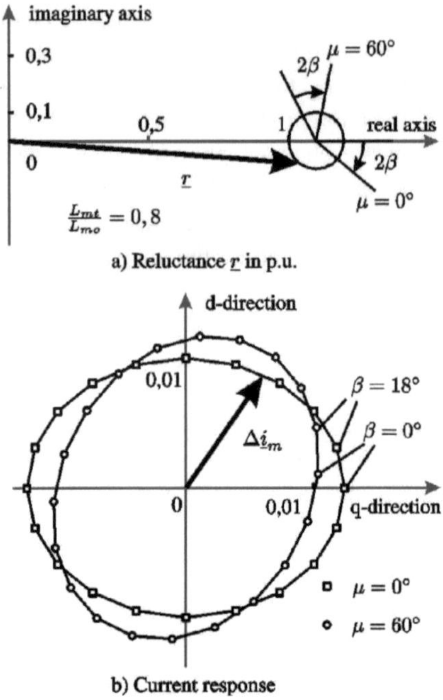

Figure 3. Reluctance $\underline{r}$ in pu and current response to small flux variations in the case of saturation with β as parameter.

If the PMSM has a uniform air-gap permeance, most controllers disregard the reluctance variation along the air gap. Consequently, the direction of the q-axis fixed to the rotor can be chosen deliberately. Furthermore, the reciprocity property, mentioned in [10], implies that

$$L_{dmo} - L_{qmo} = L_{dmt} - L_{qmt} = 0 \tag{17}$$

As a result, the difference in (14) becomes zero and the inductance (15) reduces to

$$\underline{L}_{sat}(\beta, \mu) = \frac{\Delta L_q + \Delta L_d}{2} e^{-j2(\beta(t)-\mu)} \tag{18}$$

This means that a noticeable phase shift between $\Delta \underline{i}_m$ and $\Delta \underline{\phi}_m$ is caused by (18) only.

In the case of a saturated smooth air-gap synchronous machine the reluctance $\underline{r}$ in pu is presented in Fig. 3(a) for a given modulus of $\underline{i}_m$. This figure shows that the direction of $\underline{i}_m$ influences the phase shift between $\Delta \underline{i}_m$ and $\Delta \underline{\phi}_m$ for the same β. The trajectory of $\Delta \underline{i}_m$ for a circular trajectory of $\Delta \underline{\phi}_m$ is shown in Fig. 3(b). The figure shows elliptical trajectories with a maximum $\Delta \underline{i}_m$ in the direction of $\underline{i}_m$, corresponding the point in Fig 3(a) with the maximum modulus of $\underline{r}$.

Multiple saliencies

Due to the construction of the rotor, the reluctance variation along the air gap can display global extrema in q- and d-axis as well as several local extrema. Such a reluctance variation

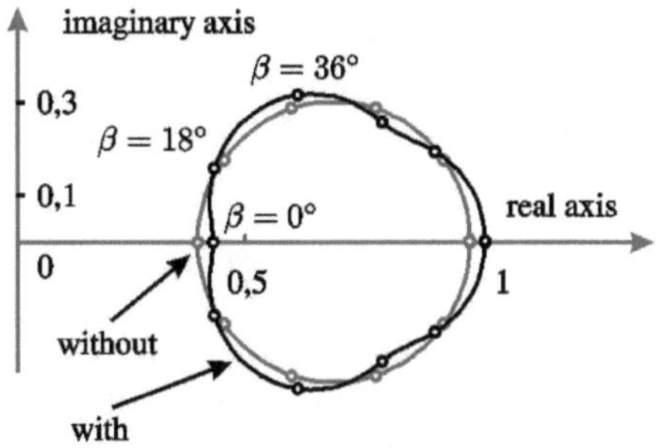

Figure 4. Reluctance $\underline{r}$ in pu of an IPMSM with and without multiple saliencies with β as parameter.

is called a multiple saliency. Assuming sinusoidal reluctance variations, these multiple saliencies can be modeled in a similar way as the reluctance variation with extrema in q- and d-axis only. As a result, by using previous discussion, the equation in (14) can be replaced for modeling multiple saliencies by

$$\underline{L}_{rel} = \sum_i \frac{L_{q,i} - L_{d,i}}{2} e^{-j(i\beta(t)+\varphi_i)}, \quad i \in IN_0 \tag{19}$$

with φ_i a possible space phase shift.

To illustrate the model with multiple saliencies, the trajectory of $\underline{r}$ for an unsaturated salient-pole machine with β as parameter is calculated by using (19) instead of (14). Two cases, with and without an extra sinusoidal reluctance variation having four extrema per pole pitch (i equals to 2 and 4 in (19)), are shown in Fig. 4. This reluctance trajectory is also observable in [7] and mentioned in [8]. In [7], by applying finite element simulations, almost the same trajectory as in Fig. 4 can be observed. In [8] the effect of multiple saliencies is measured as a variation in the stator current instead of an inductance. However, these results are not modeled such as in (19).

Recently proposed estimators

Approximated small signal dynamic flux model

In an IPMSM the magnetizing current is mainly generated by the permanent magnets. For this reason, in some recently proposed sensorless drives, such as [1–8], the angle of $\underline{i}_m$ is approximated by $\pi/2$. For μ equal to $\pi/2$, the equation (10) results in

$$\Delta\underline{i}_m^{dq}(t) = \frac{1}{L_{qmo}L_{dmt}}\left(\frac{L_{qmo}+L_{dmt}}{2} - \frac{L_{qmo}-L_{dmt}}{2}e^{-j2\beta(t)}\right)\Delta\underline{\phi}_m^{qd}(t) \tag{20}$$

By defining

$$L = \frac{L_{qmo}-L_{qmt}}{2}, \quad \Delta L = \frac{L_{qmo}-L_{dmt}}{2} \tag{21}$$

the relationship (20) can also be written as

$$\Delta\underline{i}_m^{dq}(t) = \left(\frac{L}{L^2-\Delta L^2} - \frac{\Delta L}{L^2-\Delta L^2}e^{-j2\beta(t)}\right)\Delta\underline{\phi}_m^{qd}(t) \tag{22}$$

In some estimators a stationary reference frame ($\alpha\beta$) is used. Furthermore, equation (16) rather than the one in (10) is considered. Transformation of a variation $\Delta\underline{x}$ from the reference frame (qd) to the stationary reference frame ($\alpha\beta$), with the real axis parallel to the α-axis, is given by

$$\Delta\underline{x}^{qd}(t) = e^{j\theta_r}\left(j\Delta\theta_r \underline{x}_0^{\alpha\beta} + \Delta\underline{x}^{\alpha\beta}\right) \tag{23}$$

with θ_r the rotor angle defined as the angle between the α-axis and the q-axis and with $\underline{x}_0$ the mean value of the vectors $\underline{x}$ at the beginning and end of the variation. Transformation of (16) to the stationary reference frame, by using (23), results in

$$\begin{aligned} j\Delta\theta_r \underline{i}_{mo}^{\alpha\beta} + \Delta\underline{i}_m^{\alpha\beta} &= \frac{L}{L^2 - L\Delta L^2}\left(j\Delta\theta_r \underline{\phi}_{mo}^{\alpha\beta} + \Delta\underline{\phi}_m^{\alpha\beta}\right) \\ &\quad - \frac{\Delta L}{L^2 - \Delta L^2}\left(j\Delta\theta_r \underline{\phi}_{mo}^{\alpha\beta} + \Delta\underline{\phi}_m^{\alpha\beta}\right) * e^{j2\theta_r} \end{aligned} \tag{24}$$

for μ equal to $\pi/2$. Inverting (24)

$$j\Delta\theta_r \underline{\phi}_{mo}^{\alpha\beta} + \Delta\underline{\phi}_m^{\alpha\beta} = L\left(j\Delta\theta_r \underline{i}_{mo}^{\alpha\beta} + \Delta\underline{i}_m^{\alpha\beta}\right) + \Delta L e^{j2\theta_r}\left(j\Delta\theta_r \underline{i}_{mo}^{\alpha\beta} + \Delta\underline{i}_m^{\alpha\beta}\right) \tag{25}$$

results in the matrix notation

$$\Delta\phi_m^{qd}(t) = \begin{pmatrix} L + \Delta L\cos(2\theta_r) & \Delta L\sin(2\theta_r) \\ \Delta L\sin(2\theta_r) & L - \Delta L\cos(2\theta_r) \end{pmatrix} \cdot \Delta\underline{i}_m^{\alpha\beta}(t) \tag{26}$$

This equation is well-known as it shows the sinusoidal variation of the magnetic reluctance along the air gap with the pole pitch as period.

Stator voltage equation

By supplying a high-frequency voltage to the motor terminals, a high-frequency stator flux linkage variation of small amplitude is generated. In an IPMSM, without damper effects e.g. due to short-circuited windings or eddy currents, this flux variation mainly occurs in the main flux instead of the leakage flux path. For this reason and by disregarding the voltage drop across the stator resistance and leakage inductance, the motor voltage equation, in a two-dimensional stationary reference system ($\alpha\beta$), can be approximated by

$$\underline{v}_s^{\alpha\beta}(t) = \frac{\Delta\underline{\phi}_m^{\alpha\beta}(t)}{\Delta t} \tag{27}$$

with $\underline{v}_s$ the complex stator voltage. Transformation of (27) to the synchronous reference frame (qd) results in

$$\underline{v}_s^{qd}(t) = \frac{\Delta\underline{\phi}_m^{qd}(t)}{\Delta t} - j\frac{\Delta\theta_r}{\Delta t}\underline{\phi}_{mo}^{qd}(t) \tag{28}$$

High-frequency voltage pulse train

In modern IPMSM drives, a pulse-width-modulated (PWM) inverter is used. This means that, at normal operation, a voltage pulse train at high frequency is supplied to the motor terminals. From equation (28) it follows that the current variation is piecewise linear. As a result, according to the model in (10), a magnetizing current variation occurs, which depends on the direction of the main flux variation in the reference frame *(qd)* and on the magnetizing current. For a current controlled drive, the current $\underline{i}_m$ can be approximated by the desired $\underline{i}_m$ calculated within the controller. Moreover, in the synchronous reference frame, the variation of $\underline{i}_m$ equals $\Delta\underline{i}_s$, with $\underline{i}_s$ the complex stator current, as the equivalent magnetizing current due to the magnets is constant.

The high-frequency flux variation, generated by using a PWM, can be used to estimate the rotor angle. Calculating the main flux variation with (28) and transforming the measured stator current to an estimated synchronous reference frame $(\widehat{qd})$, it follows from (10) that an estimation of the reluctance $\underline{r}$ can be obtained. Substituting $\underline{i}_m$ with its desired value calculated within the controller, the angle μ and the inductances in (7)–(8) can be approximated, which result, together with the estimated $\underline{r}$, in an estimation of the angle β. As the angle of the main flux variation can also be calculated in the stationary reference frame by using (27), a new estimation of the q-axis is obtained.

If the motion-induced voltage is known, $\Delta\underline{\phi}_m$ can be calculated by using (28). However, for a slow rotor motion, a back-EMF of small amplitude has to be measured or calculated which is strongly disturbed by noise produced by normal operation of the PWM and motor. As for most drives the mechanical time constant is higher than the electrical one, the rotor speed and the motion-induced voltage can be assumed to be constant during a sufficiently small time period. Therefore, by subtracting the stator voltage generated by two successive PWM pulses, back-EMF measurements are avoided. Together with (10), this results in the following system

$$\frac{\Delta\underline{\phi}_{m,2}^{qd} - \Delta\underline{\phi}_{m,1}^{dq}}{\Delta t} = \underline{v}_{s,2}^{qd} - \underline{v}_{s,1}^{qd} \tag{29}$$

$$\frac{\Delta\underline{i}_{s,2}^{qd} - \Delta\underline{i}_{s,1}^{dq}}{\Delta t} = \underline{r}(\beta_2)\frac{\Delta\underline{\phi}_{m,2}^{qd}}{\Delta t} - \underline{r}(\beta_1)\frac{\Delta\phi_{m,1}^{qd}}{\Delta t} \tag{30}$$

In the sensorless drive presented in [1], the reluctance $\underline{r}$ is estimated by using equation (30). This method is called indirect flux detection by online reactance measurement (INFORM) as introduced by Schrödl. However, in such an estimator μ is approximated by $\pi/2$, which means that the reluctance $\underline{r}$ is estimated by using equation (22) instead of (10). Furthermore, a β_2 value is used that is equal to $\beta_1 + \pi$. As a result, the system of (29) and (30) together with the relationship in (22), results in

$$\underline{v}_{s,1}^{qd} = -\underline{v}_{s,2}^{qd} \tag{31}$$

$$\frac{\Delta\underline{i}_{s,2}^{qd} - \Delta\underline{i}_{s,1}^{dq}}{\Delta t} = \left(\frac{L}{L^2 - \Delta L^2} - \frac{\Delta L}{L^2 - \Delta L^2}e^{-j2\beta_2}\right)2\underline{v}_{s,2}^{qd} \tag{32}$$

as the reluctance $\underline{r}$ in (22) varies periodically with 2β. Furthermore, in the INFORM method the estimation of $\underline{r}$ is repeated in the two other stator phases. The average reluctance of the three phases approximately coincides with

$$\frac{\underline{r}(\beta) + \underline{r}(\beta + \frac{2\pi}{3}) + \underline{r}(\beta - \frac{2\pi}{3})}{3} = \frac{L}{L^2 - \Delta L^2} \tag{33}$$

As a consequence by subtracting (33) from an estimation of the reluctance $\underline{r}$ in (32), calculating 2β is done by using the inverse tangent function only. Clearly, this method requires no knowledge about the inductances. However, it requires successive flux variations in opposite directions, which can disturb proper motor operation [2]. Note that the higher the inductance difference between the two orthogonal magnetic axes, the smaller the rotor variation that can be detected.

Sinusoidal high-frequency voltage

Instead of using the PWM generated pulse train, in some estimators, such as in [3–5,8], the current response is measured on a sinusoidal high-frequency voltage within the stator voltage. In most of these strategies, calculations are performed in a stationary reference frame by using the equations in (24) and (27). The stator voltage is given by

$$\underline{v}_s^{\alpha\beta}(t) = \underline{V}_s^{\alpha\beta}(t) + \underline{v}_{s,i}^{\alpha\beta}(\omega_i t) \tag{34}$$

with $\underline{V}_s$ a complex voltage and with ω_i the pulsation of the injected high-frequency voltage $\underline{v}_{s,i}$. With the voltage in (34), the voltage equation in (27), at a high frequency, is written as

$$\underline{v}_{s,i}^{\alpha\beta}(\omega_i t) = \frac{d\underline{\phi}_{m,i}^{\alpha\beta}(\omega_i t)}{dt} \tag{35}$$

Furthermore, at a sufficiently high frequency, the rotor angle can be assumed to be constant. Consequently, equation (24), for a sufficiently high frequency, results in

$$\frac{d\underline{i}_{s,i}^{\alpha\beta}(\theta_r, \omega_i t)}{dt} = \frac{L}{L^2 - \Delta L^2}\underline{v}_{s,i}^{\alpha\beta}(\omega_i t) - \frac{\Delta L}{L^2 - \Delta L^2} e^{j2\theta_r}\left(\underline{v}_{s,i}^{\alpha\beta}(\omega_i t)\right)* \tag{36}$$

In some methods a voltage rotating at a high pulsation ω_i is added to the stator voltage

$$\underline{v}_{s,i}^{\alpha\beta}(\omega_i t) = V_i e^{j\omega_i t} \tag{37}$$

The high-frequency current response, obtained by using (36), will include a positive and negative rotating component

$$\underline{i}_{s,i}^{\alpha\beta}(\theta_r, \omega_i t) = I_0 e^{j\omega_i t} + I_1 e^{j(2\theta_r - \omega_i t)} \tag{38}$$

with

$$I_0 = \frac{L}{L^2 - \Delta L^2} \cdot \frac{V_i}{\omega_i} \quad I_1 = \frac{\Delta L}{L^2 - \Delta L^2} \cdot \frac{V_i}{\omega_i} \tag{39}$$

Transformation of the current response to a reference frame that rotates at ω_i results in

$$\underline{i}_{s,i}^{\omega_i}(\theta_r, \omega_i t) = I_0 + I_1 e^{j2(\theta_r - \omega_i t)} \tag{40}$$

In other estimation methods a high-frequency voltage in an estimated quadrature axis is added to the fundamental voltage

$$\underline{v}_{s,i}^{\alpha\beta}(\hat{\theta}_r, \omega_i t) = V_i \cos(\omega_i t)\, e^{j\hat{\theta}_r} \tag{41}$$

By using (36), the high-frequency current vector, observed from the estimated system $(\widehat{qd})$, is given by

$$\underline{i}_{s,i}^{\widehat{dq}}(\theta_r, \widehat{\theta}_r, \omega_i t) = \left(I_0 + I_1 \cdot e^{j2(\theta_r - \hat{\theta}_r)}\right) \cos(\omega_i t) \tag{42}$$

By removing the offset I_0 in (40) or in the modulus of the current in (42) and by using the inverse tangent function, the argument of the exponential function in (40) or in (42) is calculated. From this result, the rotor angle can be estimated. It follows from I_1 in (39) that the higher the value of ΔL, the higher the resolution of a position estimation, which is already concluded for an estimator that uses PWM generated pulses.

Estimation errors

As mentioned before, in most position estimators the magnetizing current direction is approximated by the d-axis direction. However, for high loads, the controller forces an important stator current along the q-axis. This means that the magnetizing current direction deviates from the d-axis. Consequently, the model in (22) or (24) introduces an estimation error. Compensating this error is done in [5,6] by measuring the error during the self-commissioning of the drive. The error can also be predicted by simulating the drive, modeled with (10), with an estimator that uses a high-frequency voltage pulse train and that is based on (22). The simulation results as a function of μ are presented in Fig. 5. The error on θ_r is zero if the magnetizing current is aligned with one of the magnetic axes. The higher the deviation of $\underline{i}_m$ from the d-axis, the higher the estimation error; an increased error is shown if saturation becomes more important. However, as permanent demagnetization of the magnets has to be avoided and the stator current has to be limited, the results are meaningful for small deviations of μ from $\pi/2$ only.

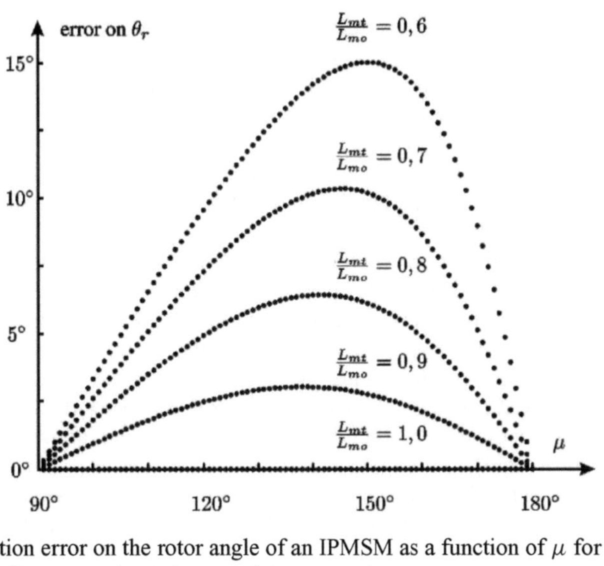

Figure 5. Estimation error on the rotor angle of an IPMSM as a function of μ for various magnetic states in the case of an approximated magnetizing current.

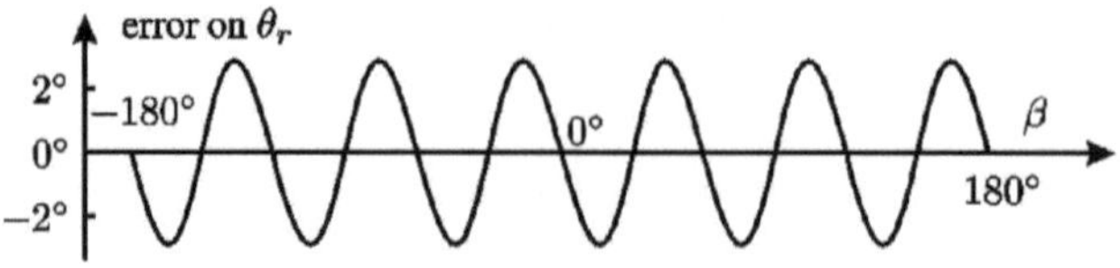

Figure 6. Estimation error on the rotor angle of an IPMSM as a function of β in the case of neglecting the presence of multiple saliencies.

In most estimators the influence of multiple saliencies on the current response is considered as a source of disturbance. Simulating an unloaded IPMSM, with (10) and (18), and a position estimator, that uses a PWM generated high-frequency voltage and equation (22), predicts the estimation error. Simulation results, based on a multiple saliency as modeled in Fig. 4, are shown in Fig. 6. Clearly, the error on θ_r oscillates as a function of β.

Conclusions

This paper discusses fundamental equations used in high-frequency signal based IPMSM position estimators. For this purpose, a small signal dynamic flux model is presented that takes into account the nonlinear magnetic condition and the magnetic interaction between the direct and the quadrature magnetic axis. An addition to the model is proposed to tackle multiple saliencies. Using the novel equations some recently proposed motion-state estimators are described. It is shown that the higher the inductance difference between the two orthogonal magnetic axes, the higher the position estimation resolution. Furthermore simulation results reveal the estimation error caused by estimators that neglect the presence of multiple saliencies or that approximate the magnetizing current angle by $\pi/2$.

References

[1] M. Schrödl, Sensorless control of permanent magnet synchronous motors, Electr. Mach. Power Syst., Vol. 22, No. 2, pp. 173–185, 1994.
[2] E. Robeischl, M. Schrödl, Optimized INFORM measurement sequence for sensorless PM synchronous motor drives with respect to minimum current distortion, IEEE Trans. Ind. Appl., Vol. 40, No. 2, pp. 591–598, 2004.
[3] M.J. Corley, R.D. Lorenz, Rotor position and velocity estimation for a salient-pole permanent magnet synchronous machine at standstill and high speeds, IEEE Trans. Ind. Appl., Vol. 34, No. 4, pp. 784–789, 1998.
[4] M. Linke, R. Kennel, J. Holtz, "Sensorless Position Control of Permanent Magnet Synchronous Machines Without Limitation at Zero Speed", Proceedings of the 28th Annual Conference of the IEEE Industrial Electronics Society, Sevilla, Spain, CD-ROM, November 5–8, 2002.
[5] C. Silva, G.M. Asher, M. Sumner, K.J. Bradley, Sensorless rotor position control in a surface mounted PM machine using HF rotating injection, EPE J., Vol. 13, No. 3, pp. 12–18, 2003.
[6] M. Schrödl, "Zuverlässigkeit sensorloser INFORM-geregelter Permanentmagnetmotor-Antriebe im Transient-betrieb bis Stillstand", Elektrotechnik und Informationtechnik, Heft 2, pp. 48–57, Febuary 2004.
[7] U.H. Rieder, M. Schrödl, "Optimization of Saliency Effects of External Rotor Permanent Magnet Synchronous Motors with Respect to Enhanced INFORM-Capability for Sensorless Control", Proc. of the 10th European Conference on Power Electronics and Applications, Toulouse, France, CD-ROM, September 2003.

[8] M.W. Degner, R.D. Lorenz, Using multiple saliencies for the estimation of flux, position, and velocity in AC machines, IEEE Trans. Ind. Appl., Vol. 34, No. 5, pp. 1097–1104, 1998.

[9] J.A.A. Melkebeek, “Small Signal Dynamic Modelling of Saturated Synchronous Machines”, Conf. Proc. Int. Conf. El. Mach., Lausanne, Switzerland, Part 2, September 18–21, 1984, pp. 447–450.

[10] J.A.A. Melkebeek, J.L. Willems, Reciprocity relations for the mutual inductances between orthogonal axis windings in saturated salient-pole machines, IEEE Trans. Ind. Appl., Vol. 26, No. 1, pp. 107–114, 1990.

II-2. SENSORLESS CONTROL OF SYNCHRONOUS RELUCTANCE MOTOR USING MODIFIED FLUX LINKAGE OBSERVER WITH AN ESTIMATION ERROR CORRECT FUNCTION

Tsuyoshi Hanamoto, Ahmad Ghaderi, Teppei Fukuzawa and Teruo Tsuji

Kyushu Institute of Technology, 2-4 Hibikino, Wakamatsu-ku, Kitakyushu 808-0196, Japan
hanamoto@life.kyutech.ac.jp, ahmad-ghaderi@edu.life.kyutech.ac.jp, tsuji@life.kyutech.ac.jp

Abstract. The modified flux observer with an estimation error correct function for the sensorless control method of synchronous reluctance motor is presented. The validity of the proposed method is verified by experiments. The experimental setup is based on the Real Time Linux for operating system and Field programmable Logic Array interface board.

Introduction

Recently, a motor control for a motion control is widely used in various industrial applications. AC motors are better to be used because they have some advantages, such as easiness of maintenance. In addition, sensorless speed control of the AC motors has been proposed for the demand of the reduction of weight, size, and total cost.

Synchronous reluctance motor (Syn.RM) is a kind of the AC motors and has the advantage that it is mechanically simple and robust because they need not the permanent magnet for a material of a rotor, then many researchers are proposed the sensorless algorithm [1–4].

In this paper, we propose a novel sensorless control method for Syn.RM. The sensorless control is based on the modified flux linkage observer, which is proposed by authors for permanent magnet synchronous motors (PMSM) [5]. The observer is able to estimate the modified flux linkage and the electromotive force (EMF) simultaneously, and the motor speed and the rotor position are calculated from these estimated values. But as same as the other method, the precision of the observer-based estimation is affected by the parameter fluctuations [7]. In this paper, we propose the new estimation method for Syn.RM using the modified flux linkage observer with an estimation error correct function. A Proportional-Integral (PI) type controller is added to the system to compensate the estimation error. It operates that the estimated magnitude of the flux corresponds to the nominal value.

The high-speed experimental system is required to achieve the proposed method because the observer matrix is changed for every control period and the gains must be recalculated.

S. Wiak, M. Dems, K. Komęza (eds.), Recent Developments of Electrical Drives, 155–164.

Thus, the experimental setup is based on the Real Time Linux (RTLinux) [8] for operating system. The RTLinux is used for achievement of the real time control and it guarantees to satisfy hard real time constraints in light of maintaining soft real time requirements. To acquire the data from sensors and to output the gate signals to the system, the interface board is accomplished designed by the Field programmable Logic Array (FPGA).

The environment of the system development is so convenient and sophisticated to combine the RTLinux operating system and the FPGA-based interface board.

The validity of the proposed method is verified by experiments.

Mathematical model of Syn.RM

Fig. 1 shows the mathematical model of Syn.RM, where d, q show the dq axes rotating at ω_e, α, β show $\alpha\beta$ axes, u, v, w show three phase axes, θ_e shows the electrical angle from α (or u) axis and $\omega_e = d\theta_e/dt$.

The voltage equations of the Syn.RM in $\alpha\beta$ axes are described as follows

$$\begin{bmatrix} v_\alpha \\ v_\beta \end{bmatrix} = \begin{bmatrix} R + PL_\alpha & 0 \\ 0 & R + PL_\alpha \end{bmatrix} \begin{bmatrix} i_\alpha \\ i_\beta \end{bmatrix} + PL_\beta \begin{bmatrix} \cos 2q_e & \sin 2q_e \\ \sin 2q_e & -\cos 2q_e \end{bmatrix} \begin{bmatrix} i_\alpha \\ i_\beta \end{bmatrix} \quad (1)$$

where, v: armature voltage, i:armature current, R: armature resistance, L: armature inductance, P: differential operation $(= d/dt)$, subscript $_{d,q}$ denotes d-axis component, q-axis component, respectively.

L_d, L_q are calculated using, L_α, L_β as follows

$$\begin{bmatrix} L_d \\ L_q \end{bmatrix} = \begin{bmatrix} 1 & 1 \\ 1 & -1 \end{bmatrix} \begin{bmatrix} L_\alpha \\ L_\beta \end{bmatrix} \quad (2)$$

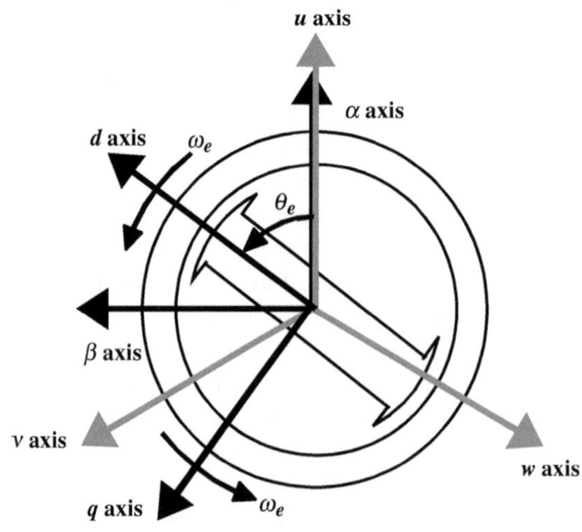

Figure 1. Analytical model of synchronous reluctance motor.

Equation (1) is rewritten as follows [2] when the de-coupling control for dq axes is achieved in the speed control system of the Syn.RM

$$\begin{bmatrix} v_\alpha \\ v_\beta \end{bmatrix} = R \begin{bmatrix} i_\alpha \\ i_\beta \end{bmatrix} + P \begin{bmatrix} y_\alpha \\ y_\beta \end{bmatrix} \tag{3}$$

where, flux linkage y_α, y_β are defined as follows,

$$\begin{bmatrix} y_\alpha \\ y_\beta \end{bmatrix} = \begin{bmatrix} L_\alpha + L_\beta \cos 2q_e & L_\beta \sin 2q_e \\ L_\beta \sin 2q_e & L_\alpha - L_\beta \cos 2q_e \end{bmatrix} \begin{bmatrix} i_\alpha \\ i_\beta \end{bmatrix} \tag{4}$$

Use the well-known relationship described in (5), y_α, y_β are calculated as follows,

$$\left.\begin{aligned} \cos 2q_e &= 2\cos^2 q_e - 1 = 1 - 2\sin^2 q_e \\ \sin 2q_e &= 2 \sin q_e \cos q_e \end{aligned}\right| \tag{5}$$

$$\begin{bmatrix} y_\alpha \\ y_\beta \end{bmatrix} = L_q \begin{bmatrix} i_\alpha \\ i_\beta \end{bmatrix} + (L_d - L_q) \begin{bmatrix} \cos q_e & 0 \\ 0 & \sin q_e \end{bmatrix} \begin{bmatrix} \cos q_e & \sin q_e \\ \cos q_e & \sin q_e \end{bmatrix} \begin{bmatrix} i_\alpha \\ i_\beta \end{bmatrix} \tag{6}$$

When we set

$$Y = (L_d - L_q) i_d, \tag{7}$$

the flux linkage are described as follows.

$$\begin{bmatrix} y_\alpha \\ y_\beta \end{bmatrix} = L_q \begin{bmatrix} i_\alpha \\ i_\beta \end{bmatrix} + Y \begin{bmatrix} \cos q_e \\ \sin q_e \end{bmatrix} \tag{8}$$

Y is able to be treated as a constant because it changes slowly compared with the sampling period. Then the derivative of (8) is

$$P \begin{bmatrix} y_\alpha \\ y_\beta \end{bmatrix} = \begin{bmatrix} PL_q i_\alpha & -Y w_e \sin q_e \\ PL_q i_\beta & Y w_e \cos q_e \end{bmatrix} \tag{9}$$

In this paper, the EMF of each axis (e_a, e_b) are determined as the following equation.

$$\left.\begin{aligned} e_\alpha &= -Y w_e \sin q_e \\ e_\beta &= Y w_e \cos q_e \end{aligned}\right| \tag{10}$$

Finally, (3) is described as follows.

$$P \begin{bmatrix} i_\alpha \\ i_\beta \end{bmatrix} = \begin{bmatrix} \dfrac{R}{L_q} & 0 \\ 0 & -\dfrac{R}{L_q} \end{bmatrix} \begin{bmatrix} i_\alpha \\ i_\beta \end{bmatrix} - \frac{1}{L_q} \begin{bmatrix} e_\alpha \\ e_\beta \end{bmatrix} + \frac{1}{L_q} \begin{bmatrix} v_\alpha \\ v_\beta \end{bmatrix} \tag{11}$$

This equation is equivalent of the equation for a PMSM [5], then we can also apply the flux linkage observer for the Syn.RM.

Sensorless speed control method of Syn.RM

Modified linkage observer with an estimation error correct function

In this chapter, we consider the estimation method of the rotor speed and the position.

The EMF is assumed that it consists the fundamental component, which rotate the constant angular speed, w_e and the DC component denoted as follows. The DC component is not necessary in the ideal case, but in the real system this term is very effective for the ripple reduction of the estimated speed calculation.

$$\begin{bmatrix} e_\alpha \\ e_\beta \end{bmatrix} = \begin{bmatrix} A_\alpha \cos q_e + B_\alpha \sin q_e + e_{\alpha 0} \\ A_\beta \cos q_e + B_\beta \sin q_e + e_{\beta 0} \end{bmatrix} = \begin{bmatrix} e_{\alpha 1} + e_{\alpha 0} \\ e_{\beta 1} + e_{\beta 0} \end{bmatrix} \tag{12}$$

From (8), the following equation are obtained.

$$\begin{bmatrix} y_\alpha \\ y_\beta \end{bmatrix} = \begin{bmatrix} y_\alpha - L_q i_\alpha \\ y_\beta - L_q i_\beta \end{bmatrix} = \begin{bmatrix} Y \cos q_e \\ Y \sin q_e \end{bmatrix} \tag{13}$$

The derivative of the EMF is given by the following equation.

$$P \begin{bmatrix} e_\alpha \\ e_\beta \end{bmatrix} = \begin{bmatrix} -w_e^2 Y \cos q_e \\ -w_e^2 Y \sin q_e \end{bmatrix} = -w_e^2 \begin{bmatrix} y_\alpha \\ y_\beta \end{bmatrix} \tag{14}$$

But in the practical case, the estimated rotor position has the estimation error, and L_d, L_q are considered as a function of an armature current. So, we propose to use the following equation instead of (14), where K_y is a compensation coefficient.

$$P \begin{bmatrix} e_\alpha \\ e_\beta \end{bmatrix} = -K_y w_e^2 \begin{bmatrix} y_\alpha \\ y_\beta \end{bmatrix} \tag{15}$$

Finally, the voltage equation of α-axis is written as

$$P\boldsymbol{x} = \boldsymbol{A}\boldsymbol{x} + \boldsymbol{b} v_\alpha, \tag{16}$$

where,

$$\begin{aligned} \boldsymbol{x} &= \begin{bmatrix} i_\alpha & y_\alpha & e_{\alpha 1} & e_{\alpha 0} \end{bmatrix}^T \\ \boldsymbol{A} &= \begin{bmatrix} \frac{R}{L_q} & 0 & -\frac{1}{L_q} & -\frac{1}{L_q} \\ 0 & 0 & 1 & 0 \\ 0 & -K_y w_e^2 & 0 & 0 \\ 0 & 0 & 0 & 0 \end{bmatrix} \\ \boldsymbol{b} &= \begin{bmatrix} \frac{1}{L_q} & \mathbf{0} & \mathbf{0} & \mathbf{0} \end{bmatrix}^T \end{aligned} \tag{17}$$

To apply the same manner for β-axis, the flux linkage and the EMF of β phase are also obtained.

How to calculate the coefficient K_y is as follows.

1. The magnitude of the flux linkage is estimated by

$$Y = \sqrt{y_\alpha^2 + y_\beta^2} \tag{18}$$

2. K_y is obtained the output of the PI controller shown in Fig. 2. In the figure, the flux linkage reference Y_n is calculated by the nominal values of the inductance and the d-axis reference i_d^*.

$$Y_n = (L_d - L_q) i_d^* \tag{19}$$

3. For every control period, K_Y in (17) is recalculated.
4. K_Y is convergence to the appropriate value after several iterations.

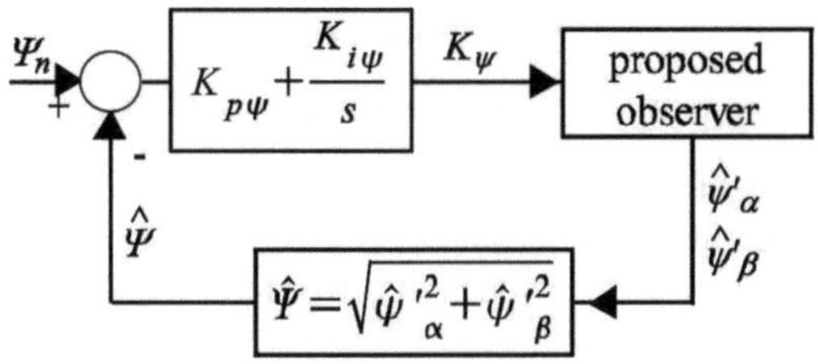

Figure 2. Estimation error correct function.

The flux linkage and the EMF are estimated directly applying the full order observer to (16). The speed reference ω^* and the voltage reference v_α^* are used instead of ω_e and v_α because these values are not measured in this system.

To convert the discrete time system at control period T_s, the following equation is obtained.

$$\boldsymbol{x}(k+1) = (\boldsymbol{A}_d - \boldsymbol{g}_d\boldsymbol{c}_d)\boldsymbol{x}(k) + \boldsymbol{b}_d v_\alpha^*(k) + \boldsymbol{g}_d i_\alpha(k) \tag{20}$$

where,

$$\begin{aligned}
\boldsymbol{A}_d &= e^{-AT_s} \\
\boldsymbol{b}_d &= \int_0^{T_s} e^{A\tau} d\tau \cdot \boldsymbol{b} \\
\boldsymbol{g}_d &= (g_1 \quad g_2 \quad g_3 \quad g_4) \\
\boldsymbol{c}_d &= (1 \quad 0 \quad 0 \quad 0)
\end{aligned}$$

$\boldsymbol{g}_d$ is the observer gain in the discrete time system. We select the observer gains to have all of the poles of $(\boldsymbol{A}_d - \boldsymbol{g}_d\boldsymbol{c}_d)$ on the real axis in unit circle as the multiple poles.

In the proposed method, $\boldsymbol{A}_d$ is changed because $\boldsymbol{K}_Y$ is calculated for every control period and it is also the function of the speed reference. Though the online calculation is required to convert the discrete time system and calculation of the observer gains, the computer technology able to achieve the calculation within 200 μs.

Speed and position estimation

The estimated speed w_e are obtained by the following equation. A low pass filter (LPF) is added after the output of w_e in the experimental system.

$$w_e = \frac{e_{\beta 1} y'_\alpha - e_{\alpha 1} y'_\beta}{y'^2_\alpha + y'^2_\beta} \tag{21}$$

Since it is enough to estimated the sin q_e and cos q_e instead of the rotor position (q_e) itself, then

$$\sin q_e = \frac{y_\beta}{\sqrt{y_\alpha^2 + y_\alpha^2}}, \quad \cos q_e \frac{y_\alpha}{\sqrt{y_\alpha^2 + y_\alpha^2}} \tag{22}$$

Experimental results

Experimental setup

To realize the validity of the control theory, there is a need for an appropriate operating system that could operate in real time. The PC-based control offers great advantages like a faster design cycle and increased productivity [6]. Here, we refer to the real time system based on the RTLinux [8].

The RTLinux is a hard real time operating system that handles time-critical tasks and runs the normal Linux as its lowest priority execution thread. Then the system includes the networking, GUI programming, and several other function.

We can construct the PC-based experimental setup which includes not only the control program but also the user GUI, for example, data entry windows of the reference, controller gains, and so on. Fig. 3 is an example of the GUI using RTiC-Lab [9], which is a semi-detached open source software designed to run on the RTLinux.

From the viewpoint of the hardware, the digital servomotor control system requires a speed detector, a position detector, both for reference, the PWM pulse generator, and the interface circuit. In this paper, we designed the interface circuit board, which includes all of the necessary functions. Fig. 4 shows the interface board that consists of the Field programmable Logic Array (FPGA), 10 MHz system clock, and an A/D converter.

An Altera FLEX10K50 is selected for the FPGA device and the circuit is designed using the VHDL, which is one of the hardware description languages. The speed and

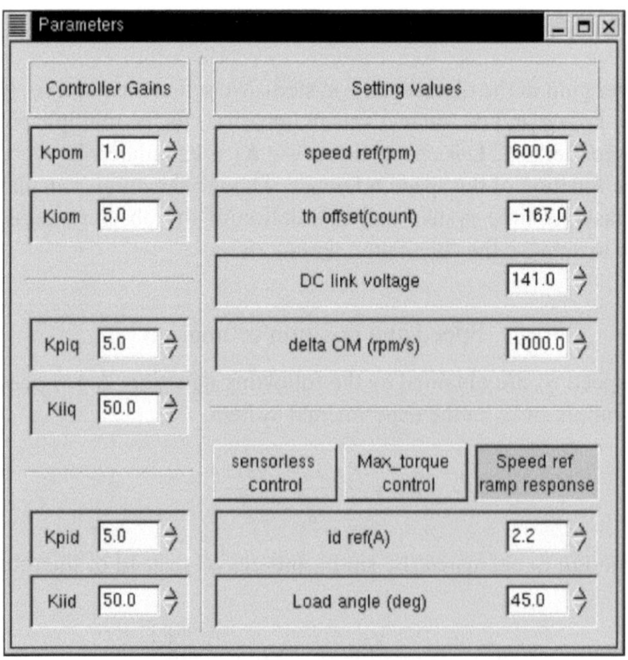

Figure 3. GUI controller using RTiC.

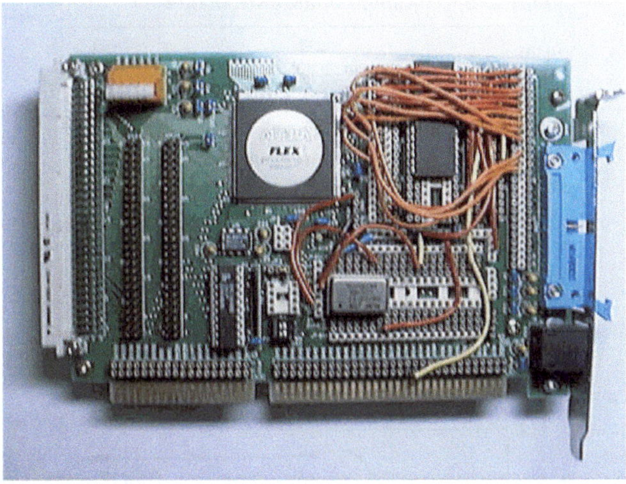

Figure 4. Interface board using the FPGA.

position detectors with speed correction function, clock generator for A/D converter, and ISA interface circuit are designed in it.

Fig. 5 shows the configuration of the experimental system consists of the RTLinux-based PC, an inverter, an interface board using the CPLD device, and the tested motor. Table 1 shows the specification of the tested motor.

Experimental results

Figs. 6 and 7 show the ramp response where the speed command is changed from 500 to 1,500/min. Fig. 6 shows the results when the sensor is used for the speed control. Here, the dashed line shows the estimated speed and the solid line shows the speed calculated by the

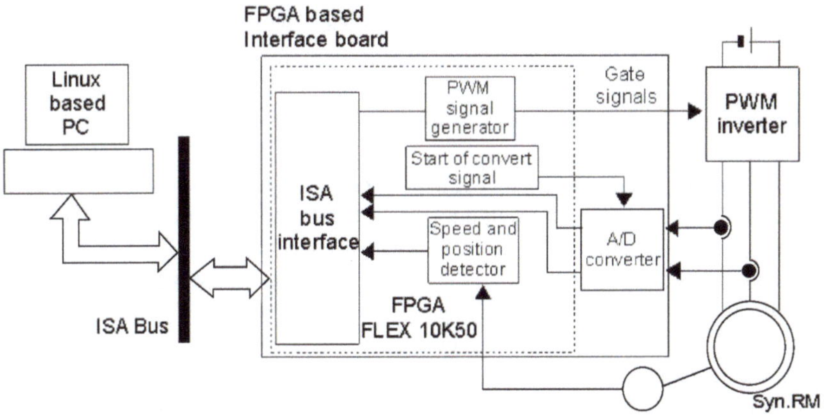

Figure 5. Configuration of the system.

Table 1. Specifications of the tested motor

Items	Value
Rated power (W) (rated/maximum)	86/355
Rated torque (Nm) (rated/maximum)	0.4/0.96
Rated current (Arms) (rated/maximum)	1.7/2.6
Inverter voltage (V)	200
Armature resistance (Ω)	1.89
d-axis armature inductance (mH)	93
q-axis armature inductance (mH)	36
Number of pole pairs	2

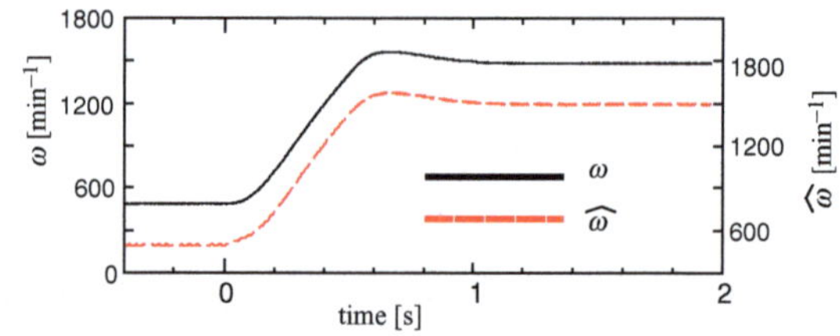

Figure 6. Experimental results using speed sensor.

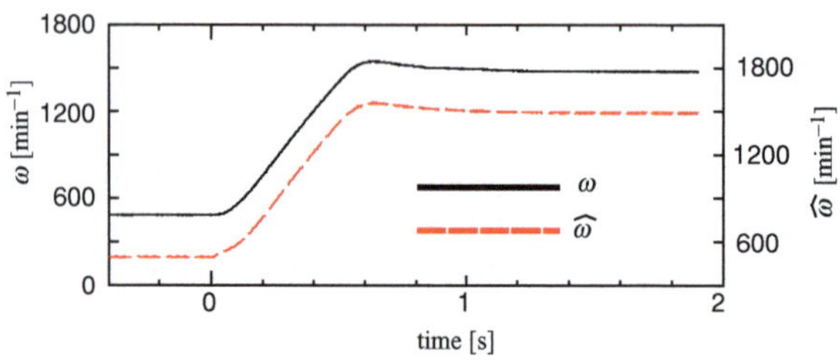

Figure 7. Experimental results using proposed method.

speed sensor. Fig. 7 shows the results using the proposed modified flux linkage observer when the same condition.

From this figure, the modified flux linkage observer gives almost the same results as them measured by a sensor. As a result our proposed methods are effective to the sensorless control of Syn.RM.

Fig. 8 illustrates the performance in the steady state. The measured and estimated rotor position of the middle speed and high-speed operation are shown. From the figure, the accuracy estimation results are obtained even if the high-speed region.

Fig. 9 show the output of the estimation error correct function K_y. From this figure K_y is decreased when the effective current i_{rms} is increased. And K_y is also the function of the motor speed.

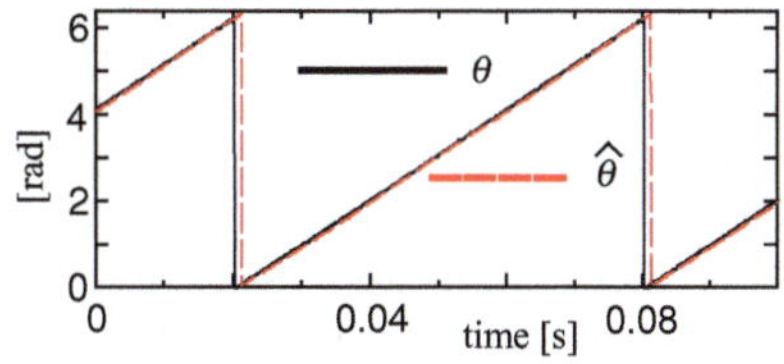

(a) ω = 500 min^{-1}

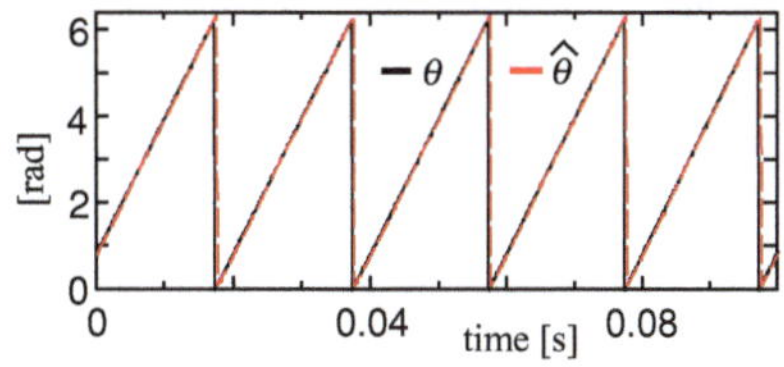

(b) ω = 1500 min^{-1}

Figure 8. Measured and estimated rotor positions in the steady state.

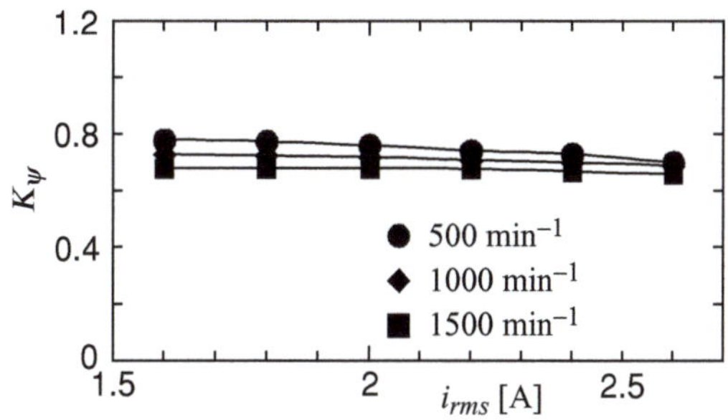

Figure 9. i_{rms} vs. K_Y.

Conclusion

In this paper, we proposed the modified flux observer with the estimation error correct function for the sensorless control method of Syn.RM. The validity of the proposed method is verified by experiments.

References

[1] T. Tamamura, Y. Honda, S. Morimoto, Y. Takeda, "Synchronous Reluctance Motor When Used Air-Condition Compressor Motor: A Comparative Study", IPEC-Tokyo 2000, 2000, pp. 654–659.

[2] T. Senju, T. Shingaki, K. Uezato, Sensorless vector control of synchronous reluctance motor with disturbance torque observer, IEEE Trans. Ind. Electron., Vol. 48, No. 2, pp. 402–407, 2001.

[3] S. Shinnaka, Mirror-phase characteristics of synchronous reluctance motor and salient-pole orientation methods for sensorless vector control, Trans. IEE Japan, Vol. 121-D, No. 2, pp. 210–218, 2001 (in Japanese).

[4] S. Saha, T. Iijima, K. Narazaki, Y. Honda, "High Speed Sensorless Control of Synchronous Reluctance Motor by Modulating the Flux-Linkage Angle", IPEC-Tokyo 2000, 2000, pp. 643–648.

[5] T. Hanamoto, T. Tsuji, Y. Tanaka, "Sensorless Speed Control of Cylindrical Type PMSM Using Modified Flux Observer", IPEC-Tokyo 2000, 2000, pp. 2104–2108.
[6] K. Yamazaki, S.P. Kommareddy, J. Liu, "Durable PC-Based Real-Time Control System for Servomotor Control in Windows NT Environment", IPEC-Tokyo 2000, 2000, pp. 355–360.
[7] T. Hanamoto, H. Ikeda, T. Tsuji, Y. Tanaka, "Sensorless Speed Control of Synchronous Reluctance Motor Using RTLinux", PCC-Osaka 2002, Vol. 2, pp. 699–703, 2002.
[8] http://www.rtlinux.org.
[9] http://rtic-lab.sourceforge.net/.

II-3. A NOVEL SENSORLESS ROTOR-FLUX-ORIENTED CONTROL SCHEME WITH THERMAL AND DEEP-BAR PARAMETER ESTIMATION

Mario J. Duran[1], Jose L. Duran[1], Francisco Perez[1] and Jose Fernandez[2]

[1]University of Málaga, Electrical Engineering Department, Plaza El Ejido S/N 29013 Malaga, Spain
mjduran@uma.es, jlduran@uma.es, fmperez@uma.es
[2]University of Jaen, Electrical Engineering Department, Alfonso X, 28, 23700 Linares (Jaén), Spain
jmoreno@ujaen.es

Abstract. In this paper a novel scheme for vector control is presented that aims to improve some of the weaknesses of the sensorless vector control. Among indirect rotor-flux-oriented control (IRFOC), some of the aspects that can be improved are the low speed behavior, current control, and parameter detuning. The present scheme includes temperature estimation to correct the deviation in steady state, a new control scheme with skin effect estimation to improve the transient accuracy, and an overcurrent protection to be able to have a control on the stator current while allowing a good performance. The proposed scheme is designed from the Matlab/Simulink environment and is experimentally tested using a 1 kW induction motor and a TMS320C31 DSP proving its validity and usefulness.

Introduction

Sensorless vector control is a mature technology whose origins go back to the early seventies[1]. However, many high performance induction motor drives are still being proposed since some problems are still not solved. Sensorless operation mode has attracted much attention and two main approaches can be considered: those based on the field orientation principles to carry out the control FOC [1] and the direct torque control DTC [2] which is inherently a sensorless method. Both have their own weaknesses and a lot of research work has been done trying to solve them.

Among the rotor field oriented schemes, the indirect approach is the most popular, but still presents problems concerning parameter detuning [3] and low speed performance [4]. In these sensorless schemes, it is necessary to estimate the speed since no encoders are used, and this can be carried out directly from the motor model, or using other approaches such as MRAS [5] or Kalman filter methods [6].

In this work the speed is estimated from the motor model but including the mechanical equation, which allows including new phenomena such as the static friction. On the other

S. Wiak, M. Dems, K. Komęza (eds.), Recent Developments of Electrical Drives, 165–176.

hand, parameter estimation is carried out to account for thermal and skin effects. Models for both effects are lumped parameter and simple enough to be included in the real-time application. Skin effect model must be calculated in every step while the thermal model could be implemented with a greater step size since the thermal constant time is higher. Parameters for both models are calculated experimentally, in the case of the motor heating, and considering the analytical case for the deep-bar effect [8].

Additionally, the problem of current control is considered proposing an overcurrent protection that allows transient currents above the rated value, which improves the drive performance.

Control scheme

As previously stated, speed estimation is required to carry out the control, and this can be done by using three from the four well-known RFOC equations [7], building a speed estimator from one of the rotor electrical equations. Nevertheless, in this paper a different proposal is made including the dynamical equation (1) into the estimator together with the four RFOC equations, obtaining a set of equations:

$$
\begin{aligned}
&T_s' p i_d + i_d = \frac{u_d}{R_s} + \omega_{mr} T_s' i_q - (T_s - T_s') p |i_{mr}| \\
&T_s' p i_q + i_q = \frac{u_q}{R_s} - \omega_{mr} T_s' i_d - (T_s - T_s') |i_{mr}| \\
&T_r p |i_{mr}| + |i_{mr}| = i_d \\
&\omega_{mr} = \omega + \frac{i_q}{T_r |i_{mr}|} \\
&\frac{P}{2} \frac{L_m^2}{L_r} i_q |i_{mr}| = T_m + J p \omega \frac{2}{P} + \alpha_f \omega \frac{2}{P}
\end{aligned}
\tag{1}
$$

Instead of obtaining the speed from an electrical equation, it is obtained from the mechanical one, including new parameters as the inertia or friction coefficient. It allows to include a variable friction that takes into account the static friction when the movement starts.

From the set of equations (1) the motor speed can be calculated using an estimator whose inputs are the voltage components and the load torque. Since these are the real inputs of an induction machine, the estimator is further called *simulated motor*.

Voltage components can be measured or reconstructed from the stator equations, but torque must be estimated since it is not a measured variable (Fig. 1).

In order to estimate the load torque, and adaptive scheme is adopted considering that the motor torque is proportional to i_q. This component can be obtained from the simulated motor as an estimated value and can also be measured. For the measurements of the currents a digital filter is used before considering the transformation matrix into the *dq* values.

The difference between them is due to the fact that the information of the load torque that the estimator is using is not correct, and so a controller can be used to update this torque value. In Fig. 2 the complete control scheme is shown.

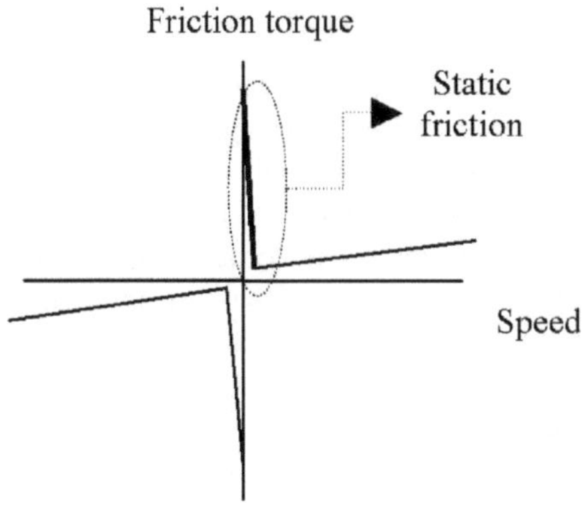

Figure 1. Evolution of the friction torque.

The speed and flux target values are compared with the estimated ones obtaining target currents that are transformed into target voltages by reconstruction from the stator equations. These voltages are, together with the estimated torque, the inputs for the simulated motor and also the outputs to generate the PWM for the VSI inverter.

Three controllers are involved in the control as it is usual in this kind of vector control, one for the flux comparison obtaining the direct component i_d and two for the speed and torque comparison obtaining the quadrature component i_q.

The rotor flux reference decreases in inverse proportion to the speed of rotation in the field-weakening region, while it is constant and equal to rated rotor flux in the base speed region.

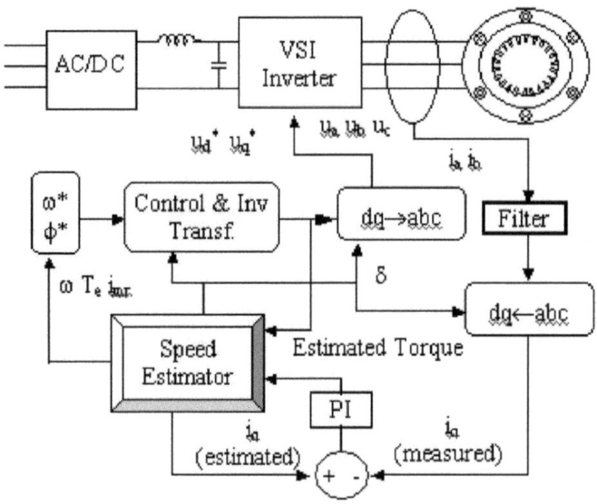

Figure 2. Control scheme using the simulated motor as a speed estimator with adaptive load torque estimation.

Overcurrent protection

Vector control provides high performance to drives, but to achieve a good transient response a high electrical torque is required, and it means high currents flowing in the machine. Apart from the achievement of decoupled and optimal control, it is also necessary to protect the motor against overcurrents. The control will need high currents to provide torque, and this can be necessary for high accelerations or load torque. In Fig. 3, it is shown the stator current evolution reference speed is increased in a rampwise manner with high accelerations, and it is clear that in these transient, currents over the rated value are required. The inverter used is VSI type and so to carry out a proper control, target voltages are supplying the motor, but without any current control. For this reason, it is convenient to include a current protection in the software design. To have control over the currents in field oriented control is relatively easy compared with other schemes such as direct torque control (DTC), since it is only necessary to control the quadrature component of the stator space vector i_q. The obvious solution is just to saturate this component in the control scheme. However, the aim here is to design a nonconservative protection that allows transient currents above the rated one.

Traditionally, for steady-state operation, manufacturers provide the maximum time for a certain value over the rated current so that the motor is not damaged. For a vector control application the motor works in transient state, but a protection based on energy considerations can also be used.

The method proposed is to use an energy counter that starts to rise then the current is over the nominal value by integrating this current. When the energy counter is over a certain energy threshold, then the protection acts limiting i_q to its nominal value.

The method proposed is to use an energy counter that starts to rise then the current is over the nominal value by integrating this current. When the energy counter is over a certain energy threshold, then the protection acts limiting i_q to its rated value.

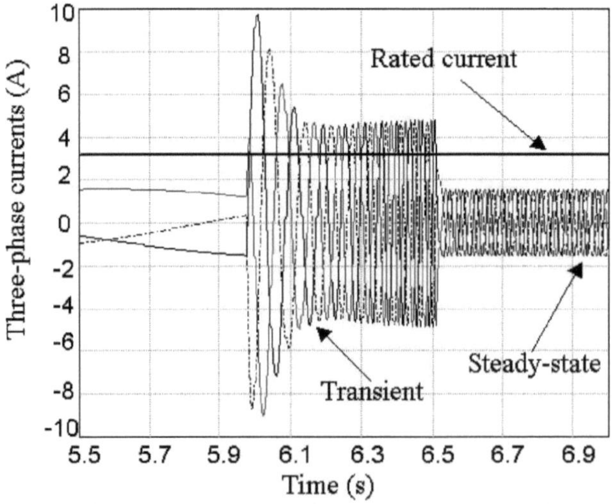

Figure 3. Three-phase currents during acceleration transient.

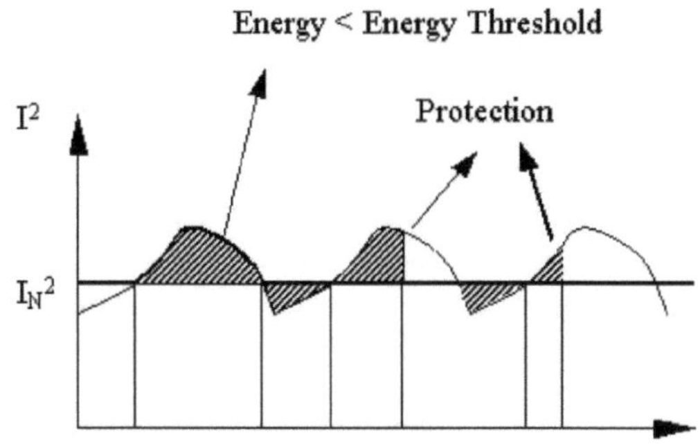

Figure 4. Proposed protection.

Some considerations have to be made in order to make the system work: the integration for the energy counter must be limited and if the current is below the nominal value the integration must continue with a negative value until the energy counter is set to zero. This makes that, if a repetitive cycle occurs (Fig. 4), the protection finally acts even if the energy of the each cycle is below the limit.

Parameter estimation

In the proposed control scheme a speed estimator was built from the motor model equations, and some parameters were involved in this estimation. Since this parameters change with the operation conditions, the problem of parameter detuning common to all vector control remains the same. In order to overcome this problem, two of the main influencing factors are considered: thermal and deep-bar effects.

In a previous paper [7] a thermal model is developed that takes electrical RFOC variables as inputs and provides stator and rotor representative temperatures. The model is simple enough not to be time consuming, but proves to be very accurate.

Considering the copper losses, hysteretic, and eddy current losses and taking into account only the stator, rotor, and environment, representative temperatures of the stator and rotor can be obtained making thermal balance.

$$\begin{aligned} R_s i_s^2 + k_{Hs}\omega_s + k_{Fs}\omega_s^2 &= G_s\theta_s + C_s\frac{d\theta_s}{dt} + G_{sr}(\theta_s - \theta_r) \\ R_r i_r^2 + k_{Hr}\omega_r + k_{Fr}\omega_r^2 &= G_r\theta_r + C_r\frac{d\theta_r}{dt} + G_{sr}(\theta_r - \theta_s) \end{aligned} \tag{2}$$

Both conduction and convection are considered in the thermal conductances.

$$G = G_0(1 + b \cdot \omega) \tag{3}$$

Model parameters are obtained from three tests: blocked shaft, DC, and AC tests. For the conductances and convection coefficients it is only necessary to consider steady-state values, while for the capacitances the thermal transient must be taken into account.

Table 1. Results for the thermal tests

Tests and simulations			Test temperature		Simulation temperature	
Variable		f (Hz)	θ_s	θ_r	θ_s	θ_r
DC	1	0	61.9	58.6	62.2	58.2
	2	0	79.4	74.0	79.8	74.3
AC Blocked shaft	1	4	51.1	63.4	50.0	60
	2	5.5	71.6	91.0	70.8	88
AC	1	0	41.25	50.8	41.4	50.8
	2	40	43.66	58.3	43.0	57.0

The results for the different test carried out to obtain the different parameters can be summarized in Table 1 showing the experimental and simulation steady-state results.

To account for the deep-bar effect, FEM solutions are not possible for a real-time application, and both analytical [9] and lumped parameters [10] of previous solutions are only valid for rectangular rotor bars. In the present work the classical analytical solution [9] is generalized starting from the same wave equation but changing the contour condition in the upper part of the bar so that there is a contribution of the sides when using Ampère's law.

The same occurs when the Poynting's flux is considered, and so there is also flux through the sides, and not only through the upper part of the bar, what is considered in the equations:

$$P + jQ = EH'(0)b(0) + 2\int_0^h EH'(z)b(z)\,dz = I_r^2 R_r + j\frac{I_r^2 L_r}{2} \tag{4}$$

All in all, a general analytical solution is presented whose main weakness is to be time-consuming due to hyperbolic functions.

Because of that an approximate solution is considered that starts from a lumped parameter π equivalent circuit (Fig. 5), and calculates the values of the different parameters by comparing the results with the previous analytical solution and minimizing the error shown in (5) using a Nelder-Mead direct search.

$$E = (1/f^{cr}) \cdot \left[R_r^{an} - R_r^{calc}\right]^2 + p_{RL}(1/f^{cl}) \cdot \left[X_r^{an} - X_r^{calc}\right]^2 \tag{5}$$

being *cr* and *cl* weight coefficients that improve the solution performance at low or high frequencies, P_{rl} a coefficient that allows a better adjustment of resistance or inductance.

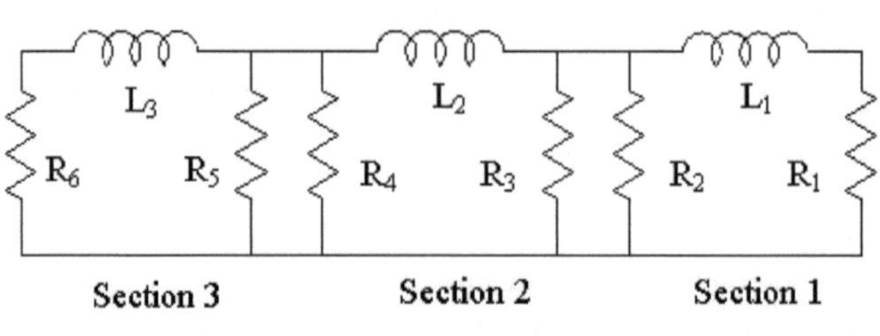

Figure 5. Lumped parameter π rotor equivalent circuit.

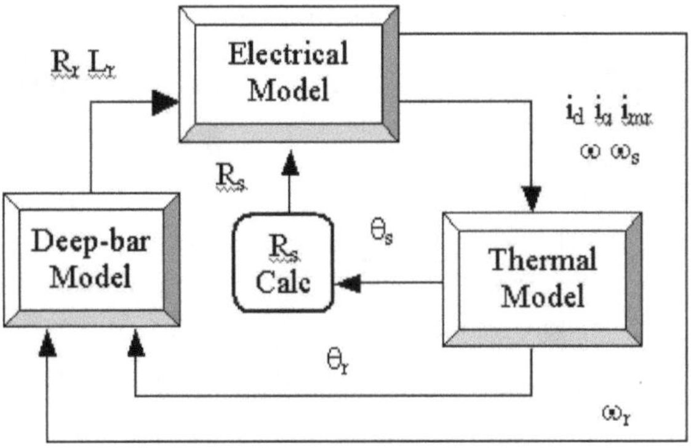

Figure 6. Global estimation scheme.

Considering a circuit with three sections, six resistances and three inductances are needed. Weight values of $cr = 2$, $cl = 2.5$, and $P_{rl} = 100$ have been chosen so that the estimation of both parameters is compensated and for a best adjustment at low frequencies, obtaining the following parameters:

$R_1 = 0.92\ \Omega/\text{cm}$, $R_2 = 0.029\ \Omega/\text{cm}$, and $R_3 = 0.11\ \Omega/\text{cm}$
$R_4 = 0.062\ \Omega/\text{cm}$, $R_5 = 0.016\ \Omega/\text{cm}$, and $R_6 = 0.123\ \Omega/\text{cm}$
$L_1 = 6.65\ \mu\text{H/cm}$, $L_2 = 27.4\ \mu\text{H/cm}$, and $L_3 = 5.1\ \mu\text{H/cm}$

Including both the deep-bar effect and thermal model into the speed estimator, the scheme shown in Fig. 6 is finally obtained.

The stator resistance is updated considering just the stator representative temperature, since the skin effect is neglected in the stator. The rotor resistance is updated thanks to the skin effect model, which already takes into account the rotor temperature changes since one of its inputs is the rotor representative temperature.

It must be noticed that the motor heating influences the deep-bar model due to the electrical conductivity variation but the motor temperature is not practically affected by the additional losses caused by the skin effect.

Experimental rig

In the experimental rig, there are a 1 kW induction motor (AEG eAM 90SY 4Ex), a Semikron Skiip with integrated rectifier and VSI inverter, a dynamo and a bank resistor for load tests, and a digital signal processor (DSP) (TMS320C31) main control board (Fig. 7). The control design is carried out in Simulink (Matlab) and compiled to be executed in the DSP.

For the acquisition data, two types can be considered: the control data and the verification data. Control data are the currents necessary to estimate the motor speed, which need PCBs specifically designed with Hall effect transducers to obtain proper voltages that can be introduced in the DSP thanks to 16-bit A/D converter. Moreover, the speed is also measured in order to have a verification tool, and so it is vital to carry out the measurement with high precision. For this purpose, a 1024 CPR encoder is used and the TTL pulses are filtered and counted into the DSP obtaining the shaft position.

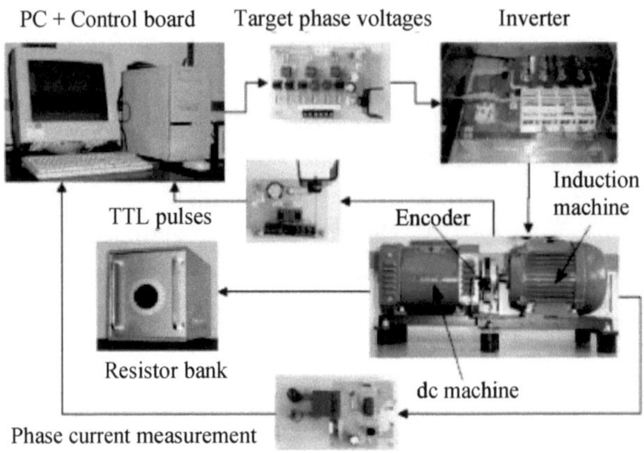

Figure 7. Experimental rig.

In order to obtain the real speed, a simple algorithm is design that allows selecting the integration step and an accurate speed measurement is achieved. Since the real speed is introduce into the Simulink design, it is immediate to compare the real speed with the reference and so the control can be rigorously tested.

Experimental results

In the experimental results, tests are carried out to see the performance in steady state, transient state, and overcurrent states. In this way the different aspects of the proposed scheme can be tested.

Concerning the proposed protection two load tests were carried out, one limiting the value of i_q to its rated value and the other with the designed protection (Fig. 8). Setting a value of the energy threshold, and applying a load torque that makes the motor consumes a current over the rated one, it can be observed that the proposed protection permits an overcurrent for a certain time, allowing the control to achieve the reference speed (1,000 rpm) in the first seconds. When the energy counter reaches the energy threshold, the protection acts and the speed falls to a value so that the current is the rated one.

The performance is improved in this way since the target values can be followed even when certain transients overcurrents are required.

In order to verify the steady-state behavior of the proposed system a constant load test was carried out during time enough to allow the motor heating. The test is carried out considering a target speed of 1,000 rpm and a load torque of approximately half the rated value, and it is made with and without the inclusion of the thermal model in the control scheme. Both tests are performed in the same conditions without changing the value of the resistance in the resistor bank. The results of both tests are shown in Fig. 9, where the evolution of the measured motor speed is displayed. The vertical lines are due to the reset of the incremental encoder position counter.

If the thermal effect is neglected and constant parameters are considered, there is a deviation due to the effect of the motor heating. On the contrary, if the thermal effect is

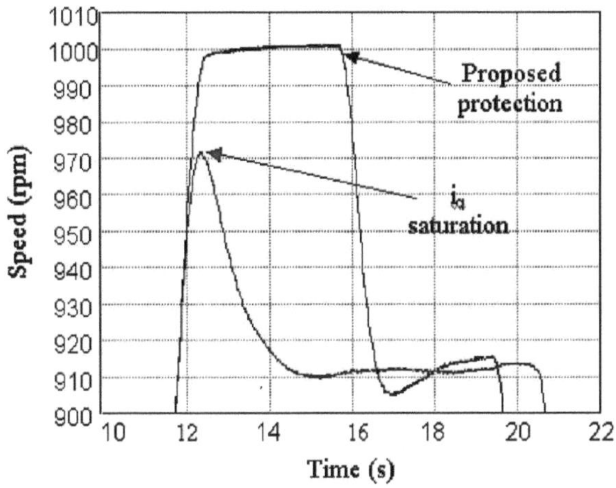

Figure 8. Protection test.

considered, parameters are updated properly as the motor temperature increases and the target speed is followed without deviation.

A requirement of the dynamical behavior is the response when sudden changes in the target speed occur. In order to be demanding with the control features, it is consider a test with a target speed going from 0 to 1,000 rpm in 1.3 s. Fig. 10 shows that even with this acceleration, the motor follows the target speed just with some oscillations in the starting (7.2 s) and braking (8.5 s). Carrying out the same test without considering the deep-bar effect leads to higher oscillations and poorer transient response and not considering the static friction in the mechanical equation also makes the motor oscillate more in the starting.

Vector control is used because of the good dynamic performance, and a normal test is also to apply a sharp load torque to verify the system response. In this case during a few seconds a nominal load torque was applied and in 28th second, approximately, the load torque was

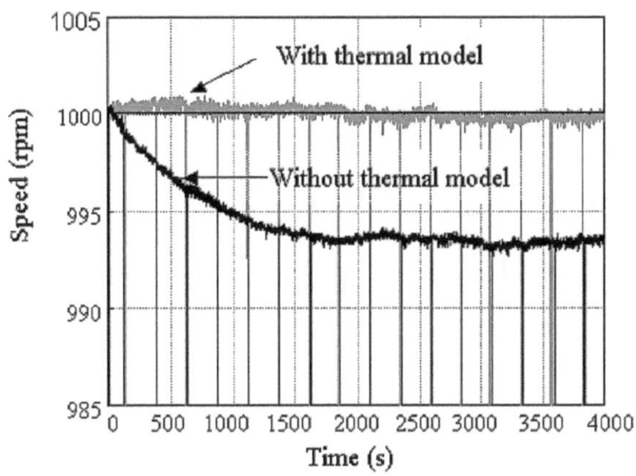

Figure 9. Constant load test.

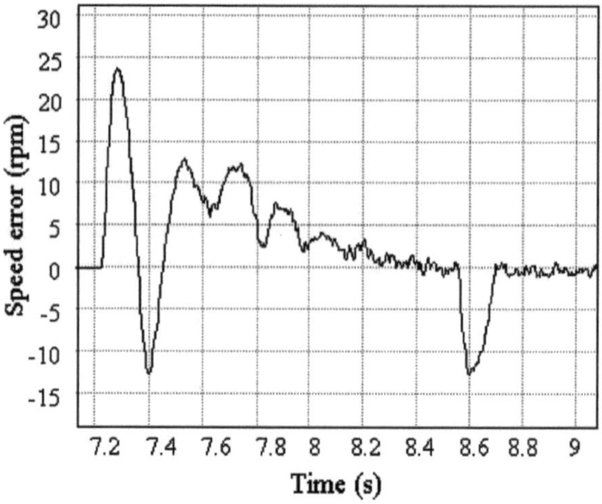

Figure 10. Speed test.

released and so the motor is instantaneously accelerated (Fig. 11). The maximum error in this test is 16 rpm and motor speed goes quickly to the target value (1,000 rpm). In the same test without the deep-bar effect model, this error was 20 rpm, what points out the relevance of including parameter variation due to this effect.

In Fig. 12 the estimated and measured quadrature current are shown. The evolution is similar, so that the correct information about the load torque is being provided to the speed estimator. The rapid change in the quadrature current when changes in the load torque occur is the clue to obtain a quick response of the control.

Until the 28th second the torque is gradually being increased and this information is introduced in the controller thank to the measured quadrature current. When the torque is released the quadrature component is also suddenly changed by the control following in this way the target speed.

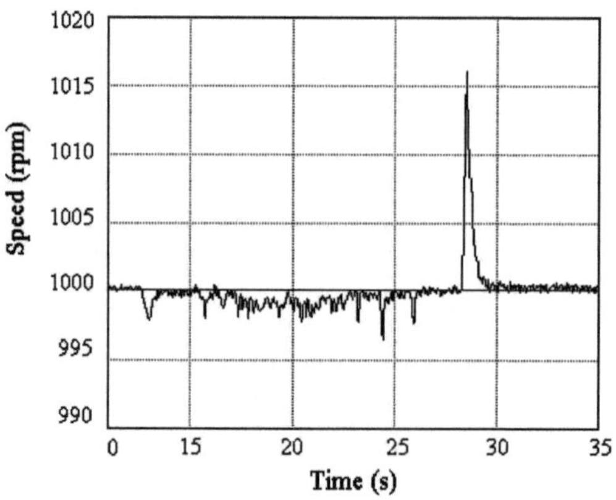

Figure 11. Load test.

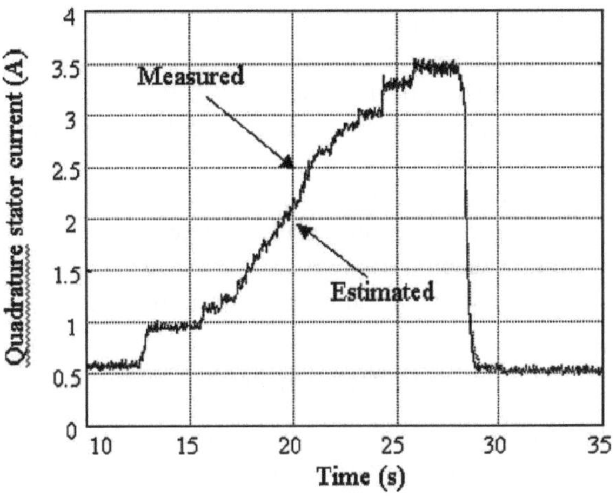

Figure 12. Evolution quadrature current in load test.

Conclusions

The scheme that has been proposed improves the performance of the drive in several ways. On the one hand the thermal state estimation can correct the steady-state deviation in the motor speed that otherwise is produced when the motor is heated and parameter detuning occur. Another improvement of the scheme is the inclusion of the skin effect estimation and the consideration of the static friction that allow to obtain a good performance both against speed reference or load torque changes. Since apart from the control characteristics it is necessary to avoid overcurrents, the proposed protection proves to permit transient currents over the rated value improving the drive performance. Not saturating directly the current helps the motor to reach the target values even with high transients torque required.

All the improvements have been tested experimentally and with high accuracy measurements, validating the effectiveness of the proposed solution.

List of symbols

abc	Three-phase values
b_s, b_r, b_{sr}	Stator-environment, rotor-environment, and stator-rotor convection coefficients
C_s, C_r	Stator and rotor thermal capacitances
dq	Field oriented values
G_s, G_r, G_{sr}	Stator-environment, rotor-environment, and stator-rotor thermal conductances
i_d, i_q	Direct and quadrature components of stator current space vector
i_{mr}, ω_{mr}	Modulus and angular speed of rotor magnetizing current space vector
J	Inertia moment
k_H, k_F	Hysteresis and eddy current coefficients
L_r, L_m	Rotor self-inductance and magnetizing inductance
P	Number of poles

p	Derivative operator
R_s, R_r	Stator and rotor resistance
T_m, T_e	Load and electrical torque
T_s, T_r, T_r'	Stator and rotor time constants and stator transient time constant, respectively
α_f	Friction coefficient
θ_s, θ_r	Stator and rotor representative temperature
ω	Motor speed
ω_s, ω_r	Angular speed of stator and rotor currents

References

[1] F. Blashke, The principle of field-orientation as applied to the new transvector closed-loop control system for rotating field machines, Siemens Rev., Vol. 34, No. 5, pp. 217–220, 1972.

[2] T. Naguchi, I. Takahashi, A new quick-response and high-efficiency control strategy of an induction motor, IEEE Trans. Ind. Appl., Vol. IA-22, pp. 820–827, 1986.

[3] E.Y.Y. Ho, P.C. Sen, Decoupling control of induction motor drives, IEEE Trans. Ind. Electron., Vol. 35, pp. 253–262, 1998.

[4] J. Holtz, J. Quan, Sensorless vector control of induction motors at very low speed using a nonlinear inverter model and parameter identification, IEEE Trans. Ind. Appl., Vol. 38, No. 4, 2002.

[5] M. Wang, E. Levi, Evaluation of steady-state and transient behaviour of a MRAS based sensorless rotor flux oriented induction machine in the presence of parameter detuning, Elect. Mach. Power Syst., Vol. 27, No. 11, pp. 1171–1190, 1999.

[6] M.N. Marwali, A. Keyhani, "A Comparative Study of Rotor Flux Based MRAS and Back EMF Based MRAS Speed Estimators for Speed Sensorless Vector Control of Induction Machines", Proc. IEEE Ind. Appl. Soc. Annu. Meet. IAS'97, New Orleans, LA, 1997, pp. 160–166.

[7] J. Fernández Moreno, F. Pérez Hidalgo, M.J. Durán Martínez, Realization of tests to determine the parameters of the thermal model of induction machine, IEE Proc. Electr. Power Appl., Vol. 148, pp. 392–397, 2001

[8] M.J. Durán, J.L. Durán, F. Pérez, J. Fernández, "Improved Sensorless Induction Machine Vector Control with On-line Parameter Estimation Taking into Account Deep-Bar and Thermal Effects", 28th Annual Conference of the IEEE Ind. Electron. Soc. IECON, Sevilla, 2002.

[9] P.L. Alger, Induction Machines, Gordon and Breach Science Publishers, New York, 2nd edition, 1970.

[10] W. Levy, C.F. Landy, M.D. McCulloch, Improved models for the simulation of deep bar induction motors, IEEE Trans. Energy Convers., Vol. EC-5, No. 2, pp. 393–400, 1990.

II-4. WIDE-SPEED OPERATION OF DIRECT TORQUE-CONTROLLED INTERIOR PERMANENT-MAGNET SYNCHRONOUS MOTORS

Adina Muntean[1], M.M. Radulescu[1] and A. Miraoui[2]

[1]Small Electric Motors and Electric Traction (SEMET) Group, Technical University of Cluj-Napoca, P.O. Box 45, RO-400110 Cluj-Napoca 1, Romania
adina.muntean@mae.utcluj.ro, mircea.radulescu@mae.utcluj.ro

[2]Laboratory of Electronics, Electrotechnics and Systems (L2ES), University of Technology of Belfort-Montbéliard, rue Thierry-Mieg, F-90010 Belfort, France
abdellatif.miraoui@utbm.fr

Abstract. In this paper, an integrated design and direct torque control (DTC) of inverter-fed interior permanent-magnet synchronous motors (IPMSMs) for wide-speed operation with high torque capability is presented. The double-layer IPM-rotor design is accounted for IPMSMs requiring a wide torque-speed envelope. A novel approach for the generation of the reference stator flux-linkage magnitude is developed in the proposed IPMSM DTC scheme to insure extended torque-speed envelope with maximum-torque-per-stator-current operation range below the base speed as well as constant-power flux-weakening and maximum-torque-per-stator-flux operation regions above the base speed. Simulation results to show the effectiveness of the proposed DTC scheme are provided and discussed.

Introduction

Due to their many positive features, including high torque-to-inertia and power-to-weight ratios, fast dynamics, compact design, and low maintenance, inverter-fed interior permanent-magnet synchronous motors (IPMSMs) are viable contenders for industrial drives with high torque capability over a wide-speed range. Indeed, PMs being completely embedded inside the steel rotor core, a mechanically robust construction of IPMSMs allowing wide speed-torque envelope is primarily obtained. Secondly, the rotor-buried PMs, covered by steel pole-pieces, significantly change the magnetic circuit of the motor, since, on the one hand, the PM cavities create flux barriers within the rotor, thus reducing the permeance in a flux direction that crosses these cavities, and, on the other hand, high-permeance paths are created for the flux across the steel rotor-poles and also in space-quadrature to the rotor-PM flux; this establishes the rotor magnetic saliency. Hence, it is a hybrid torque production mechanism in IPMSMs, because in addition to the magnet (or field-alignment) torque due to the interaction of rotor-PM flux and the armature (stator) mmf, there is also a reluctance torque component due to rotor magnetic saliency. Thirdly, IPMSM having a small effective

S. Wiak, M. Dems, K. Komęza (eds.), Recent Developments of Electrical Drives, 177–186.

airgap, the armature reaction is quite important, and can be conveniently used for airgap flux-weakening in order to extend the motor torque capability toward high speeds.

Several current vector control schemes were earlier proposed for wide-speed range control of IPMSMs, the motor torque being indirectly controlled via subordinated stator-current loops [1–3]. All these control schemes are based on steady-state motor characteristics, whereas the IPMSM dynamic behaviour is implicitly solved by the current controller.

Recently, the direct torque control (DTC) has been proposed for high-performance wide-speed operation of IPMSMs [4–7]. In principle, the IPMSM DTC involves the direct and independent control of the stator flux-linkages and the electromagnetic torque by selecting proper voltage switching vectors of the voltage-source inverter (VSI) supplying the motor. This selection is made to restrict the differences between the references of stator flux-linkage magnitude and electromagnetic torque and their actual (estimated) values. The advantages of the IPMSM DTC over conventional current control schemes include the elimination of current controller, coordinate transformation, and PWM signal generator, the lesser dependence on motor parameters as well as the fast torque response in steady-state and transient operating conditions.

In this paper, an integrated design and DTC of VSI-fed IPMSMs for wide-speed operation with high torque capability is presented. Hence, the paper is organized as follows. In "IPMSM Design for Wide-Speed Operation," the double-layer IPM-rotor design is adopted for IPMSMs requiring a wide torque-speed envelope. In "DTC of VSI-fed IPMSM for Wide-Speed Operation," an IPMSM DTC scheme incorporating both the optimized constant-torque and flux-weakening controllers for wide-speed range operation is developed. Simulation results to validate the proposed IPMSM DTC scheme are presented and discussed in "Simulation Results." Conclusions are drawn in section "Conclusions."

IPMSM design for wide-speed operation

The stator of the considered VSI-fed IPMSM is a typical AC design accommodating a three-phase distributed winding in slots to produce the synchronously-rotating, quasi-sinusoidal armature-mmf wave. Conversely, the IPMSM rotor can be designed in different configurations. However, only two of them with radially-magnetized buried-type IPMs have been accounted as being advantageous for wide-speed operation [8–10]. The high-energy rotor-PMs usually consist of sintered-NdFeB blocks inserted after magnetization into the rotor cavities.

Fig. 1 shows the cross-sectional configurations of both IPMSMs in conjunction with their rated-load magnetic flux distribution obtained from finite element analysis. The first IPMSM rotor topology has only one (single-layer) PM per rotor-pole, whereas in the second one, each rotor-PM is splitted up in two layers with iron separation in the radial direction of the rotor core.

The well-known coordinate system (d,q) bounded to the rotor (i.e. rotating at synchronous speed ω_r) is defined hereafter with the d-axis aligned with the stator PM flux-linkage vector $\underline{\psi}_{s0} = \underline{\psi}_{PM}$ and the orthogonal q-axis aligned with the back-emf vector $\omega_r \underline{\psi}_{PM}$ (Fig. 2). By noticing that the (total) stator flux-linkage vector can be splitted into the flux-linkage (with the stator winding) due to the excitation rotor-PMs, $\underline{\psi}_{PM,}$ and the armature-reaction flux, which entails the self-inductances L_{sd} and L_{sq} $(L_{sd} \leq L_{sq})$ of the

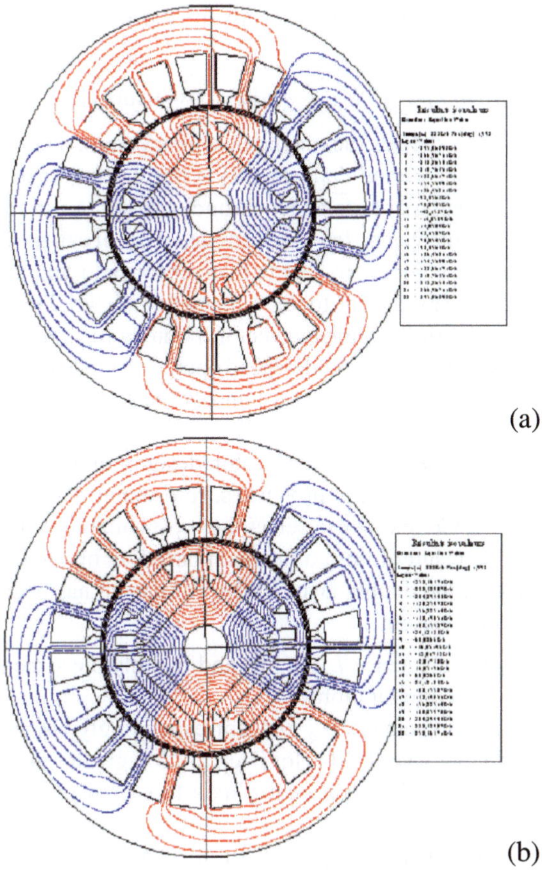

Figure 1. IPMSM cross-sectional design and magnetic flux distribution under rated-load condition for (a) single- and (b) double-layer IPM-rotor topology, respectively.

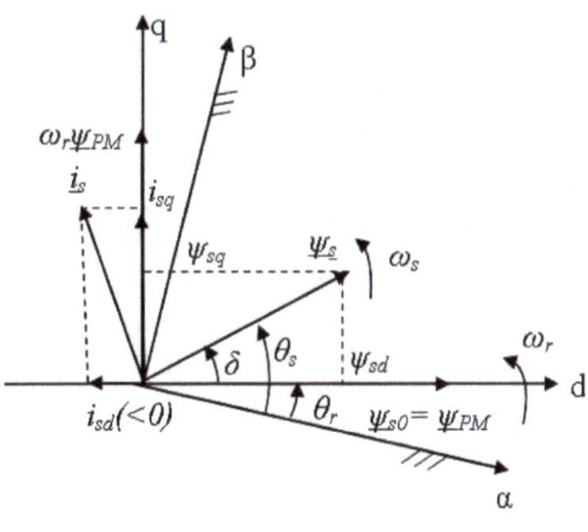

Figure 2. Different coordinate systems for vector representation of IPMSM quantities.

stator winding along the d- and q-axis, respectively, the IPMSM electromagnetic torque may be expressed as [5,6]

$$m_e = (3p/2)|\underline{\psi}_s|(|\underline{\psi}_{PM}| - \xi|\underline{\psi}_s|cos\delta)\ \sin\delta/L_{sq}(1-\xi) \tag{1}$$

where $\xi = (L_{sq} - L_{sd})/L_{sq}$ defines the magnetic saliency ratio, and δ represents the angle between flux-linkage vectors $\underline{\psi}_s$ and $\underline{\psi}_{PM}$ (Fig. 2); δ is constant for steady-state operation, hence both $\underline{\psi}_s$ and $\underline{\psi}_{PM}$ vectors rotate at synchronous speed ω_r; in transient operation, δ varies, hence $\underline{\psi}_s$ and $\underline{\psi}_{PM}$ rotate at different speeds, $\omega_s \neq \omega_r$. It can be identified in equation (1) the first IPMSM torque component, as the magnet (or field-alignment) torque and the second one, as the reluctance torque due to rotor magnetic saliency. From equation (1), it also results that, for a certain stator flux-linkage vector modulus $|\underline{\psi}_s|$, the IPMSM rotor design achieving high torque capability over a wide-speed operation range requires increased values of the rotor-PM linkage flux magnitude $|\underline{\psi}_{PM}|$ and of the stator self-inductance difference $L_{sq} - L_{sd}$.

A comparison between the two IPMSM rotor designs of Fig. 1, for constant rotor-PM volume and for identical magnetic properties, rotor outer diameter, airgap, and stator specifications, has been made in order to select the most suitable structure for high-torque wide-speed operation. As result of this comparison based on finite-element magnetic field analysis of both IPMSMs, the double-layer IPM-rotor design has been adopted for motor prototype by the following reasons.

1. The d-axis stator self-inductance L_{sd} is low and roughly the same for both single- and double-layer PM-rotor configurations.
2. The q-axis stator self-inductance L_{sq} and, correspondingly, the inductance difference $L_{sq} - L_{sd}$ for the double-layer IPM rotor is up to 20% greater than for the single-layer IPM rotor, mainly due to the additional q-axis flux path provided between the two rotor-PM layers.
3. The q-axis stator self-inductance L_{sq} for the rotor topology with only one PM per pole decreases greatly with the stator-current rising, because of the magnetic saturation, whereas for the double-layer PM-rotor topology this effect is less significant.
4. The stator flux-linkage due to the double-layer of rotor-PMs is about 10% greater than in the case of single-layer IPM rotor.
5. The electromagnetic torque developed up to the rated rotor speed by the double-layer IPMSM is about 10% increased in comparison with that produced by a single-layer IPMSM, for the same armature mmf. However, the torque performances using flux-weakening at high speeds for both IPMSMs are quite similar.

DTC of VSI-fed IPMSM for wide-speed operation

In the DTC scheme for VSI-fed IPMSM, the inner torque controller is based on the expression of the electromagnetic torque given by equation (1). Hence, torque is controlled by regulating (through inverter voltages) the amplitude $|\underline{\psi}_s|$ and the angle δ of the stator flux-linkage vector.

The d- and q-axis stator flux-linkages are

$$\psi_{sd} = L_{sd}i_{sd} + |\underline{\psi}_{PM}| \tag{2}$$

$$\psi_{sq} = L_{sq}i_{sq} \tag{3}$$

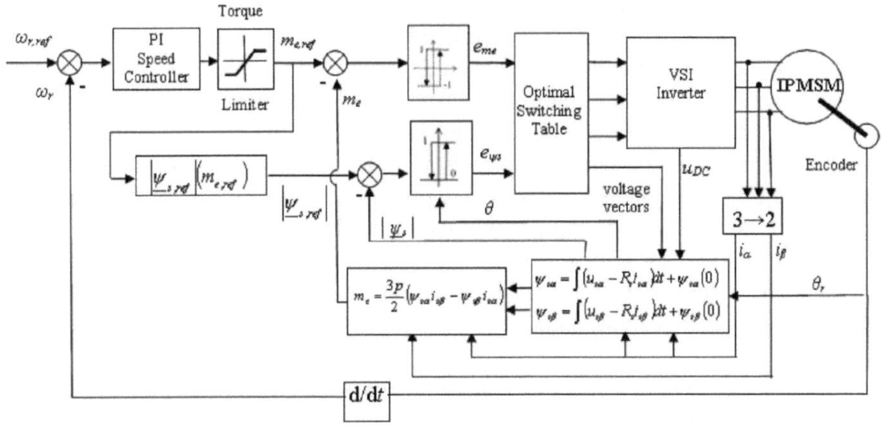

Figure 3. Block diagram of the VSI-fed IPMSM DTC scheme for wide-speed operation.

From equations (2) and (3), the stator flux-linkage vector modulus can be expressed as

$$|\underline{\psi}_s| = (\psi_{sd}^2 + \psi_{sq}^2)^{1/2} = [(L_{sd}i_{sd} + |\underline{\psi}_{PM}|)^2 + (L_{sq}i_{sq})^2]^{1/2} \tag{4}$$

By differentiating equation (1) with respect to time, for constant stator flux-linkage magnitude, one obtains

$$\mathrm{d}m_e/\mathrm{d}t = (3p/2)|\underline{\psi}_s|(|\underline{\psi}_{PM}|\cos\delta - \xi|\underline{\psi}_s|\cos 2\delta)(\mathrm{d}\delta/\mathrm{d}t)/L_{sq}(1-\xi) \tag{5}$$

Equation (5) emphasizes that the electromagnetic torque can be dynamically controlled by means of controlling the rate of change of the angle δ.

There are upper limits of variation for both control quantities, $|\underline{\psi}_s|$ and δ, to achieve stable IPMSM DTC. Firstly, since according to equation (1), $m_e = 0$ for $\delta = 0$, the condition for positive slope $\mathrm{d}m_e/\mathrm{d}\delta$ around $\delta = 0$ leads to

$$|\psi_s| < |\underline{\psi}_{PM}|/\xi \tag{6}$$

Secondly, by differentiating equation (1) with respect to δ and equating it to zero, the maximum allowable angle δ_{lim} can be found as

$$\delta_{lim} = \cos^{-1}\{|\underline{\psi}_{PM}|/4\xi|\underline{\psi}_s| - [(|\underline{\psi}_{PM}|/4\xi|\underline{\psi}_s|)^2 + 1/2]^{1/2} \tag{7}$$

that is

$$\delta \leq \delta_{lim} \tag{8}$$

The block diagram of the proposed DTC scheme for wide-speed operation with high torque capability of a VSI-fed IPMSM is shown in Fig. 3. The three-phase stator variables are transformed to the α,β-axes variables of the (α,β) stationary coordinate system shown in Fig. 2.

The α, β stator currents, obtained from current sensors, and the stator voltages $u_{s\alpha}$ and $u_{s\beta}$, calculated from the measured DC-link voltage, are then used for stator flux-linkage vector and electromagnetic torque estimation. Some methods of compensation for the effect of stator-resistance variation and for the DC offset in the measurements, particularly at low speed, have been recently reported [11,12]. The initial angular position of the stator flux-linkage vector $\underline{\psi}_s$ may be obtained from a low-resolution encoder. Subsequently, this encoder is not needed under the DTC scheme.

Electromagnetic torque and stator flux-linkage magnitude errors, generated by comparison between estimated and reference values, are inputs to the respective flux and torque hysteresis regulators. The discretized outputs of these regulators are inputs to the optimum voltage switching selection table. It is used to properly choose the VSI-fed voltage vectors to regulate the stator flux and torque within their error bands.

In the IPMSM DTC scheme of Fig. 3, the reference electromagnetic torque, $m_{e,ref}$, is obtained as the output of the rotor-speed controller from the outer loop, and is limited at a certain value, which guarantees the stator current not to exceed its maximum admissible value.

In its turn, the reference value of the stator flux-linkage vector modulus, $|\underline{\psi}_{s,ref}|$, is generated in the proposed IPMSM DTC scheme as a function of the electromagnetic torque reference, i.e. $|\underline{\psi}_{s,ref}|(m_{e,ref})$, by maximizing the IPMSM torque over the wide-speed operation range in the presence of current and voltage constraints.

The stator-current limit, $I_{s,lim}$, is an IPMSM thermal rating or a VSI maximum available current. The stator-voltage limit, $U_{s,lim}$, is the VSI maximum available output voltage, depending on the DC-link voltage. Hence, the current and voltage constraints establish the following operating limits for the VSI-fed IPMSM:

$$\underline{i}_s| = (i_{sd}^2 + i_{sq}^2)^{1/2} \le I_{s,lim} \tag{9}$$

$$|\underline{u}_s| = (u_{sd}^2 + u_{sq}^2)^{1/2} \le U_{s,lim} \tag{10}$$

In the speed operation range I from standstill up to the base rotor speed ω_{rb}, current constraint of equation (9) is dominant, preventing the IPMSM overheating, whereas voltage constraint of equation (10) can be met, since the back-emf is rather low. Thus, the required function $|\underline{\psi}_{s,ref\,\mathrm{I}}|(m_{e,ref\,\mathrm{I}})$ for the reference value of the stator flux-linkage magnitude in the speed range I can be obtained by ensuring the IPMSM constant-torque operation in which the maximum torque-to-stator current ratio is achieved at the stator-current limit $I_{s,lim}$, i.e. the motor is accelerated by the maximum available torque below the base speed; it results

$$(i_{sd,\mathrm{I}}^2 + i_{sq,I}^2)^{1/2} = I_{s,lim} \tag{11}$$

$$m_{e,max\mathrm{I}} = (3p/4)|\underline{\psi}_{PM}|i_{sq,\mathrm{I}}\{1 + [1 + (2\xi L_{sq} i_{sq,\mathrm{I}}/|\underline{\psi}_{PM}|)^2]^{1/2}\} \tag{12}$$

For the currents $i_{sd,\mathrm{I}}$ and $i_{sq,\mathrm{I}}$, equation (5) can be written as

$$|\psi_{s,ref\,\mathrm{I}}| = [(L_{sd} i_{sd,\mathrm{I}} + |\underline{\psi}_{PM}|)^2 + (L_{sq} i_{sq,\mathrm{I}})^2]^{1/2} \tag{13}$$

Considering in equation (12) $m_{e,max\,\mathrm{I}} = m_{e,ref\,\mathrm{I}}$, solving equations (11) and (12) for the currents $i_{sd,\mathrm{I}}$ and $i_{sq,\mathrm{I}}$, and then substituting in equation (13), one obtains the function $|\underline{\psi}_{s,ref\,\mathrm{I}}|(m_{e,ref\,\mathrm{I}})$ requested in the IPMSM DTC scheme over the speed range I, i.e. from standstill up to the base rotor speed ω_{rb}. If one defines ω_{rb} as the highest speed for the constant-torque operation mode with the maximum torque subject to the stator-current limit, and, at the same time, as the lowest speed for which the stator-voltage limit is reached,

ω_{rb} can be readily deduced from the steady-state IPMSM stator-voltage equations in the (d,q) coordinate system (neglecting the stator-resistance voltage drop):

$$u_{sd} = -\omega_r L_{sq} i_{sq} \tag{14}$$

$$u_{sq} = \omega_r L_{sd} i_{sd} + \omega_r |\underline{\psi}_{PM}| \tag{15}$$

one obtains from equations (13)–(15)

$$\omega_{rb} = U_{s,lim}/[(L_{sd} i_{sd,\mathrm{I}} + |\underline{\psi}_{PM}|)^2 + (L_{sq} i_{sq,\mathrm{I}})^2]^{1/2} \tag{16}$$

The IPMSM speed operation range II, just above the base rotor speed, is a flux-weakening constant-power region. The highest attainable IPMSM torque subject to both stator-current and -voltage limits of equations (9) and (10) yields

$$m_{e,max\mathrm{II}} = (3p/2)[|\underline{\psi}_{PM}| i_{sq,\mathrm{II}} - (L_{sq} - L_{sd}) i_{sd,\mathrm{II}} i_{sq,PM}] \tag{17}$$

where

$$(i_{sd,\mathrm{II}}{}^2 + i_{sq,\mathrm{II}}{}^2)^{1/2} = I_{s,lim} \tag{18}$$

$$i_{sd,\mathrm{II}} = (|\underline{\psi}_{PM}| L_{sd} - \{(|\underline{\psi}_{PM}| L_{sd})^2 + (L_{sq}{}^2 - L_{sd}{}^2) \times [|\underline{\psi}_{PM}|^2 + (L_{sq} I_{s,lim})^2 - (U_{s,lim}/\omega_r)^2]\}^{1/2})/(L_{sq}^2 - L_{sd}^2) \tag{19}$$

Rewriting equation (13) for $i_{sd,\mathrm{II}}$ and $i_{sq,\mathrm{II}}$, accounting in equation (17) $m_{e,max\mathrm{II}} = m_{e,ref\mathrm{II}}$, and eliminating the currents $i_{sd,\mathrm{II}}$ and $i_{sq,\mathrm{II}}$ between equations (13) and (17)–(19), one obtains the required function $|\underline{\psi}_{s,ref\,\mathrm{II}}|(m_{e,ref\,\mathrm{II}})$ for the IPMSM DTC scheme over the flux-weakening constant-power speed range II.

Since for the considered IPMSM drive $|\underline{\psi}_{PM}|/L_{sd} < I_{s,lim}$, there is a high-speed flux-weakening region III, where IPMSM constant-power operation is no more achievable. However, the torque capability can be insured by the maximum torque-to-stator flux ratio subject to the stator-voltage limit alone. The rotor speed, at which IPMSM constant-power operation ceases, is termed as base power speed, ω_{rbp}, and can be simply determined by

$$\omega_{rbp} = U_{s,lim}/(L_{sd} I_{s,lim} - |\underline{\psi}_{PM}|) \tag{20}$$

Beyond ω_{rbp}, IPMSM flux-weakening operation is still available up to theoretically infinite speed.

The IPMSM maximum available torque, $m_{e,max\,\mathrm{III}}$, as previously defined for the high-speed flux-weakening operation range III, is determined by introducing the upper-limit angle δ_{lim} of equation (8) into equation (1) expressing the IPMSM torque, thus leading to

$$m_{e,max\,\mathrm{III}} = (3p/2)|\underline{\psi}_s|(|\underline{\psi}_{PM}| - \xi|\underline{\psi}_s|\{|\underline{\psi}_{PM}|/4\xi|\underline{\psi}_s| - [(|\underline{\psi}_{PM}|/4\xi|\underline{\psi}_s|)^2 + 1/2]^{1/2}\}) \times (1 - \{|\underline{\psi}_{PM}|/4\xi|\underline{\psi}_s| - [(|\underline{\psi}_{PM}|/4\xi|\underline{\psi}_s|)^2 + 1/2]^{1/2}\}^2)/L_{sq}(1-\xi) \tag{21}$$

Equation (21) with $m_{e,max\,\mathrm{III}} = m_{e,ref\mathrm{III}}$, yields the required function $|\underline{\psi}_{s,ref\,\mathrm{III}}|(m_{e,ref\,\mathrm{III}})$ for the IPMSM DTC scheme over the high-speed flux-weakening operation range III.

For the three IPMSM operation modes that have been previously identified over the wide-speed range (below and above the base speed) the specific reference relationships $|\underline{\psi}_{s,ref}|(m_{e,ref})$ can be computed off-line, and subsequently incorporated into the IPMSM DTC scheme as a simple look-up table.

Table 1. Specifications of prototype IPMSM

Number of pole-pairs, p	3
Stator phase resistance, R_s	0.895 Ω
PM flux-linkage magnitude, $\lvert \underline{\psi}_{PM} \rvert$	0.2979 Wb
d-axis stator self-inductance, L_{sd}	12.16 mH
q-axis stator self-inductance, L_{sq}	21.3 mH
Stator-current limit, $I_{s,lim}$	6.75 A
Stator-voltage limit, $U_{s,lim}$	400 V
Base rotor speed, ω_{rb}	2,500 rpm

Simulation results

Extensive dynamic simulations using Matlab/Simulink software are carried out on a prototype IPMSM having the specifications given in Table 1 in order to validate and assess the performance of the proposed VSI-fed IPMSM DTC scheme over wide-speed operation range.

Fig. 4 shows the simulated dynamic responses of DTC IPMSM speed, torque, and stator flux-linkage with respect to a step change in speed reference from 0 to 4,000 rpm under

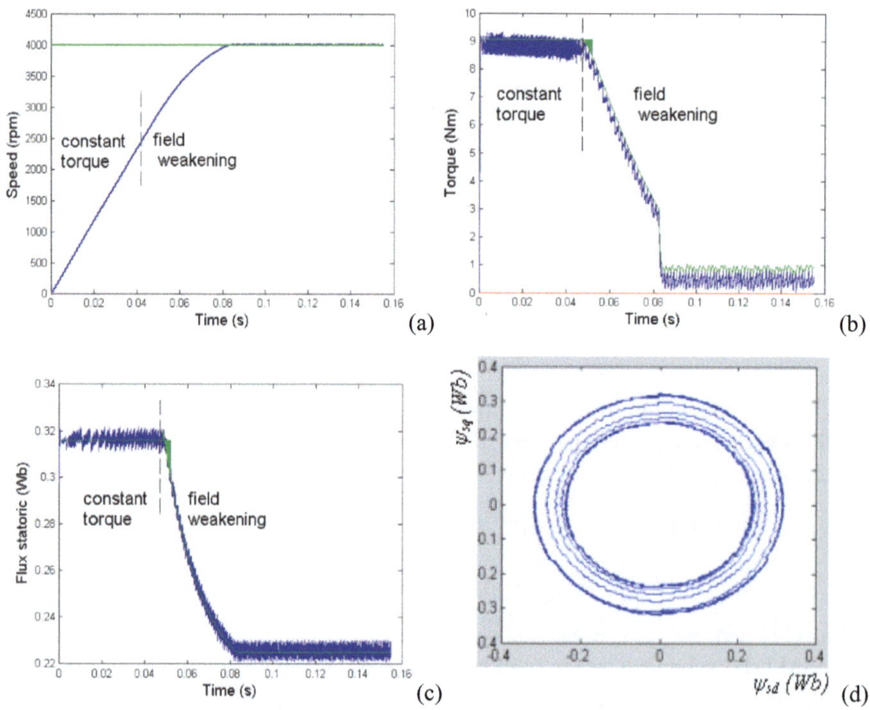

Figure 4. Dynamic simulation results for prototype IPMSM DTC over constant-torque and flux-weakening wide-speed operation ranges: (a) rotor-speed response; (b) torque response; (c) response of the stator flux-linkage magnitude; (d) locus of the stator flux-linkage vector.

no-load condition and subject to current and voltage constraints. It is seen from Fig. 4, that a smooth transition between the constant-torque and flux-weakening speed operation regions occurs when the rotor speed exceeds the base speed. With the proposed DTC scheme, IPMSM is accelerated by the maximum available torque in both constant-torque and flux-weakening operation modes over the wide-speed range in the presence of current and voltage constraints. Fig. 4(d) displays the dynamic locus of the stator flux-linkage vector, which is almost a circle in both constant-torque and flux-weakening wide-speed operation ranges.

Conclusions

An integrated approach to the proper design and DTC of VSI-fed IPMSMs requiring wide speed-torque envelope has been proposed.

The relationship between the reference electromagnetic torque and stator flux-linkage has been derived to be used in IPMSM DTC insuring maximum-torque-per-stator-current operation below the base speed as well as constant-power flux-weakening and maximum-torque-per-stator-flux operations above the base speed.

The simulated dynamic response in step speed command has confirmed the effectiveness of the proposed IPMSM DTC scheme over wide-speed operation range.

References

[1] S. Morimoto, M. Sanada, Takeda, Y. Wide-speed operation of interior permanent magnet synchronous motors with high-performance current regulator, IEEE Trans. Ind. Appl., Vol. 30, No. 4, pp. 920–926, 1994.

[2] J.-M. Kim, S.-K. Sul, Speed control of interior permanent magnet synchronous motor drive for the flux weakening operation, IEEE Trans. Ind. Appl., Vol. 33, No.1, pp. 43–48, 1997.

[3] M.N. Uddin, T.S. Radwan, M.A. Rahman, Performance of interior permanent magnet motor drive over wide speed range, IEEE Trans. Energy Convers., Vol. 17, No. 1, pp. 79–84, 2002.

[4] M.F. Rahman, L. Zhong, K.W. Lim, A direct torque-controlled interior permanent magnet synchronous motor drive incorporating field weakening, IEEE Trans. Ind. Appl., Vol. 34, No. 6, pp. 1246–1253, 1998.

[5] P. Vas, Sensorless Vector and Direct Torque Control, Oxford, UK: Oxford University Press, 1998, pp. 223–237 (Ch. 3).

[6] J. Luukko, "Direct Torque Control of Permanent Magnet Synchronous Machines—Analysis and Implementation", Ph.D. dissertation, Lappeenranta University of Technology, Finland, 2000, 172 p.

[7] L. Qinghua, A.M. Khambadkone, A. Tripathi, M.A. Jabbar, "Torque Control of IPMSM Drives Using Direct Flux Control for Wide Speed Operation", Proc. IEEE Int. Conf. Electr. Mach. Drives Conf. (IEMDC 2003), Vol. 1, Madison, Wisconsin, USA, June 1–4, 2003, pp. 188–193.

[8] Y. Honda, T. Higaki, S. Morimoto, Y. Takeda, Rotor design optimization of a multi-layer interior permanent-magnet synchronous motor. IEE Proc. Electr. Power Appl., Vol. 145, No. 2, pp. 119–124, 1998.

[9] L. Qinghua, M.A. Jabbar, A.M. Khambadkone, "Design Optimization of a Wide-Speed Permanent Magnet Synchronous Motor", Proc. IEE Int. Conf. Power Electr. Mach. Drives (PEMD 2002), Bath, UK, April 16–18, 2002, pp. 404–408.

[10] F. Rahman, R. Dutta, "A New Rotor of IPM Machine Suitable for Wide Speed Range", Rec. 29th Ann. Conf. IEEE Ind. Electron. Soc. (IECON 2003), Roanoke, Virginia, USA, November 2–6, 2003, CD-ROM.

[11] J. Luukko, M. Niemelä, J. Pyrhönen, Estimation of the flux linkage in a direct-torque-controlled drive, IEEE Trans. Ind. Electron., Vol. 50, No. 2, pp. 283–287, 2003.

[12] L. Tang, F. Rahman, M.E. Haque, "Low speed performance improvement of a direct torque-controlled interior permanent magnet synchronous machine drive", Rec. 19th IEEE Ann. Appl. Power Electron. Conf. (APEC 2004), Anaheim, CA, USA, February 22–26, 2004, pp. 558–564.

II-5. OPTIMAL SWITCHED RELUCTANCE MOTOR CONTROL STRATEGY FOR WIDE VOLTAGE RANGE OPERATION

F. D'hulster[1], K. Stockman[1], I. Podoleanu[2] and R. Belmans[2]

[1]Hogeschool West-Vlaanderen, Dept. PIH, Graaf Karel de Goedelaan 5, B-8500 Kortrijk, Belgium
frederik.dhulster@howest.be, kurt.stockman@howest.be
[2]KU Leuven, Dept. ESAT, Div. ELECTA, Kasteelpark Arenberg 10, B-3001 Leuven, Belgium
isan.podoleanu@esat.kuleuven.ac.be, ronnie.belmans@esat.kuleuven.ac.be

Abstract. This paper describes a technique to obtain optimal torque control parameters of a switched reluctance motor (SRM). A relationship between dc-link voltage and rotor speed is used, reducing the number of control parameters. Using a nonlinear motor model, surfaces are created describing torque, torque ripple, and efficiency as function of rotor speed and the main control parameters. Next, optimization software generates optimal control parameter combinations out of these surfaces for equidistant torque-speed performance. The advantage of this technique is an offline optimization platform and the simplicity to create additional surfaces (e.g., acoustic noise, vibrations, ...).

Introduction

Due to the ever increasing application demands put on switched reluctance motor drives, a flexible control strategy is gaining importance. Some applications demand a low acoustic noise or vibration level, others feature high efficiency. This paper deals with the design and implementation of an optimal control strategy for an 8/6 SRM, operating in a broad supply voltage range. Robust control must be applied for a dc-link voltage range of 115–325 V and a speed range of 0–2,000 rpm. At full motor load, a maximum torque control strategy must be used to obtain maximum mechanical power at the motor shaft. At medium load, different combinations of phase current and control angles are possible for a given reference torque. This degree of freedom enables optimization of the torque control parameters.

A complete optimization of machine geometry—converter—control of a SRM is proposed in [1] using genetic algorithms (GA) as an optimization tool. In many applications, the use of standard motor designs is preferred rather than developing a motor geometry for every new application.

The motor behavior as function of its torque control parameters is calculated only once and can serve as input for an offline optimization platform. Through a weighted sum of objective functions, the control of a standard SRM can be optimized for different applications. Fig. 1 illustrates the flowchart of this procedure.

First, the basic equations for the nonlinear SRM model are explained. Next, the main parameters (N) for the torque control are derived, taking into account the relation between

S. Wiak, M. Dems, K. Komęza (eds.), Recent Developments of Electrical Drives, 187–200.

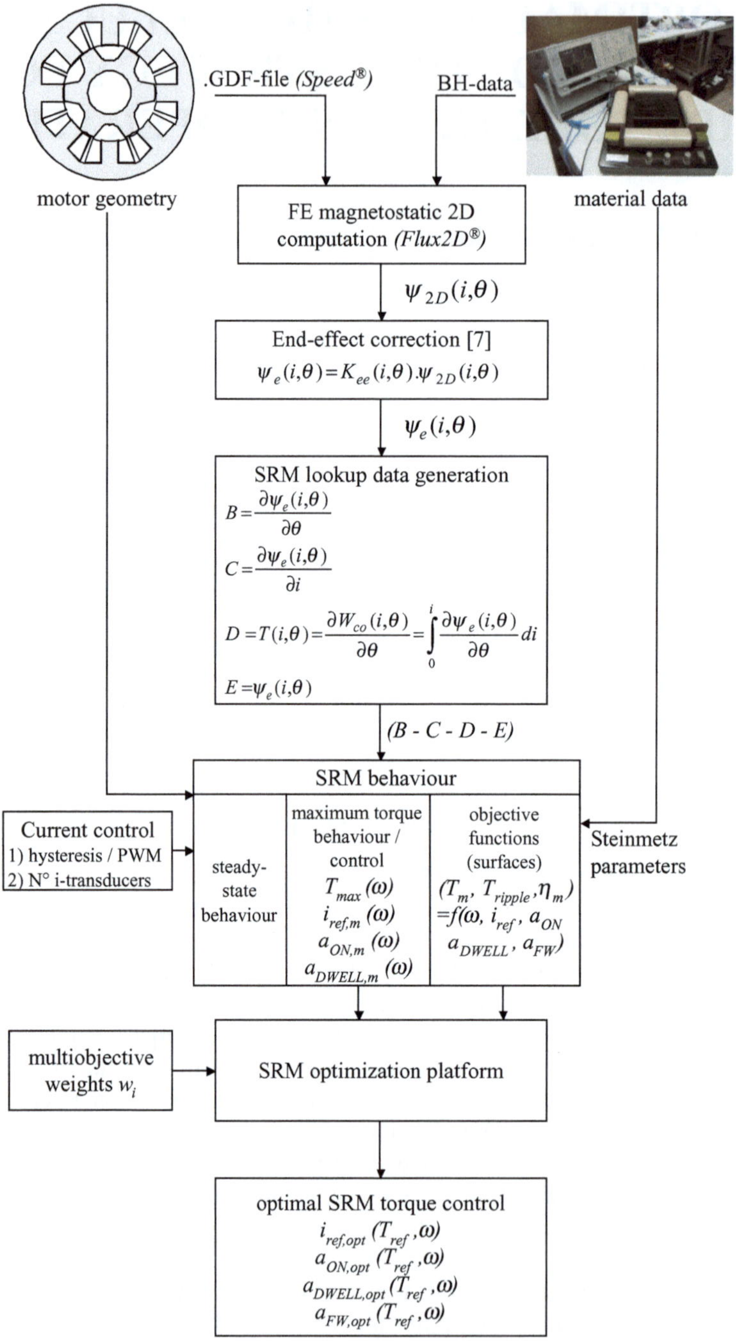

Figure 1. General flowchart of the optimal torque control of SRMs.

the dc-link voltage and the rotor speed. Then, N-dimensional surfaces are created, representing the SRM behavior as function of the torque control parameters. Finally, the optimal control parameters for the complete torque-speed range are determined using a genetic algorithm (GA) search tool or alternatively a "search for all" tool.

SRM system equations and drive model

The static behavior of a SRM can be explained by two equations, describing the current in a stator phase (1) and the instantaneous electromagnetic torque T, produced by a stator phase (2). Both equations depend on the partial derivatives of the flux-linkage $\psi(i,\theta)$.

$$\frac{di}{dt} = \frac{1}{\frac{\partial\psi(i,\theta)}{\partial i}}\left(u - Ri - \frac{\partial\psi(i,\theta)}{\partial\theta}\omega\right) \tag{1}$$

$$T(i,\theta) = \left.\frac{\partial W_{co}(i,\theta)}{\partial\theta}\right|_{i=cst} = \int_0^i \frac{\partial\psi(i,\theta)}{\partial\theta}\cdot di \tag{2}$$

with:

$-\dfrac{\partial\psi(i,\theta)}{\partial i} = p_i(i,\theta)$: phase inductance [H]

$-\dfrac{\partial\psi(i,\theta)}{\partial\theta} = p_\theta(i,\theta)$: back-emf coefficient

$-\omega$: rotor speed (rad/s)

$-u$: phase voltage (V)

$-i$: phase current (A)

$-R$: phase resistance (Ω).

This single-phase behavior, represented by four matrices as function of rotor position and phase current, is deducted from a magnetostatic finite element analysis (Fig. 2). The unaligned rotor position is set to 30° and aligned to 60°. Fig. 3 shows the single-phase static behavior of the motor, further used in this paper.

SRM control optimization is only possible using an accurate dynamic motor model, including saturation, iron loss estimation, and torque ripple calculation, combined with a drive model using the appropriate torque and current control (hysteresis or PWM). Both motoring and generating mode are supported, for different phase current sensing. Superposition of single-phase SRM-modeling, using lookup tables with 2D magnetostatic finite element

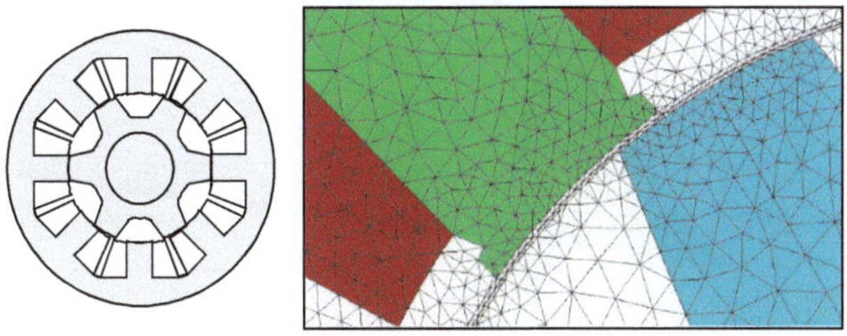

Figure 2. Geometry and 2D finite element model (Flux2D®).

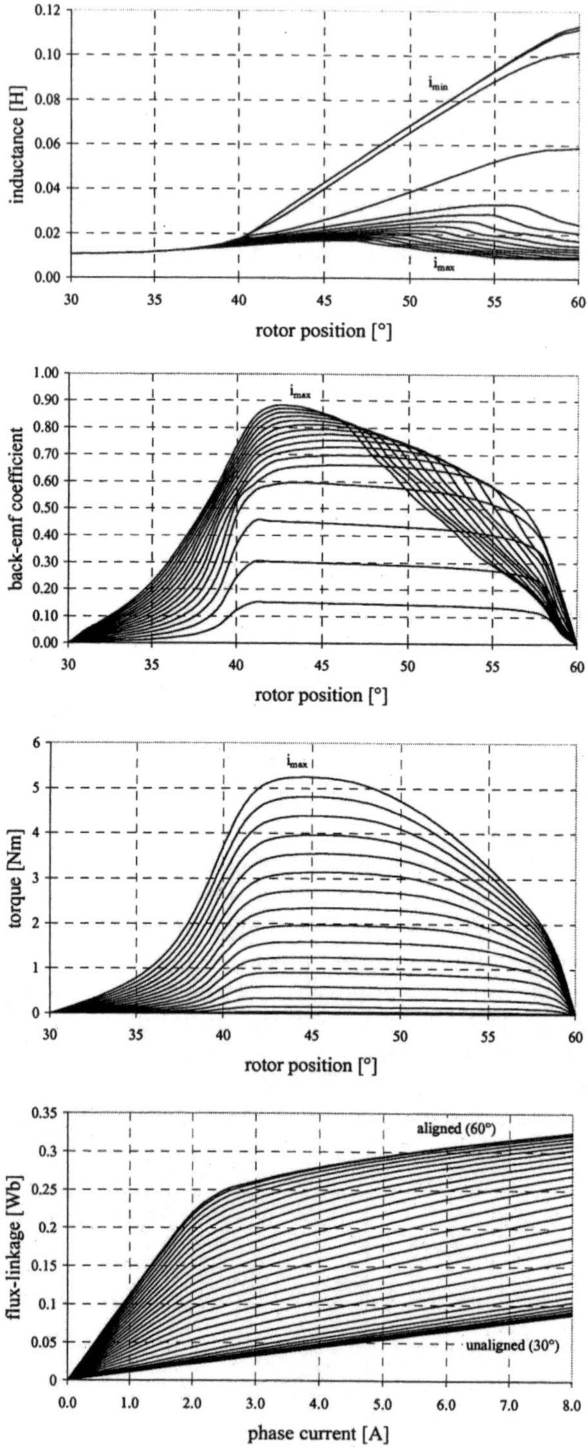

Figure 3. Single-phase SRM lookup data.

flux-linkage data, is described in [2]. This model is extended with iron loss calculation, based on the modified Steinmetz equation [3]. The Steinmetz parameters, describing the iron losses function for sinusoidal excitation are measured on a standard Epstein frame. Further in this paper, only motoring operation is considered.

If ventilation and friction losses are neglected, efficiency and torque ripple for motoring operation are:

$$\eta_m = \frac{P_m}{P_m + P_{Cu} + P_{Fe}} \tag{3}$$

$$T_{ripple} = \frac{\max(T) - \min(T)}{T_m} \tag{4}$$

with:

– P_m: mechanical power (W)
– P_{Cu}: Joule losses (W)
– P_{Fe}: iron losses (W)
– T_m: average torque (Nm).

SRM torque control

Unlike dc-machines or rotating field machines, in SRMs no direct link exists between torque and current, in this way complicating its control. This is linked to the fact that even in steady state the stored magnetic energy in the machine is not constant. A basic torque controller (Fig. 4) consists of lookup tables with the control parameters (turn-on angle a_{ON}, dwell angle $a_{DWELL} = a_{OFF} - a_{ON}$, freewheeling angle a_{FW}, and reference current i_{ref}),

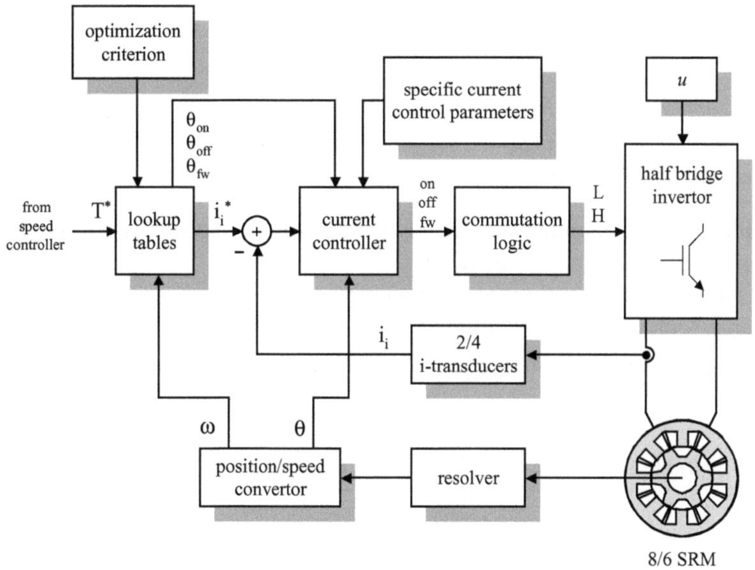

Figure 4. Basic SRM torque control structure.

determined according to an optimization criterion. The current can be controlled using a hysteresis or PWM control technique. With a PWM current controller the system produces less acoustic noise due to the fixed switching period, but its PI control parameters must be selected carefully. In addition to this basic structure, also different torque ripple reduction principles can be implemented, of which examples in [4,5].

Voltage-speed relationship

So far, not much research is done on the control of SRMs under different supply voltage conditions. Reference [6] proposes a maximum torque control strategy during short disturbances in the dc-link voltage due to voltage sags or load transients. For steady state behavior, [7] describes the similarity between supply voltage decrease and rotor speed increase on the current waveform of SR generators. This reduces the number of parameter sets in the drive. Equation (1) can be rewritten to:

$$\frac{di}{d\theta} = \frac{1}{\frac{\partial\psi(i,\theta)}{\partial i}} \left(\frac{u - Ri}{\omega} - \frac{\partial\psi(i,\theta)}{\partial\theta} \right) \tag{5}$$

Relation (5) states that, for a given phase current behavior, a relation exists between the phase voltage and the rotor speed:

$$\frac{u - Ri}{\omega} = cst \tag{6}$$

For a given voltage u, speed $\omega(u)$ and reference torque, the optimal control parameters can be obtained from the parameter set, defined for u_{ref}, using an equivalent rotor speed $\omega(u_{ref})$:

$$\omega(u_{ref}) = \frac{(u_{ref} - R \cdot i_{ref})}{(u - R \cdot i_{ref})} \cdot \omega(u) \tag{7}$$

with:

$-i_{ref}$: reference phase current (A)
$-u_{ref}$: reference phase voltage (V).

An example of this relation between supply voltage and rotor speed is given in Fig. 5 with the numerical simulation results in Table 1.

If the control is to be optimized for a supply voltage range $[u, u_{ref}]$ in a motor speed range $[0, \omega_{max}]$, then the equivalent optimal control parameters must be calculated for the supply voltage u_{ref} in a speed range $\left[0, \omega_{max} \cdot \frac{u_{ref} - R \cdot i_{ref}}{u - R \cdot i_{ref}}\right]$

Table 1. Numerical steady state simulation results

ω (rad/s)	u (V)	T_m (Nm)	P_m (W)	P_{Cu} (W)	P_{Fe} (W)	η_m
432	290	2.28	928	119	59	0.839
200	145	2.34	450	122	20	0.760

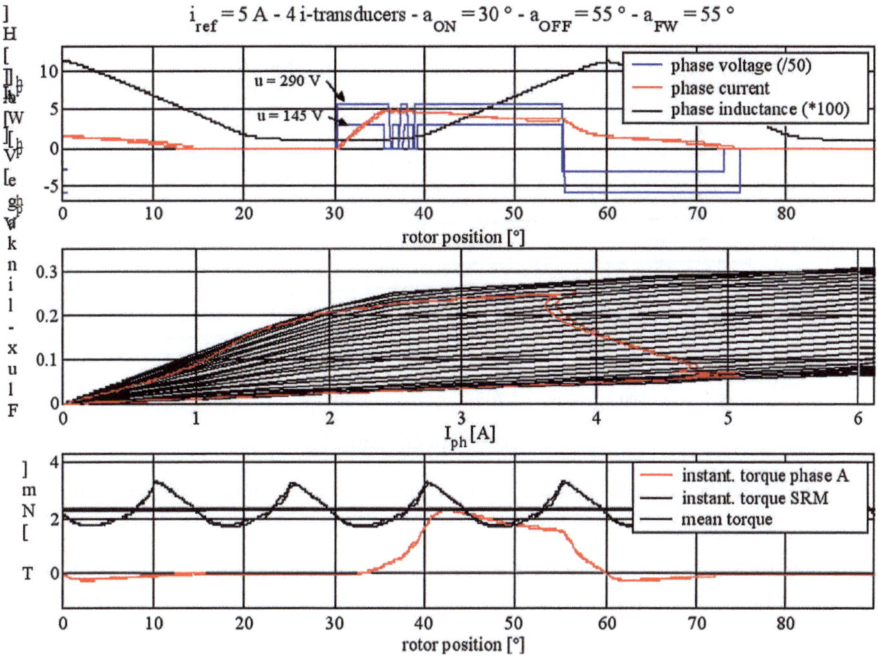

Figure 5. Comparison of current and torque behavior for voltage-speed combinations of the same parameter set ($\omega_1 = 432$ rad/s, $u_1 = 290$ V, $\omega_2 = 200$ rad/s, $u_2 = 145$ V).

SRM maximum torque control

When the speed or position controller demands maximum torque performance from the motor, no freedom is left for optimization. Both turn-on and dwell angle are determined to maximize the loop-surface during energy conversion [6]. In this paper only motoring operation is elaborated. The turn-on angle is calculated to reach the reference current at the start of pole-overlap:

$$a_{ON} = a_{ref} - \frac{\omega \cdot L_u \cdot i_{max}}{u_{ref} - p_\theta(i_{ref}, a_{ref}) \cdot \omega} \tag{8}$$

with:

$-a_{ref}$: start of inductance increase (pole-overlap)
$-L_u$: inductance at unaligned rotor position [H].

For the full rotor speed range, maximum torque control parameters are obtained, using the maximum available phase current i_{max}. Based on this maximum available torque at every rotor speed, the torque-speed plane is divided into equidistant torque-speed reference curves (Fig. 6). An important feature is the equidistance between torque references. This enables to design a stable speed or position controller. Intersection between different reference torque-speed lines would inevitably result in unstable operating points. The control angles and the

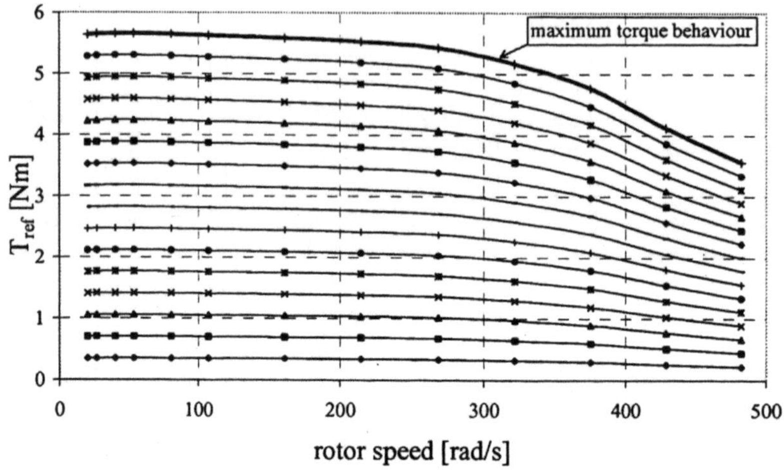

Figure 6. Equidistant reference torque curves, related to the maximum torque behavior for u_{ref}.

SRM behavior for the maximum torque control are illustrated in Fig. 7. Maximum torque control parameters are not obtained using the optimization algorithm because finding the parameters for the unique peak value of a surface is not an obvious task for any search tool.

SRM objective functions (surfaces)

Objective functions, describing the SRM behavior as a function of the control parameters, are the input functions of the optimization platform. Different functions or surfaces can

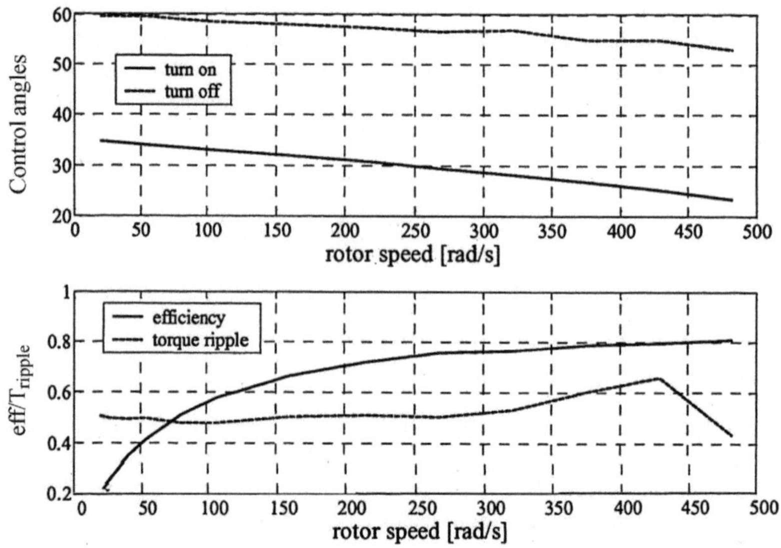

Figure 7. SRM maximum torque control angles and behavior for u_{ref} and i_{max}.

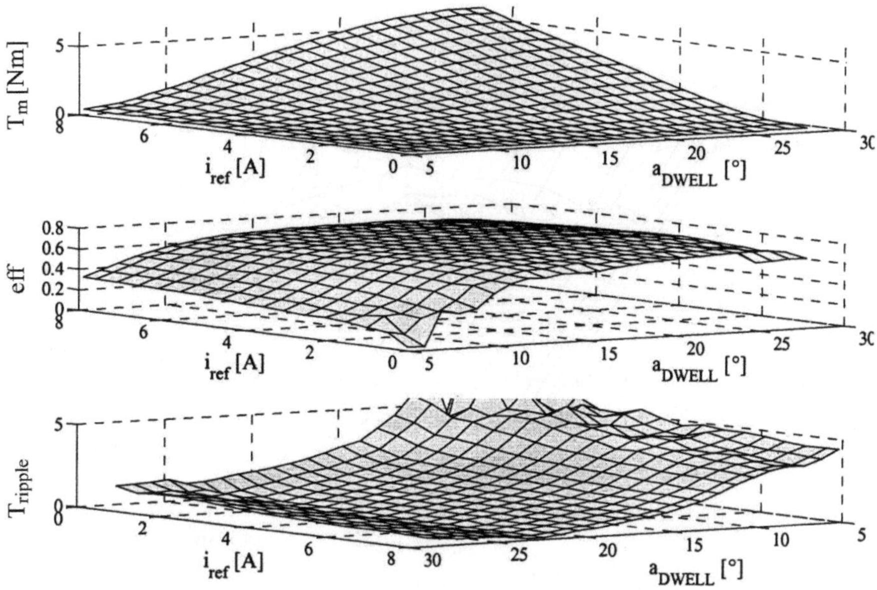

Figure 8. Torque, efficiency, and torque ripple for constant turn-on angle and rotor speed ($a_{ON} = 30°$; $\omega_{ref} = 160.85$ rad/s).

be calculated, using the nonlinear motor model, e.g., efficiency, torque ripple, acoustic noise.... Besides those surfaces, allowing an optimization criterion, the torque surface is also needed as a constraint function to satisfy the reference torque demand. Fig. 8 shows the surfaces of the torque, efficiency, and torque ripple for a fixed turn-on angle and rotor speed. Different combinations of the parameters can result in the same torque production, allowing optimization of the parameters for a given reference torque constraint.

Optimal control parameters determination

As pointed out, for each speed and reference torque, the appropriate input variables $i_{ref,opt}$, $a_{ON,opt}$, and $a_{DWELL,opt}$ must be determined in such a way that the overall performance matches an optimization criterion. For SRMs, the optimality condition is in general determined by straightforward requirements with regard to the efficiency, torque ripple, or acoustic noise. The efficiency should be maximized, the torque ripple and acoustic noise minimized.

All objective functions are combined into a single value function, called generic cost (c). For example, the generic cost function of efficiency and torque ripple is:

$$c = w_1(1 - \eta_m) + w_2 \left(\frac{T_{ripple}}{\max(T_{ripple})} \right) \tag{9}$$

The optimal solution is a combination of input variables for which the cost function is minimized, for a given speed and reference torque.

Although the surfaces of Fig. 8 seem relatively smooth, this is not the general behavior. In practice, noise on the surface results in many combinations of input variables with the

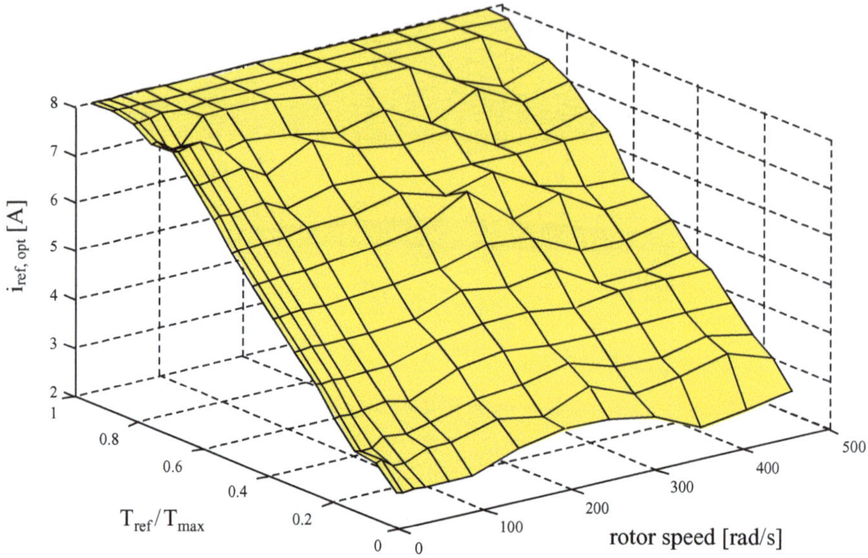

Figure 9. Optimal reference $i_{ref,opt}$ for u_{ref} ($w_1 = 0.5$; $w_2 = 0.5$).

same value for the cost function. As a direct consequence, only numerical algorithms able to find a global solution can be used, avoiding local minima. As a general constraint with regard to the final implementation, the chosen algorithm should always find the solution within a reasonable time. Moreover, the solution should be found from any initial starting point.

There are several algorithms to determine the desired minimum, but only two were implemented. The first attempt uses a genetic algorithm (GA) as it is characterized by a high probability to find a global minimum. However, for a few operating points, no useful solution is found. A second algorithm ("search for all") takes all possible combinations of input variables with a constant step and determines the constrained minimum. This method is straightforward to implement and a solution is found for every operating point. The calculation time is function of the number of parameters and the step size. A GA search method has the disadvantage that a solution is not guaranteed and that one particular solution is searched, without taking into account that this combination could be useful for other torque reference values. The direct "search for all" method calculates the objective and constraint function values for a parameter combination and tests the cost for all torque reference values. This strongly reduces the computation time.

With a weight of 0.5 for efficiency and 0.5 for torque ripple, the optimal control parameters are presented in Figs. 9–11, using the "search for all" algorithm.

Measurement results

The optimal control parameters, are programmed into a SRM drive and its behavior is measured on a test setup with load machine. Validating if the control is really optimal is

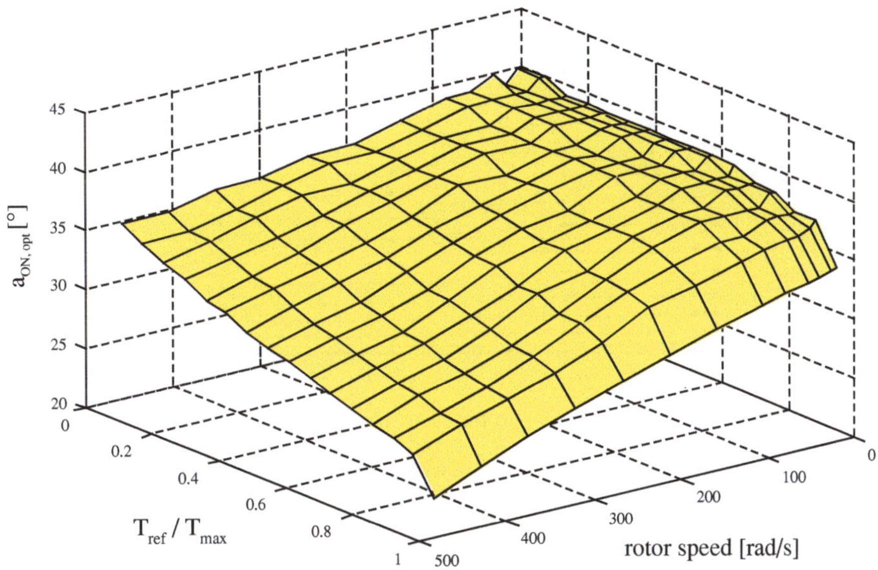

Figure 10. Optimal turn-on angle $a_{ON,opt}$ for $u_{ref}(w_1 = 0.5; w_2 = 0.5)$.

not easy. The model accuracy is verified by means of torque and efficiency measurements. Fig. 12 represents the measured torque-speed performance, according to the reference torque values for every rotor speed. No intersection between the lines occurs, resulting in a stable position or speed controller. Efficiency is determined by measuring the electrical power,

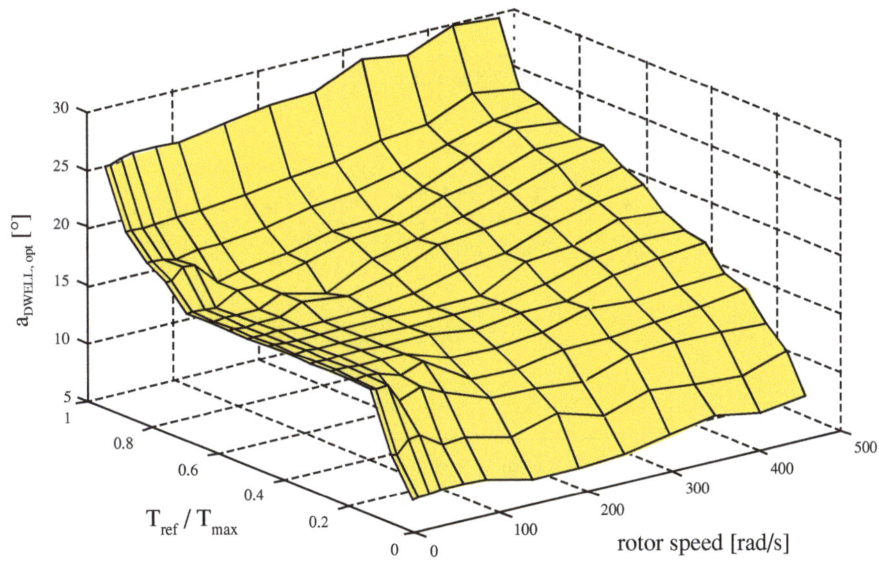

Figure 11. Optimal dwell angle $a_{DWELL,opt}$ for $u_{ref}(w_1 = 0.5; w_2 = 0.5)$.

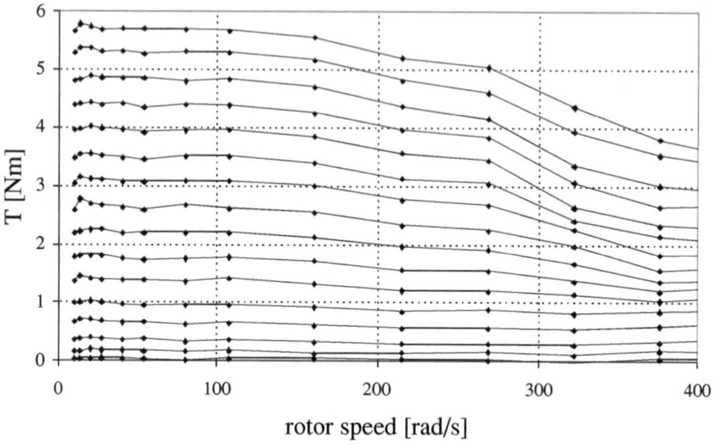

Figure 12. Measured torque-speed performance with optimal control parameters for $u_{ref}(w_1 = 0.5; w_2 = 0.5)$.

supplied to the motor, and the mechanical shaft torque. Efficiency measurements as function of reference torque and rotor speed are compared with simulations in Figs. 13 and 14.

Conclusions

Different torque control strategies can be implemented in SRM drives, operating at varying supply voltage conditions. A technique is presented to obtain optimal SRM torque control parameters, according to a weighted optimization criterion. The dc-link voltage is not considered as a fundamental parameter due to its analogy with rotor speed. Using a nonlinear

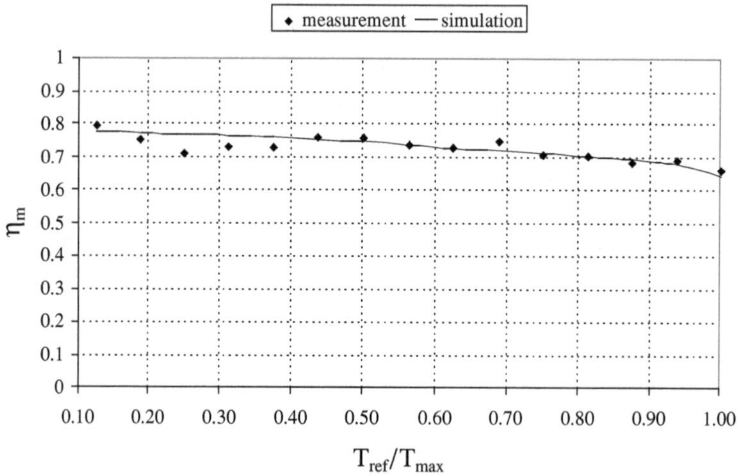

Figure 13. Measured and simulated motor efficiency as function of reference torque (u_{ref}; $\omega = 214$ rad/s).

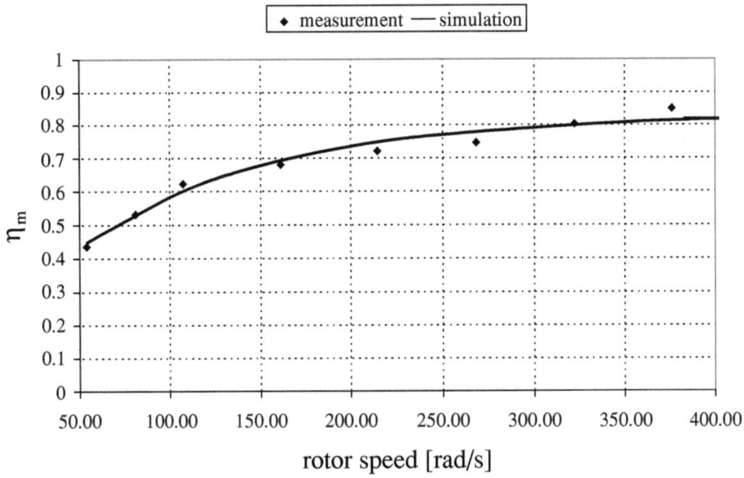

Figure 14. Measured and simulated motor efficiency as function of rotor speed (u_{ref}; $T_{ref} = 0.5T_{max}$).

SRM drive model, the behavior is stored in N-dimensional surfaces, serving as objective and constraint functions for the optimization platform. The objective functions in this paper are limited to motor efficiency and torque ripple but can easily be extended with acoustic noise or temperature. The surfaces are calculated only once for each motor geometry and different control parameter sets can be obtained for different application demands.

Acknowledgments

The authors wish to thank the Flemish Government (IWT) for granting the research project "Bepaling van de optimale stuur- en regelparameters voor systemen met SR-motor aandrijving. Ontwerp van een ontwikkelingsplatform." (IWT 020343). The general optimization work is part of the IUAP/PAI P4/20 project "Coupled problems" sponsored by the Belgian Federal Government.

References

[1] E. Lomonova, A. Matveev, "Application of genetic algorithm for design of switched reluctance drives," Proceedings of the European Conference on Power Electronics and Applications (EPE 2003), Toulouse, France, p. 12, 2003.

[2] F. D'hulster, K. Stockman, J. Desmet, R. Belmans, "Advanced nonlinear modelling techniques for switched reluctance machines," IASTED International Conference on Modelling, Simulation and Optimization (MSO 2003), Banff, Alberta, Canada, pp. 44–51, July 2–4, 2003.

[3] J. Reinert, R. Inderka, R. W. De Doncker, "A novel method for the prediction of losses in switched reluctance machines," Proceedings of the European Conference on Power Electronics and Applications (EPE 1997), Trondheim, Norway, pp. 3608–3612, 1997.

[4] I. Husain, M. Ehsani, "Torque ripple minimization in switched reluctance motor drives by PWM current control," IEEE Transactions on Power Electronics, Vol. 11, No. 1, pp. 83–88, 1996.

[5] R.B. Inderka, R.W. De Doncker, "DITC – Direct instantaneous torque control of switched reluctance drives," Proceedings of the IEEE-IAS Annual Meeting, Pittsburgh, Pennsylvania, USA, pp. 1605–1609, October 13–18, 2002.
[6] F. D'hulster, K. Stockman, R. Belmans, "Maximum torque control strategy for switched reluctance motors during dc-link disturbances," Proceedings of the European Conference on Power Electronics and Applications (EPE 2003), Toulouse, France, p. 6, 2003.
[7] R.B. Inderka, M. Menne, R.W. De Doncker, "Generator operation of a switched reluctance machine drive for electric vehicles", EPE journal, Vol. 11, No. 3, August 2001.

II-6. EFFECT OF STRESS-DEPENDENT MAGNETOSTRICTION ON VIBRATIONS OF AN INDUCTION MOTOR

A. Belahcen

Laboratory of Electromechanics, Helsinki University of Technology, P.O. Box 3000, FIN-02015 HUT, Finland

anouar.belahcen@hut.fi

Abstract. A model for the magnetoelastic coupling in electrical machines is presented. It couples transient electromagnetic field equations with dynamic elastic ones. Computations are made to show the effect of stress-dependent magnetostriction on the vibrations of the stator core of an induction machine. It is shown that the magnetostriction changes the amplitude of vibrations velocity up to 800%. A relative difference of more than 6,000% is found between calculation with stress-dependent and stress-independent magnetostriction. Measurements are made for validation.

Introduction

The effect of magnetostriction and inverse magnetostriction (Villary effect) on the vibrations and acoustic noise of rotating electrical machines is still a subject of controversy. Indeed, different authors [1–3] presented different models for the magnetostriction and came up with different results. Some authors believe that the magnetostriction affects the vibrations of rotating electrical machines [1,2]; others claim that the magnetostriction can be ignored [3]. We investigate the problems of magnetoelasticity and magnetostriction by means of coupled transient and dynamic FE analysis.

Models for static analysis with current-supplied systems have been presented by Ren et al. [4] and further developed by Mohammed et al. [5]. Uncoupled dynamic models for the vibrations of rotating electrical machines also have been presented [6–8]. The model we purpose is developed from both the static coupled and dynamic uncoupled models. It is a model that handle transient dynamic systems with voltage-supply.

The goal of this study is to establish the effect of magnetostriction and magnetoelastic coupling on the vibrations and noise of rotating electrical machines. Results about the effect of coupling on the vibrations have already been presented in [9]. The data about the magnetic properties of the materials used in this work and its measurements have been presented in [10].

This paper presents the effect of magnetostriction and stress-dependent magnetostriction on the vibrations of an induction motor.

S. Wiak, M. Dems, K. Komęza (eds.), Recent Developments of Electrical Drives, 201–210.

Methods

Magnetic and elastic fields

The $\boldsymbol{A}-\phi$ formulation of the magnetic field in two dimensions and the displacement based formulation of the elastic field are used.

$$
\begin{bmatrix}
\begin{array}{l}\boldsymbol{S}(\boldsymbol{A}_{k+1},\boldsymbol{U}_{k+1})+\\ \dfrac{\partial \boldsymbol{S}(\boldsymbol{A}_{k+1},\boldsymbol{U}_{k+1})}{\partial \boldsymbol{A}}\boldsymbol{A}_{k+1}\end{array} & [\boldsymbol{D}^{\mathrm{r}}]^{\mathrm{T}} & [\boldsymbol{L}\boldsymbol{D}^{\mathrm{s}}]^{\mathrm{T}} & \dfrac{\partial \boldsymbol{S}(\boldsymbol{A}_{k+1},\boldsymbol{U}_{k+1})}{\partial \boldsymbol{U}}\boldsymbol{A}_{k+1} \\
\boldsymbol{D}^{\mathrm{r}} & \boldsymbol{C}^{\mathrm{r}} & 0 & 0 \\
\boldsymbol{L}\boldsymbol{D}^{\mathrm{s}} & 0 & \boldsymbol{G}^{\mathrm{s}} & 0 \\
-\dfrac{\partial \tilde{\boldsymbol{F}}_{k+1}}{\partial \boldsymbol{A}} & 0 & 0 & \tilde{\boldsymbol{K}}-\dfrac{\partial \tilde{\boldsymbol{F}}_{k+1}}{\partial \boldsymbol{U}}
\end{bmatrix}
\begin{bmatrix}
\Delta \boldsymbol{A}_{k+1}^{n} \\
\Delta \boldsymbol{u}_{k+1}^{\mathrm{r}n} \\
\Delta \boldsymbol{i}_{k+1}^{\mathrm{s}n} \\
\Delta \boldsymbol{U}_{k+1}^{n}
\end{bmatrix}
=
\begin{bmatrix}
\boldsymbol{S}'(\boldsymbol{A}_k)\boldsymbol{A}_k+[\boldsymbol{D}^{\mathrm{r}}]^{\mathrm{T}}\boldsymbol{u}_k^{\mathrm{r}}+[\boldsymbol{D}^{\mathrm{s}}]^{\mathrm{T}}\boldsymbol{K}^{\mathrm{T}}\boldsymbol{i}_k^{\mathrm{s}}- \\
(\boldsymbol{S}(\boldsymbol{A}_{k+1}^{n})\boldsymbol{A}_{k+1}^{n}+[\boldsymbol{D}^{\mathrm{r}}]^{\mathrm{T}}\boldsymbol{u}_{k+1}^{\mathrm{r}n}+[\boldsymbol{D}^{\mathrm{s}}]^{\mathrm{T}}\boldsymbol{K}^{\mathrm{T}}\boldsymbol{i}_{k+1}^{\mathrm{s}n}) \\
\boldsymbol{L}\boldsymbol{D}^{\mathrm{s}}\boldsymbol{A}_k-\boldsymbol{H}^{\mathrm{s}}\boldsymbol{i}_k^{\mathrm{s}}-\boldsymbol{C}^{\mathrm{s}}(\boldsymbol{V}_{k+1}^{\mathrm{s}}+\boldsymbol{V}_k^{\mathrm{s}})- \\
(\boldsymbol{L}\boldsymbol{D}^{\mathrm{s}}\boldsymbol{A}_{k+1}^{n}-\boldsymbol{H}^{\mathrm{s}}\boldsymbol{i}_{k+1}^{\mathrm{s}n}-\boldsymbol{C}^{\mathrm{s}}(\boldsymbol{V}_{k+1}^{\mathrm{s}}+\boldsymbol{V}_k^{\mathrm{s}})) \\
\boldsymbol{D}^{\mathrm{r}}\boldsymbol{A}_k-\boldsymbol{C}^{\mathrm{r}}\boldsymbol{u}_k^{\mathrm{r}}-\boldsymbol{G}^{\mathrm{r}}\boldsymbol{i}_k^{\mathrm{r}}- \\
(\boldsymbol{D}^{\mathrm{r}}\boldsymbol{A}_{k+1}^{n}-\boldsymbol{C}^{\mathrm{r}}\boldsymbol{u}_{k+1}^{\mathrm{r}n}-\boldsymbol{G}^{\mathrm{r}}\boldsymbol{i}_{k+1}^{\mathrm{r}n}) \\
\tilde{\boldsymbol{F}}_{k+1}^{n}-\tilde{\boldsymbol{K}}\boldsymbol{U}_{k+1}^{n}
\end{bmatrix}
\tag{1}
$$

The 2D finite element (FE) equations for the magnetic, and elastic field are coupled through the displacements and the forces. These equations are solved together with the circuit equations of the windings of the machine as described in [9]. The system of equations to be solved at each iteration is written as (1), where $\boldsymbol{A}$ is the magnetic vector potential; $\boldsymbol{C}$, $\boldsymbol{D}$, $\boldsymbol{G}$, and $\boldsymbol{H}$ are the coupling matrices between the magnetic vector potential and the electric parameters in the windings parts of the machine; $\boldsymbol{L}$ is a matrix for the connections of the stator windings; $\boldsymbol{S}$ and $\boldsymbol{K}$ are the magnetic and mechanical stiffness matrices; $\boldsymbol{u}$ and $\boldsymbol{i}$ are respectively voltages and currents and $\boldsymbol{U}$ is the nodal displacements vector.

The superscripts r and s refer respectively to rotor and stator. The subscripts k and $k+1$ refer respectively to previous and present steps in the time stepping method. The sign $\sim$ over the matrix $\boldsymbol{K}$ and the force $\boldsymbol{F}$ means that they are replaced by their dynamic counterparts. A full description of the above matrices is given in [9,11].

Magnetic and magnetostrictive forces

The forces, which are the load for the elastic field, are separated into magnetic force (also called reluctance forces in some works [4]) and magnetostriction forces.

The magnetic forces are calculated, at any iteration, from the calculated magnetic vector potential, based on the local application of the virtual work principle:

$$\boldsymbol{F}^{\mathrm{T}} = -\frac{\partial W}{\partial \boldsymbol{U}}\bigg|\phi = \text{constant} \tag{2}$$

With the energy W per element $\boldsymbol{S}_{\mathrm{e}}$ given by:

$$W = \int_{S_{\mathrm{e}}} \int_0^{\boldsymbol{B}} \boldsymbol{H} \cdot \mathrm{d}\boldsymbol{B}\, \mathrm{d}\boldsymbol{S}_{\mathrm{e}}, \tag{3}$$

we obtain the contribution of one element to the nodal magnetic forces calculated on the reference element $\hat{S}_{\mathrm{e}}$ as:

$$\boldsymbol{F}^{\mathrm{T}} = -\int_{\hat{S}_{\mathrm{e}}} \left[\frac{1}{\mu_{\mathrm{e}}} \nabla A \cdot \frac{\partial}{\partial \boldsymbol{U}} (\nabla A) |\boldsymbol{J}| + \int_0^{\boldsymbol{B}} \boldsymbol{H} \cdot \mathrm{d}\boldsymbol{B} \frac{\partial}{\partial \boldsymbol{U}} (|\boldsymbol{J}|) \right] \mathrm{d}\hat{S}_{\mathrm{e}} \tag{4}$$

where, $|\boldsymbol{J}|$ is the determinant of the Jacobian matrix for the transformation from the reference element to the actual one. These individual contributions from elements to nodal forces are added to each other to obtain the global nodal magnetic forces. The nodal magnetostriction forces are calculated also at element level and assembled in the same manner as the magnetic forces.

The calculation of the magnetostrictive forces is based on an original method called the *method of magnetostrictive stress*. This method is explained hereafter.

Let's consider an element of iron in a magnetic field $\boldsymbol{H}$. Due to magnetostriction of iron, this element will shrink or stretch depending on the sign of its magnetostriction. This change in dimensions is described by a magnetostrictive strain tensor $\{\varepsilon_{ms}\}$. Corresponding to this strain, a magnetostrictive stress tensor $\{\sigma_{ms}\}$ can be calculated using Hook's law. The nodal magnetostrictive forces are the set of nodal forces due to this stress.

The measurements presented in [10] give the component of magnetostrictive stress $\sigma_{ms\parallel}$ in a direction parallel to that of the magnetic field. The other component of magnetostrictive stress orthogonal to the direction of the magnetic field $\boldsymbol{H}$ can be calculated within two assumptions. First, there is no magnetostrictive shear stress in the frame defined by the direction parallel to $\boldsymbol{H}$ and the one orthogonal to it. Second, there is no volume magnetostriction; which is a good assumption in the range of flux density occurring in electrical machines. The latter assumption means that the magnetostriction strain in the direction orthogonal to that of the magnetic field is opposite and has half the amplitude compared to the strain parallel to the direction of the magnetic field.

If $\sigma_{ms\perp}$ is the magnetostrictive stress in the direction orthogonal to the magnetic field, using the first assumption, we can write:

$$\begin{bmatrix} \sigma_{ms\parallel} \\ \sigma_{ms\perp} \\ 0 \end{bmatrix} = \boldsymbol{E} \begin{bmatrix} \varepsilon_{ms\parallel} \\ \varepsilon_{ms\perp} \\ 0 \end{bmatrix} \tag{5}$$

where $\boldsymbol{E}$ is the stress-strain matrix. In the case of plane stress, use of the second assumption leads to:

$$\sigma_{ms\perp} = \frac{2v - 1}{2 - v} \sigma_{ms\parallel} \tag{6}$$

The magnetostrictive nodal forces are calculated for each element as follows. Let θ be the angle defined by the direction of the magnetic field and the x-axis. The projections of each edge of the element on the directions parallel and orthogonal to the magnetic field are respectively

$$s_{\parallel} = \cos(\theta) s_x + \sin(\theta) s_y \tag{7}$$

and

$$s_{\perp} = -\sin(\theta)s_x + \cos(\theta)s_y \tag{8}$$

where s_x and s_y are respectively the projection of the considered edge of the element on the x- and y-axis. The forces, per unit length, parallel and orthogonal to the direction of the magnetic field are respectively

$$F_{ms\|} = \sigma_{ms\|}s_{\|} \tag{9}$$

and

$$F_{ms\perp} = \sigma_{ms\perp}s_{\perp} \tag{10}$$

These forces are distributed equally between the two nodes of the given edge. The forces in the original Cartesian coordinate system are obtained as the projection of $F_{ms\|}$ and $F_{ms\perp}$ on the axis of that system:

$$F_{msx} = \cos(\theta)F_{ms\|} - \sin(\theta)F_{ms\perp} \tag{11}$$

and

$$F_{msy} = \sin(\theta)F_{ms\|} + \cos(\theta)F_{ms\perp} \tag{12}$$

When the stress dependency of magnetostriction is taken into account, only the data of magnetostrictive stress are changed. All the rest is the same.

Vibrations

The solution of the magnetoelastic FE analysis produces among others the nodal displacements as a function of time. The displacements of a node on the outer surface of the stator core are transformed with Discrete Fourier Transform (DFT) and numerically differentiated to obtain the frequency components of the velocity of vibrations of the node considered.

In the following calculations, a total of 3,000 time steps are calculated with 300 time steps per period of the line voltage (20 ms). This leads to a sampling frequency of 15 kHz, a frequency resolution of 5 Hz and a maximum frequency of 7.5 kHz. The amplitudes of these vibrations are the quantities under consideration in the result section.

Results

Validation

An induction machine-like test device has been built to verify the presented model for magnetostriction. The test device is shown in Fig. 1.

The test device is constructed in a way that simulates the flux path in an induction machine, meanwhile it minimizes the reluctance forces. The latter are due to the presence of the air gap while the test device has no air gap. The only magnetic forces in the test device are the magnetostrictive forces and the Lorentz forces. The effect of the latter ones on the vibrations of the test device can be neglected due to the low currents in the windings and also due to the high relative mass-ratio between the iron core of the device and its windings. Thus, the only cause of vibrations in the test device is the magnetostriction.

Computations and measurements have been made for the test device. The simulated flux lines in the cross section of the device are shown in Fig. 2. The flux density in the back

Figure 1. Picture of the test device. The search coil for measurement of the back iron magnetic flux density can be seen. Rubber tubes separate both mechanically and electrically the windings from the iron core.

iron core of the test device has also been measured by a search coil. This flux density is compared to the simulated one in Fig. 3. The vibrations of the outer surface of the test device have been measured with a laser vibrometer at different points. The same values have been calculated. The simulated and measured displacements at a point on the outer surface are shown in Fig. 4.

Application

The method developed in this work is applied to a small size (37 kW) induction machine. Different computational approaches are used to establish both the effect of stress-independent and stress-dependent magnetostriction on the vibrations of such a machine. The parameters of the simulation machine are given in Table 1.

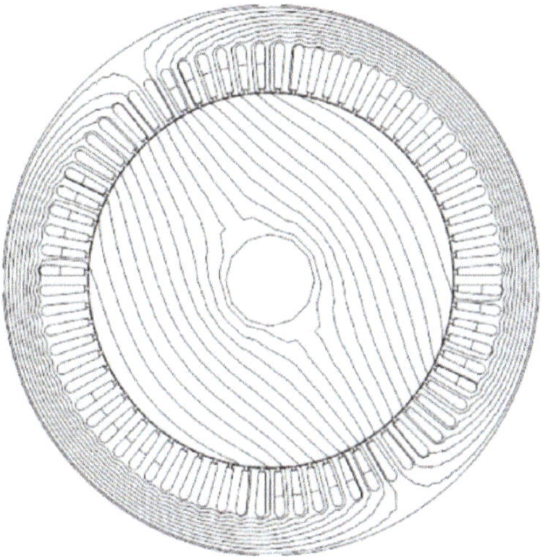

Figure 2. Plot of the calculated flux lines in the test device.

Table 1. Parameters of the induction machine

Rated power	37 kW
Rated voltage	380 V
Connection	Star
Slip	1.7%
Frequency	50 Hz
Number of phases	3
Number of poles	4
Stator outer diameter	310 mm
Stator inner diameter	200 mm
Stack length	289 mm
Number of stator slots	48
Rotor outer diameter	198.4 mm
Number of rotor slots	40

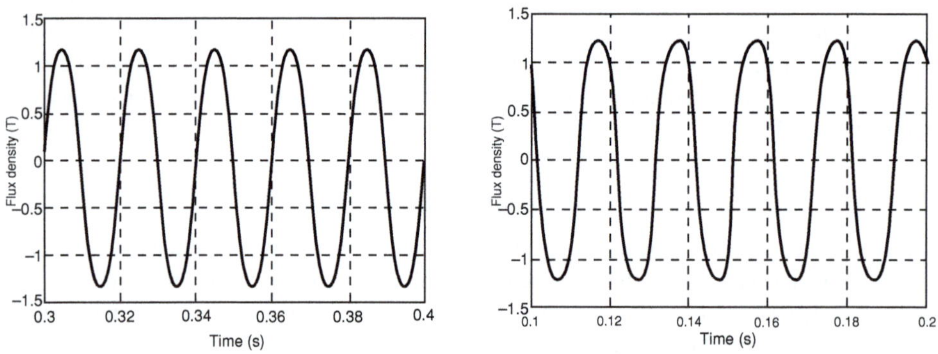

Figure 3. Measured and simulated flux density in the back iron core of the test device. (a) Measured. (b) Simulated.

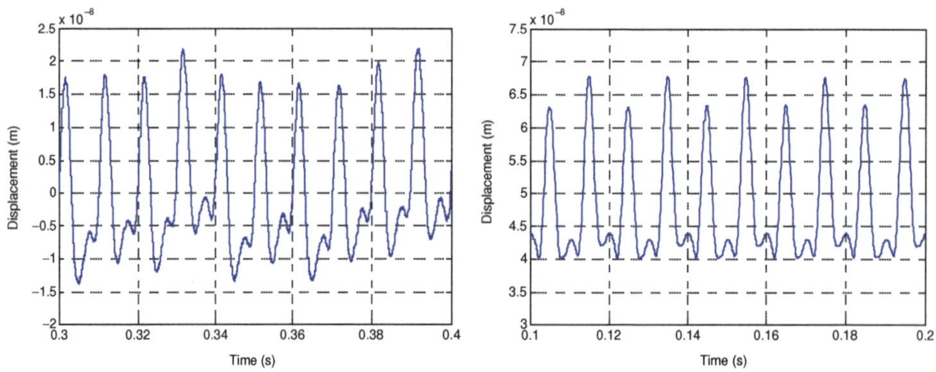

Figure 4. Measured and simulated displacements at the surface of the test device (in measurement the DC-component is omitted). (a) Measured. (b) Simulated.

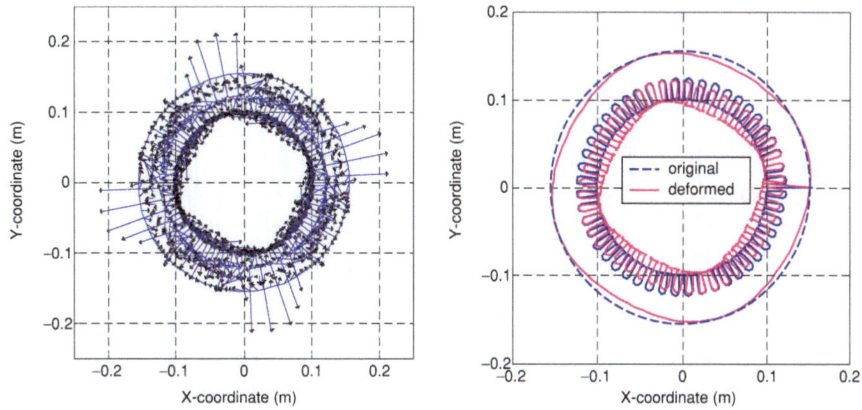

Figure 5. Calculated magnetostrictive force (normalized to 50,870 N/m) and deformation (magnified 20,000 times) of the stator core of the induction machine. (a) Forces. (b) Deformation.

The calculated magnetostrictive forces acting on the stator core of the induction machine at the last time step and the corresponding deformation are shown in Fig. 5(a,b) respectively. These are forces and deformations calculated with stress-independent magnetostriction. The reluctance forces are not shown.

Calculations with no magnetostriction and these with stress-dependent magnetostriction have also been undertaken. The velocity of vibrations of a node on the outer surface of the stator core of the machine (point P in Fig. 5(b)) calculated with different approaches are compared.

Fig. 6 shows the relative difference in the amplitudes of velocity between the cases no magnetostriction and stress-independent magnetostriction. Fig. 7 shows the relative

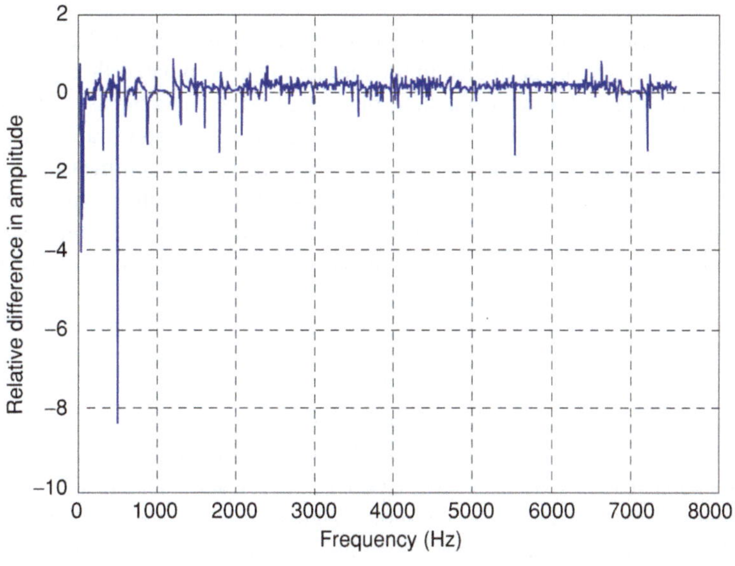

Figure 6. Calculated relative difference in the amplitude of velocities of point P (Fig. 5b). Differences between the cases no magnetostriction and stress-independent magnetostriction.

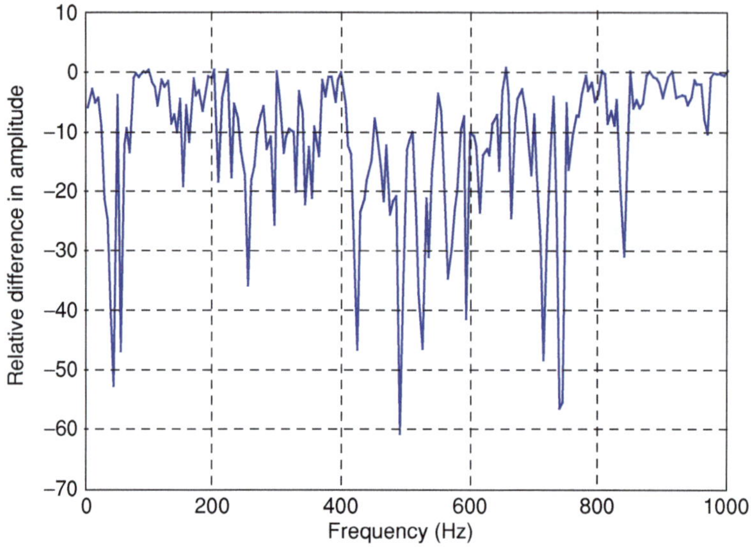

Figure 7. Calculated relative difference in the amplitude of velocities of point P (Fig. 5b). Differences between the cases stress-independent and stress-dependent magnetostriction.

difference in the amplitudes of velocity between the cases stress-independent and stress-dependent magnetostriction. The relative differences are calculated as $(|v_1| - |v_2|)/|v_1|$.

Analysis and discussion

Validation

The measured displacements from the test device are slightly higher than the simulated ones. This difference can be seen also in the measured and simulated flux densities. They are due mainly to differences in the magnetic properties of the materials used in simulations. Indeed, the manufacturing process of the test device slightly deteriorated the magnetic properties of the iron sheets. However, the correspondences between measured quantities and these simulated with the presented model for magnetostriction are rather good from both the amplitudes and wave forms points of view. In the future, better magnetic properties of the manufactured machine can be introduced into the simulation software for better results.

Simulations

The vibrations of the induction machine are affect by the magnetostriction. The amplitudes of most of the frequency components are increased. The most increased frequency components are these at 490 Hz (840%), 40 Hz (400%), 60 Hz (270%), and 50 Hz (320%). Some other frequency components are damped due to the magnetostriction. Among these the ones at 1,190 and 1,470 Hz damped respectively 80% and 70%. The 100 Hz component is damped only by 7%.

The stress dependency of magnetostriction adds to the effect of stress-independent magnetostriction so that the amplitudes of almost all the frequencies are increased. The increase reaches some 7,000% for the frequency component at 1,595 Hz e.g. However, the accuracy of the simulations with stress-dependent magnetostriction cannot be established due to the effect of magnetostrictive stress. Indeed, in the FE iteration process, the stress from magnetostrictive forces cannot be separated from the stress due to other forces (reluctance and Lorentz forces). Thus the stress state of the material is not accurately estimated, leading to inaccuracies in calculation of the stress-dependent magnetostriction. Although, we can say that both stress-independent and stress-dependent affect the vibrations of rotating electrical machines.

Conclusions

A model for the magnetoelastic coupling is presented and used in the simulations of an induction machine. The goal of these simulations is to establish the effect of the magnetostriction on the vibrations of rotating electrical machines. For this purpose an original method for the calculation of magnetostrictive forces is presented.

It is shown that the magnetostriction affects the vibrations of rotating electrical machines by increasing or decreasing the amplitudes of velocities measured at the outer surface of the stator core of the machine. These velocity are the ones responsible for acoustic noise. Furthermore, The stress dependency of the magnetostriction adds to the increase of the above amplitudes.

The modeling of vibrations and noise of electrical machines should take into account the effect of magnetostriction and its stress dependency.

References

[1] P. Witczak, Calculation of force densities distribution in electrical machinery by means of magnetic stress tensor, Arch. Electr. Eng., Vol. XLV, No. 1, pp. 67–81, 1996.

[2] L. Låftman, "The Contribution to Noise from Magnetostriction and PWM Inverter in an Induction Machine", Doctoral thesis, IEA Lund Institute of Technology, Sweden, 94 p, 1995.

[3] K. Delaere, "Computational and Experimental Analysis of Electrical Machine Vibrations Caused By Magnetic Forces and Magnetostriction", Doctoral thesis, Katholieke Universiteit Leuven, Belgium, 224 p, 2002.

[4] Z. Ren, B. Ionescu, M. Besbes, A. Razek, Calculation of mechanical deformation of magnetic material in electromagnetic devices, IEEE Trans. Magn. Vol. 31, No. 3, pp. 1873–1876, 1995.

[5] O. Mohammed, T. Calvert, R. McConnell, "A Model for Magnetostriction in Coupled Nonlinear Finite Element Magneto-elastic Problems in Electrical Machines", International Conference on Electric Machines and Drives IEMD '99, Seattle, Washington, USA, pp. 728–735, May 1999.

[6] F. Ishibashi, S. Noda, M. Mochizuki, "Numerical simulation of electromagnetic vibration of small induction motor", IEE Proc. Electr. Power Appl., Vol. 145, No. 6, pp. 528–534, November 1998.

[7] G.H. Jang, D.K. Lieu, "The effect of magnetic geometry on electric motor vibration", IEEE Trans. Magn., Vol. 27, No. 6, pp. 5202–5204, November 1991.

[8] C.G. Neves, R. Carlson, N. Sadowski, J.P.A. Bastos, N.S. Soeiro, "Forced Vibrations Calculation in Switched Reluctance Motor Taking into Account Viscous Damping", International Conference on Electric Machines and Drives IEMD'99, May 1999.
[9] A. Belahcen, "Magnetoelastic Coupling in Rotating Electrical Machines", IEEE Tran. Mag., Vol. 41, No. 5, pp. 1624–1627, May 2005.
[10] A. Belahcen, M. El Amri, "Measurement of Stress-Dependent Magnetisation and Magnetostriction of Electrical Steel Sheets", Internation Conference on Electrical Machines, Cracow, Poland, CD-ROM Paper No. 258, September 5–8, 2004.
[11] A. Arkkio, "Analysis of Induction Motors Based on the Numerical Solution of the Magnetic Field and Circuit Equations", Doctoral thesis, Acta Polytechnica Scandinavica, Electrical Engineering Series No. 59, 97 p. Available at http://lib.hut.fi/Diss/198X/isbn951226076X/.

II-7. COMPARISON OF STATOR- AND ROTOR-FORCE EXCITATION FOR THE ACOUSTIC SIMULATION OF AN INDUCTION MACHINE WITH SQUIRREL-CAGE ROTOR

C. Schlensok and G. Henneberger
Institute of Electrical Machines (IEM), RWTH Aachen University, Schinkelstraße 4, D-52056 Aachen, Germany
christoph.schlensok@iem.rwth-aachen.de, henneberger@iem.rwth-aachen.de

Abstract. In this paper the structure- and air-borne noise of an induction machine with squirrel-cage rotor are estimated. For these, different types of surface-force excitations and rotational directions are regarded for the first time. The comparison of the different excitations shows, that it is necessary to take the rotor excitation into account, and that the direction of the rotation has a significant effect on the noise generation.

Introduction

The drivers of passenger cars nowadays make great demands on the acoustics of the technical equipment such as the electrical power steering. Therefore, it is of high interest to estimate the audible noise radiation of these components. The induction machine with squirrel-cage rotor used as power-steering drive is computed in three steps: coupled to the casing caps by the bearings. For this, the rotor excitation has to be taken into account as well for comparison reasons.

1. electromagnetic simulation,
2. structural-dynamic computation, and
3. acoustic estimation.

The theory is briefly described in [1] and therefore not repeated. In the case of an induction machine with skewed squirrel-cage rotor the location of the maximum force excitation of the stator teeth depends on the rotational direction. So far, only stator-teeth excitation has been regarded in literature [2–4]. Further on the impact of the force exciting the rotor is taken into account. Therefore, four different cases of electromagnetic surface-force excitation are compared and discussed in this paper as listed in Table 1.

Since the rotor of the induction machine is skewed (skewing angle $= 10^{\circ}$) the stator teeth are excited very asymmetrically. The location of the maximal tooth excitation depends on the direction of rotation. In case of right-hand rotation the highest excitation

S. Wiak, M. Dems, K. Komęza (eds.), Recent Developments of Electrical Drives, 211–223.

Table 1. Cases for different force excitations

Type of excitation	Rotational direction
Stator-teeth	Left-hand
Rotor-teeth	Right-hand
Stator-teeth	Right-hand
Rotor- and stator-teeth	Right-hand

values are positioned on the side of the mounting-plate. Left-hand rotation results in maximal excitation locations on the opposite side of the machine. For this, both directions are computed and the audible acoustic-noise radiation is compared. Fig. 1 defines the rotational direction.

Usually it is sufficient to simply take the force excitation of the stator in to consideration to make good predictions of the radiated noise. The stator of the regarded machine is weakly coupled to the casing mechanically spoken by hard rubber rings around the casing caps and steel-spring pins in the notches of the stator and casing. The rotor on the other hand is strongly coupled to the casing caps by the bearings. For this, the rotor excitation has to be taken into account as well for comparison reasons.

Electromagnetic simulation

The first step of the computational process is the electromagnetic simulation. The induction machine is simulated with a three-dimensional magnetostatic model, which uses stator and rotor currents as excitations. Due to computational timesaving reasons the rotor-bar currents are derived from a two-dimensional, transient computation [5]. The 2D model consists of 6,882 first order triangular elements and the computation of one time step in 2D takes

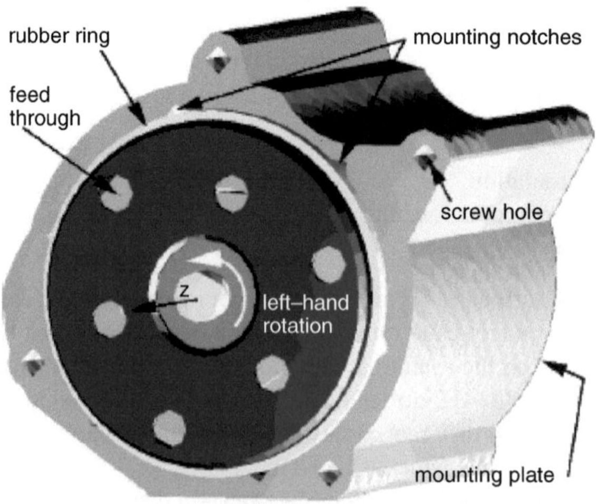

Figure 1. Definition of rotational direction; location of the mounting plate, the mounting notches, the screw holes, and the rubber rings.

$t_{2D} = 24.7$ s. A 3D time-step simulation takes $t_{3D} = 494$ min due to 288,782 first order tetrahedral elements in the 3D model. The duration of the transient phenomenon t_{tp} equals the rotor-time constant t_r:

$$t_{tp} = t_r \approx 0.1 \text{ s} \tag{1}$$

Depending on the time step Δt ($\Delta t_{3D} = 416.6$ μs) the number of time steps "lost" N_{lost} for analysis is:

$$N_{lost} = \frac{t_{tp}}{\Delta t} = 240 \tag{2}$$

In the case of transient 3D simulation the extra simulation time would approximately sum up to:

$$t_{extra} = N_{lost} \cdot \bar{t}_{3D} - N_{lost} \cdot \bar{t}_{2D} = 3,576h = 149d \tag{3}$$

The 3D static simulation can be performed simultaneously on several computers. So that the effective computational time is reduced drastically to about 3 weeks in total, for both: the 2D transient and 3D static simulation.

Two global results are provided:

1. the net force onto the rotor and
2. the torque of the machine for each time step.

Due to the symmetry of the machine only a half model has to be applied and the radial and tangential components of the net force cannot be computed. Therefore, only the axial component of the net force and the torque are analyzed. All electromagnetic simulations are performed employing the open-source software iMOOSE of the IEM [6].

The studied point of operation is at nominal speed $n_N = 1{,}200$ rpm and $f_1 = 48.96$ Hz. $N = 120$ time steps are computed and analyzed with the 3D model. For the 2D model the equivalent time steps are taken into regard.

The time behavior of the axial component of the net force, which depends only on the skewing angle is depicted in Fig. 2. The average value is $F_z = 9.22$ N. The direction of the force depends on the rotational direction. In the case of left-hand rotation the force acts

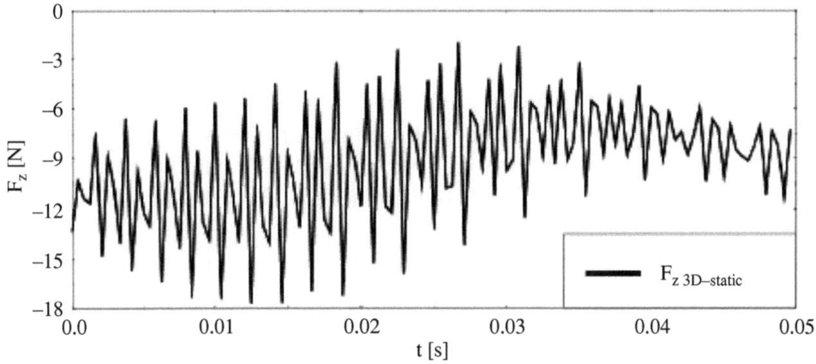

Figure 2. Net-force behavior for 3D model.

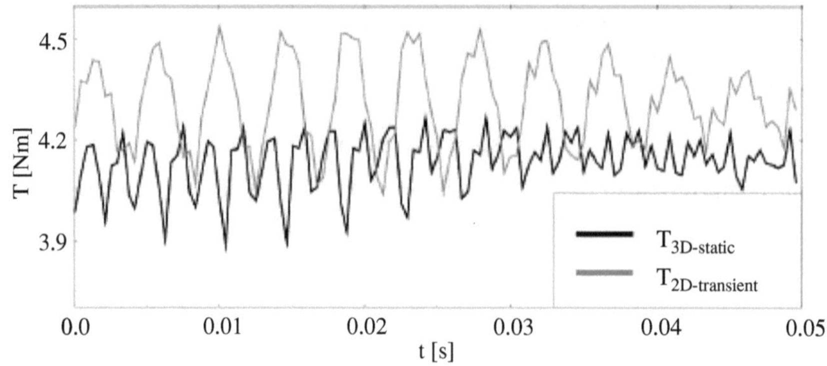

Figure 3. Torque behavior for 2D and 3D model.

in negative z-direction (see Fig. 1). For right-hand rotation the rotor is dragged into the opposite direction.

Fig. 3 shows the time behavior of the torque for 2D transient and 3D static computation. The average value of the 3D torque is lower because of the rotor skewing and the front leakage: $M_{3\text{D}} = 4.13$ Nm $<$ 4.31 Nm $= M_{2\text{D}}$. Both effects are neglected by the 2D model. The net-force and the torque behavior are analyzed using the Fast-Fourier Transformation (FFT) [7]. Due to the smaller time step in the case of the two-dimensional simulation $\Delta t_{2\text{D}} = 1/3\Delta t_{3\text{D}}$ the cut-off frequency $f_{co,2\text{D}}$ in the spectrum is three times $f_{co,3\text{D}} = 1{,}200$ Hz. The time step $\Delta t_{3\text{D}}$ and the number of time steps $N_{3\text{D}}$ in case of the 3D simulation are chosen in such a way that the resolution in the frequency domain is exactly the rotor speed $f_R = 20$ Hz. The resolution of the 2D spectrum is 10 times that of the 3D spectrum because of the high number of time steps $N_{2\text{D}} = 3{,}600$. With the Criteria of Nyquist [8] and

$$\Delta f = \frac{2 \cdot f_{co}}{N} \tag{4}$$

the resolutions of the spectra result in:

$$\Delta f_{3\text{D}} = 20 \text{ Hz} \quad \text{and} \quad \Delta f_{2\text{D}} = 2 \text{ Hz} \tag{5}$$

Fig. 4 shows the spectrum of the axial net-force component of the 3D simulation.

The main orders found are at intervals of $f_{int} = 240$ Hz. The same orders are found in the torque spectrum in Fig. 5. This reflects the very close link of the axial net-force component and the torque: The torque vector points into the axial direction. Structure-borne sound-measurements show the highest values at 720 and 940 Hz next to others. These two significant orders might be caused by the axial force and torque excitation. The spectrum of the 2D simulation is similar to Fig. 5.

Next to these two global values the electromagnetic computation also provides the flux-density distribution for each time step. From this the surface-force density is derived at the interface of air and the lamination of the machine [9].

Only the 3D model is regarded for the investigations in the following. For each time step the surface-force density of the stator and rotor lamination are computed of the 3D model.

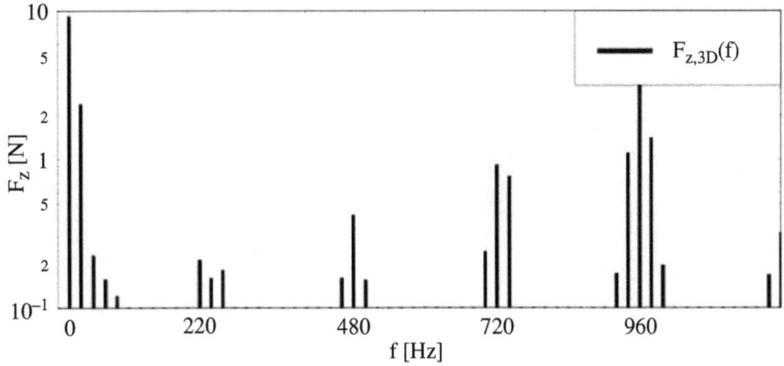

Figure 4. Spectrum of the axial net-force component of the 3D model.

The excitation of each considered element is analyzed by using the FFT and transforming the forces to the frequency domain.

The surface-force density-excitation for one time step is depicted in Fig. 6. The highest values are reached at the up-running edges of the stator teeth.

The surface-force excitation is transformed to the frequency domain as well. The FFT is again used. In a first step the values of the three components (x, y, and z) of all $N = 120$ time steps of each stator-surface element are collected. There are $E_{stator} = 20{,}602$ shell elements. Then the FFT is performed for each of these elements. Finally the transformed values for the three components are rearranged into two files (real and imaginary part) for each of the frequencies in the spectrum (number of frequencies: $N_f = 61$).

Structural-dynamic simulation

The next step in the computational process is the simulation of the deformation of the entire machine structure due to the surface-force density-excitation derived from the electromagnetic simulation. For this, an extra model of the entire machine is generated consisting of the stator and rotor laminations including the winding and the squirrel cage, the shaft, the bearings, the casing, and the casing caps. The model is described more detailed in [1].

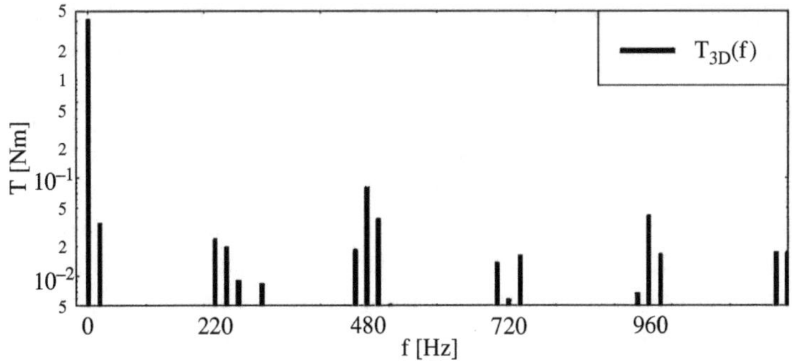

Figure 5. Spectrum of the torque of the 3D model.

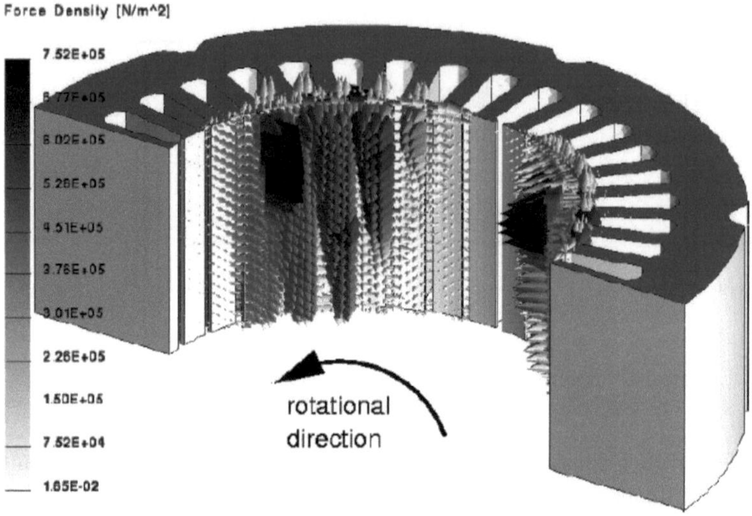

Figure 6. Surface-force density-excitation for one time step at left-hand rotation.

Four different types of force excitations are object of investigation as listed in the introduction section of this paper. The locations of the highest force excitations depend on the rotational direction. In the case of right-hand rotation the maximal forces arise at the side of the mounting-plate of the machine (see Figs. 1 and 6). For left-hand rotation the highest excitation is located on the opposite side.

Exemplarily, Fig. 7 shows the real and the imaginary part of the surface-force density-excitation for $f = 420$ Hz for left-hand rotation. There is a phase shift between both parts. This will result in a pulsating deformation behavior for the excitation at this frequency. The maximal forces in both, the real and imaginary part, are positioned at four locations. The order of deformation does not depend on the rotational direction.

The frequencies analyzed and the resulting mechanical orders of deformation r are listed in Table 2. $r = 2$ is the most often order found. Mainly second and fourth order deformations are detected. Orders higher than $r = 6$ usually do not produce strong deformation and are not critical in respect of noise radiation. The most important order is the elliptical second order [10].

The resulting deformation of the stator and the casing of the machine in the case of pure stator-teeth excitation is shown in Fig. 8 in an overemphasized representation for the frequency of $f = 620$ Hz.

The spring pins keeping the stator fixed in the casing damp the deformation and decouple both parts very well. The deformation of the casing is much lower than that of the stator and cannot be sensed in the figure. Some deformation orders are shown in Fig. 9.

Fig. 10 depicts the real part of the deformation of the entire machine structure in scalar representation at $f = 720$ Hz. Although the order of deformation of the stator deformation is $r_{stator} = 2$ the deformation order of the casing of the machine is $r_{casing} = 4$. This effect stems from the skewing and the mounting of the machine. The skewing results in torsional vibrations [5]. The machine is mounted on one front plate. This way the deformation is

Table 2. Mechanical orders of deformation found for all analyzed frequencies

r	f (Hz)
2	420, 520, 720, 940
4	100, 1,040, 1,140
6	620

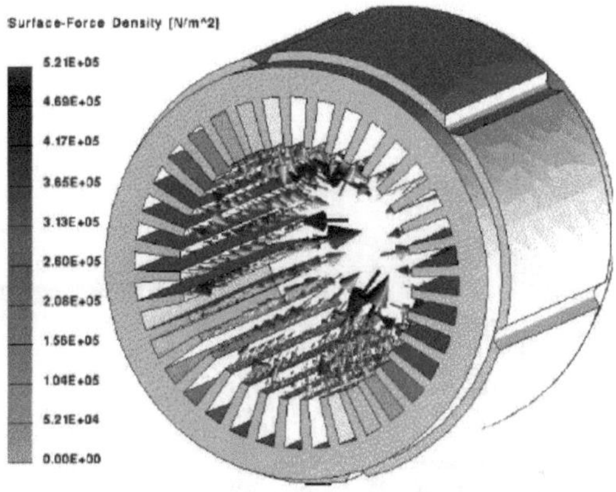

(a) Surface-Force Density-Excitation, Left-Hand Rotation, Real Part.

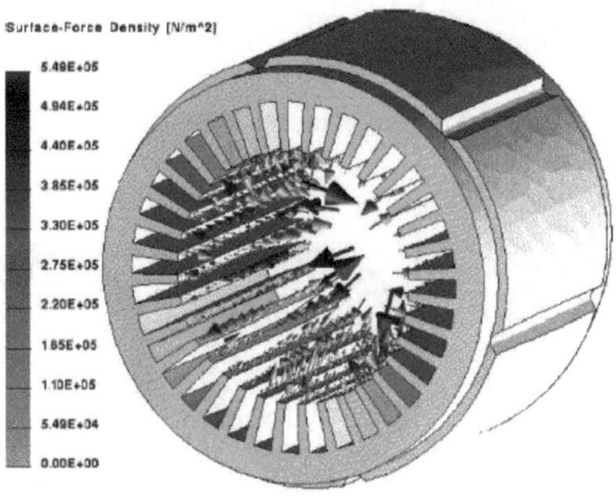

(b) Surface-Force Density-Excitation, Right-Hand Rotation, Imagnary Part.

Figure 7. Location of the maximal surface-force density-excitation for $f = 420$ Hz at left-hand rotation: (a) real part. (b) Imaginary part.

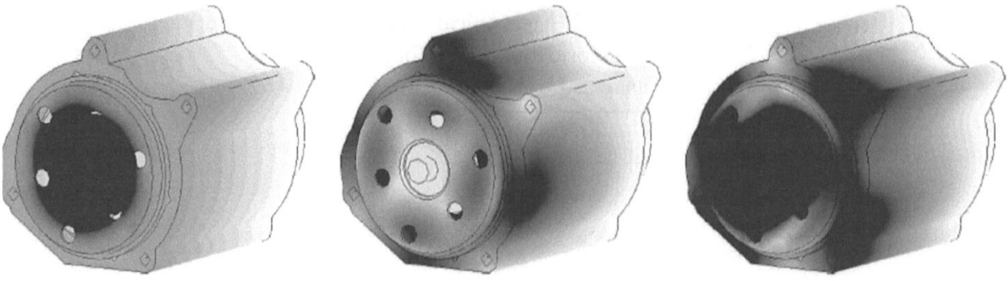

(a) Rotor-Teeth Excitation, Right-Hand Rotation, $f = 940\,\mathrm{Hz}$ (Real Part).

(b) Stator-Teeth Excitation, Right-Hand Rotation, $f = 940\,\mathrm{Hz}$ (Real Part).

(c) Rotor- and Stator-Teeth Excitation, Right-Hand Rotation, $f = 940\,\mathrm{Hz}$ (Real Part).

Figure 8. Deformation of the stator at $f = 620$ Hz.

Figure 9. Mechanical orders of the deformation.

(a) $r = 2$. (b) $r = 4$. (c) $r = 6$.

Figure 10. Deformation of the entire machine in the case of pure stator-teeth excitation, real part.

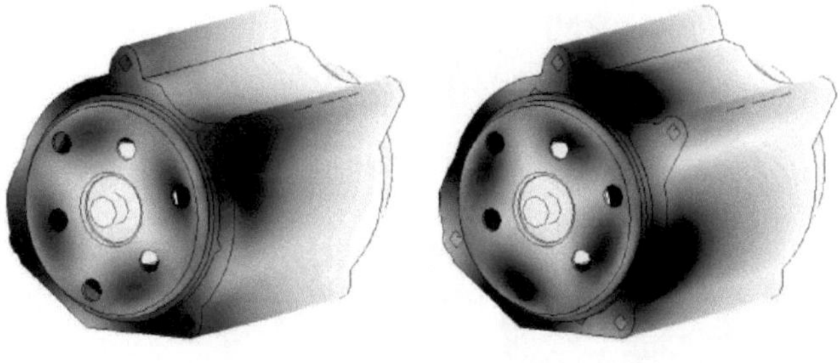

(a) Stator-Teeth Excitation, Left-Hand Rotation, $f = 720$ Hz.

(b) Stator-Teeth Excitation, Right-Hand Rotation, $f = 720$ Hz.

Figure 11. Deformation of the entire machine structure: pure rotor, pure stator, and combined rotor-stator excitation.

"reflected" at this stiff front plate and produces the double order on the opposite front plate ("open end"). In the case of right-hand rotation the maximal deformation arises on the "open end" of the structure. This is the same location of the maximal force excitation. Therefore, the deformation of the machine depends strongly on the rotational direction.

In a next step the deformation stemming from the combined rotor- and stator-excitation at right-hand motion is simulated. Exemplarily Fig. 11 depicts the real part of the deformation for pure rotor excitation (Fig. 11(a)), pure stator-excitation (Fig. 11(c)), and for combined excitation (Fig. 11(c)) at $f = 940$ Hz. All three pictures show the same scaling (black strong, white weak deformation).

Pure rotor excitation results in strong deformation of the shaft and the casing caps. This deformation mainly deforms these parts in radial and axial direction of the machine. If pure stator-excitation is regarded mainly the casing at the "open end" and the outer parts of the casing caps are deformed. Finally the combination of both excitations results in strong deformation of the casing and the casing caps. These observations are stated for all studied frequencies listed in Table 2.

Next to the possibility of using the deformation of the structure for acoustic simulation the structure-borne sound at certain locations can be derived. For this reason the node nearest to the location of the accelerometer for measurements on the flank of the casing and one node on the casing cap where the converter is mounted are chosen. Fig. 12 shows the locations of the nodes regarded with their IDs. The definition of the tangential, radial, and axial direction is displayed as well.

The structure-borne sound-level L_S is calculated as follows for radial, tangential, and axial direction:

$$L_S = 20 \cdot \log \frac{a}{1\ \mu\text{m/s}^2}\ \text{dB} \tag{6}$$

a is the acceleration of the specific node at the regarded frequency f [1].

Fig. 13 depicts the structure-borne sound-levels for some selected frequencies in the case of left-hand rotation and pure stator-teeth force-excitation. The axial sound-level is about $\Delta L_S = 20$ dB lower throughout the spectrum. This fact can be stated for right-hand rotation

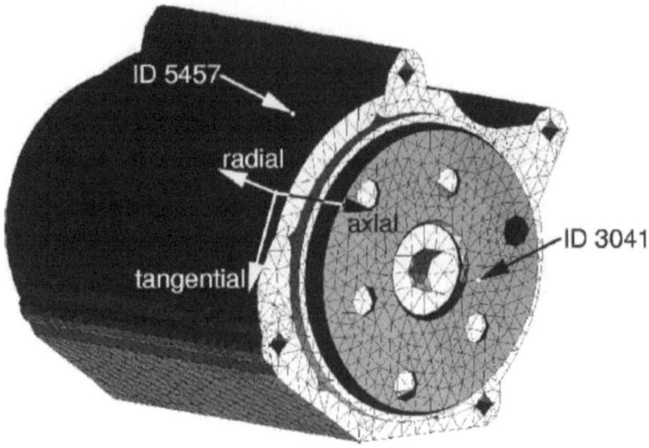

Figure 12. Locations of the nodes for structure-borne sound-analysis.

and both pure stator and combined rotor- and stator-excitation as well. The same levels are reached for the tangential and radial components for all regarded frequencies.

Referring to the deformation plot in Fig. 11(a) Fig. 14 shows exemplarily the results for the levels calculated on the casing cap (Node 3041) in the case of pure rotor-force excitation for Node 3041. The three components reach nearly the same levels. The highest values are detected for $f = 620$ Hz and $f = 720$ Hz.

Fig. 15 depicts the levels in the case of pure stator-excitation (see Fig. 11(c)) for Node 3041. Except for much lower levels at $f = 100$ Hz and $f = 620$ Hz slightly lower values are reached for $f = 720$ Hz. The axial component is the lowest for $f = 620$ Hz and the higher frequencies similar to the results in Fig. 13.

Finally the combined stator- and rotor-excitation is regarded (Fig. 16). At Node 3041 the spectrum is similar to pure stator-excitation with the exception of the significantly higher levels at $f = 620$ Hz and $f = 720$ Hz. This spectrum suits the result of the acceleration measurements very well.

It can be stated that pure stator-excitation results in fairly good estimations concerning the structure-borne sound-levels. Only very few orders are affected significantly

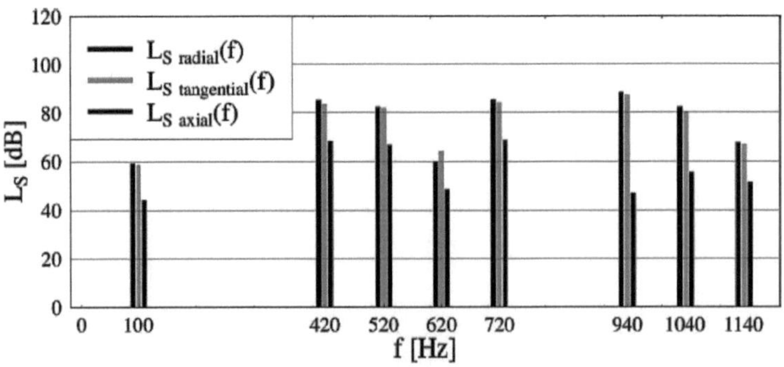

Figure 13. Structure-borne sound-levels, left-hand rotation, pure stator excitation, ID 5457.

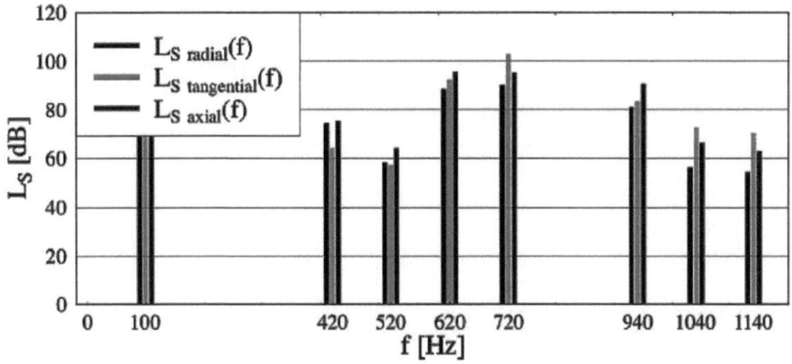

Figure 14. Structure-borne sound-levels, right-hand rotation, pure rotor excitation, ID 3041.

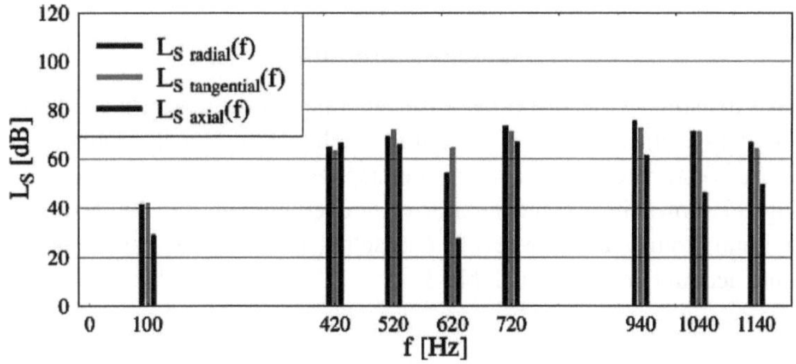

Figure 15. Structure-borne sound-levels, right-hand rotation, pure stator excitation, ID 3041.

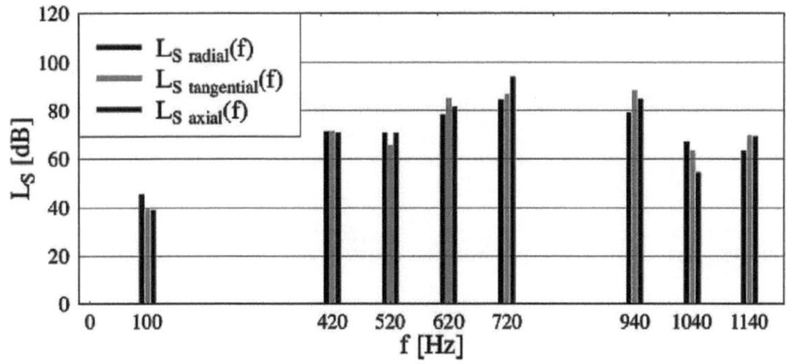

Figure 16. Structure-borne sound-levels, right-hand rotation, combined stator-rotor excitation, ID 3041.

Table 3. Maximal levels of the estimated air-borne sound-pressure (sl: stator excitation, left-hand rotation; sr: stator excitation, right-hand rotation; srr: stator + rotor excitation, right-hand rotation)

f(Hz)	L_{sl}(dB)	L_{sr}(dB)	L_{srr}(dB)
420	16	26	9
520	23	23	23
620	9	6	14
720	27	25	27
940	28	29	28
1,040	11	21	22
1,140	9	10	10

by the forces amplify the axial component. Therefore, it is of advantage to take the rotor excitation into account to get more exact results concerning the structure-borne sound.

Acoustic simulation

The last step is to estimate the air-borne noise generated by the different excitations. For this reason a boundary-element model of the entire machine structure is applied. The air-borne sound-pressure is estimated on an analysis hemisphere around the machine at a distance of $d = 1$ m. Fig. 17 shows the result for stator-rotor excitation at $f = 420$ Hz.

The maximum sound-pressure levels L reached for the three cases taking the stator excitation into account are listed in Table 3.

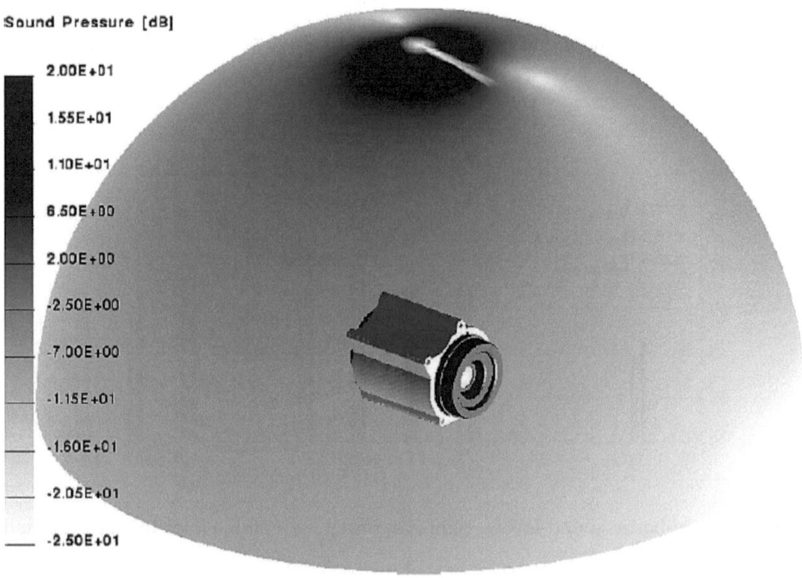

Figure 17. Sound-pressure distribution at $f = 420$ Hz for stator-rotor excitation and right-hand rotation.

The results show that the direction of the rotation has a significant effect on the noise generation. Except for $f = 720$ Hz and $f = 620$ Hz all orders are amplified up to $\Delta L =$ 10 dB. If the rotor-force excitation is taken into account some orders become louder and some quieter. The air-borne sound-levels do not suit the acceleration measurements as well as those of the structure-borne sound.

Conclusion

In this paper the structure- and air-borne noise of an induction machine with squirrel-cage rotor are estimated.

For this, different types of surface-force excitations and rotational directions are regarded for the first time. In general the calculated structure-borne sound-levels suit the acceleration measurements of the industrial partner very well. The acoustic-noise levels differ from those.

The comparison of the different excitations show, that it is necessary to take the rotor excitation into account. In case of pure stator-excitation e.g. the first stator-slot harmonic at 720 Hz does not reach as significantly high levels as expected although it is one of the strongest orders measured.

References

[1] C. Schlensok, T. Küest, G. Henneberger, "Acoustic Calculation of an Induction Machine with Squirrel Cage Rotor", 16th International Conference on Electrical Machines, ICEM, Crakow, Poland, September 2004.

[2] B.-T. Kim, B.-I. Kwon, Reduction of electromagnetic force harmonics in asynchronous traction motor by adapting the rotor slot number, IEEE Trans. Magn., Vol. 35, No. 5, pp. 3742–3744, 1999.

[3] T. Kobayashi, F. Tajima, M. Ito, S. Shibukawa, Effects of slot combination on acoustic noise from induction motors, IEEE Trans. Magn., Vol. 33, No. 2, pp. 2101–2104, 1997.

[4] L. Vandevelde, J.J.C. Gyselinck, F. Bokose, J.A.A. Melkebeek, Vibrations of magnetic origin of switched reluctance motors, COMPEL, Vol. 22, No. 4, pp. 1009–1020, 2003.

[5] G. Arians, Numerische Berechnung der elektromagnetischen Feldverteilung, der strukturdynamischen Eigenschaften und der Geräusche von Asynchronmaschinen, Aachen: Shaker Verlag, 2001. Dissertation, Institut fur Elektrische Maschinen, RWTH, Aachen.

[6] G. Arians, T. Bauer, C. Kaehler, W. Mai, C. Monzel, D. van Riesen, C. Schlensok, iMOOSE, www.imoose.de.

[7] I.N. Bronstein, K.A. Semendjajew, Taschenbuch der Mathematik. 25. Auflage, Leipzig, Stuttgart: B.G. Teubner Verlagsgesellschaft, 1991.

[8] H.D. Lke, Signalübertragung, Berlin, Heidelberg, and New York: Springer-Verlag, 1999.

[9] I.H. Ramesohl, S. Küppers, W. Hadrys, G. Henneberger, Three dimensional calculation of magnetic forces and displacements of a claw-pole generator, IEEE Trans. Magn., Vol. 32, No. 3, pp. 1685–1688, 1996.

[10] Jordan, H., Geräuscharme Elektromotoren, Essen: Verlag W. Girardet, 1950.

II-8. A CONTRIBUTION TO DETERMINE NATURAL FREQUENCIES OF ELECTRICAL MACHINES. INFLUENCE OF STATOR FOOT FIXATION

J-Ph. Lecointe, R. Romary and J-F. Brudny

Laboratoire Systèmes Electrotechniques et Environnement, Université d'Artois, Technoparc Futura, 62400 Béthune, France

jp.lecointe@fsa.univ-artois.fr, jfrancois.brudny@univ-artois.fr, raphael.romary@univ-artois.fr

Abstract. In this paper, four methods to determine the mechanical characteristics (natural frequencies, mode numbers) of electrical machine stators are developed. Result comparison concerns analytical laws, a finite element software, a modal experimental procedure and a method based on analogies between mechanic and electric domains. Simple structures are studied in order to analyze the validity of each method with accuracy. The fixation of a stator yoke allows to observe the modifications of the mechanical behavior.

Introduction

The study of electrical machine noise always leads to mechanical resonance problems. The noise origins are generally divided into three sources which are mechanic, aerodynamic, and magnetic [1]. The noise of magnetic origin is produced by the electromagnetic radial forces between the stator and the rotor. The noise resulting from these forces can be particularly severe when a force of magnetic origin is close to a natural frequency because circumferential modes of the stator are excited [2]. It is particularly the case of switched reluctance machines [3] but it could be also problematic for classical alternative current machines supplied by converters. That is why an accurate knowledge of the mechanical behavior of the machine—especially the natural frequencies—is important in noise and vibration prediction.

Most studies often use finite element software. These last ones give accurate and usable results if model is well fitted to the studied structure. Indeed, materials constituting the machine have to be correctly estimated; otherwise the advantages offered by FE are reduced. In this paper, different methods are studied in order to estimate which ones can give accurate values of natural frequencies of simple structures, as fast as possible in order to establish a rapid diagnosis. Four methods are performed. The first one uses analytical expressions based on Jordan's work, but the proposed laws are improved thanks to fewer restrictive hypotheses. The second method uses a FE software (Ansys). The third method is original because it is based on analogies between mechanic and electrical domains. Consequently, the mechanical problem is transformed into an electrical circuit resonance determination. The last method is experimental: a modal hammer allows to verify the calculated values.

S. Wiak, M. Dems, K. Komęza (eds.), Recent Developments of Electrical Drives, 225–236.

Table 1. Used variables

Symbol	Quantity
R_c	Average radius of the yoke
e_c	radial thickness of the yoke
L_e	length of the cylinder
E	Young modulus
P	Mass density
N	Poisson ratio

Usual studies consider machines in free conditions. It allows the influence of different parameters to be quantified; several papers have already discussed the effects of the feet, the cooling ribs, the windings, or the end-bells [4,5]. In this paper, the influence of the fixation on a rigid chassis is studied. The purpose is to evaluate the fixation impact on the natural frequencies and on the shape of the mode numbers (Table 1).

Methods of natural frequency determination

Four methods are performed to determine the natural frequencies and the associated mode number. Technologies and principles for each of them are quite different. The older one is entirely based on analytic beam theory [6]–[7] whereas another one is totally numeric (finite element). The third developed method considers analogies between mechanical; electrical quantities and the identification of mechanical parameters allows to transform the mechanical problem into an electrical circuit study. The experimental method uses a modal station and gives the reference results. Table 1 presents the used variables.

Analytical method

The presented laws have been rewritten [8] more accurately about smooth free rings. Considering that the stator is the most responsive part compared to the rotor, the ball bearings or the flanges, this analytical method gives a fast determination of stator natural radial frequencies. This method allows to determine only the frequencies in two dimensions. They are noted F_i, where "i" is the mode number:

$$m = 0: F_0 = \Omega \tag{1}$$

$$m = 1: F_1 = \Omega\sqrt{2} \tag{2}$$

$$m = 1: F_1 = \Omega\sqrt{\frac{\Gamma a + m_+ - \sqrt{\Theta}}{2\left(3\Gamma^2 m_- + 1\right)}} \tag{3}$$

where

$$\Theta = m_+^2 + \Gamma^2\left(2am_+ - 4m^2m_-^2\right) + \Gamma^4\left(a^2 - 12m^2m_-^3\right) \tag{4}$$

$$a = 4m^4 - m^2 - 3 \tag{5}$$

$$m_+ = m + 1 \quad m_- = m - 1 \tag{6,7}$$

$$\Omega^2 = \frac{E}{\rho R_c^2} \tag{8}$$

$$\Gamma = \frac{e_c^2}{12R_c^2} \tag{9}$$

Numerical determination with a FE software

The finite element software (Ansys) solves the conventional eigenvalue equation:

$$\{[H] - \omega^2 [M]\} [\chi] = \{0\} \tag{10}$$

where $[H]$ and $[M]$ are, respectively, the stiffness and the mass matrixes. The solutions $\omega/(2\pi)$ and $[\chi]$ are the natural frequencies and the nodal displacements. Resolution uses the block Lanczos algorithm; values of Young modulus, mass density, and Poisson ratio are required.

Equivalent electric circuit

As the first method allows to find only the natural resonances of structures in free conditions, a second method based on analogies between mechanic and electric domains has been developed. The equivalences are presented at Table 2.

The stator is divided into M levels, each of them containing N cells (Fig. 1). Each cell is characterized by its mass. The deformation of the structure is represented by the relative displacement of a cell compared to the others cells. From a mechanical point of view, the rigid linkages can be taken into account with springs and, from an electrical point of view, with capacitors. A resistor allows to take into account the energy lost in the movement by viscous friction.

The equivalent scheme of the structure is presented at Fig. 2. The voltages applied on the internal part of the first level represent the forces supported by the stator. The voltage fluctuations at the external periphery give the evolution of the deformations. Consequently,

Table 2. Equivalences

	Mechanical quantities		Electrical quantities
Mass	M	Inductance	L
Rigid linkages	K	Capacitor	C
Viscous friction	F_v	Resistor	R
Force	x	Voltage	U
Speed	dx/dt	Current	i

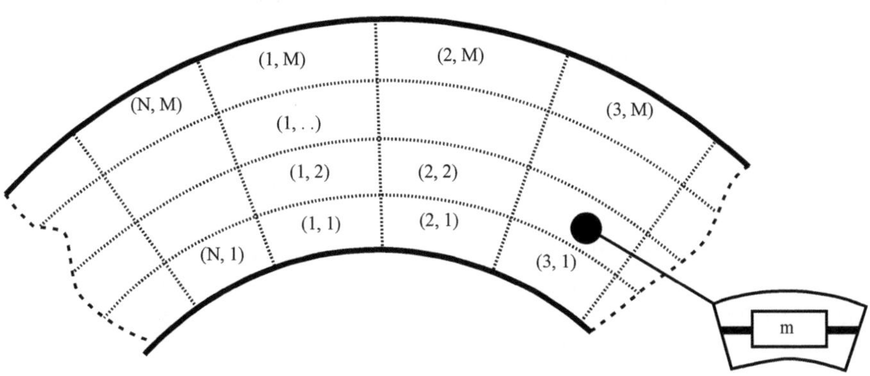

Figure 1. Stator division.

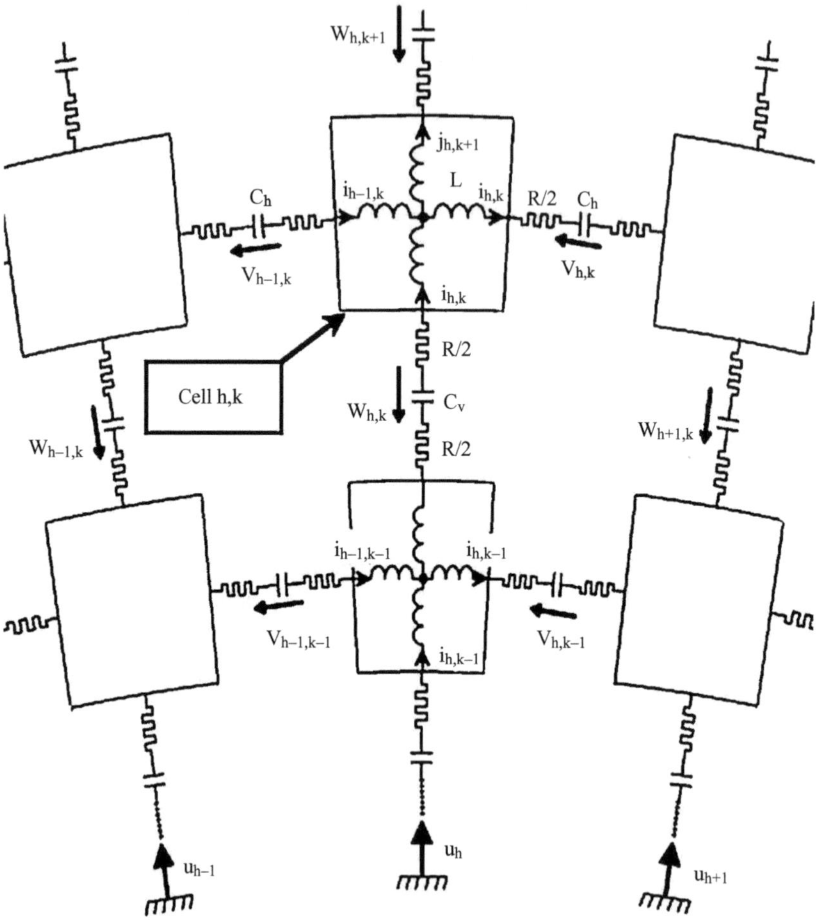

Figure 2. Equivalent electric circuit.

such a model gives the possibility to model different excitations: sinusoidal or pulsed. Therefore, the validity of the model can be verified with modal experimentations performed with an impact hammer.

For a sinusoidal excitation characterized by a frequency ω_e and a mode number m, it becomes:

$$u_h(t) = U_m \cos\left(\omega_e t - hm\frac{2\pi}{N}\right) \tag{11}$$

where h gives the position of the force along the internal periphery. Successive calculations give the frequency value for which the vertical response is maximal and thus the radial frequency can be determined. The fixation of the machine can be studied by imposing a potential zero in chosen points of the external periphery.

Next step consists in determining the values of the equivalent parameters.

The equivalent inductance is given by the elementary cell mass which is given by the expression:

$$L = \pi \frac{r_{k+1}^2 - r_k^w}{N} \rho L_e \tag{12}$$

where r_{k+1} and r_k are, respectively, the external and the internal radius of the level k.

Then, the equivalence between potential and electrical energies gives the relation $K = 1/C$. The Hooke law and the classical capacitor calculation relations allows to determine the expressions of capacitors C_v and C_h, according to the considered geometry:

$$C_v = \frac{(r_{k+1} - r_k)\, N}{(r_{k+1} + r_k)\, E_y} \tag{13}$$

$$C_h = \frac{(r_{k+1} + r_k)\pi}{(r_{k+1} - r_k) N E_y} \tag{14}$$

The equivalent resistance is the most difficult to determine. As it does not influence the frequency response, this coefficient is arbitrary chosen.

The equations of the electric circuit lead to a second order differential system composed of $4 \times M \times N$ lines. Consequently, the response of the structure is determined in the state space. The state vector is composed of $4 \times M \times N$ elements: vertical and horizontal currents and capacitor voltages. Computation is realized with Matlab⇔ and Simulink⇔. Such a process has a double advantage. First, computation time is lower than FE software because the matrix size is smaller. Secondly, it could be set up on any computer without any specific software.

Modal experimental device: impact hammer test

The modal test using a hammer is the least expensive. The examined structure is excited by an impact given with a specific hammer (Meggitt Endevco, model 2302-5) which allows to measure the characteristics of the shock. A piezoelectric accelerometer (Bruël & Kjaer, model 4384) allows to observe the response of the structure. A spectrum analyzer (Bruël & Kjaer, 2035, 2 channels) and a modal analysis software (Star System™) provide the processing and the analysis of the measures. The Fig. 3 presents a scheme of the experimental device. The frequency limit of the used hammer is around 8 kHz.

Studied structures

Three elementary structures are studied. Two are perfectly sleek rings (Fig. 4) whereas the third one is composed of a statoric yoke made of steel which is equipped of two welded feet (Fig. 5). Table 3 presents the dimensions of the structures. The geometry of the cylinders is quite different. Indeed, the first one is elongated whereas the second one presents an important diameter. Studying such different configurations allows the method accuracy to be quantified. A massive structure equipped of feet is deliberately studied in order to avoid the perturbations generated by coils, cooling ribs, or stack lamination. In this way, the main phenomenon observed is the foot fixation.

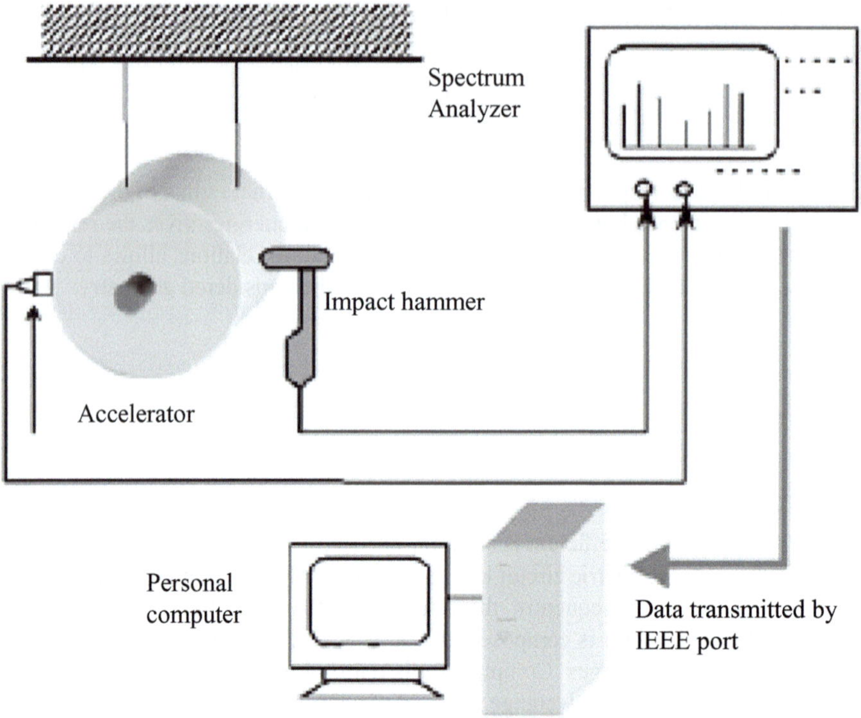

Figure 3. Experimental device.

Results and experimental validation

Structures in free conditions

First, all the methods are applied to the two cylinders in free conditions to check their accuracy. Analytical method, FE software, and analogy method require the same parameters: Young modulus, mass density, and Poisson ratio are, respectively, equal to $E = 2, 1.10^{11}$ Pa,

Figure 4. Picture of the smooth cylinders.

Figure 5. Picture of the statoric yoke.

Table 3. Dimensions

	Cyl 1	Cyl 2	Statoric ring
R_c (mm)	53	133.5	120.5
e_c (mm)	10	23	42
L_e (mm)	136	104	260

Meshing and 2D deformations (cylinder 2)

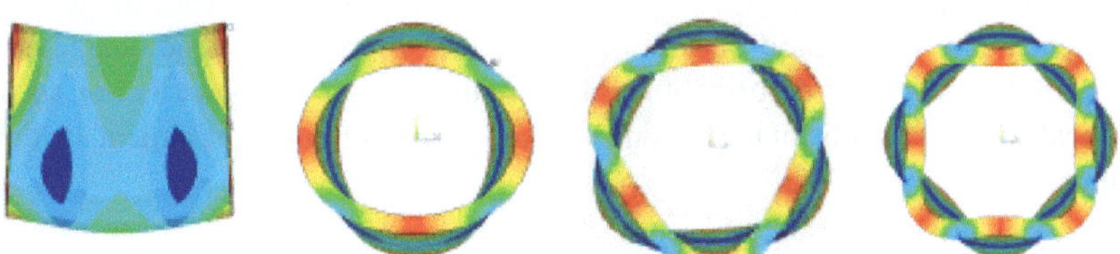

3D deformations (cylinder 2)

Figure 6. FE results.

$\rho = 7{,}850$ kg/m^3, and $\upsilon = 0, 3$. Fig. 6 presents the FE meshing of the cylinder 2 and the usual 2D deformation shape. Additional 3D frequencies and bending appear.

Fig. 7 shows the cylinder 2 response. It is obtained by the method using analogies with $M = 5$ and $N = 16$. The excitation is sinusoidal and the current in the last vertical branch of the circuit presents a transient state and then becomes sinusoidal. The division of the cylinder

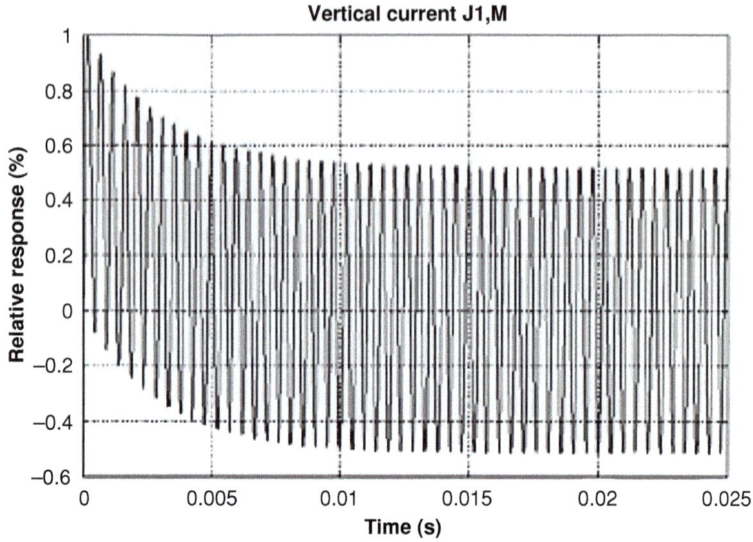

Figure 7. Vertical current $j_{M,N}$ response (sinusoidal excitation).

can be noticed on Fig. 8 which represents the special evolution of the external surface; the different shapes correspond to the excitations (modes 2, 3, and 4). A shock simulation leads to the response presented at Fig. 9 whereas Fig. 10 presents the FFT of the signal.

In free conditions, the structures are suspended with rigid rubber bands or with elastic rubber bands if the weight of the structure is not too important. A meshing is constituted of 192 points drawn on each external surface and shared out four planes (Fig. 11). Each point is excited with the modal hammer four times. Mode shapes given by the modal software can be perfectly identified. Fig. 12 presents the cylinder 2 shapes.

Results for each method applied to both cylinders are presented at Table 4 and hammer test is chosen as reference. Results show that, independently of the considered geometry, the maximum relative error for analytical method is 2.3% for the four first modes. The accuracy of such analytical process is noticeable in spite of the particular geometry of the cylinder 2. The FE method does not give results so precise but the characterization is more complete with 3D deformations. The method using analogies is not so precise as analytical laws. The advantage of this method is its future development in the third dimension.

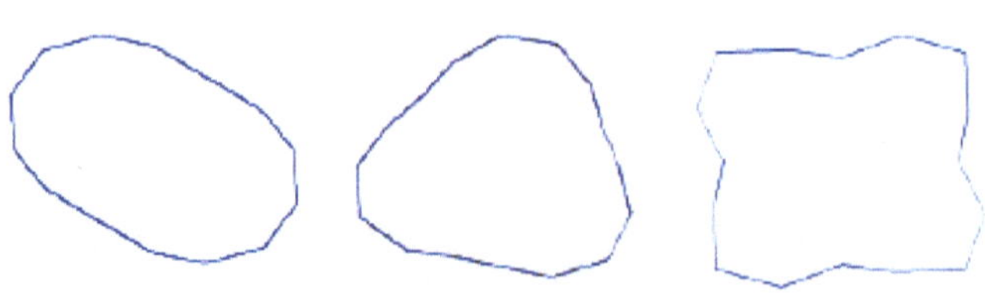

Figure 8. Shapes of deformation.

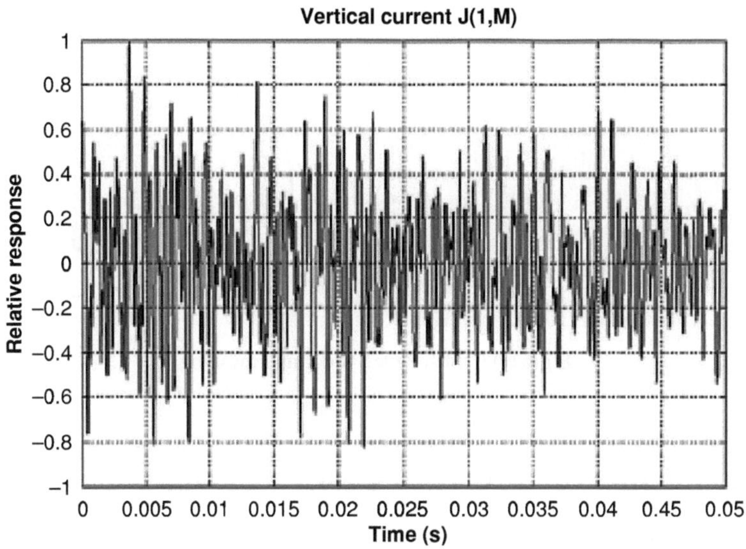

Figure 9. Vertical current $j_{M,N}$ response (impact excitation).

Fixed structure

Figure 13 presents the mode shapes and the natural frequencies of the statoric ring first in free condition and then when it is attached.

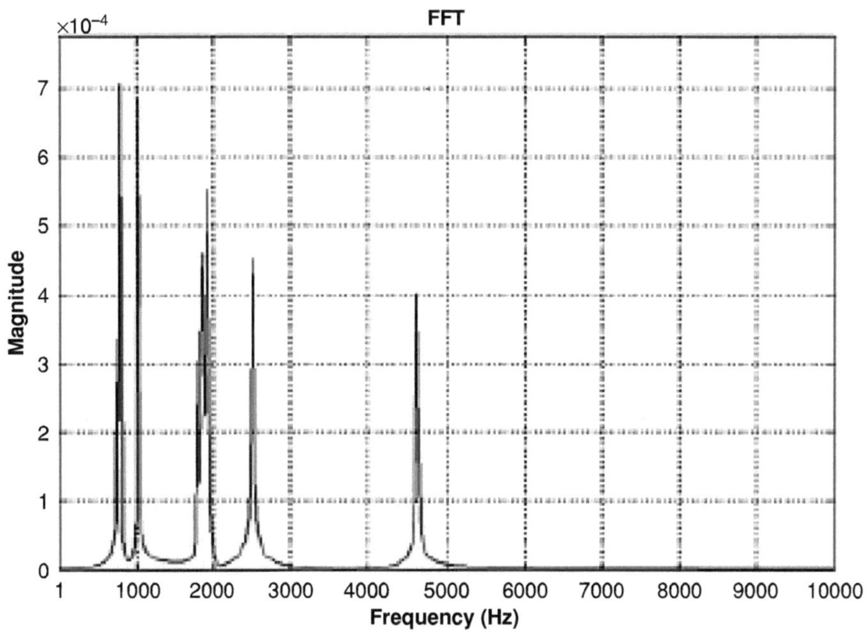

Figure 10. FFT.

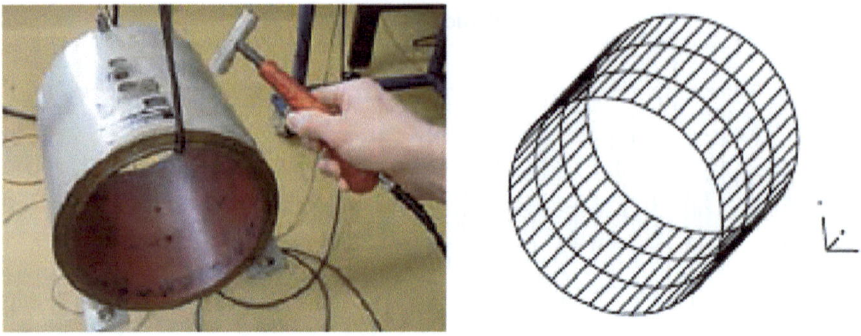

Figure 11. Suspended stator and meshing.

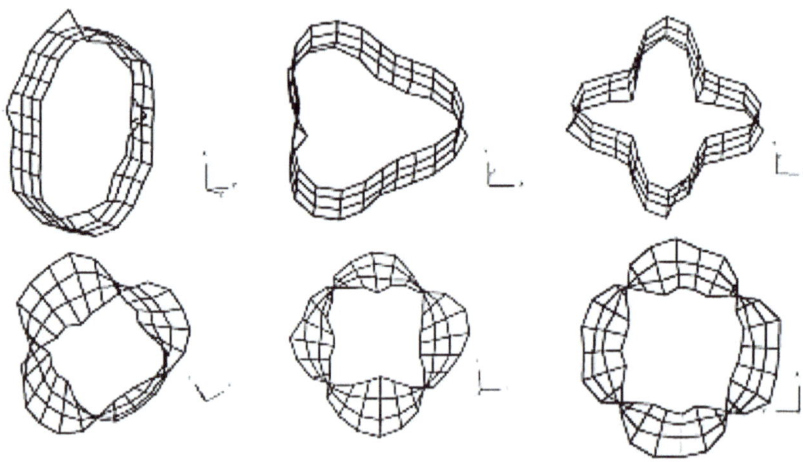

Figure 12. Experimental results: modes 2-2D, 3-2D, 4-2D and modes 2-3D, 3-3D, 4-3D.

Table 4. Results for the sleek cylinders

	Natural frequencies (Hz)			
Mode	Analytical laws	FE Software	Analogies	Hammer test
Cyl 1				
2-2D	2,244	2,232	2,095	2,280
2-3D	x	2,808	x	2,740
3-2D	6,208	6,169	6,395	6,280
3-3D	x	7,155	x	6,930
Cyl 2				
2-2D	756	824	799	742.6
2-3D	x	1,454	x	1,350
3-2D	2,102	2,295	1,899	2,060
3-3D	x	3,752	x	3,470
4-2D	3,940	4,301	3,700	3,860
4-3D	x	6,278	x	5,810
5-2D	6,200	6,761	6,318	6,060
5-3D	x	8,976	x	8,230

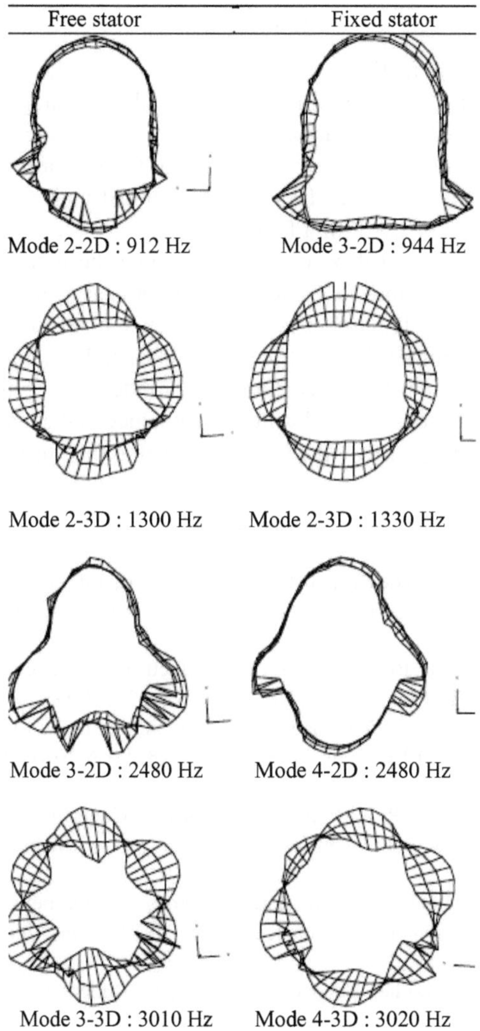

Figure 13. Mode shapes and natural frequencies.

In free mode, the three first modes are perfectly identified. However, some modes are not as perfectly defined as for the sleek cylinders. For example, the shape of the mode 2 in two dimensions is undoubtedly deformed by the welded feet.

When the stator yoke is fixed to the chassis, the natural frequencies are increased but the maximal relative variation is equal to 5.2%. This evolution is attributed to the fixation of the structure. If the natural frequencies are not modified a lot, the mode shapes differ from the shapes observed in free mode. For example, the modes 3-2D and 3-3D in free mode become almost modes 4-2D and 4-3D when the structure is fixed. The fixation prevents the structure from moving. It is clearly visible at presented pictures.

Conclusion

In this paper, four different methods to determine the mechanical behavior of cylinders—sleek or equipped of feet—have been studied. Analytical and analogy methods give accurate results but the FE software keeps the advantage of the 3D mode determination. The extended Jordan's laws allow to find very quickly the 2D radial frequencies with a simple computer whereas others methods requires specific software or modal analysis equipment. In this way, the analytical method can be used for a first diagnosis.

Thus, analytical laws can still be used to determine the natural frequencies of fixed electrical rotating machines. Indeed, frequencies in both cases are very close. However, it has been shown that the mode number is different. Such a modification can influence the determination of noise emitted by electrical machines. Indeed, analytical theories [9] of noise prediction consider non-fixed structure as a cylinder of infinite length or a vibrating sphere. Consequently, the next step will study the influence of the machine fixation on noise emission. At last, the analogy method will be extended to the third dimension.

References

[1] P. François, "La génération des bruits et la réponse des structures dans les moteurs asynchrones, en particulier en ce qui concerne les écoulements", Revue Générale de l'Electricité, Avril 1968, pp. 377–392.

[2] B. Cassoret, R. Corton, D. Roger, J.-F. Brudny, Magnetic noise reduction of induction machines, IEEE Trans. Power Electron., Vol. 18, No. 2, pp. 570–579, 2003.

[3] D.E. Cameron, J.H. Lang, S.D. Umans, The origin and reduction of acoustic noise in doubly salient variable-motors, IEEE Trans. Ind. App., Vol. 28, No. 6, November/December, pp. 1250–1255, 1992.

[4] C. Couturier, B. Cassoret, P. Witczak, J.-F. Brudny, "A Contribution To Study The Induction Machine Stator Resonance Frequencies", ICEM 98, Vol. 1, Istanbul, Turkey, September 2–4, 1998, pp. 485–489.

[5] W. Cai, P. Pillay, Z. Tang, Impact of stator windings and end-bells on resonant frequencies and mode shapes of switched reluctance motors, IEEE Trans. Ind. App., Vol. 38, No. 4, pp. 1027–1036, 2002.

[6] S.P. Timoshenko, J.N. Goodier, Theory of Elasticity, 3rd student edition, McGraw-Hill Companies, New York, 1970.

[7] H. Jordan, Geräuscharme Electromotoren, Essen: W. Girardet, 1950.

[8] J.-Ph. Lecointe, R. Romary, T. Czapla, J.-F. Brudny, Five methods of stator natural frequencies determination. Case of induction and switched reluctance machines, Mechanical Systems and Signal Processing, Elsevier, Vol. 18, pp. 1133–1159, 2004.

[9] Ph.L. Alger, The magnetic noise of polyphase induction motors, Trans. Amer. IEEE, Pt. III A, No. 73, pp. 118–125, 1954.

II-9. DIAGNOSIS OF INDUCTION MACHINES: DEFINITION OF HEALTH MACHINE ELECTROMAGNETIC SIGNATURE

D. Thailly, R. Romary and J.F. Brudny

CNRT Futurelec 2—Réseaux et Machines Electriques du Futur, Laboratoire Systèmes Electrotechniques et Environnement—Université d'Artois, Faculté des Sciences Appliquées, Technoparc Futura, 62400 Béthune, France

delphine.thailly@fsa.univ-artois.fr, romary@fsa.univ-artois.fr, brudny@fsa.univ-artois.fr

Abstract. This paper deals with the diagnosis of induction machines using data contained in the radial external magnetic field. This work presents a theoretical approach which permits to study the evolution of each flux density air-gap component through the stator. The aim of this method is to find, by computation, the magnitude of measured spectral lines. The study is made on the couples of toothing spectral lines and justifies why these couples do not have the same magnitude, what is not obvious in a first approach where the practical spectrum is directly compared with this one of the air-gap flux density.

Introduction

From an economical point of view, it appears that, for a factory, predictive maintenance of electrical machines is essential. Various procedures have been brought [1,2], and it should be noted a recent orientation in the exploitation of data enclosed in magnetic field (MF) which surrounds the machine. It was display that is advisable to dissociate two cases: the axial MF [3], which principally corresponds to the coil end leakage of armature winding, and the radial MF [4], which is the subject of our research.

Previous studies [5] have shown that it is necessary to distinguish two frequency domains of the radial MF according to the skin effect which appears in the magnetic sheets. This paper is restricted to low frequency components induced by slot effect in induction machines. The aim is to investigate the flux density evolution through the stator frame to determine a correspondence between the flux density in the air-gap and the one which appears at the measured point outside the machine.

This paper presents first, the theoretical approach, then, the principle of measurement of the radial MF, and finally, the comparison between results deduced from experimental tests and these ones obtained by analytical computations.

S. Wiak, M. Dems, K. Komęza (eds.), Recent Developments of Electrical Drives, 237–246.

Analytical expression of the air-gap flux density

The analytical expression of the normal air-gap flux density b^g is achieved by multiplying the expression of the magneto-motive force (m.m.f.) which appears at the limit of the air-gap by this one of the air-gap permeance per surface unit.

To reduce the developments, it is assumed that the magnetic effects are only created by the three phase p pole pair stator which windings are crossed by the currents i_q $(1 \leq q \leq 3)$. Taking the stator phase 1 axis as spatial reference and locating any point of the air-gap by the variable α, the m.m.f. expression $\varepsilon(\alpha)$, assuming first infinite the iron permeability, is given by the following relation:

$$\varepsilon(\alpha) = \sum_{q=1}^{3} i_q \sum_{\substack{hs=1 \\ hs\, uneven}}^{+\infty} A_{hs} \cos\left[hs\left(p\alpha - (s-1)\frac{2\pi}{3}\right)\right] \tag{1}$$

where s is the slip.

Considering that the m.m.f. evolutes in a linear way on the slot width leads to defines A_{hs} by the expression:

$$A_{hs} = \frac{2z}{\pi hs}(-1)^{\frac{hs-1}{2}} \frac{\sin\left[\frac{(1-rds)\pi}{pN^s} hs\right]}{\frac{(1-rds)\pi}{pN^s} hs} \frac{\sin\left[\frac{hs\pi}{6}\right]}{m \sin\left[\frac{hs\pi}{6m}\right]} \tag{2}$$

where rds is the ratio of the slot width to the slot step, N^s is the number of stator slots per pole pair, z is the total number of turns for one stator phase under a pole pair, and m is the number of slots per pole per phase.

The used expression of the permeance per surface unit, takes the stator and rotor slotting effects into account separately but, also the interaction between the both [6]. So this quantity, function of α but also of θ which represents the angular displacement between the stator and rotor references, can be written as:

$$p(\alpha, \theta) = \sum_{ks=0}^{+\infty} \sum_{kr=-\infty}^{+\infty} P_{kskr} \cos\left[(ks N^s - kr N^r)\, p\alpha + kr N^r p\theta\right] \tag{3}$$

N^r represents the bars number per pole pair, P_{kskr} is a term which depends on the machine geometry, and ks and kr are two integers.

Let us assume that the stator is supplied by a balanced three phase sine currents of I r.m.s value and ω angular frequency. It results that:

$$\theta = \theta_0 + (1-s)\frac{\omega}{p} t \tag{4}$$

Choosing a temporal origin such as, for $t = 0$, $i_1 = I\sqrt{2}$ and $\theta_0 = 0$, leads to

$$b^g = \sum_{hs} \sum_{ks=-\infty}^{+\infty} \sum_{kr=-\infty}^{+\infty} \hat{b}^g_{\substack{hs \\ ks,kr}} \cos\{[1 + kr Nr(1-s)]\omega t - (hs + ks Ns + kr Nr) p\alpha\} \tag{5}$$

with:

$$\hat{b}^g_{\substack{hs \\ ks,kr}} = \frac{3I\sqrt{2}}{2} A_{hs} P_{kskr} \tag{6}$$

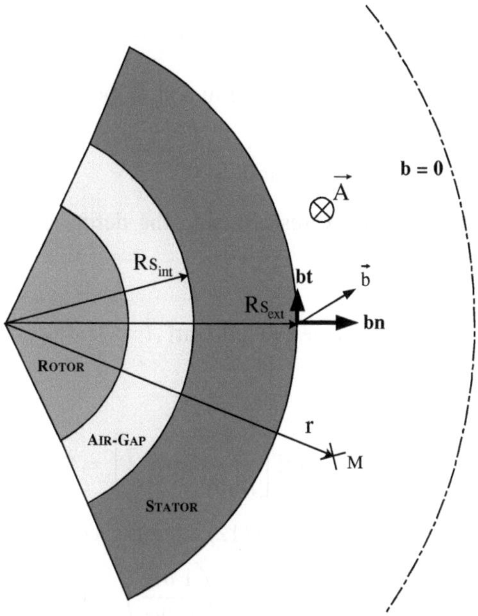

Figure 1. Section of the machine.

hs is an integer equal to $6\,\nu + 1$ where ν is a positive, negative, or null integer. Expression (5) can also be expressed with the following form:

$$b^g = \sum_{h=-\infty}^{+\infty} \sum_{k=-\infty}^{+\infty} \hat{b}^g_{h,k} \cos(k\omega t - hp\alpha) \tag{7}$$

where:

$$k = 1 + kr\,Nr(1 - s) \tag{8}$$

$$h = (hs + ks\,Ns + kr\,Nr) \tag{9}$$

It has been shown [7] that this method defines precisely the frequencies and gives satisfactory results in the magnitudes.

Evolution of the flux density through the stator

In order to establish the analytical expression of the flux density $\vec{b}$, in the surrounding of the machine, at M point, it is necessary to study its evolution through the stator frame. Fig. 1 presents a section of a machine.

Rs_{int} and Rs_{ext} are the interior and exterior stator radii, and r is the distance of the sensor from the motor shaft. The relative magnetic permeability of the stator iron is denoted μ_r.

To determine the diffusion of the flux density through the stator [8], it is required to exploit the Maxwell equations:

$\mathrm{Div}\,\vec{b} = 0$, which describes the conservative aspect of the flux density,
$\mathrm{Curl}\,\vec{h} = \vec{j}$ which is the local expression of the Ampere theorem

where $\vec{b}$, $\vec{h}$, and $\vec{j}$ are, respectively, the flux density, the magnetic field, and the current density vectors.

The equations will be solved in the stator iron and in the external air. So, as there is no current in these areas, it comes:

$$\text{Curl}\,\vec{h} = \vec{0} \tag{10}$$

The flux density vector $\vec{b}$ can be expressed using the definition of the magnetic vector potential $\vec{A}$, as well as:

$$\vec{b} = \left| \begin{matrix} bn \\ bt \\ 0 \end{matrix} \right. = \text{Curl}\,\vec{A} \tag{11}$$

so:

$$\text{Curl}\,\vec{h} = \text{Curl}\left[\frac{1}{\mu_0\mu_r}\text{Curl}\,\vec{A}\right] = 0 \tag{12}$$

As $\vec{A}$ is only in the axial direction, equation (12) amounts, consequently, to resolve:

$$\frac{\partial}{\partial r}\left(r\frac{\partial A}{\partial r}\right) + \frac{\partial}{\partial \alpha}\left(\frac{1}{r}\frac{\partial A}{\partial \alpha}\right) = 0 \tag{13}$$

This equation admits one solution which has the following form [9]:

$$A = \sum_{\varsigma=0}^{+\infty}\left(\lambda_1 r^{\varsigma} + \lambda_2 r^{-\varsigma}\right)\sin\varsigma(\alpha - \alpha') \tag{14}$$

Identification with equation (7) leads to $\varsigma\alpha' = k\omega t$ and, as the sign of the flux density pole number is just significant of the rotating sense of the component: $\varsigma = |h|p$.

It is also possible to obtain the normal and tangential components of the flux density vector:

$$\begin{cases} bn_{h,k} = \frac{1}{r}\dfrac{\partial A}{\partial \alpha} = \hat{b}n_{h,k}\cos\left(|h|\,p\alpha - k\omega t\right) \\ bt_{h,k} = -\dfrac{\partial A}{\partial r} = \hat{b}t_{h,k}\sin\left(|h|\,p\alpha - k\omega t\right) \end{cases} \tag{15}$$

So:

$$\left| \begin{array}{l} \hat{b}n_{h,k} = \dfrac{|h|\,p}{r}\left(\lambda_1 r^{|h|p} + \lambda_2 r^{-|h|p}\right) \\ \hat{b}t_{h,k} = -\,|h|\,p\left(\lambda_1 r^{|h|p-1} - \lambda_2 r^{-|h|p-1}\right) \end{array} \right. \tag{16}$$

It can be noticed that these equations are valid whatever the area.

In the following, the upper indexes "*g*," "*i*," or "*a*" will be used to distinguish different variables related to, respectively, air-gap, iron and external air areas.

In the air

According to the flux density equals zero when r tends toward infinity, leads to:

$$\lambda_1^a = 0 \tag{17}$$

Equation system (15) becomes:

$$\begin{cases} bn^a_{h,k} = |h|\, p\lambda^a_2 r^{-|h|p-1}\cos(|h|\, p\alpha - k\omega t) \\ bt^a_{h,k} = |h|\, p\lambda^a_2 r^{-|h|p-1}\sin(|h|\, p\alpha - k\omega t) \end{cases} \tag{18}$$

It can be noticed that the normal component magnitude of the flux density is equal to the tangential one.

In order to determine the second coefficient λ^a_2, it is necessary to know the magnitude of the flux density at the boundary $r = Rs_{\text{ext}}$. At this point, if the value of the flux density components is noted

$$\hat{b}^a_{h,k_{(r=Rs_{ext})}}$$

it comes:

$$\lambda^a_2 = \frac{\hat{b}^a_{h,k_{(r=Rs_{ext})}}}{|h|\, p(Rs_{ext})^{-|h|p-1}} \tag{19}$$

Then, the equations of the flux density in the air are given by:

$$\begin{cases} bn^a_{h,k} = \hat{b}^a_{h,k_{(r=Rs_{ext})}}(r')^{-|h|p-1}\cos(|h|\, p\alpha - k\omega t) \\ bt^a_{h,k} = \hat{b}^a_{h,k_{(r=Rs_{ext})}}(r')^{-|h|p-1}\sin(|h|\, p\alpha - k\omega t) \end{cases} \tag{20}$$

with:

$$r' = \frac{r}{Rs_{ext}} \tag{21}$$

In the stator

On the iron-air boundary ($r' = 1$), the normal component of the flux density is preserved as well as the tangential one of the magnetic field:

$$\left| \begin{array}{l} \hat{b}n^i_{h,k_{(r=Rs_{ext})}} = \hat{b}^a_{h,k_{(r=Rs_{ext})}} \\ \hat{b}t^i_{h,k_{(r=Rs_{ext})}} = \mu_r \hat{b}^a_{h,k_{(r=Rs_{ext})}} \end{array} \right. \tag{22}$$

Going back to the developed Maxwell equations (15), it can be written, for $r\prime = 1$:

$$\begin{cases} \hat{b}^a_{h,k_{(r=Rs_{ext})}} = \frac{|h|p}{Rs_{ext}}\left(\lambda^i_1 Rs^{|h|p}_{ext} + \lambda^i_2 Rs^{-|h|p}_{ext}\right) \\ \mu_r \hat{b}^a_{h,k_{(r=Rs_{ext})}} = -\,|h|\, p\left(\lambda^i_1 Rs^{|h|p-1}_{ext} - \lambda^i_2 Rs^{-|h|p-1}_{ext}\right) \end{cases} \tag{23}$$

The resolution of the system (23) leads to the following expressions of λ^i_1 and λ^i_2.

$$\lambda^i_1 = \frac{(1-\mu_r)\hat{b}^a_{h,k_{(r=Rs_{ext})}}}{2\,|h|\, pRs^{|h|p-1}_{ext}} \tag{24}$$

$$\lambda^i_2 = \frac{(\mu_r+1)\,\hat{b}^a_{h,k_{(r=Rs_{ext})}}}{2\,|h|\, pRs^{-|h|p-1}_{ext}} \tag{25}$$

Assuming that $\mu_r \gg 1$, the evolution equations of the flux density through the stator are finally express as following:

$$\begin{cases} bn^i_{h,k} = \dfrac{\mu_r}{2}\hat{b}^a_{h,k_{(r=Rs_{ext})}} \left[(r')^{-|h|p-1} - (r')^{|h|p-1}\right]\cos(|h|\,p\alpha - k\omega t) \\ bt^i_{h,k} = \dfrac{\mu_r}{2}\hat{b}^a_{h,k_{(r=Rs_{ext})}} \left[(r')^{-|h|p-1} - (r')^{|h|p-1}\right]\sin(|h|\,p\alpha - k\omega t) \end{cases} \tag{26}$$

As, in this approach, it is the magnetic field radial component which is considered, it is just necessary to continue the study on the normal component of the flux density.

In this case, as the normal flux density is preserved on the boundary between the air-gap and the stator and since this one can be obtained by the equation (7), it is consequently possible to establish a link between the flux density which exists in the air-gap and this which flows at the outside of the machine.

$$\text{If } r = Rs_{int}, \quad \text{then } \hat{b}n^i_{h,k_{(r=Rs_{int})}} = \hat{b}^g_{h,k}$$

So, it can be deduced:

$$\hat{b}^a_{h,k_{(r=Rs_{ext})}} = \frac{\hat{b}^g_{h,k}}{\frac{\mu_r}{2}\left[\left(\frac{Rs_{int}}{Rs_{ext}}\right)^{-|h|p-1} - \left(\frac{Rs_{int}}{Rs_{ext}}\right)^{|h|p-1}\right]} \tag{27}$$

From equation (20), it can be defined the $K_{h(r)}$ coefficient which takes into account the decreasing of the magnitude of each component, regarding to its number of pole pair:

$$K_{h_{(r)}} = \frac{\hat{b}n^a_{h,k}}{\hat{b}^g_{h,k}} \tag{28}$$

$$K_{h_{(r)}} = \mathcal{K} r'^{-|h|p-1} \tag{29}$$

where $\mathcal{K}$ is a constant coefficient equals to:

$$\mathcal{K} = \frac{2}{\mu_r\left[\left(\frac{Rs_{int}}{Rs_{ext}}\right)^{-|h|p-1} - \left(\frac{Rs_{int}}{Rs_{ext}}\right)^{|h|p-1}\right]} \tag{30}$$

It can be noticed that the higher $|h|$ is, the more rapidly K_h, and consequently, the corresponding component magnitude, decreases regarding its values in the air-gap.

Measured flux

In order to measure the radial magnetic field in the surrounding of the machine, a coil, used as a flux sensor, is set, in a plan parallel to the shaft, as it is shown on Fig. 2.

The 2β angle corresponds to the angular size of the sensor from the motor shaft ($ls \approx 2\beta r$). The sensor length Ls is sufficiently small with regard to this of the machine L, so that the developed theory remains valid even the bars are skewed.

The flux $\Psi_{h,k}$, linked by this sensor, is tied to the external normal flux density component depending on its geometry.

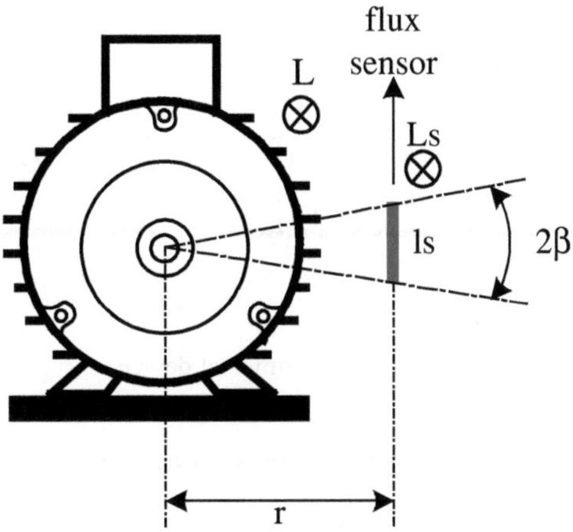

Figure 2. Measured flux.

If the sensor has n turns and it has a rectangular shape of S area, and considering that $l \ll r$, then the $\Psi_{\mathrm{h,k}}$ flux components regarding to the external flux density ones, are expressed by:

$$\Psi_{h,k_{(r)}} = nS\frac{\sin(hp\beta)}{hp\beta}\hat{b}n^{a}_{h,k_{(r)}}\cos k\omega t \tag{31}$$

Then, to obtain the component at the $k\omega$ angular frequency, it is necessary to add the different terms which have this same frequency.

$$\Psi_{k_{(r)}} = \sum_{h=-\infty}^{+\infty} \Psi_{h,k_{(r)}} \tag{32}$$

The terms which compose $\Psi_{k_{(r)}}$ are very numerous but many do not contribute significantly. Actually, relation (30) shows that only the terms with a small value of h will have a prominent effect. This property is enhanced by the quantity

$$\frac{\sin(hp\beta)}{hp\beta}$$

which appears in expression (31).

The data given by the coil is a voltage which results of the Lenz law. Consequently, the $e_{\mathrm{k(r)}}$ components of the corresponding measured e.m.f. are obtained by the following relation:

$$e_{k_{(r)}} = -\frac{d\Psi_{k_{(r)}}}{dt}$$

$$e_{k_{(r)}} = -k\omega nS\sum_{h=-\infty}^{+\infty}\frac{\sin(hp\beta)}{hp\beta}K_{h_{(r)}}\hat{b}^{g}_{h,k'}\sin k\omega t \tag{33}$$

It can be noticed that the electro-motive force spectra amplify the high frequencies, what permits to distinguish the slot spectral lines more easily.

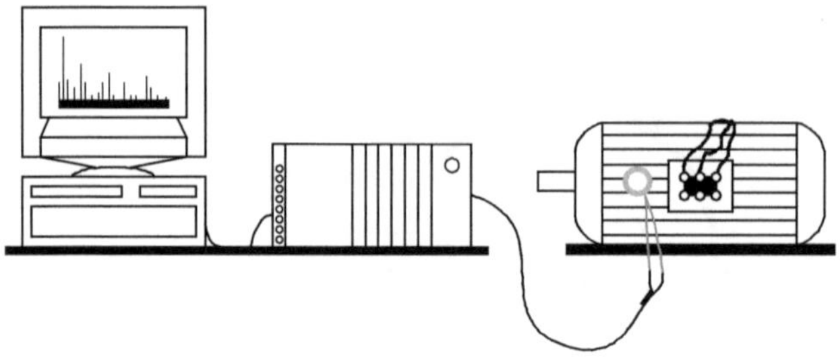

Figure 3. Experimental device.

Principle of measurements

The synoptic of the Fig. 3 shows the experimental device that allows performing the measurement of the radial magnetic field component. The principle is to pick up the sensor signal with an analyzer via an acquisition card.

A small winded flux sensor, with $nS = 30 \times 10^{-3}$, has been realized. It is located at a distance r equals to 100×10^{-3} m.

The tests are made considering a star connected 4 poles, 4 kW, 380/660 V, 50 Hz squirrel-cage induction machine with 36 stator slots and 44 rotor bars. For this machine, $Rs_{int} = 60 \times 10^{-3}$ m, $Rs_{ext} = 90 \times 10^{-3}$ m.

The stator is supplied by 380 r.m.s. phase to phase voltage. The supply current is equal to $I = 1.5$ A. In order to neglect the magnetic effects generated by the rotor, the slip is close to 0.

Results

In the measured spectrum of the e.m.f. only the spectral lines corresponding to $kr = 0$ (fundamental) and $kr = \pm 1$ will be considered for the comparison with the theory.

Concerning the fundamental, it is measured 5×10^{-3} V (r.m.s. value). As the machine is underfed, it will be supposed that the maximum of the flux density in the air-gap is 0.6 T. In that case, assuming that the flux density component at 50 Hz is mainly composed of one p pair pole wave ($h = 1$, $\hat{b}^g_{1,1} = 0.5$ T), μ_r have to be equal to 600. This low value can be justified taking into account that the frame is of cast iron and that the relative permeability of this material is about 300.

For the harmonics, Table 1 gives the relative magnitudes of the frequencies of the couple at $[1 \pm N^r(1-s)]f$ regarding to their fundamental.

Table 1. Relative magnitudes of the frequencies of the couple at $[1 \pm N^r(1-s)]f$ regarding to the fundamental

	$[1 - N^r(1-s)]f$	$[1 + N^r(1-s)]f$
Calculation without K_h correction coefficient	−5.49 db	−5.93 db
Calculation with K_h correction coefficient	−31.72 db	−22.31 db
Calculation with K_h and with prominent components	−31 db	−22.7 bd
Practical spectral lines	−27.06 db	−18.84 db

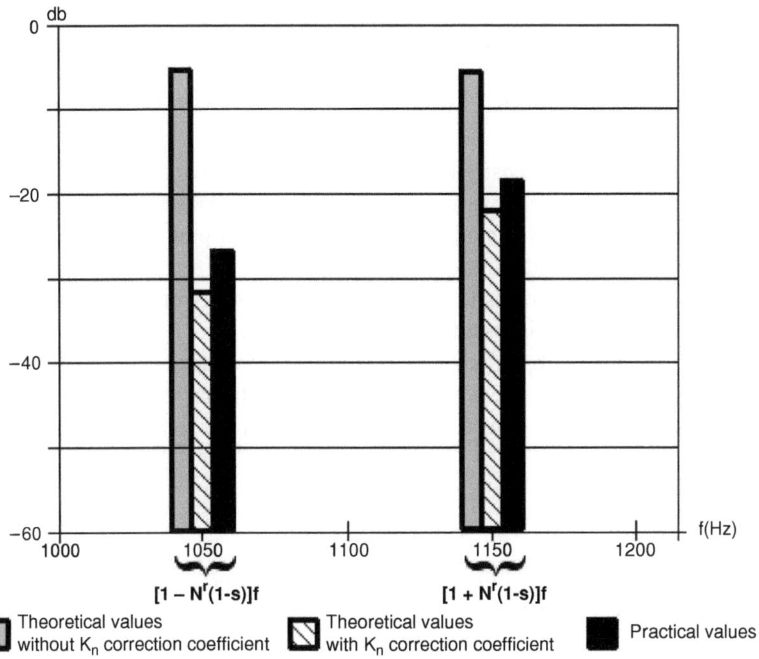

Figure 4. Comparison between practical and theoretical toothing spectral lines in relative values regarding to the fundamental.

The two first table lines correspond to computation results where *kr* and *ks* vary from -5 to 5 and *hs* varies from 1 to 13.

The third table line gives the computation results where only the prominent components depending on their number of pole pairs are considered.

For the spectral line at $[1 - N^r(1 - s)]f$ frequency ($kr = -1$), the lowest number of pole pairs is obtained for $hs = 1$ and $ks = 1$; in that case, equation (8) gives $|h|p = 2$.

For this at $[1 - N^r\ (1 - s)]f$ frequency ($kr = +1$), $ks = -1$ and $hs = -5$ leads to consider a prominent component such $|h|p = 6$.

The latest table line gives the relative magnitudes of this couple of frequencies which are obtained by experimentation.

The results show that considering only the prominent component which composes one spectral line is sufficient in the study of the external magnetic field.

Fig. 4 gives a graphic illustration of the results. It can be observed that when only the analytical expression of the electro-motive force is computed, without take the K_h correction coefficient into account, the frequencies of the couple at $[1 \pm N^r(1 - s)]f$ have similar magnitudes which are high toward the magnitude of the fundamental.

When the K_h coefficient is introduced to modify the magnitudes of the components which compose the spectral lines, it can be noticed, in one hand, that the frequencies of this same couple do not have yet the same magnitudes, and on the other hand, that the difference with practical results is reduced (the error is only about 4 dB).

Let us also precise that the b^g expression given by (7) shows that current harmonics are induced in the stator windings and the rotor bars. These currents have been neglected in our study but some of them contribute to define the magnitude of the considered spectral lines. That can partially justify the deviations.

Conclusion

It is shown that all the flux density components do not evolve, through the stator, in the same way according to the pole pair number *hp* of this component. So, it is determined the correction to apply on the calculated air-gap flux density components in order to appraise the measured spectrum outside of the machine.

The presented method permits to analyze a practical spectrum toward a reference one obtained by calculation, that is specific for each machine.

This quantitative approach is interesting if faults are introduced in the modeling of the air-gap flux density. In fact, the effects of the faults in the external magnetic field, considering the pole number of the flux density components which are generated, can be precisely predetermined.

Acknowledgment

This work is part of the project “Futurelec 2” within the “Centre National de Recherche Technologique (CNRT) of Lille.”

References

[1] W.T. Thomson, M. Fenger, Current signature analysis to detect induction motor faults, IEEE Ind. Appl. Soc. Meet. (IAS Magazine), Vol. 7, No. 4, pp. 26–34, 2001.

[2] M.E.H. Benbouzid, Bibliography on induction motors faults detection and diagnosis, IEEE Trans. Energy Convers., Vol. 14, No. 4, pp. 1065–1074, 1999.

[3] H. Hénao, T. Assaf, G.A. Capolino, “Detection of Voltage Source Dissymmetry in an Induction Motor Using the Measurement of Axial Leakage Flux”, Proceedings of International Conference of Electrical Machines (ICEM 2000), Espoo Finland, August 2000, pp. 1110–1114.

[4] D. Belkhayat, R. Romary, M. El Adnani, R. Corton, J.F. Brudny, Fault diagnosis in induction motors using radial field measurement with an antenna, Inst. Phys. Meas. Sci. Technol., Vol. 14, No. 9, pp. 1695–1700, 2003.

[5] D. Roger, O. Ninet, S. Duchesne, Wide frequency range characterization of rotating machine stator-core laminations, Eur. Phys. J. Appl. Phys., pp. 103–109, 2003.

[6] J.F. Brudny, Etude quantitative des Harmoniques du Couple du Moteur asynchrone triphasé d’Induction, Thèse d’Habilitation, No. H29, Lille, France, 1991.

[7] J.F. Brudny, Modélisation de la Denture des Machines asynchrones. Phénomène de resonance, J. Phys. III, pp. 1009–1023, 1997.

[8] K.J. Bins, P.J. Lawrenson, Analysis and Computation of Electric and Magnetic Field Problems, Oxford: Pergamon press, 1973.

[9] Ph.L. Alger, The Nature of Induction Machines, 2nd edition, New York: Gordon & Breach Publishers, 1970.

II-10. IMPACT OF MAGNETIC SATURATION ON THE INPUT-OUTPUT LINEARIZING TRACKING CONTROL OF AN INDUCTION MOTOR

D. Dolinar[1], P. Ljušev[2] and G. Štumberger[1]

[1]Faculty of Electrical Engineering and Computer Science, University of Maribor, Smetanova 17, SI-2000 Maribor, Slovenia
dolinar@uni-mb.si
[2]Ørsted DTU Automation, Technical University of Denmark, Elektrovej, Building 325, DTU, DK-2800 Kgs. Lyngby, Denmark
pl@oersted.dtu.dk

Abstract. This paper deals with the tracking control design of an induction motor, based on input-output linearization with magnetic saturation included. Magnetic saturation is accounted for by the nonlinear magnetizing curve of the iron core and is used in the control design, the observer of state variables, and in the load torque estimator. Experimental results show that the proposed input-output linearizing tracking control with the included saturation behaves better than the one without saturation. It also introduces smaller position and speed errors, and better motor stiffness.

Introduction

The magnetic saturation phenomenon, which occurs in the induction motor (IM) iron core, has been recognized from the very beginning of IM use. The models of saturated induction machines presented in the literature [1–3] show that magnetic saturation is most frequently considered by nonlinear variable mutual inductance, while the additional magnetic cross-couplings are neglected. In general, leakage flux path saturation is rarely included in the models, although its role in the squirrel-cage IM with closed rotor slots can be rather high. The level of idealization, introduced by neglecting magnetic cross-saturation, depends on the type of model, i.e. on the selected state variables [4].

Although magnetic saturation of the IM is very important for the performances of the controlled drive, it has hardly ever been properly considered in control design because of its mathematical complexity. Pioneering efforts to address the perturbing effects of magnetic saturation on the field oriented control are found in [5].

Saturated induction motor models found in the literature [1,3] are used primarily for the analysis of an induction machine operation. Some exceptions in which linear cascade rotor

S. Wiak, M. Dems, K. Komęza (eds.), Recent Developments of Electrical Drives, 247–260.

field oriented control is used can be found in [6–8] but they present an incomplete solution for the control algorithm. An observer for state variables that includes magnetic saturation is found in [9]. Some nonlinear control approaches, based on modified IM models that include saturation, are given in [10–13] as well.

In this paper an input-output linearizing tracking control design is presented. It is based on the mixed "stator current vector $\mathbf{i}_s$, rotor flux linkage vector $\mathbf{\Psi}_r$" saturated IM model introduced by [3]. The effect of magnetic cross-saturation is totally included in the main flux path saturation. The saturation of the stator and rotor leakage flux path is neglected. In addition, a state variable observer and a load torque estimator were designed with included magnetic saturation.

An $\mathbf{i}_s$, $\mathbf{\Psi}_r$ saturated IM model is presented in this paper. The input-output linearization based on the introduced model is carried out, with respect to the rotor position. To obtain the required unmeasurable state variables, an observer based on the inverse saturated IM model was designed. The experimental results of the proposed input-output linearizing tracking control of IM with included magnetic saturation show better dynamic performances of the drive than the classical control, where magnetic saturation is not considered. The main improvements are the smaller rotor position and speed errors, as well as higher stiffness and better load torque rejection, which results in a smaller stator current when the motor is loaded with a step change in the load torque [14].

Saturated induction machine model

The general approach to IM modeling incorporating magnetic saturation with different selections of state variables is presented in [3]. The standard two-phase model of an IM in the general reference frame with stator current vector $\mathbf{i}_s$ and rotor flux linkage vector $\mathbf{\Psi}_r$ as state variables, is given by a nonlinear model (1)

$$\begin{aligned} \mathbf{u} &= \mathbf{A}(\mathbf{x})\frac{d\mathbf{x}}{dt} + \mathbf{B}(\mathbf{x})\mathbf{x} \\ J\frac{d\omega_r}{dt} &= (t_e - t_l) - f\omega_r \end{aligned} \tag{1}$$

where

$$\mathbf{u} = \begin{bmatrix} u_{sd} & u_{sq} & 0 & 0 \end{bmatrix}^T \quad \mathbf{x} = \begin{bmatrix} i_{sd} & i_{sq} & \psi_{rd} & \psi_{rq} \end{bmatrix}^T$$

$$\mathbf{A}(\mathbf{x}) = \begin{bmatrix} L_l - \dfrac{L_{rl}^2}{L_{dd}} & -\dfrac{L_{rl}^2}{L_{dq}} & 1 - \dfrac{L_{rl}}{L_{dd}} & -\dfrac{L_{rl}}{L_{dq}} \\ -\dfrac{L_{rl}^2}{L_{dq}} & L_l - \dfrac{L_{rl}^2}{L_{qq}} & -\dfrac{L_{rl}}{L_{dq}} & 1 - \dfrac{L_{rl}}{L_{qq}} \\ 0 & 0 & 1 & 0 \\ 0 & 0 & 0 & 1 \end{bmatrix}$$

$$\mathbf{B}(\mathbf{x}) = \begin{bmatrix} R_s & -\omega_g\left(L_l - \dfrac{L_{rl}^2}{L_r}\right) & 0 & -\omega_g\dfrac{L_m}{L_r} \\ \omega_g\left(L_l - \dfrac{L_{rl}^2}{L_r}\right) & R_s & \omega_g\dfrac{L_m}{L_r} & 0 \\ -R_r\dfrac{L_m}{L_r} & 0 & \dfrac{R_r}{L_r} & -\omega_{sl} \\ 0 & -R_r\dfrac{L_m}{L_r} & \omega_{sl} & \dfrac{R_r}{L_r} \end{bmatrix}$$

$$t_e = p\frac{L_m}{L_r}(i_{sq}\psi_{rd} - i_{sd}\psi_{rq})$$

u_{sd}, u_{sq} and i_{sd}, i_{sq} are the d- and q-axis stator voltages and currents, ψ_{rd} and ψ_{rq} are the rotor flux linkages, R_s and R_r are the stator and rotor resistances, L_m is the mutual static inductance, L_s and L_{sl} are the stator self-inductance and the stator leakage inductance, L_r and L_{rl} are the rotor self-inductance and the rotor leakage inductances, $L_l = L_{sl} + L_{rl}$ is the total leakage inductance, L_{dq} is the cross-coupling inductance, L_{dd} and L_{qq} are the inductances along d- and q-axis, ω_g is the angular speed of the general reference frame, ω_r is the rotor angular speed, $\omega_{sl} = \omega_g - \omega_r$ is the slip angular frequency, J is the drive moment of inertia, f is the coefficient of viscous friction, t_e and t_l are the electrical and the load torque, and p is the number of pole pairs. Inductances L_{dd}, L_{qq}, and L_{dq} are given as follows:

$$\begin{aligned} \frac{1}{L_{dd}} &= \frac{1}{L_{rl}+L}\cos^2\rho_m + \frac{1}{L_{rl}+L_m}\sin^2\rho_m \\ \frac{1}{L_{qq}} &= \frac{1}{L_{rl}+L_m}\cos^2\rho_m + \frac{1}{L_{rl}+L}\sin^2\rho_m, \qquad \Psi_m = |\boldsymbol{\Psi}_m| \\ \frac{1}{L_{dq}} &= \left(\frac{1}{L_{rl}+L} - \frac{1}{L_{rl}+L_m}\right)\cos\rho_m\sin\rho_m, \qquad i_m = |\mathbf{i}_m| \\ L_m &= \frac{\Psi_m}{i_m}, \quad L = \frac{d\Psi_m}{di_m}, \quad \cos\rho_m = \frac{\psi_{md}}{\Psi_m}, \quad \sin\rho_m = \frac{\psi_{mq}}{\Psi_m} \end{aligned} \tag{2}$$

where i_m is the modulus of the magnetizing current vector. L represents the so-called dynamic inductance, ψ_{md} and ψ_{mq} are the d- and q-components of the magnetizing flux linkage vector $\boldsymbol{\Psi}_{\mathrm{m}}$, ρ_m is the angle of magnetizing flux linkage vector $\boldsymbol{\Psi}_{\mathrm{m}}$, while Ψ_m is the modulus of $\boldsymbol{\Psi}_{\mathrm{m}}$. L_m and L are given by equations (2) and are obtained from the measured magnetizing curve presented in Fig. C1, Appendix C.

Two-phase $\mathbf{i}_{\mathrm{s}}$, $\boldsymbol{\Psi}_{\mathrm{r}}$ model (1) can be written in the state-space form as:

$$\begin{aligned} \frac{d\mathbf{x}}{dt} &= -\mathbf{A}(\mathbf{x})^{-1}\mathbf{B}(\mathbf{x})\mathbf{x} + \mathbf{A}(\mathbf{x})^{-1}\mathbf{u} = \mathbf{C}\mathbf{x} + \omega_g\mathbf{Z}\mathbf{x} + \omega_r\mathbf{W}\mathbf{x} + \mathbf{D}\mathbf{u} \\ \frac{d\omega_r}{dt} &= \frac{p}{J}\frac{L_m}{L_r}(i_{sq}\psi_{rd} - i_{sd}\psi_{rq}) - \frac{1}{J}t_l - \frac{f}{J}\omega_r \end{aligned} \tag{3}$$

where:

$$\mathbf{C} = \begin{bmatrix} c_{11} & c_{12} & c_{13} & c_{14} \\ c_{21} & c_{22} & c_{23} & c_{24} \\ c_{31} & 0 & c_{33} & 0 \\ 0 & c_{42} & 0 & c_{44} \end{bmatrix} \quad \mathbf{Z} = \begin{bmatrix} z_{11} & z_{12} & z_{13} & z_{14} \\ z_{21} & z_{22} & z_{23} & z_{24} \\ 0 & 0 & 0 & z_{34} \\ 0 & 0 & z_{43} & 0 \end{bmatrix}$$

$$\mathbf{W} = \begin{bmatrix} 0 & 0 & w_{13} & w_{14} \\ 0 & 0 & w_{23} & w_{24} \\ 0 & 0 & 0 & w_{34} \\ 0 & 0 & w_{43} & 0 \end{bmatrix} \quad \mathbf{D} = \begin{bmatrix} d_{11} & d_{12} & 0 & 0 \\ d_{21} & d_{22} & 0 & 0 \\ 0 & 0 & 0 & 0 \\ 0 & 0 & 0 & 0 \end{bmatrix}$$

Elements of matrices **C**, **Z**, **W**, **D** are given in Appendix A.

Input-output linearization

The two-phase IM model (3) is transformed to the stationary reference frame $\alpha\beta$ where $\omega_g = 0$. It is written in the compact form (4)

$$\dot{\mathbf{x}} = \mathbf{f}(\mathbf{x}) + \mathbf{G}\mathbf{u} = \mathbf{f}(\mathbf{x}) + \mathbf{g}_\alpha u_{s\alpha} + \mathbf{g}_\beta u_{s\beta} \tag{4}$$

where $\mathbf{x} = \begin{bmatrix} \theta_r & \omega_r & i_{s\alpha} & i_{s\beta} & \psi_{r\alpha} & \psi_{r\beta} \end{bmatrix}$

$$\mathbf{f}(\mathbf{x}) = \begin{bmatrix} \omega_r \\ p\frac{L_m}{L_r}\frac{1}{J}\left(i_{s\beta}\psi_{r\alpha} - i_{s\alpha}\psi_{r\beta}\right) - \frac{1}{J}t_l - \frac{f}{J}\omega_r \\ c_{11}i_{s\alpha} + c_{12}i_{s\beta} + c_{13}\psi_{r\alpha} \\ +c_{14}\psi_{r\beta} + w_{13}\omega_r\psi_{r\alpha} + w_{14}\omega_r\psi_{r\beta} \\ c_{21}i_{s\alpha} + c_{22}i_{s\beta} + c_{23}\psi_{r\alpha} \\ +c_{24}\psi_{r\beta} + w_{23}\omega_r\psi_{r\alpha} + w_{24}\omega_r\psi_{r\beta} \\ c_{31}i_{s\alpha} + c_{33}\psi_{r\alpha} + w_{34}\omega_r\psi_{r\beta} \\ +c_{42}i_{s\beta} + c_{44}\psi_{r\beta} + w_{43}\omega_r\psi_{r\alpha} \end{bmatrix} \tag{5}$$

$$\mathbf{G} = \begin{bmatrix} \mathbf{g}_\alpha & \mathbf{g}_\beta \end{bmatrix} = \begin{bmatrix} 0 & 0 & d_{11} & d_{21} & 0 & 0 \\ 0 & 0 & d_{12} & d_{22} & 0 & 0 \end{bmatrix}^T$$

θ_r is the rotor angle, $u_{s\alpha}$, $su_{s\beta}$ and $i_{s\alpha}$, $i_{s\beta}$ are the stator voltages and currents and $\psi_{r\alpha}$ and $\psi_{r\alpha}$ are the rotor flux linkages. Coefficients $c_{(\cdot)}$, $w_{(\cdot)}$, and $d_{(\cdot)}$ are given in Appendix A. The output vector $\mathbf{y} = \begin{bmatrix} \theta_r & \Psi_r^2 \end{bmatrix}^T$ is chosen by (6), where $\Psi_r^2 = \psi_{r\alpha}^2 + \psi_{r\beta}^2$

$$\mathbf{y} = \phi(\mathbf{x}) = \begin{bmatrix} \phi_1(\mathbf{x}) & \phi_2(\mathbf{x}) \end{bmatrix}^T = \begin{bmatrix} \theta_r & \Psi_r^2 \end{bmatrix}^T \tag{6}$$

Since the load torque t_l is not the system control input and cannot be directly measured, it is excluded from the nominal part of the motor model and is later considered as an external disturbance. The load torque is obtained by a simple load torque estimator described in [14].

The input-output linearization technique is based on exact cancellation of the system's nonlinearities in order to obtain a linear relationship between the control inputs and the system outputs in the closed loop. The nonlinear control law is deduced to a successive differentiation of each output until at least one input appears in the derivative. The derivative of the first output $\phi_1(\mathbf{x}) = \theta_r$ is:

$$\dot{y}_1 = \frac{\partial \phi_1(\mathbf{x})}{\partial \mathbf{x}} \frac{d\mathbf{x}}{dt} = L_f \phi_1 = \omega_r \tag{7}$$

The second and the third derivatives are given by (8) and (9).

$$\begin{aligned} \ddot{y}_1 &= \frac{\partial}{\partial \mathbf{x}} \dot{y}_1 \frac{d\mathbf{x}}{dt} = \frac{\partial}{\partial \mathbf{x}} (L_f \phi_1) \frac{d\mathbf{x}}{dt} = L_f^2 \phi_1 \\ &= p \frac{L_m}{L_r} \frac{1}{J} \left(i_{s\beta} \psi_{r\alpha} - i_{s\alpha} \psi_{r\beta} \right) - \frac{1}{J} t_l - \frac{f}{J} \omega_r \end{aligned} \tag{8}$$

$$\dddot{y}_1 = \frac{\partial}{\partial \mathbf{x}} \ddot{y}_1 \frac{d\mathbf{x}}{dt} = L_f^3 \phi_1 + L_{g\alpha} L_f^2 \phi_1 u_{s\alpha} + L_{g\beta} L_f^2 \phi_1 u_{s\beta} \tag{9}$$

Lie derivatives $L_f^3 \phi_1$, $L_{g\alpha} L_f^2 \phi_1$, and $L_{g\beta} L_f^2 \phi_1$ are given in Appendix B. The input voltages $u_{s\alpha}$ and $u_{s\beta}$ appeared in $\dddot{y}_1$, therefore, the relative degree of the first subsystem is three.

The second output to be differentiated is the square of the rotor flux linkage modulus Ψ_r^2. Its first derivative is given by equation (10).

$$\begin{aligned} \dot{y}_2 &= \frac{\partial \phi_2(\mathbf{x})}{\partial \mathbf{x}} \frac{d\mathbf{x}}{dt} = L_f \phi_2 \\ &= 2R_r \frac{L_m}{L_r} \left(i_{s\alpha} \psi_{r\alpha} + i_{s\beta} \psi_{r\beta} \right) - 2R_r \frac{L_m}{L_r} \left(\psi_{r\alpha}^2 + \psi_{r\beta}^2 \right) \end{aligned} \tag{10}$$

The second derivative of Ψ_r^2 is given by (11).

$$\begin{aligned} \ddot{y}_2 &= \frac{\partial}{\partial \mathbf{x}} \dot{y}_2 \frac{d\mathbf{x}}{dt} = \frac{\partial}{\partial \mathbf{x}} (L_f \phi_2) \frac{d\mathbf{x}}{dt} \\ &= L_f^2 \phi_2 + L_{g\alpha} L_f \phi_2 u_{s\alpha} + L_{g\beta} L_f \phi_2 u_{s\beta} \end{aligned} \tag{11}$$

Lie derivatives $L_f^2 \phi_2$, $L_{g\alpha} L_f \phi_2$, and $L_{g\beta} L_f \phi_2$ are given in Appendix B. The input voltages $u_{s\alpha}$ and $u_{s\beta}$ appeared in $\ddot{y}_2$, therefore, the relative degree of the second subsystem is two. Consequently, the total relative degree of the system is unequal to the system order $n = 6$, which reveals the existence of uncontrollable internal dynamics. It is easy to prove (see Ref. [14]) that the dynamics of the third subsystem is stable if the third output ϕ_3 is selected as $\phi_3 = \arctan\ \psi_{r\beta}/\psi_{r\alpha}$, as in [11].

In the next step the input-output linearization of the nominal system is done [11]. The nominal part of the IM model is written in the form of higher derivatives of outputs y_1 and y_2

$$\begin{bmatrix} \dddot{y}_1 \\ \ddot{y}_2 \end{bmatrix} = \begin{bmatrix} L_f^3 \phi_1 \\ L_f^2 \phi_2 \end{bmatrix} + \mathbf{E}(\mathbf{x}) \begin{bmatrix} u_{s\alpha} \\ u_{s\beta} \end{bmatrix} = \mathbf{D}(\mathbf{x}) + \mathbf{E}(\mathbf{x})\mathbf{u} \tag{12}$$

where

$$\mathbf{E}(\mathbf{x}) = \begin{bmatrix} L_{g\alpha} L_f^2 \phi_1 & L_{g\beta} L_f^2 \phi_1 \\ L_{g\alpha} L_f \phi_2 & L_{g\beta} L_f \phi_2 \end{bmatrix} \tag{13}$$

is the decoupling matrix. The decoupling matrix $\mathbf{E}(\mathbf{x})$ is nonsingular, except at the motor start-up, where $\psi_{r\alpha}^2 + \psi_{r\beta}^2 = \Psi_r^2$ [14].

The obtained model (12) is still nonlinear and coupled. The linearized and decoupled IM model (15) is obtained by an appropriate selection of the control input $\mathbf{u} = \lfloor u_{s\alpha} \quad u_{s\beta} \rfloor^T$ (see equation (14)).

$$\begin{bmatrix} u_{s\alpha} \\ u_{s\beta} \end{bmatrix} = \mathbf{E}^{-1}(\mathbf{x}) \left(- \begin{bmatrix} L_f^3 \phi_1 \\ L_f^2 \phi_2 \end{bmatrix} + \begin{bmatrix} v_\alpha \\ v_\beta \end{bmatrix} \right) \tag{14}$$

Note that $\mathbf{v} = \lfloor v_\alpha \quad v_\beta \rfloor^T$ in equation (15) is the new system input.

$$\begin{aligned} \begin{bmatrix} \dddot{y}_1 \\ \ddot{y}_2 \end{bmatrix} &= \mathbf{D}(\mathbf{x}) + \mathbf{E}(\mathbf{x}) \left(\mathbf{E}^{-1}(\mathbf{x}) \mathbf{D}(\mathbf{x}) + \mathbf{E}^{-1}(\mathbf{x}) \mathbf{v} \right) \\ &= \mathbf{D}(\mathbf{x}) + \mathbf{E}(\mathbf{x}) \mathbf{u} = \begin{bmatrix} v_\alpha \\ v_\beta \end{bmatrix} = \mathbf{v} \end{aligned} \tag{15}$$

The input-output behavior of the system (15) is linear, but the relationship between the control input $\mathbf{v}$ and the states $\mathbf{x}$ is still nonlinear. This nonlinear relationship is eliminated by selecting a new set of state variables, $\mathbf{z} = [z_1\ z_2\ z_3\ z_4\ z_5]$ introduced by the nonlinear transformation $\mathbf{z} = \mathbf{T}(\mathbf{x})$, defined as:

$$\mathbf{y} = \mathbf{T}(\mathbf{x}) = \begin{bmatrix} \theta_r \\ \omega_r \\ p \frac{L_m}{L_r} \frac{1}{J} \left(i_{s\beta} \psi_{r\alpha} - i_{s\alpha} \psi_{r\beta} \right) - \frac{1}{J} t_l - \frac{f}{J} \omega_r \\ 2R_r \frac{L_m}{L_r} \left(i_{s\alpha} \psi_{r\alpha} + i_{s\beta} \psi_{r\beta} \right) - 2R_r \frac{L_m}{L_r} \left(\psi_{r\alpha}^2 + \psi_{r\beta}^2 \right) \end{bmatrix} \tag{16}$$

The block diagram of the decoupled and linearized IM model is given in Fig. 1. This system is linearized, decoupled and unstable. Both stabilization and tracking can be achieved without any concern about the stability of the internal dynamics using the linear tracking controllers designed by the pole placement [14].

The reference output vector $\mathbf{y}^*$ is given by (17)

$$\mathbf{y}^* = \begin{bmatrix} y_1^* \\ y_2^* \end{bmatrix} = \begin{bmatrix} \Theta_r^* \\ \Psi_r^{2*} \end{bmatrix} \tag{17}$$

where Θ_r^* and Ψ_r^{2*} represent the reference trajectories of the rotor position and the square of rotor flux linkage modulus. The differences between the reference and the actual values of the controlled outputs are tracking errors (18).

$$e_1 = y_1^* - y_1\,, \quad e_2 = y_2^* - y_2 \tag{18}$$

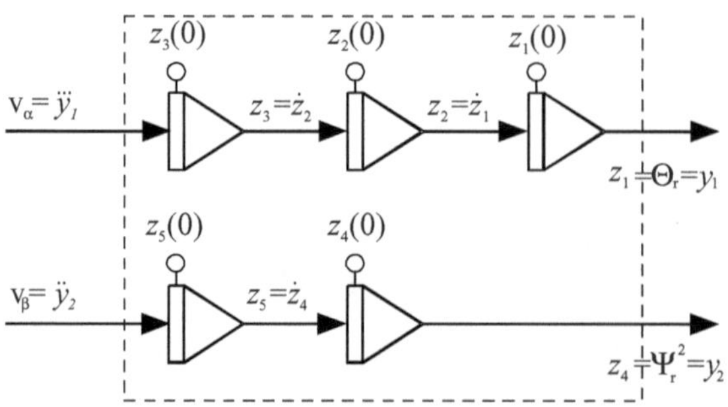

Figure 1. Linearized and decoupled IM model.

Taking into account equations (18), the position controller and the rotor flux linkage controller are given by (19).

$$\begin{aligned} v_\alpha &= k_{\alpha 0}e_1 + k_{\alpha 1}\dot{e}_1 + k_{\alpha 2}\ddot{e}_1 + \dddot{y}_1^* \\ v_\beta &= k_{\beta 1}e_2 + k_{\beta 2}\dot{e}_2 + \ddot{y}_2^* \end{aligned} \tag{19}$$

After inserting (19) in the linearized system (15), the tracking error dynamics of the closed loop system is given by (20)

$$\begin{aligned} &\dddot{e}_1 + k_{\alpha 2}\ddot{e}_1 + k_{\alpha 1}\dot{e}_1 + k_{\alpha 0}e_1 = 0 \\ &\ddot{e}_2 + k_{\beta 2}\dot{e}_2 + k_{\beta 1}e_2 = 0 \end{aligned} \tag{20}$$

with $k_{(\cdot)}$ being positive constants. The desired dynamics of the tracking errors e_1 and e_2 is assured by selecting the corresponding eigenvalues $\lambda_{(\cdot)}$ of the characteristic equations (21).

$$\begin{aligned} \lambda^3 + k_{\alpha 2}\lambda^2 + k_{\alpha 1}\lambda + k_{\alpha 0} &= 0 \\ \lambda^2 + k_{\beta 2}\lambda + k_{\beta 1} &= 0 \end{aligned} \tag{21}$$

Observer design

The state variables of the selected IM model are necessary to realize control, based on the described input-output linearization. The corresponding observer, similar to the one presented in [15], is given by (22). It is based on the electromagnetic subsystem of the two-phase $\mathbf{i}_s$, $\mathbf{\Psi}_r$ state-space IM model (3) in the $\alpha\beta$ reference frame. The coefficients $k_{(\cdot)}$ are determined in the literature [14].

$$\frac{d}{dt}\begin{bmatrix} \hat{i}_{s\alpha} \\ \hat{i}_{s\beta} \\ \hat{\psi}_{r\alpha} \\ \hat{\psi}_{r\beta} \end{bmatrix} = (\mathbf{C} + \omega_r \mathbf{W})\begin{bmatrix} \hat{i}_{s\alpha} \\ \hat{i}_{s\beta} \\ \hat{\psi}_{r\alpha} \\ \hat{\psi}_{r\beta} \end{bmatrix} + \mathbf{D}\begin{bmatrix} u_{s\alpha} \\ u_{s\beta} \end{bmatrix} + \begin{bmatrix} k_1 & -\omega_r k_2 \\ -\omega_r k_2 & k_1 \\ k_3 & -\omega_r k_4 \\ -\omega_r k_4 & k_3 \end{bmatrix}\left(\begin{bmatrix} \hat{i}_{s\alpha} \\ \hat{i}_{s\beta} \end{bmatrix} - \begin{bmatrix} i_{s\alpha} \\ i_{s\beta} \end{bmatrix}\right) \tag{22}$$

The symbol $(\hat{\cdot})$ denotes the observed values.

Experimental results

The experiments have been performed to test the proposed input-output linearizing tracking control. The elements of the experimental system are the three-phase Semikron IGBT inverter, the three-phase 3 kW IM Sever with wound rotor, whose parameters are given in Appendix C, and the DC motor Mavilor Mo2000 with an Infranor DC power converter, as the dynamic load. The control algorithm was executed on the dSPACE DS1103 microcontroller board. A block diagram of the proposed IM drive's tracking control that includes magnetic saturation is presented in Fig. 2.

Experiments were done using the reference value $\Psi_r^{2*} = 1.6\ (Vs)^2$. The smooth reference trajectories for the position θ_r and the speed ω_r were generated from the kinematic model and are shown in Figs. 3(a) and 4(a). The step changes of the load torque t_l vs. time are shown in Fig. 3(d). The results of the input-output linearizing tracking control with the included saturation were compared with the results obtained with the same type of control

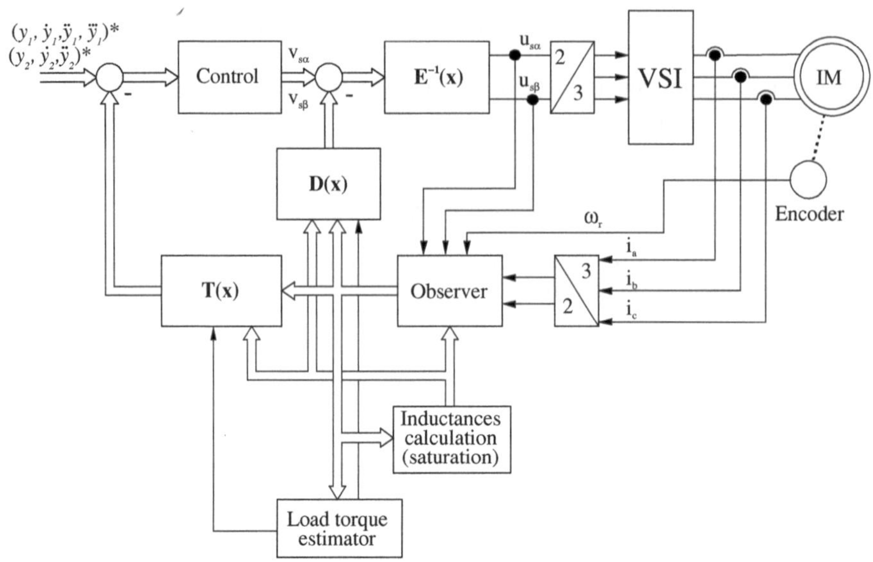

Figure 2. Block diagram of the IM's input-output linearizing tracking control.

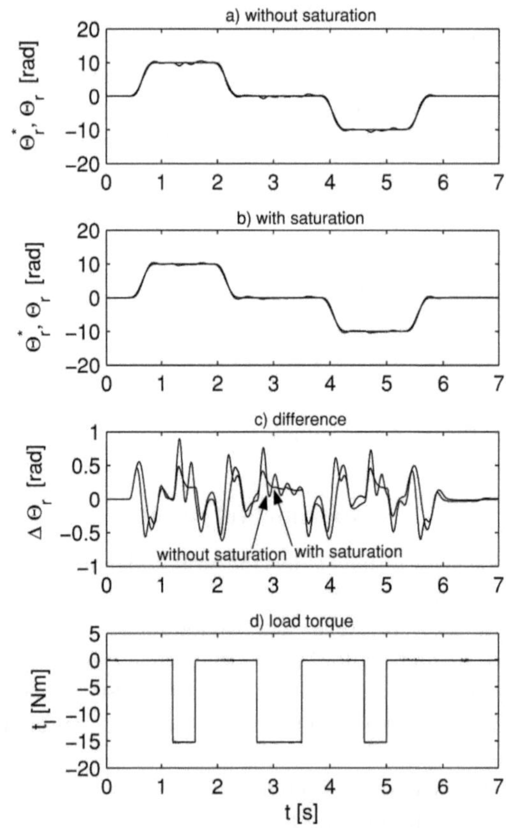

Figure 3. Reference and measured rotor position trajectory Θ_r^* and Θ_r: (a) saturation is not included, (b) saturation is included, (c) difference $\Delta\Theta_r = \Theta_r^* - \Theta_r$ without and with saturation, and (d) load torque t_l.

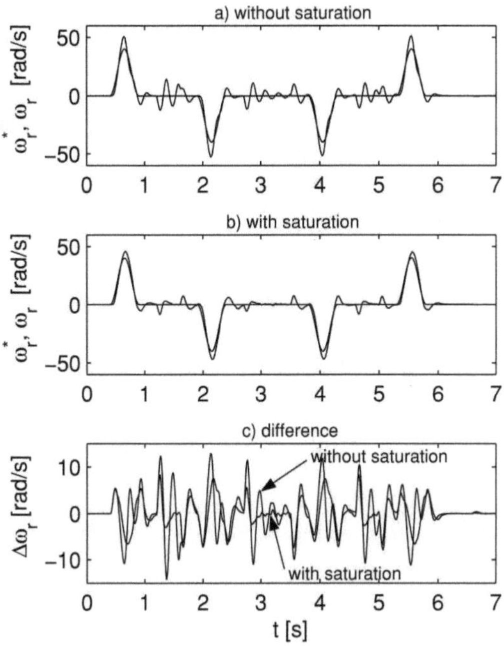

Figure 4. Reference and measured rotor speed trajectory $\omega_r^*$and ω_r: (a) saturation is not included, (b) saturation is included, and (c) difference $\Delta\omega_r = \omega_r^* - \omega_r$ without and with saturation.

without included saturation [11]. The settings of the controllers were equal in both cases: $k_{\alpha 0} = 750{,}000, k_{\alpha 1} = 25{,}000, k_{\alpha 2} = 275, k_{\beta 1} = 90{,}000$, and $k_{\beta 2} = 600$.

An analysis of the results showed that the position error $\Delta\theta_r$ in Fig. 3 and the rotor speed error $\Delta\omega_r$ in Fig. 4 are considerably smaller when magnetic saturation is included in the control algorithm, observer, and load torque estimator than in the case when magnetic saturation is neglected.

It is obvious from the results in Fig. 5 that tracking control with the included magnetic saturation performed the position task with a slightly higher stator current $i_s = \sqrt{i_{s\alpha}^2 + i_{s\beta}^2}$, than the one without saturation. In contrast, tracking control without any included magnetic saturation required smaller stator current to perform the same task at no-load, but it responded with a much higher increase in stator current i_s, when the motor was loaded with step changes of the load torque (Fig. 5a).

The reason for the described behavior of the controlled IM in Fig. 5 can be explained if the controlled system is analyzed together with the observer and, if only for explanation, the stator currents $i_{s\alpha}$, $i_{s\beta}$ are transformed to the common dq reference frame, i.e. to the stator currents i_{sd}, i_{sq} (Fig. 6). The observer of electromagnetic state variables, with included magnetic saturation yields a smaller rotor flux linkage module Ψ_r for equal stator current value than the linear observer introduced in [15]. Accordingly, the input-output linearizing tracking control with included magnetic saturation increases the magnetizing stator current i_{sd} in the direction of the rotor flux linkage vector, to achieve the reference value of the rotor flux linkage module. Therefore, the IM with the proposed input-output linearizing control is going to be magnetized in the best possible way to ensure the proper stiffness and optimal dynamic response. When the step changes of the load torque are applied on

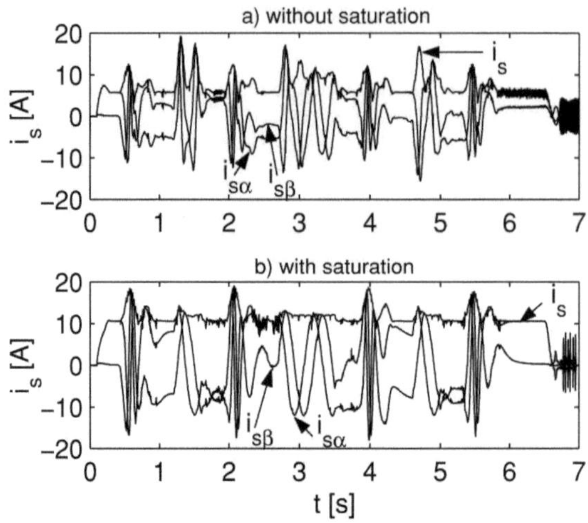

Figure 5. Stator currents $i_{s\alpha}$, $i_{s\beta}$, and $i_s = \sqrt{i_{s\alpha}^2 - i_{s\beta}^2}$: (a) saturation is not included and (b) saturation is included.

the shaft, the input-output linearizing control with included magnetic saturation performs much better than the control with neglected saturation, requiring smaller stator current i_{sq} to produce the necessary torque with the rotor flux linkage vector. The transformed currents are shown in Fig. 6. The measured stator current i_{sd} in the case of the input-output linearizing tracking control with and without included magnetic saturation agrees with the corresponding value of i_{sd} determined from the nonlinear and linearized magnetizing curve of the IM, used in the observer with and without included magnetic saturation (Fig. C1, Appendix C).

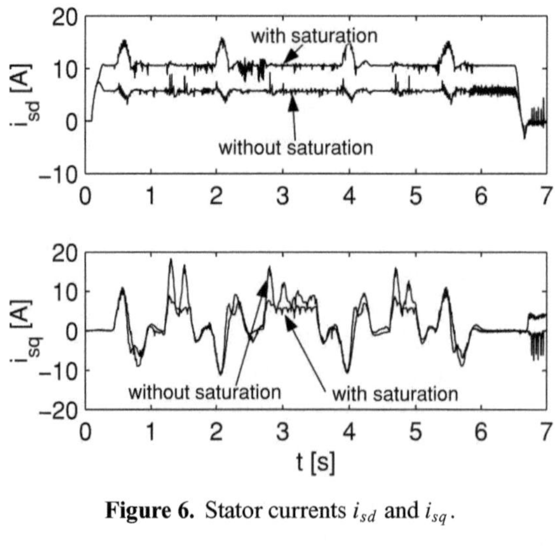

Figure 6. Stator currents i_{sd} and i_{sq}.

Conclusion

Consideration of magnetic saturation in the IM model substantially improves its accuracy, leading to a more efficient and consistent synthesis of the control algorithm, observer, and estimator of load torque. The proposed input-output linearizing control of IM with included magnetic saturation improves the dynamic performance of the drive. It gives smaller rotor position and speed errors, as well as a higher stiffness and a better load torque rejection, which results in a smaller stator current, when the load torque is introduced. An important reason for the improved behavior of the controlled IM is more adequately calculated value of the rotor flux linkage when magnetic saturation is considered in the observer design.

Appendix A

Elements of matrices C, Z, W, D

$$L_{im} = L_l^2 - L_l L_{rl}^2\left(\frac{1}{L_{rl}+L_m} + \frac{1}{L_{rl}+L}\right) + \frac{L_{rl}^{4}}{(L_{rl}+L_m)(L_{rl}+L)}$$

$$c_{11} = -\frac{1}{L_{im}}\left\{R_s\left(L_l - \frac{L_{rl}^2}{L_{qq}}\right) + R_r\frac{L_m}{L_r}\left[L_l - L_{rl}^2\left(\frac{1}{L_{rl}+L_m} + \frac{1}{L_{rl}+L}\right)\right.\right.$$
$$\left.\left. - \frac{L_{sl}L_{rl}}{L_{dd}} + \frac{L_{rl}^3}{(L_{rl}+L_m)(L_{rl}+L)}\right]\right\}$$

$$c_{12} = \frac{1}{L_{im}}\left\{R_s\frac{L_{rl}^2}{L_{dq}} - R_r\frac{L_m}{L_r}\frac{L_{sl}L_{rl}}{L_{dq}}\right\}$$

$$c_{13} = \frac{1}{L_{im}}\left\{\frac{R_r}{L_r}\left[L_l - L_{rl}^2\left(\frac{1}{L_{rl}+L_m} + \frac{1}{L_{rl}+L}\right) - \frac{L_{sl}L_{rl}}{L_{dd}} + \frac{L_{rl}^3}{(L_{rl}+L_m)(L_{rl}+L)}\right]\right\}$$

$$c_{14} = -\frac{1}{L_{im}}\left\{\frac{R_r}{L_r}\frac{L_{sl}L_{rl}}{L_{dq}}\right\}, \qquad c_{21} = c_{12}, \qquad c_{23} = c_{14}$$

$$c_{22} = -\frac{1}{L_{im}}\left\{R_s\left(L_l - \frac{L_{rl}^2}{L_{dd}}\right) + R_r\frac{L_m}{L_r}\left[L_l - L_{rl}^2\left(\frac{1}{L_{rl}+L_m} + \frac{1}{L_{rl}+L}\right)\right.\right.$$
$$\left.\left. - \frac{L_{sl}L_{rl}}{L_{qq}} + \frac{L_{rl}^3}{(L_{rl}+L_m)(L_{rl}+L)}\right]\right\}$$

$$c_{44} = -\frac{R_r}{L_r} = c_{33}$$

$$c_{24} = \frac{1}{L_{im}}\left\{\frac{R_r}{L_r}\left[L_l - L_{rl}^2\left(\frac{1}{L_{rl}+L_m} + \frac{1}{L_{rl}+L}\right) - \frac{L_{sl}L_{rl}}{L_{qq}}\right.\right.$$
$$\left.\left. + \frac{L_{rl}^3}{(L_{rl}+L_m)(L_{rl}+L)}\right]\right\}$$

$$c_{31} = R_r\frac{L_m}{L_r} = c_{42}, \qquad c_{33} = -\frac{R_r}{L_r} = c_{44}, \qquad c_{42} = R_r\frac{L_m}{L_r} = c_{31}$$

$$z_{12} = \frac{1}{L_{im}}\left\{\left(L_l - \frac{L_{rl}^2}{L_r}\right)\left(L_l - \frac{L_{rl}^2}{L_{qq}}\right)\right\}$$

$$z_{13} = -\frac{1}{L_{im}}\left\{\frac{L_m}{L_r}\frac{L_{rl}^2}{L_{dq}} + \frac{L_{sl}L_{rl}}{L_{dq}}\right\}$$

$$z_{21} = -\frac{1}{L_{im}}\left\{\left(L_l - \frac{L_{rl}^2}{L_r}\right)\left(L_l - \frac{L_{rl}^2}{L_{dd}}\right)\right\}, \quad z_{22} = z_{11}$$

$$z_{14} = -\frac{1}{L_{im}}\left\{-\frac{L_m}{L_r}\left(L_l - \frac{L_{rl}^2}{L_{qq}}\right) + \left[L_l - L_{rl}^2\left(\frac{1}{L_{rl}+L_m} + \frac{1}{L_{rl}+L}\right)\right.\right.$$
$$\left.\left. -\frac{L_{sl}L_{rl}}{L_{dd}} + \frac{L_{rl}^3}{(L_{rl}+L_m)(L_{rl}+L)}\right]\right\}$$

$$z_{23} = -\frac{1}{L_{im}}\left\{\frac{L_m}{L_r}\left(L_l - \frac{L_{rl}^2}{L_{dd}}\right) + \left[-L_l + L_{rl}^2\left(\frac{1}{L_{rl}+L_m} + \frac{1}{L_{rl}+L}\right)\right.\right.$$
$$\left.\left. +\frac{L_{sl}L_{rl}}{L_{qq}} - \frac{L_{rl}^3}{(L_{rl}+L_m)(L_{rl}+L)}\right]\right\}$$

$$z_{24} = z_{13}, \quad z_{34} = 1, \quad z_{43} = -1, \quad w_{13} = \frac{1}{L_{im}}\left\{\frac{L_{sl}L_{rl}}{L_{dq}}\right\}, w_{13} = \frac{1}{L_{im}}\left\{\frac{L_{sl}L_{rl}}{L_{dq}}\right\}$$

$$w_{14} = \frac{1}{L_{im}}\left\{L_l - L_{rl}^2\left(\frac{1}{L_{rl}+L_m} + \frac{1}{L_{rl}+L}\right) - \frac{L_{sl}L_{rl}}{L_{dd}} + \frac{L_{rl}^3}{(L_{rl}+L_m)(L_{rl}+L)}\right\}$$

$$w_{23} = -\frac{1}{L_{im}}\left\{L_l - L_{rl}^2\left(\frac{1}{L_{rl}+L_m} + \frac{1}{L_{rl}+L}\right) - \frac{L_{sl}L_{rl}}{L_{qq}} + \frac{L_{rl}^3}{(L_{rl}+L_m)(L_{rl}+L)}\right\}$$

$$w_{24} = -w_{13}, \quad w_{34} = -1, \quad w_{43} = 1$$

$$d_{11} = \frac{1}{L_{im}}\left\{L_l - \frac{L_{rl}^2}{L_{qq}}\right\}, \quad d_{12} = \frac{1}{L_{im}}\left\{\frac{L_{rl}^2}{L_{dq}}\right\} d_{21} = d_{12}, \quad d_{22} = \frac{1}{L_{im}}\left\{L_l - \frac{L_{rl}^2}{L_{dd}}\right\}$$

Appendix B

Lie derivatives

$$L_f^3\phi_1 = \frac{\partial}{\partial \mathbf{x}}\left(L_f^2\phi_1\right)\frac{d\mathbf{x}}{dt} = \nabla\left(L_f^2\phi_1\right)[\mathbf{f}+\mathbf{Gu}]$$
$$= p\frac{L_m}{L_r}\frac{1}{J}\left[\left(c_{22}+c_{33}-\frac{f}{J}\right)i_{s\beta}\psi_{r\alpha} - \left(c_{11}+c_{44}-\frac{f}{J}\right)i_{s\alpha}\psi_{r\beta}\right.$$
$$+ c_{12}\left(i_{s\alpha}\psi_{r\alpha} - i_{s\beta}\psi_{r\beta}\right) - \omega_r\left(i_{s\alpha}\psi_{r\alpha} + i_{s\beta}\psi_{r\beta}\right)$$
$$+ (c_{24}+c_{13}-2w_{13}\omega_r)\psi_{r\alpha}\psi_{r\beta}$$
$$\left. + (c_{23}+w_{23}\omega_r)\psi_{r\alpha}^2 - (c_{14}+w_{14}\omega_r)\psi_{r\beta}^2\right] + \frac{f}{J^2}t_l + \frac{f^2}{J^2}\omega_r$$

$$L_f^2\phi_2 = \frac{\partial}{\partial \mathbf{x}}\left(L_f\phi_2\right)\frac{d\mathbf{x}}{dt} = \nabla\left(L_f\phi_2\right)[\mathbf{f}+\mathbf{Gu}]$$
$$= 2R_r\frac{L_m}{L_r}\left[\left(c_{11}+c_{33}-2\frac{c_{31}}{L_m}\right)i_{s\alpha}\psi_{r\alpha} + \left(c_{22}+c_{44}-2\frac{c_{31}}{L_m}\right)i_{s\beta}\psi_{r\beta}\right.$$

$$+c_{12}\left(i_{s\beta}\psi_{r\alpha}+i_{s\alpha}\psi_{r\beta}\right)+\omega_r\left(i_{s\beta}\psi_{r\alpha}-i_{s\alpha}\psi_{r\beta}\right)$$
$$+\left(c_{14}+c_{23}+w_{14}\omega_r+w_{23}\omega_r\right)\psi_{r\alpha}\psi_{r\beta}+\left(c_{13}-2\frac{c_{33}}{L_m}\right)\psi_{r\alpha}^2$$
$$\left.+\left(c_{24}-2\frac{c_{33}}{L_m}\right)\psi_{r\beta}^2+w_{31}\omega_r\left(\psi_{r\alpha}^2-\psi_{r\beta}^2\right)+c_{31}\left(i_{s\alpha}^2+i_{s\beta}^2\right)\right]$$

$$L_{g\alpha}L_f^2\phi_1=\frac{\partial}{\partial x}\left(L_f\phi_1\right)g_\alpha=p\frac{L_m}{L_r}\frac{1}{J}\left(d_{21}\psi_{r\alpha}-d_{11}\psi_{r\beta}\right)$$
$$L_{g\beta}L_f^2\phi_1=\frac{\partial}{\partial x}\left(L_f\phi_1\right)g_\beta=p\frac{L_m}{L_r}\frac{1}{J}\left(d_{22}\psi_{r\alpha}-d_{12}\psi_{r\beta}\right)$$
$$L_{g\alpha}L_f\phi_2=\frac{\partial}{\partial x}\left(L_f\phi_2\right)g_\alpha=2R_r\frac{L_m}{L_r}\left(d_{11}\psi_{r\alpha}+d_{21}\psi_{r\beta}\right)$$
$$L_{g\beta}L_f\phi_2=\frac{\partial}{\partial x}\left(L_f\phi_2\right)g_\beta=2R_r\frac{L_m}{L_r}\left(d_{12}\psi_{r\alpha}+d_{22}\psi_{r\beta}\right)$$

Appendix C

Table 1. Parameters of the 3 kW induction motor with wound rotor Sever ZPD112MK4:

R_s	1.976 Ω
L_m	0.223 H
L_s	0.2335 H
f	0.0007 Nms/rad
R_r	2.91 Ω
L_r	0.2335 H
J	0.031 kgm^2
T_n	15 Nm

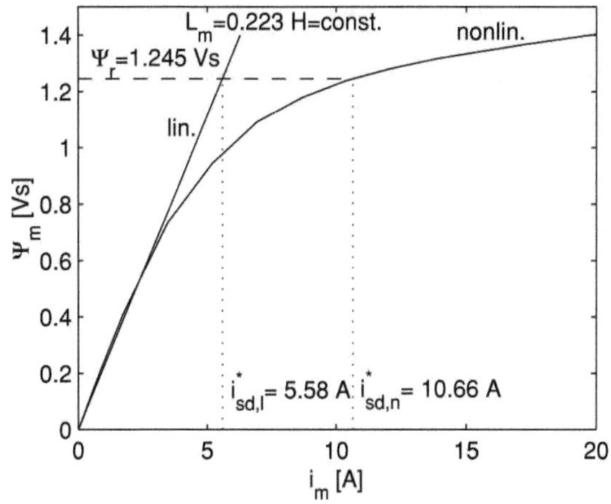

Figure C 1. Rotor flux linkage and corresponding stator current in the case of linear and nonlinear magnetizing curve.

Acknowledgment

This work was supported in part by the Slovene Ministry of Education, Science and Sport, Project No. P2-0115.

References

[1] P. Vas, Electrical Machines and Drives: A Space-Vector Theory Approach, Oxford: Oxford University Press, 1992.
[2] J.C. Moreira, T.A. Lipo, Modelling of saturated ac machines including air gap flux harmonic components, IEEE Trans. Ind. Appl., Vol. 28, No. 2, pp. 343–349, 1997.
[3] E. Levi, A unified approach to main flux saturation modelling in d-q axis models of induction machines, IEEE Trans. Energy Convers., Vol. 10, No. 3, pp. 455–461, 1995.
[4] E. Levi, Impact of cross-saturation on accuracy of saturated induction machine models, IEEE Trans. Energy Convers., Vol. 12, No. 3, pp. 211–216, 1997.
[5] R.D. Lorenz, D.W. Novotny, Saturation effects in field-oriented induction machines, IEEE Trans. Ind. Appl., Vol. 26, No. 5, pp. 283–289, 1990.
[6] E. Levi, S. Vukosavíc, V. Vučkovíc, "Saturation Compensation Schemes for Vector Controlled Induction Motor Drives", PESC'90 Record, San Antonio, TX, USA, pp. 591–598, 1990.
[7] P. Vas, Sensorless Vector and Direct Torque Control, Oxford: Oxford University Press, 1998.
[8] E. Levi, M. Sokola, S.N. Vukosavíc, A method for magnetizing curve identification in rotor flux oriented induction machines, IEEE Trans. Energy Convers., Vol. 15, No. 2, pp. 157–162, 2000.
[9] Z. Krzeminski, A. Jaderko, "A Speed Observer System of Induction Motor with Magnetizing Curve Identification", EPE-PEMC 2000, Kosice, Slovakia, 2000.
[10] Z. Krzeminski, "Nonlinear Control of Induction Motor", Proc. 10th IFAC World Congress, Munchen, Germany, 1987, pp. 349–354.
[11] R. Marino, S. Peresada, P. Valigi, Adaptive nonlinear control of induction motors via extended matching, Lect. Note Contr. Inform. Sci., Vol. 160, pp. 1435–1454, 1991.
[12] T. von Raumer, J.M. Dion, L. Dugard, "Adaptive Nonlinear Speed and Torque Control of IM", Proceedings of European Control Conference, Gronigen, June 1993, pp. 592–596.
[13] R.T. Novotnak, J. Chiasson, M. Bodson, High-performance motion control of an induction motor with magnetic saturation, IEEE Trans. Contr. Syst. Technol., Vol. 7, No. 3, pp. 315–327, 1999.
[14] P. Ljušev, "Analysis of Induction Motor Control Taking into Account Magnetic Saturation", Master thesis, University of Maribor, Slovenia, 2002.
[15] G.C. Verghese, S.R. Sanders, Observers for flux estimation in induction machines, IEEE Trans. Ind. Electron., Vol. 35, No. 1, pp. 85–94, 1988.

II-11. DIRECT POWER AND TORQUE CONTROL SCHEME FOR SPACE VECTOR MODULATED AC/DC/AC CONVERTER-FED INDUCTION MOTOR

M. Jasinski, M. P. Kazmierkowski and M. Zelechowski

Warsaw University of Technology, Institute of Control & Industrial Electronics,
ul. Koszykowa 75, 00-662 Warszawa,
mja@isep.pw.edu.pl, mpk@isep.pw.edu.pl,
zelechom@isep.pw.edu.pl *WWW: http://www.ee.pw.edu.pl/icg*

Abstract. A novel control scheme for PWM rectifier-inverter system is proposed. Fast control strategies such as line voltage Sensorless Virtual Flux (VF) based Direct Power Control with Space Vector Modulator (DPC-SVM) for rectifier and Direct Torque Control with Space Vector Modulator (DTC-SVM) for inverter side are used. These strategies lead to good dynamic and static behaviour of the proposed control system—Direct Power and Torque Control- Space Vector Modulated (DPTSVM). Simulations and experiment results obtained show good performance of the proposed system. Additional power feedforward loop from motor to rectifier control side improved dynamic behaviours of the power flow control. As a result, better input-output energy matching allows decreasing the size of the dc-link capacitor

Introduction

The adjustable speed drives (ASD) with diode rectifier nowadays is the most popular on the marked. Large electrolytic capacitor is used as an energy-storing device to decouple rectifier and the inverter circuits. The capacitors have some drawbacks: low reliability, high size, weight and cost. Hence, reliability of the dc-link capacitor is the major factor limiting the lifetime of the ASD systems [1].

Development of control methods for Pulse Width Modulated (PWM) boost rectifier (active rectifier) was possible thanks to advances in power semiconductors devices and Digital Signal Processors (DSP). Therefore, the Insulated Gate Bipolar Transistors (IGBT) AC/DC/AC converter controlled by PWM is used in motor drive systems (Fig.1). Thanks to active rectifier the dc-link capacitor can be reduced [2]. Farther reduction of the capacitor can be achieved by power feedforward loop from motor side to the control of the PWM rectifier. A lot of works are given attention to reduce the dc-link capacitor. However, a small capacitance leads to a high dc-voltage fluctuation. To avoid this drawback various dc-voltage control schemes have been proposed. Some of them take into account the inverter dynamics

S. Wiak, M. Dems, K. Komęza (eds.), Recent Developments of Electrical Drives, 261–274.

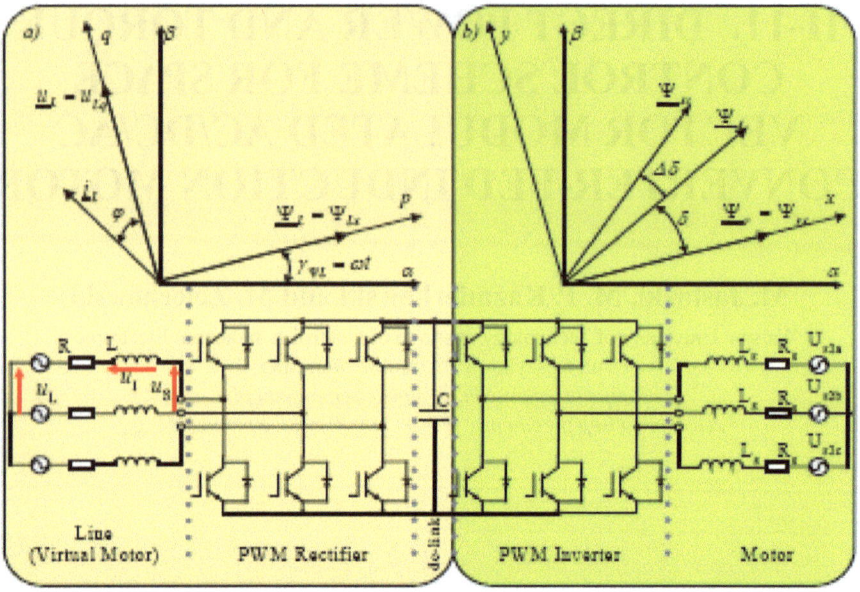

Figure 1. Representation of three-phase PWM rectifier—inverter system; vector diagram and coordinate system for: a) PWM rectifier side b) inverter side.

to improve the PWM rectifier current control by feedback linearization [3] and master-slave [1] manner. Another control methodology proposed a fast dc-link voltage controller which works with dc-voltage and motor variables as inputs [4]. Moreover, various methods of the output power estimation have been discussed in [5].

In the mentioned methods active and reactive powers of the PWM rectifier are indirectly controlled via current control loops. Besides, stator current controllers control the torque and flux of the motor too.

In this paper a line voltage sensorless Virtual Flux (VF) based Direct Power Control with Space Vector Modulator (DPC-SVM) is applied to control of the PWM rectifier.

The inverter with induction motor is controlled via Direct Torque Control with Space Vector Modulator (DTCSVM). Contrary to the scheme proposed in [6], our solution includes not stator flux controller but space vector modulator.

Hence, an AC/DC/AC converter of Fig. 1, is controlled by Sensorless Direct Power and Torque Control-Space Vector Modulated (DPT-SVM) scheme. In comparison to methods that control an active and reactive power, torque and flux in indirect manner the coordinates transformation and decoupling are not required. Moreover, the current control loops are avoided.

In respect of dynamic, of dc-voltage control the power balance between line and motor is very important. Therefore, to improve instantaneous input/output power matching, the additional feedforward power control loop is introduced.

Thanks to better control of the power flow the fluctuation of the dc-link voltages will be decrease. So the size of the dc-link capacitor can be reduced.

Direct power and direct torque control space vector modulated (DPT-SVM) scheme

Direct Power Control (DPC) for PWM rectifier is based on instantaneous control of active p and reactive q power flow from/to the line and to/from active load. In classical approach [7, 8] of DPC there are two power control loops with hysteresis comparators and switching table.

Therefore, the key point of the DPC implementation is sufficiently precise and fast estimation of the instantaneous line powers. The most significant drawbacks of the hysteresis-based DPC are variable switching and high sampling frequency. Introducing a Space Vector Modulator (SVM) in control strategy [9,10] allows to eliminate the both mentioned problems. Moreover, the line voltage sensors can be replaced by Virtual Flux (VF) estimator, which introduces technical and economical advantages to the system (simplification, reliability, galvanic isolation, cost reduction). Such control system is called: Virtual Flux Based Direct Power Control Space Vector Modulated (DPC-SVM) scheme [11]. Summarised, in this method linear PI controllers with Space Vector Modulator replace hysteresis comparators and switching table (Fig. 3).

Similarly like DPC, for the classical Direct Torque Control (DTC) [12], the command stator flux Ψ_{sc} and commanded torque Mec values are compared with the actual stator flux Ψ_s and electromagnetic torque M_{ec} values in hysteresis flux and torque controllers, respectively. Therefore, the well known disadvantages of DTC are: variable switching frequency, ± 1 switching over dc-link voltage U_{dc}, current and torque distortion caused by sector changes as well as high sampling frequency requirement for digital implementation of the hysteresis controllers. All above difficulties can be eliminated when, instead of the switching table, a SVM is used. Hence, the DTC-SVM strategy [13] for control of the inverter/motor part is proposed (Fig. 3).

Simplified mathematical model of the system in stationary α, β coordinates is shown in Fig. 2

DPC-SVM with virtual flux (VF)

A line current i_L is controlled by voltage drop on the input inductance L that placed between two voltage sources (line on the one side and the converter on the other). From Kirchhoff's law the input equations can be wrote:

$$U_L = U_I + U_{s1} \tag{1}$$

where

$$U_I = L\frac{d}{dt}I_L\text{—voltage drop on the inductance}$$

$$U_{sk} = \begin{bmatrix} U_{sk\alpha} \\ U_{sk\beta} \end{bmatrix} = \begin{bmatrix} \frac{2}{3}U_{dc}\left\{DA_k - \frac{1}{2}(D_{Bk} + D_{Ck})\right\} \\ \frac{\sqrt{3}}{3}U_{dc}(D_{Ak} - D_{Bk}) \end{bmatrix} \tag{2}$$

where $k = 1, 2$; 1—for the PWM rectifier, 2—for the PWM inverter.

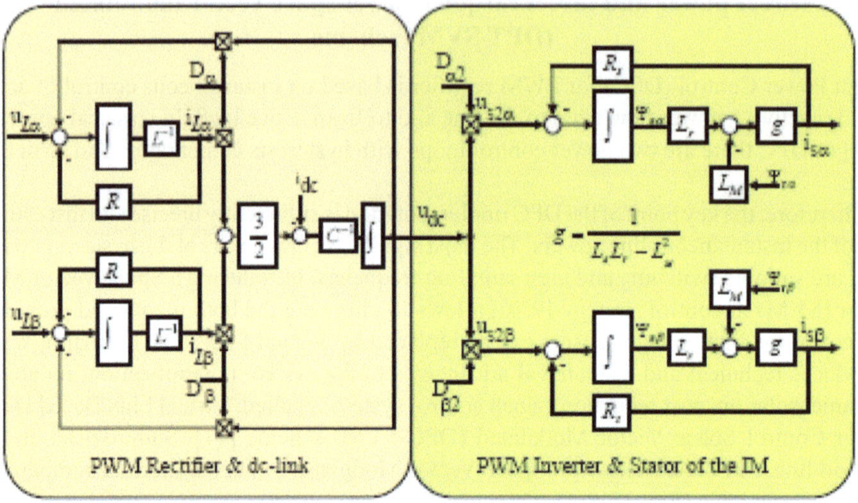

Figure 2. Modified model of AC/DC/AC converter in α, β coordinates.

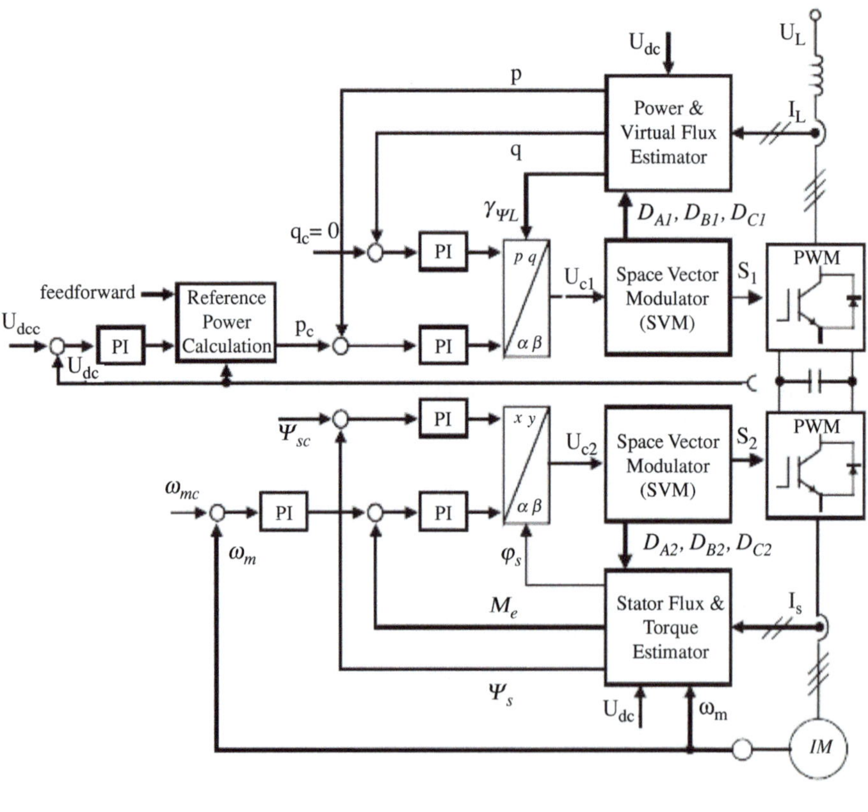

Figure 3. Basic structure of unified direct power and torque control with space vector modulator (DPT- SVM).

Voltage on the input of the converter can be calculated from measured dc-link voltage U_{dc} and duty cycles from PWM rectifier's modulator D_{A1}, D_{B1}, D_{C1} (2). Therefore, proposed DPC-SVM is sensorless line voltage control strategy. Based on assumption that line voltage U_L with input inductances can be related as quantities of virtual AC motor (Fig. 1) and the integration of the line voltage gives Virtual flux linkage of the virtual AC motor, the VF estimator with low pass filter is used.

$$\mathbf{\Psi}_L = \int \left[(\mathbf{U}_S + \mathbf{U}_L) - \frac{1}{T_{f1}} \Psi_L \right] dt \tag{3}$$

The measured line currents and virtual flux linkage obtained from (3) can be used for power calculations [11]. With assumptions that line voltages is sinusoidal and balanced, simple equations are obtained:

$$\begin{aligned} p &= \omega \left(\Psi_{L\alpha} I_{L\beta} - \Psi_{L\beta} I_{L\alpha} \right), \\ q &= \omega \left(\Psi_{L\alpha} I_{L\alpha} + \Psi_{L\beta} I_{L\beta} \right) \end{aligned} \tag{4}$$

Both estimated powers are compared with commanded values p_c, q_c respectively, were q_c is set to zero for fulfilling the unity power factor conditions. The command active power p_c is provided from outer PI dc-link voltage controller. The obtained errors are dc quantities. These signals are delivered to PI controllers that eliminate the steady state error. The PI controllers generate dc-values voltage commands U_{pc}, U_{qc}. After coordinate transformation (5) $pq/\alpha\beta$, $U_{\alpha c}$ and $U_{\beta c}$ are delivered to SVM block, which generates switching signals (Fig. 3).

$$\mathbf{U}_c = \begin{bmatrix} U_{c\alpha} \\ U_{c\beta} \end{bmatrix} = \begin{bmatrix} -U_{qc} \cos \gamma_{\Psi L} - U_{pc} \sin \gamma_{\Psi L} \\ -U_{qc} \cos \gamma_{\Psi L} + U_{pc} \sin \gamma_{\Psi L} \end{bmatrix} \tag{5}$$

DTC-SVM

In control of an induction motor (IM) drive, supplied by a voltage source inverter, there is a possibility to control directly the electromagnetic torque and stator flux linkage by the selection of the optimum inverter switching modes. That control manner is called direct power and torque control (DTC). DTC allows very fast torque responses and flexible control of an IM. To avoid the drawbacks (variable switching frequency, voltage polarity violation) of DTC instead hysteresis controllers and switching table the space vector modulator (SVM) with PI controllers were introduced.

However, it should be noted that DTC with SVM (DTC-SVM) has all advantages of the DTC, and mathematical as well as physical principles are the same. Generally in IM, the instantaneous electromagnetic torque is proportional to the vector product of the stator flux linkage and stator current space vectors (6) in stationary $\alpha\beta$ reference frame.

$$M_e = \frac{1}{2} m_s P_b \mathbf{\Psi}_s \times \mathbf{I}_s \tag{6}$$

where $\Psi_s = \Psi_{sej\varphi s}$—stator flux linkage space vector, $I_s = I_{sej\varphi i}$—stator current space vector φ_s, φ_i—angle of the stator flux linkage space vector and angle of the stator current space vector respectively, in relation to the α axis of the stationary (stator) reference frame.

Therefore, eq. (6) can be converted into (7)

$$M_e = \frac{1}{2} m_s P_b \Psi_s \sin\gamma \tag{7}$$

where $\gamma = \phi_i - \phi_s$—angle between the stator current and stator flux linkage space vectors. Assuming, that modules (amplitudes) of the stator flux linkage is constant, and the angle φ_s is varying quickly, then M_e can be changed with very high dynamics. The rate of change of the increasing M_e is almost proportional to the rate of change $d\varphi_s/dt$ [18]. Summarizing, fast torque control is obtained when stator voltage is on the level, which kept amplitude of the stator flux constant (the voltage drop on stator resistance is neglected), and which rapidly moving the stator flux linkage space vector to demanded position (required by the torque). Therefore, by using appropriate stator voltages the stator flux linkage space vector can be controlled. It is useful to consider another expression for control of the electromagnetic torque (8):

$$M_e = \frac{L_m}{L_r L_\sigma} \Psi_r \Psi_s \sin\delta \tag{8}$$

It base on assumption, that amplitudes of stator and rotor flux linkage are constant. The rotor one because of time constant is large (eg. 0.1 s). Therefore, with this conditions follows from eq. (8) that the Mecan be controlled by changing δ in suitable direction. The δ is called a torque angle and depends on the commanded torque. It should be pointed, that accuracy of the flux calculation is indispensable. That goal can be obtained with a $\mathbf{U}_s$, $\mathbf{I}_s$ (“voltage”) model based estimator, with low pass filter (9) or by $\mathbf{I}_s$, γ_m model (10):

$$\mathbf{\Psi}_s = \int \left[(\mathbf{U}_{s2} - Rs\mathbf{I}_s) - \frac{1}{TF}\mathbf{\Psi}_s \right] dt, \tag{9}$$

or

$$\mathbf{\Psi}_s = \frac{L_m}{L_r}\mathbf{\Psi}_r + \sigma L_s \mathbf{I}_s \tag{10}$$

Equation (10) ensures better accuracy over the entire frequency range, but it require the angle γ_m of motor shaft position for dq transformation.

Power feedforward control loop

Instantaneous power supplied to an m_s – phase winding can be expressed in terms of complex space vectors as:

$$P = \frac{1}{2} m_s \mathrm{Re}\left[\mathbf{U}_{s2}\mathbf{I}_s^*\right] \tag{11}$$

Taking into consideration the overall power supplied to the stator and rotor windings from (11) can be wrote:

$$P = \frac{1}{2} m_s \left(\mathrm{Re}\left[\mathbf{U}_{s2}\mathbf{I}_s^*\right] + \mathrm{Re}\left[\mathbf{U}_r\mathbf{I}_r^*\right]\right) \tag{12}$$

The losses in resistances can be neglected, thus the internal power is:

$$P_i = P_{mag} + P_e \tag{13}$$

where P_{mag}—is the power stored in the magnetic fields, P_e is the electromagnetic power. From the assumption that only active power is derived from the dc-link to an electric motor and reactive power is derived from the inverter only electromagnetic power can be taken into consideration. In a general way P_e expressed as:

$$P_e = M_e \Omega m \tag{14}$$

where Ω_m—mechanical angular rotor speed, M_e electromagnetic torque.

Hence, active power feedforward can be realized based on Eq. 14. The electromagnetic power is the part of the power supplied to the electrical terminals of an AC motor, that is neither stored nor lost. It corresponds to the voltages induced in rotor windings and to the currents flowing into them [11]. For prediction of the power state of the motor (motoring, regenerating, loaded or unloaded) the commanded values of the electromagnetic torque and mechanical speed can be taken:

$$P_{ec} = M_{ec} \Omega_{mc} \tag{15}$$

Such calculated power can be simply added to the referenced active power of the PWM rectifier. To fulfil the stability conditions of the system the T_w delay should be introduced:

$$P_e = \frac{1}{1 + T_w s} P_{ec} \tag{16}$$

where T_w—time constant of the M_e dynamics.

Thanks to the predictive abilities of motor power feedforward loop a better dc-link voltage stabilization can be obtained. Also, fluctuations of dc-voltages may be reduced.

Dc-link capacitor design

In AC/DC/AC converter with diode rectifier there is no control of the dc-link voltage in particular during transients (Fig. 10). So that, the size of the capacitor should be grater than in a converter with PWM rectifier. A dc-link voltage control accuracy depends on the time constant of the dclink voltage controller. This time constants can be reduced by additional power feedforward control loop.

Having the maximum allowed dc-link voltage fluctuations ΔUdc, the required capacity can be calculated as :

$$C_{PWMm} = P_{out} \frac{\sqrt{2} + \sqrt{3} U_{LLrms} / U_{DC}}{2\sqrt{3} f_s U_{LLrms} \Delta U_{DC}} \tag{17}$$

where P_{out}—rated output power, U_{LLrms}—line to line voltage, f_s—sampling frequency.

Moreover, the general capacitor life time is:

$$L = L_B \times f(T_M - T_C) \times f_1(U_{dc}) \tag{18}$$

where, L is the life estimate in hours, L_B is the base life elevated maximum temperature T_M, T_C is the actual core temperature and U_{dc} is the applied dc-voltage. The voltage multiplier f_1 at higher stress level may reduce the life of the capacitor [14]. Therefore, the stabilization of the dc-voltage at the required level is important.

Table 1. Parameters of the model

Sampling and switching frequency	5 kHz
Resistance of reactors *R*	80 mΩ
Inductance of reactors *L*	10 mH
DC-link capacitor	470 μF
Phase voltage *V*	230 RMS
Source voltage frequency	50 Hz
DC-link voltage	560 V

Simulation and experimental results

Proposed approach has been tested using Saber simulations packed software. The main data and parameters of the model are shown in Table 1.

An experimental investigation was conducted on a laboratory setup (Fig. 4). The setup consists of: input inductance, two PWM converter (VLT5005, serially produced by Danfoss with replaced control interfaces) controlled by dSPACE DS1103 and induction motor set. The computer is used for software development and process visualization.

Converters, motor and input inductance parameters are shown in Table 2.

In below figures are shown different states of the DPTSVM operation. In Fig. 5a and Fig. 6a the system operates in motoring mode, with power factor near to unity (the current is in phase with the line voltage) and almost sinusoidal waveform of the line current (low Total Harmonic Distortion – THD factor).

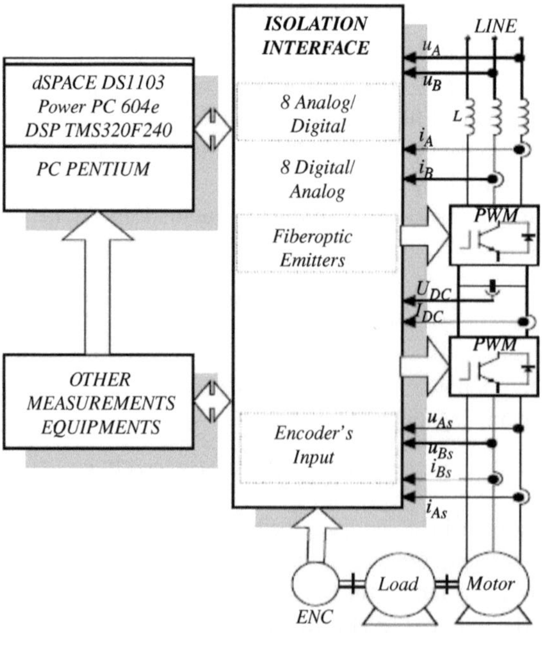

Figure 4. Laboratory setup.

Table 2. Main parameters of the laboratory setup

AC motor	
Stator winding resistance	1.85 Ω
Rotor winding resistance	1.84 Ω
Stator inductance	170 mH
Rotor inductance	170 mH
Mutual inductance	160 mH
Number of pole pairs	2
Moment of inertia	0.019 kgm^2
Phase voltage	230 V(rms)
Phase current I	6.9 A(rms)
Nominal torque MN	20 Nm
Base speed ω_b:	1415 rpm
Input inductance	
Resistance of reactors R	100 mΩ
Inductance of reactors L	10 mH
VLT5005 Converters	
Sampling and switching frequency	5 kHz
DC-link capacitor	470 μF
Nominal power P_N	5,5 kVA
Measurement conditions	
Phase voltage V	150 RMS
Source voltage frequency	50 Hz
DC-link voltage	560 V

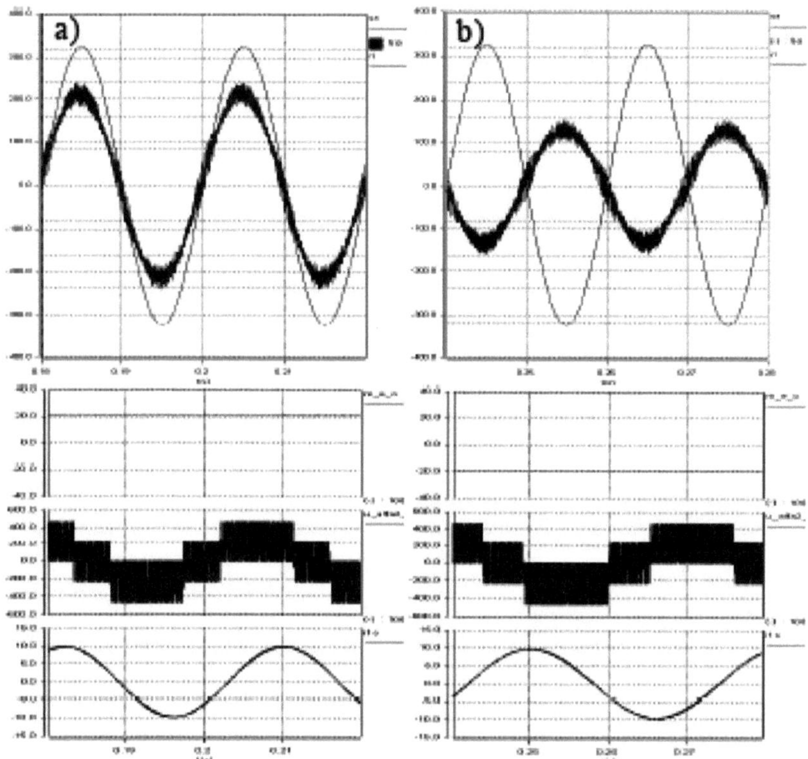

Figure 5. Steady state from the top: i_L—line current 2A/div, U_L—line voltage, M_e electromagnetic torque, $U_{s\alpha}$ component of stator voltage, $i_{s\alpha}$—stator current; a) motoring mode, b) regenerating mode.

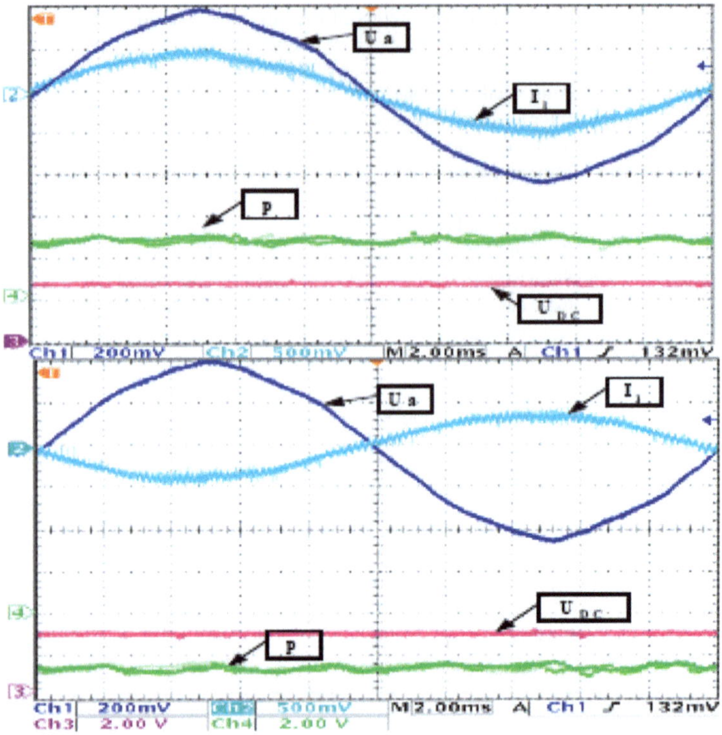

Figure 6. Experimental results—steady state. From the top: line voltage 100 V/div, line current 5 A/div, active power, dc-link voltage, a) for acceleration, b) for regeneration mode.

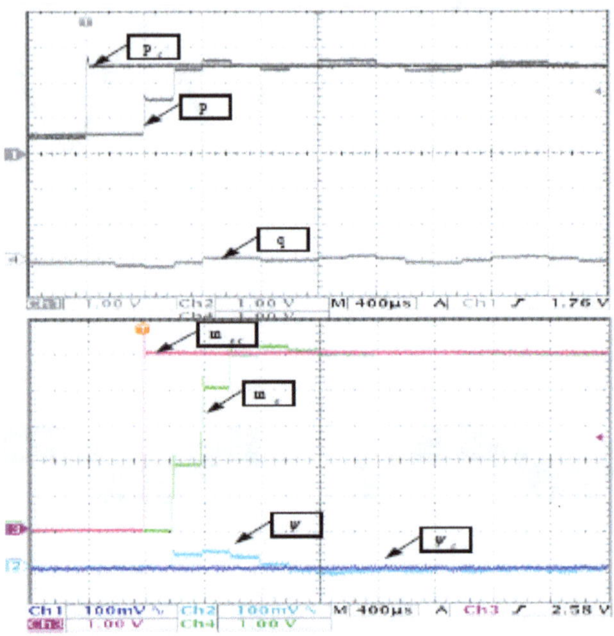

Figure 7. Experimental results. Small signal behaviour of the: a) power control loop ($p_c = 0.1 \rightarrow 0.5$ PN, p—actual active power, q—reactive power; b) torque control loop ($M_{ec} = 0 \rightarrow 1$ $\mathrm{M_N}$), M_e—actual electromagnetic torque, commanded and actual stator flux.

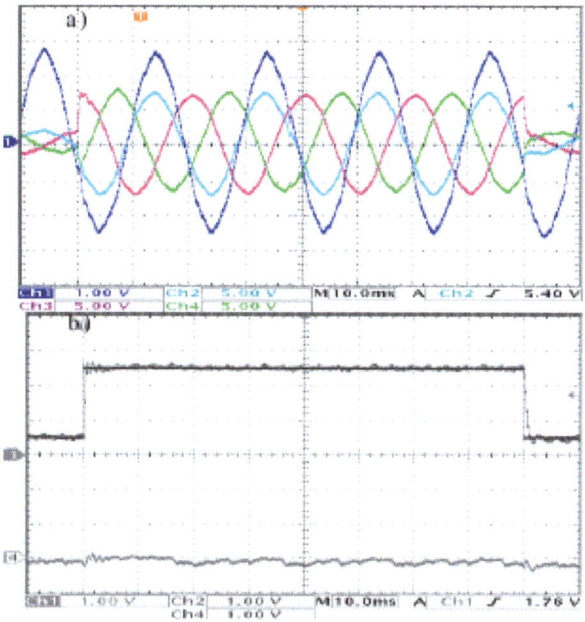

Figure 8. Experimental results. Transient in commanded active power (300–1300 W) a) ch1—line voltage, ch2,3,4—line currents, b) from the top: commanded active power, active power, reactive power.

Oscillograms of Fig. 5b and Fig. 6b illustrates operation of an AC/DC/AC converter in regenerating mode (as a transmitter of the energy from the motor to the line). Note that current is shifted by 180 degree in respect to the line voltage. In Fig. 7 experimental waveforms of the small signal test a) commanded active power and b) electromagnetic torque are presented. Power tracking performance of the PWM rectifier in back-to-back converter is shown in Fig. 8. In Fig. 9 and Fig. 10 are shown the responses to step change of the commanded electromagnetic torque from –5 do 5 Nm. That test was conduced for ac/dc/ac converter with diode rectifier (Fig. 9) as well as for back-to-back converter (Fig. 10). The behaviour of the dc-link voltage can be observed. From Fig. 9a it can be seen that the overshoot in dc-link voltage is significantly bigger then for back-to-back converter (Fig. 10a).

CONCLUSION

Virtual Flux Based Direct Power Control with Space Vector Modulator (DPC-SVM) and Direct Torque Control with Space Vector Modulator (DTC-SVM) are applied to a PWM AC/DC/AC converter. The power of the PWM rectifier and torque of the induction motor is controlled in direct manner. It means that control system operates with end-user quantities. Hence, obtained Direct Power and Torque Control- Space Vector Modulated (DPT-SVM).

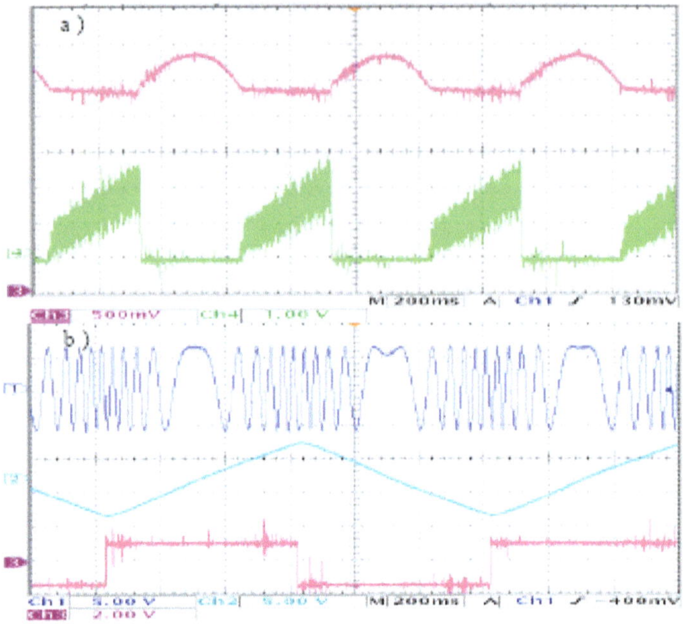

Figure 9. Experimental results with diode rectifier. Transients to commanded torque changes (−5 to 5 Nm). From the top: a) dc-link voltage 100 V/div, active power at the input of the ac/dc/ac converter b) stator current, mechanical speed, electromagnetic torque.

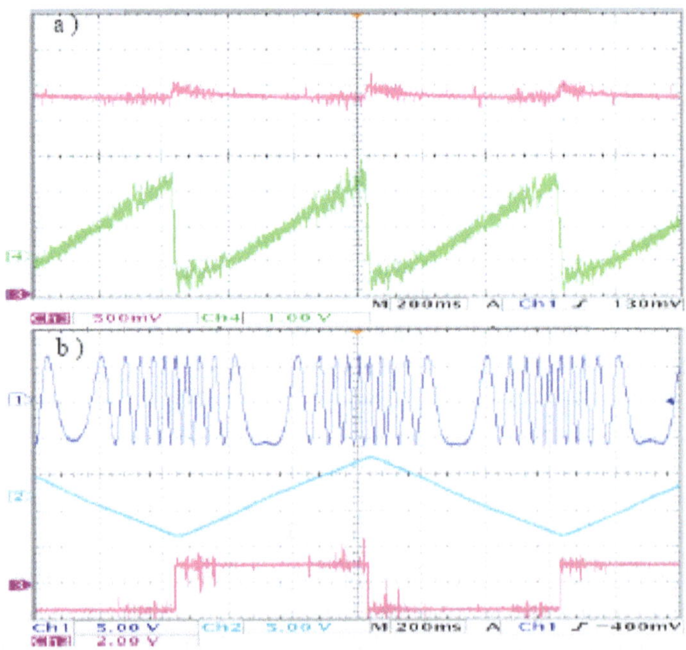

Figure 10. Experimental oscillograms with PWM rectifier. Transients to commanded torque changes (−5 to 5 Nm) From the top: a) dc-link voltage 100 V/div, active power at the input of the ac/dc/ac converter b) stator current, mechanical speed, electromagnetic torque.

Moreover, additional power feedforward control loop was implemented and tested. Proposed control system assures:

- four quadrant operation (energy saving),
- good stabilization of the dc-voltage (allows to reduce a dc-link capacitor),
- constant switching frequency,
- almost sinusoidal line current (low THD) for ideal and distorted line voltage,
- noise resistant power estimation algorithm,
- high dynamics of power and torque control,
- low motor current and torque ripple

Power feedforward loop from the motor side to the PWM rectifier improved control dynamics of the dc-link voltage. It allows fulfilling power matching conditions under transient for PWM rectifier and inverter/motor system.

References

[1] H. Hur, J. Jung, K. Nam, "A Fast Dynamics DC-link Power-Balancing Scheme for a PWM Converter-Inverter System". IEEE Trans. on Ind. Elect., vol. 48, No. 4, August 2001, pp. 794–803.

[2] L. Malesani, L. Rossetto, P. Tenti and P. Tomasin, "AC/DC/AC Power Converter with Reduced Energy Storage in the DC Link," IEEE Trans. on Ind. Appl., vol. 31, No. 2, March/April 1995, pp. 287–292.

[3] J. Jung, S. Lim, and K. Nam, "A Feedback Linearizing Control Scheme for a PWM Converter-Inverter Having a Very Small DClink Capacitor," IEEE Tran. on Ind. App., vol. 35, No. 5, September/October 1999, pp. 1124–1131.

[4] J. S. Kim and S. K. Sul, "New control scheme for ac–dc–ac converter without dc link electrolytic capacitor," in Proc. of the IEEE PESC'93, 1993, pp. 300–306.

[5] R. Uhrin, F. Profumo "Performance Comparison of Output Power Estimators Used in AC/DC/AC Converters," IEEE, 1994, pp. 344–348.

[6] A. Tripathi, A.M. Khambadkone, S.K. Panda, "Space-vector based, constant frequency, direct torque control and dead beat stator flux control of AC machines," Proc. of the IECON '01, Vol.: 2, pp. 1219–1224 vol. 2.

[7] T. Noguchi, H. Tomiki, S. Kondo, I. Takahashi, "Direct Power Control of PWM converter without power-source voltage sensors," IEEE Trans. on Ind. Appl. Vol. 34, No. 3, 1998, pp. 473–479.

[8] T. Ohnishi, "Three-phase PWM converter/inverter by means of instantaneous active and reactive power control," In Proc. of the IEEE-IECON Conf., 1991, pp. 819–824.

[9] J. Holtz "Pulsewidth Modulation for Electronics Power Conversion," In Proc. of The IEEE, vol. 82, no. 8, August 1994, pp.1194–1214.

[10] M. P. Kazmierkowski, R. Krishnan and F. Blaabjerg, Control in Power Electronics, Academic Press, 2002, p. 579.

[11] M. Malinowski, M. Jasinski, M.P. Kazmierkowski, "Simple Direct Power Control of Three-Phase PWM Rectifier Using Space Vector Modulation," in IEEE Trans. on Ind. Elect., vol. 51, No. 2, April 2004, pp. 447–454.

[12] I. Takahashi, and T. Noguchi, "A New Quick-Response and High Efficiency Control Strategy of an Induction Machine," IEEE Trans. on Ind. Appl., vol. IA-22, No. 5, September/October 1986, pp. 820–827.

[13] D. Swierczynski, M.P. Kazmierkowski, "Direct Torque Control of Permanent Magnet Synchronous Motor (PMSM) Using Space Vector Modulation (DTC-SVM),"—Simulation and Experimental Results", IECON 2002, Sevilla, Spain, on-CD.
[14] S.G. Perler, "Deriving Life Multipliers for Electrolytic Capacitors," IEEE PES Newsletter, First Quarter 2004, pp. 11–12.
[15] H. Tajima, and Y. Hori, "Speed Sensorless Field-Oriented Control of the Induction Machine". IEEE Trans. on Ind. Appl., vol. 29, No. 1, 1993, pp. 175–180.
[16] T.G. Habatler, "A space vector-based rectifier regulator for AC/DC/AC converters". IEEE Trans. on Power Electr., vol. 8, February 1993, pp. 30–36.
[17] J.Ch. Liao and S.N. Yen, "A Novel Instantaneous Power Control Strategy and Analytic Model for Integrated Rectifier/Inverter Systems". IEEE Trans. on Power Electr., vol. 15, No. 6, November 2000, pp. 996–1006.
[18] P. Vas, "Sensorless Vector and Direct Torque Control," Oxford University Press, 1998, p. 729.

II-12. EXPERIMENTAL VERIFICATION OF FIELD-CIRCUIT FINITE ELEMENTS MODELS OF INDUCTION MOTORS FEED FROM INVERTER

K. Komęza, M. Dems and P. Jastrzabek

Institute of Mechatronics and Information Systems, Technical University of Lodz, Stefanowskiego 18/22, 90-924 Lodz, Poland

komęza@p.lodz.pl, mdems@p.lodz.pl, piastrza@posejdon.wpk.p.lodz.pl

Abstract. The main aim of the paper is the presentation of the different methods that can be used during experimental verification of the validity of the field-circuit model of an induction machine for inverter feeding simulation. The second aim is to discuss, based on the DC feeding method, whether field-circuit methods or circuit methods with changeable parameters should be used to simulate transient characteristics of induction machines.

Introduction

The paper presents different methods used for experimental verification of field-circuit finite elements models of induction motors. The field-circuit models can be used in the modeling of transient states of induction motors by taking advantage of the real shape of voltage generated by the inverter [1–4]. The current and speed curves vs. time of the induction motors during transient state can be simply compared with the calculated curves to indicate the validity of the simulation. The torque curve vs. time, especially for inverter feeding, is very distorted. It is widely known that only part of torque harmonics can be obtained from measurements. The measurement of the torque during transient state is very difficult because the measured signals are the results of the mechanical systems response.

According to this problem, it is very important to work out different methods to verify the validity of used field-circuit models of induction motors.

Examined motor

The object of investigation was the three-phase induction squirrel-cage motor of 380 V (star connected) rated output power 0.37 kW. Table 1 shows the specification of the motor.

Field-circuit model

Electromechanical transients of the examined induction motor have been computed using the program Opera 2D based on the field-circuit model.

S. Wiak, M. Dems, K. Komęza (eds.), Recent Developments of Electrical Drives, 275–289.

Table 1. Specification of analyzed motor

Diameter of rotor and stator	60.5 mm, 106 mm
Air gap length, core length	0.25 mm, 56 mm
Number of phase and poles	3 phases, 4 poles
Primary winding pitch	Single layer, 5/6 short pitch
Number of series turns in stator winding	612
Rotor winding	Aluminum cage
Number of stator and rotor slots	24, 18
Depth of secondary slot	10.56 mm

The **field-and-circuit** model [1,5] is made by the assumption of a 2D electromagnetic field. In this model, coil outhangs and shorting rings of the rotor were taken into account by joining properly lumped parameters to several circuits. The application of the described method to model the magnetic field distribution in an induction motor, taking into account the movement of the rotor, required the introduction of a special element to the model which properly joined the unmoving and moving parts.

In the applied module RM [6] of the software package Opera 2D this element took the form of a gap-element. The gap region is divided uniformly on 3,168 elements along the circumference of the gap (Fig. 1). It gives time of displacement of one element equal to about 2.5×10^{-5} s at synchronous speed, comparable with the average time step of computation. The gap region division is fundamental for avoiding erroneous oscillation generations of computed electromagnetic torque.

The comparison of the calculated and measured values of rotational speed, current, and torque for starting state feeding by soft-starting (Figs. 2–4) and frequency starting devices (Figs. 5–7) can be observed.

Verification Methods

Traditional methods

The traditional methods, which are used to measure induction machine parameters, are: no-load test and short-circuit test. No-load test is useful for comparing the value of the magnetizing current measured and that calculated. Specifically in low-powered machines we focused on the problem of the air gap width estimation due to the effects of the cutting process and its influence on the sheet borders. Because dynamic field-circuit models of induction motors usually do not incorporate eddy currents, hysteresis, and mechanical losses in the stator core, it is necessary to obtain experimentally only the magnetization

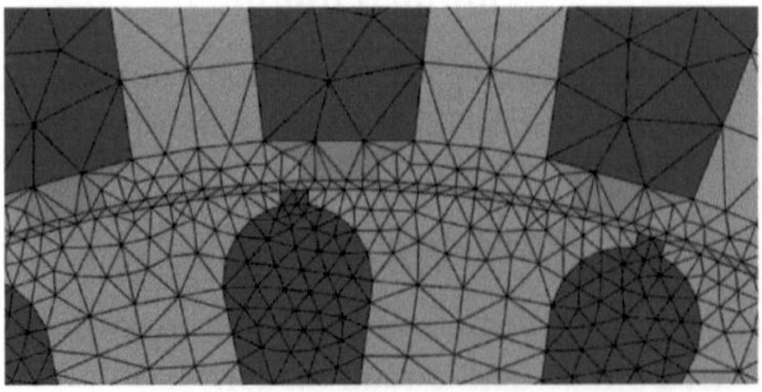

Figure 1. The gap region division.

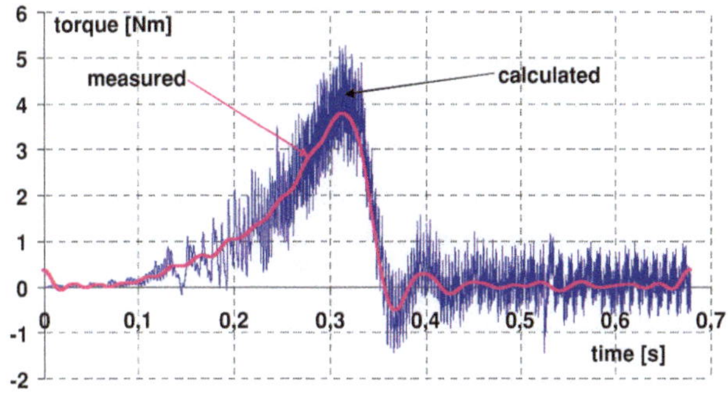

Figure 2. Torque vs. time during soft-starting starting.

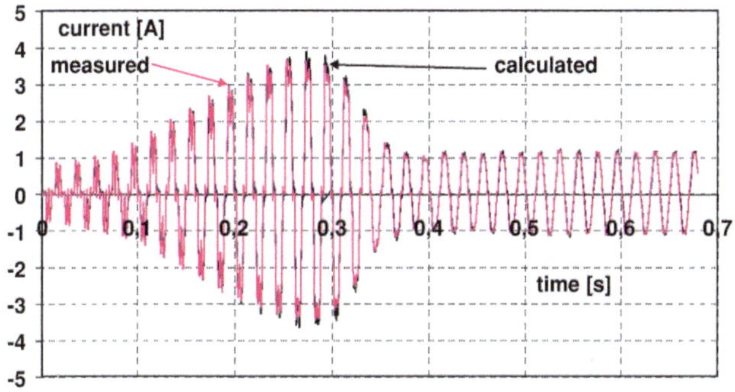

Figure 3. Comparison of calculated and measured current curves vs. time during soft starting.

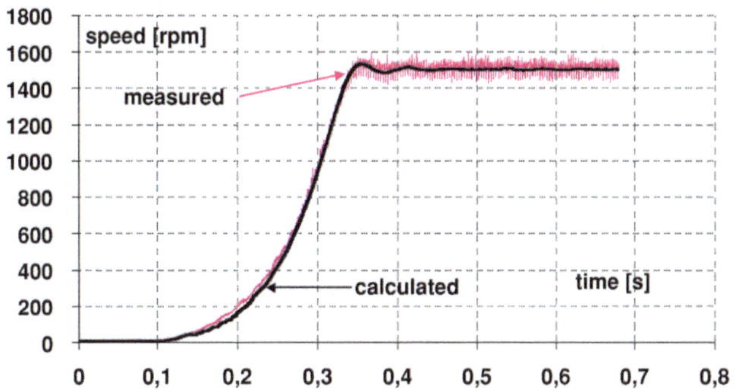

Figure 4. Comparison of calculated and measured speed curves vs. time during soft starting.

part of the no-load current. The quasi-static solvers calculate element permeability using amplitude of the magnetic flux density. This can introduce some errors in highly saturated small machines despite the transient calculation of the magnetization current needed [7-10]. In the presented motor, a comparison of measured and calculated values of the magnetizing current is made. The second test concerns the shape of calculated and measured currents at no-load. Comparing the shape of the two currents informs whether the flux distribution in the different part of the examined motor is near to the real one. The maximum value is mainly dependent on the air gap representation and saturation of the main parts of the

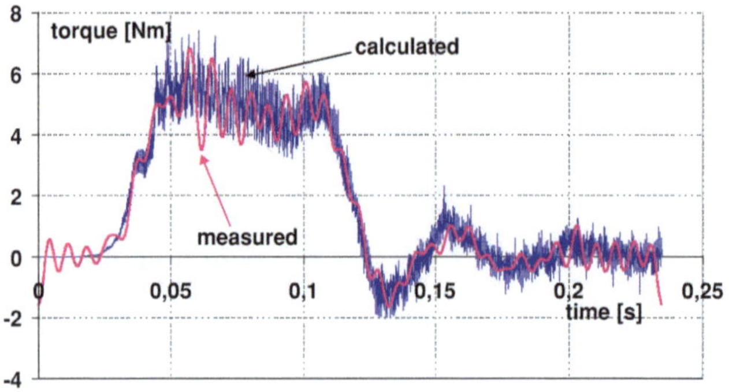

Figure 5. Torque vs. time during frequency starting.

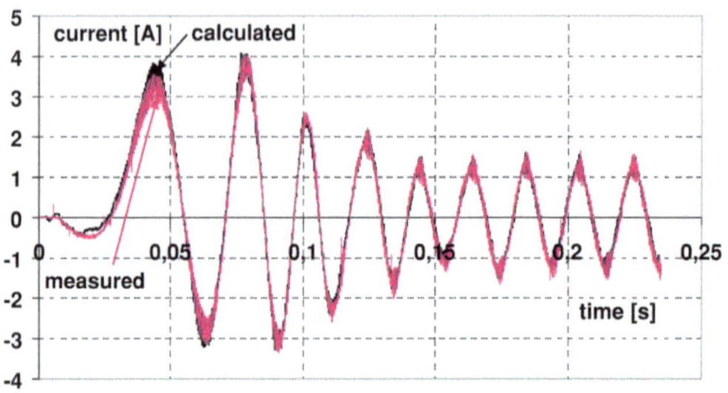

Figure 6. Comparison of calculated and measured current curves vs. time during frequency starting.

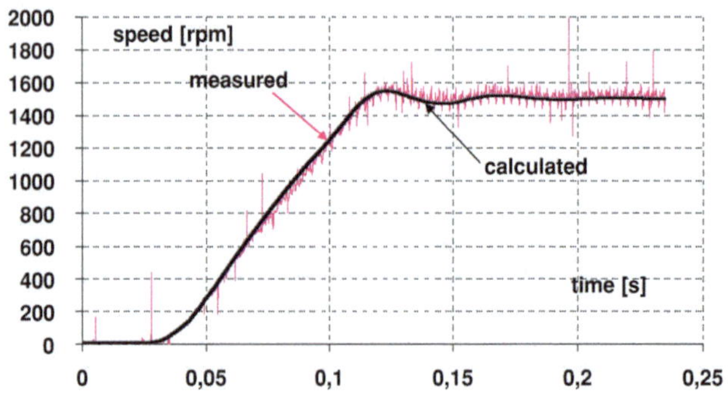

Figure 7. Comparison of calculated and measured speed curves vs. time during frequency starting.

magnetic core. Fig. 8 shows the comparison of the current vs. time calculated with transient and quasi-static solvers.

The results of comparison between two methods (AC and TR) and measurement are summarized in Table 2.

The best results are obtained by TR method. It is very difficult in practice to obtain accuracy better then 5% especially for small power motors due to inaccuracies in the production process and the results of the die-casting of the rotor cage and mechanical processing.

Table 2. The relative error between calculation and measurement

Relative error			
Phase A	Phase B	Phase C	Average
AC			
9,544	9,512	7,912	8,989
RT			
3,535	8,524	8,376	6,812

Short-circuit test examines the accuracy of the leakage reactance estimation (end parts reactance are included as lumped parameters) and the skin effect in the rotor bars. The main problem of the short-circuit test is the level of test current because of the local saturation effects of the leakage fluxes. Therefore, if possible, only a test with a nominal voltage will be really satisfactory. The measurement of the torque during this test is very helpful (Fig. 9).

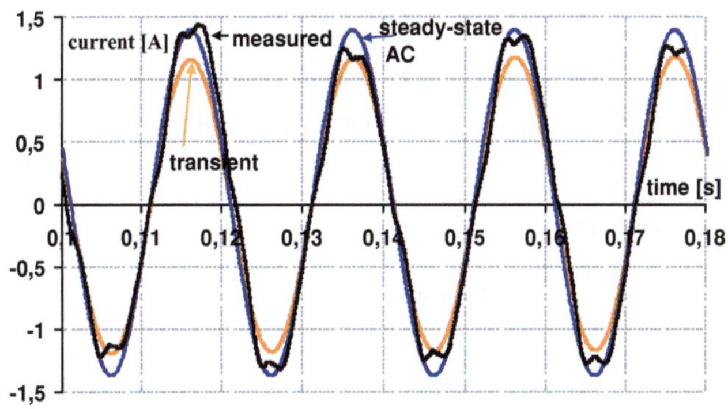

Figure 8. Current vs. time curves for steady-state, transient calculation, and measurement.

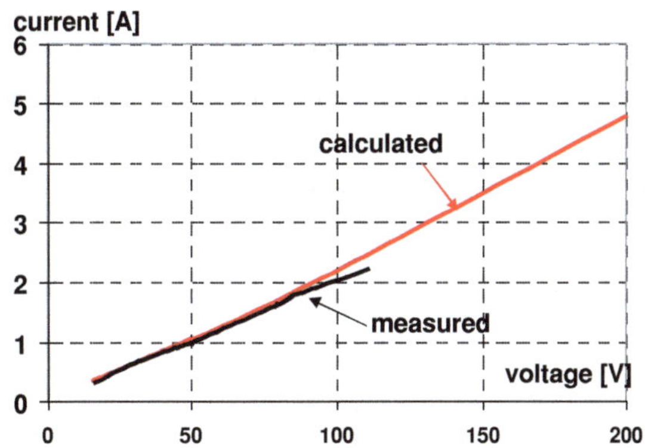

Figure 9. The short-circuit current vs. voltage.

Using the impulse DC supply test

Using this method we use the DC supply of one or two windings of the motor. With DC impulse method it is possible to test many aspects of the motor's behavior: the nonlinearity of the main flux path, influence of saturation due to leakage flux of the windings and skin effect of the currents induced in the rotor bar as well. Figs. 10 and 11 show the comparison between the measured and calculated values of input current at different DC voltage value.

It is also possible to have a look on the classical equivalent circuit of the motor (Fig. 12).

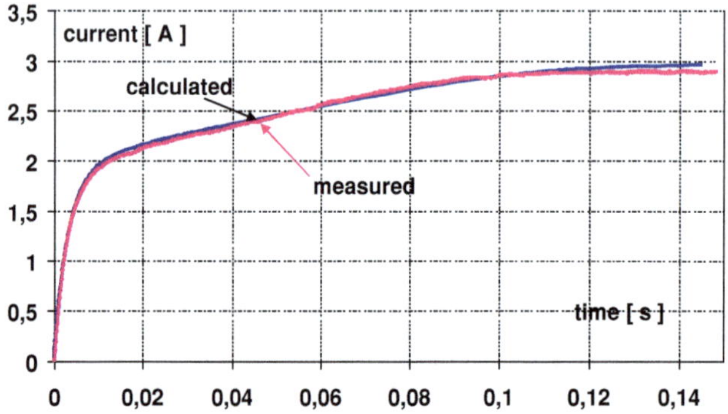

Figure 10. Current vs. time curves for DC supply at DC voltage value 63.05 V.

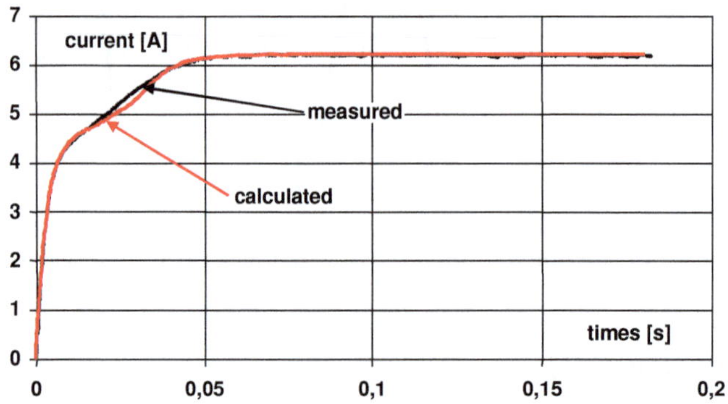

Figure 11. Current vs. time curves for DC supply at DC voltage value 138 V.

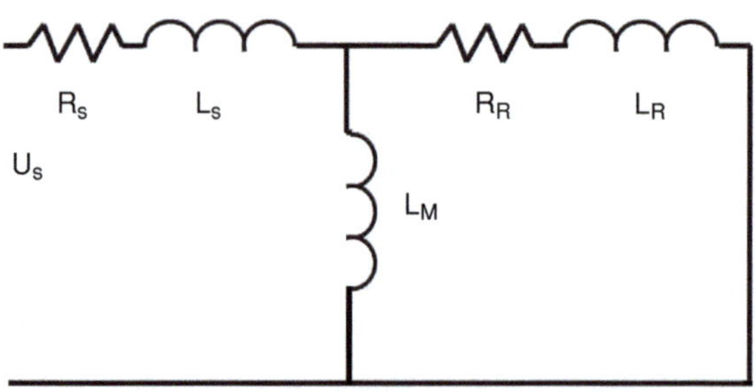

Figure 12. The classical equivalent circuit of the motor.

Using the simplified, without current induced in stator and rotor cores, model of the induction motor, the transfer function under linear condition is

$$Z(s) = R_s + sL_s + \frac{sL_m(R_r + sL_r)}{R_r + s(L_r + L_m)} \tag{1}$$

When the DC impulse signal (step) is applied to the one phase terminals of the motor the transient current response will be

$$I_s(s) = \frac{U_s(s)}{Z(s)} = \frac{U_c}{s\left(R_s + sL_s + \dfrac{sL_m(R_r + sL_r)}{R_r + s(L_r + L_m)}\right)} \tag{2}$$

where U_c is the value of applied DC voltage

$$I_s(s) = \frac{U_c(R_r + s(L_r + L_m))}{s((R_s + sL_s)(R_r + s(L_r + L_m)) + sL_m(R_r + sL_r))} \tag{3}$$

$$I_s(s) = \frac{U_c(R_r + s(L_r + L_m))}{s(s - s_1)(s - s_2)(L_sL_r + L_sL_m + L_rL_m)} \tag{4}$$

where s_1 and s_2—simple poles of the current function are the roots of the equation

$$s^2(L_sL_r + L_sL_m + L_rL_m) + s(R_sL_r + R_sL_m + R_rL_s + R_rL_m) + R_sR_r = 0 \tag{5}$$

The current vs. time function can be obtain using Heaviside's formula

$$I_s(t) = A_1e^{s_1t} + A_2e^{s_2t} + A_3e^{s_3t} \quad s_3 = 0 \tag{6}$$

where

$$A_1 = \frac{U_c(R_r + s_1(L_r + L_m))}{s_1(s_1 - s_2)(L_sL_r + L_sL_m + L_rL_m)} \tag{7}$$

$$A_2 = \frac{U_c(R_r + s_2(L_r + L_m))}{s_2(s_2 - s_1)(L_sL_r + L_sL_m + L_rL_m)} \tag{8}$$

$$A_3 = \frac{U_cR_r}{s_2s_2(L_sL_r + L_sL_m + L_rL_m)} = \frac{U_c}{R_s} \tag{9}$$

When the time constant s_1 and s_2 differs significantly it is possible to separate them from the measured current curve.

On the accuracy of the motor representation is shown by the values of the voltages induced in open windings vs. time (Figs. 13 and 14).

Upon examining the obtained results it was obvious that separation of the current curve exponential components would only be possible for small values of the instantaneous DC current, for higher value current curve vs. time differs significantly from the curve described by equation (3) (Figs. 15–18).

The parameters calculated from measured curves are shown in Table 3.

Even when approximation was possible, the obtained values changed with voltage value. Explanation of this result can be found easily by observing the field and current density distributions calculated using field-circuit method.

In Figs. 19 and 20 the distribution of the relative permeability for DC voltage equals 138 V for two different time instances are shown.

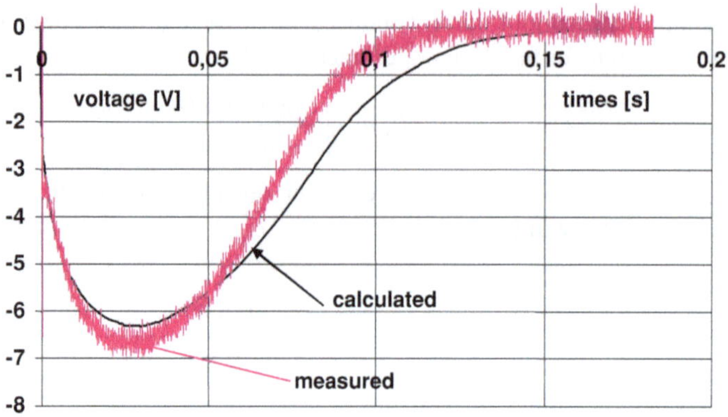

Figure 13. The voltage induced in open winding vs. time at DC voltage value 63.05 V.

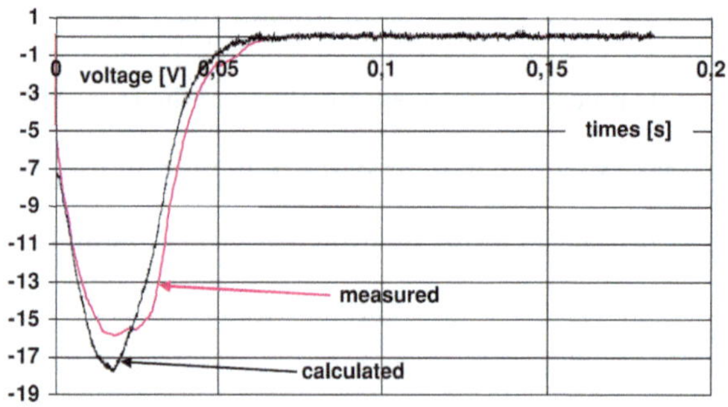

Figure 14. The voltage induced in open winding vs. time at DC voltage value 138 V.

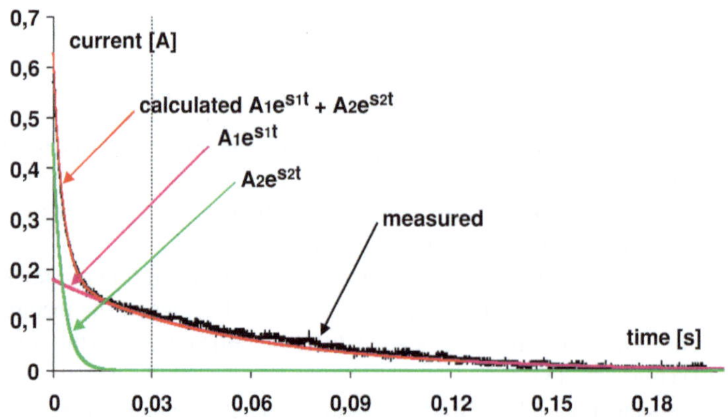

Figure 15. Decomposition of measured current curve vs. time into exponential components for DC voltage 13.4 V.

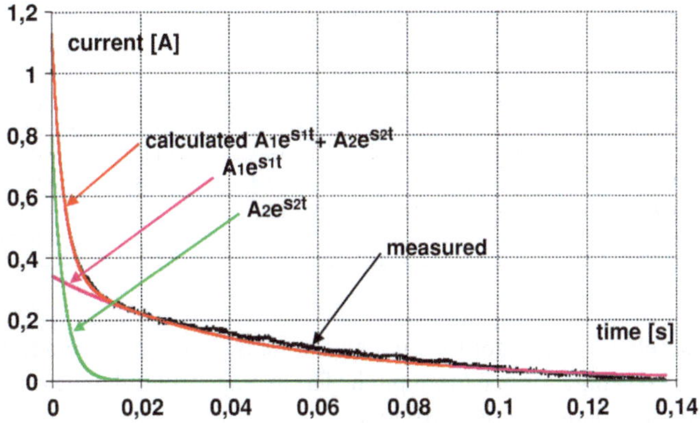

Figure 16. Decomposition of measured current curve vs. time into exponential components for DC voltage 23.82 V.

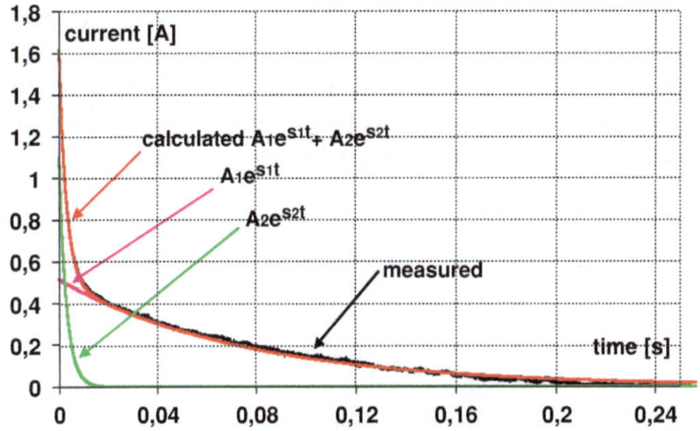

Figure 17. Decomposition of measured current curve vs. time into exponential components for DC voltage 32.49 V.

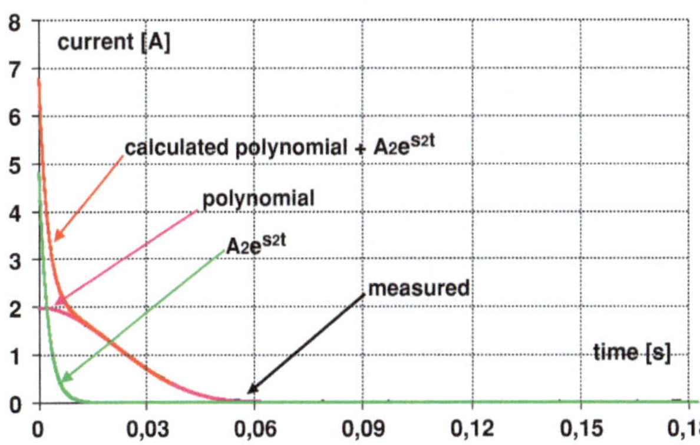

Figure 18. Decomposition of measured current curve vs. time into exponential components for DC voltage 138 V.

Table 3. Values of solution coefficients obtained from measured curves

DC supply U_s (V)	Steady-state current (A)	A_1 (A)	A_1/U_s (Ω)	$-s_1$ (1/s)	A_2 (A)	A_2/U_s (Ω)	$-s_2$ (1/s)
13.41	0.58	0.18	0.0134	18.3	0.4481	0.0334	335.9
23.31	1.01	0.34	0.0146	22.4	0.7883	0.0338	385.3
32.49	1.48	0.515	0.0159	13.3	1.1	0.0339	341.2
138.01	6.18				4.8	0.0348	422.0

In Fig. 21 the relative permeability curves vs. time is shown: a—average for both stator and rotor core, b—average for one tooth pitch. The average value changes significantly and the permeability distribution is also different.

The second important reason for the observed effect is a very well known skin effect in the rotor bars.

As can be expected the current density distribution in the rotor bar changes considerably during the time (Fig. 22). The resulting value of equivalent rotor bar resistance and inductance changes too (Figs. 23–25).

It should be emphasized that all described changes occur during the initial period when the stator current is compensated by rotor current. The main flux and magnetizing current do not grow as they occur in the final period when stator current goes to a constant value

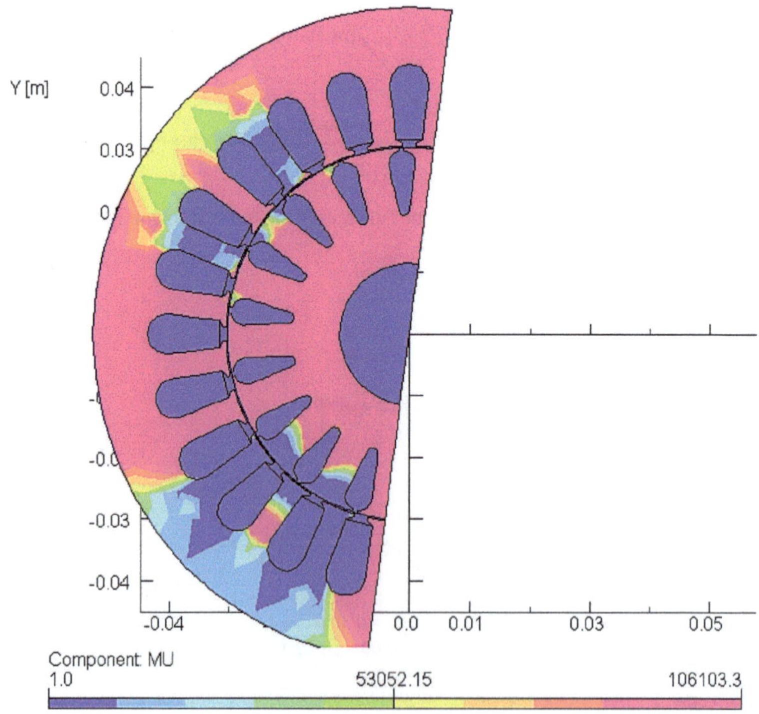

Figure 19. Relative permeability for time 0.005 s.

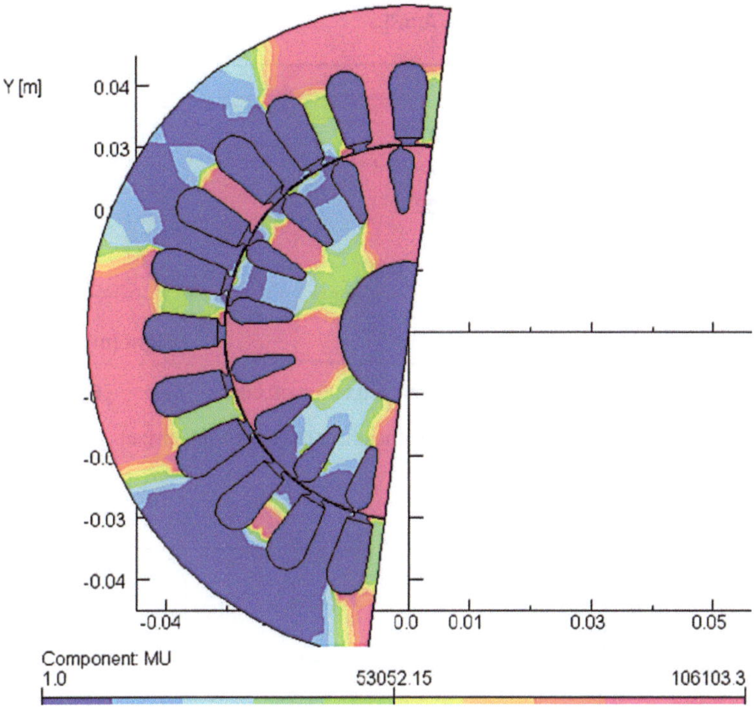

Figure 20. Relative permeability for time 0.01 s.

and rotor current disappears. According to these results, it should be clear why constant parameter equivalent circuits, sometimes used especially by Matlab Simulink users, are not usable for transient simulation of induction machines. Therefore noticeable growth of scientific reports according to adjustable parameters can be observed.

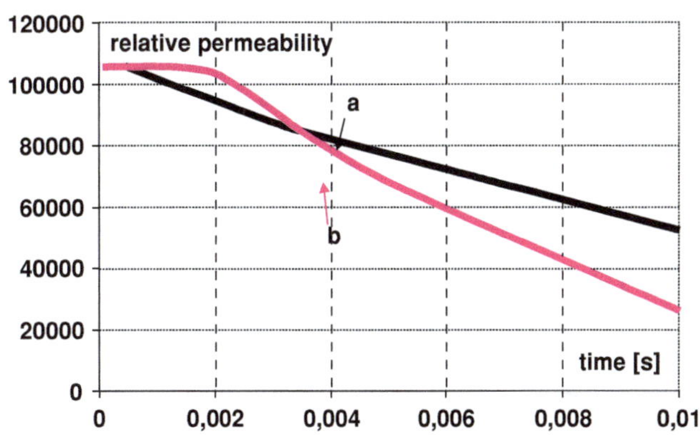

Figure 21. The relative permeability curves vs. time. (a) Average for both stator and rotor core. (b) Average for one tooth pitch.

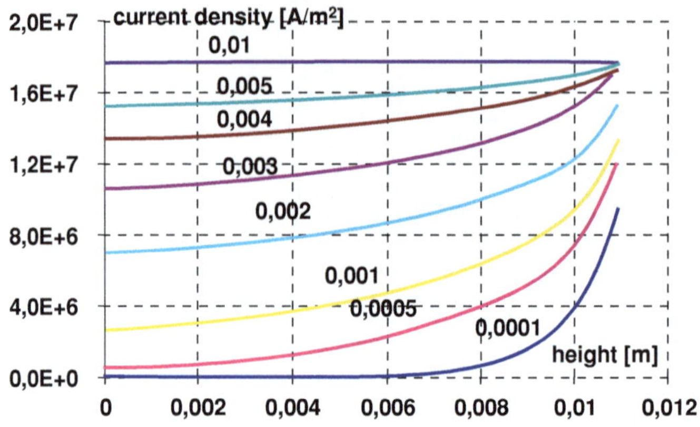

Figure 22. Current density distributions vs. bar height at different time instants for DC voltage equal 138 V.

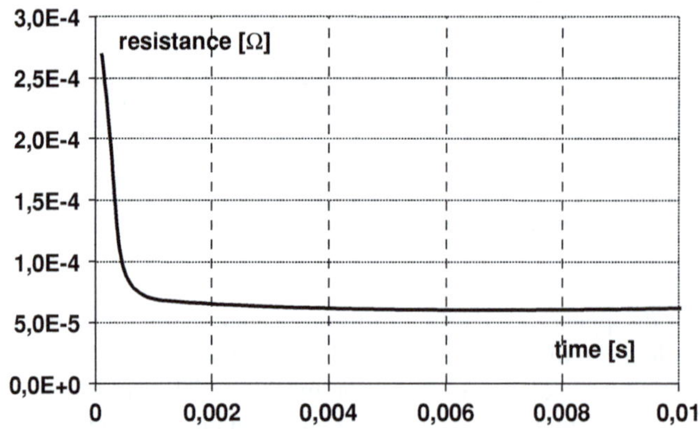

Figure 23. Equivalent resistance of the rotor bar vs. time.

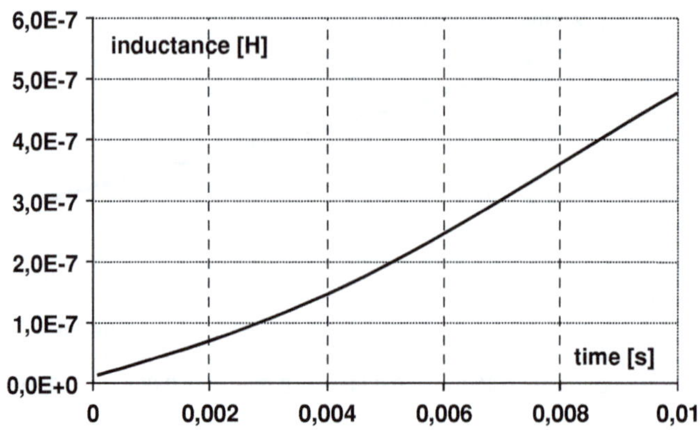

Figure 24. Equivalent inductance of the rotor bar vs. time.

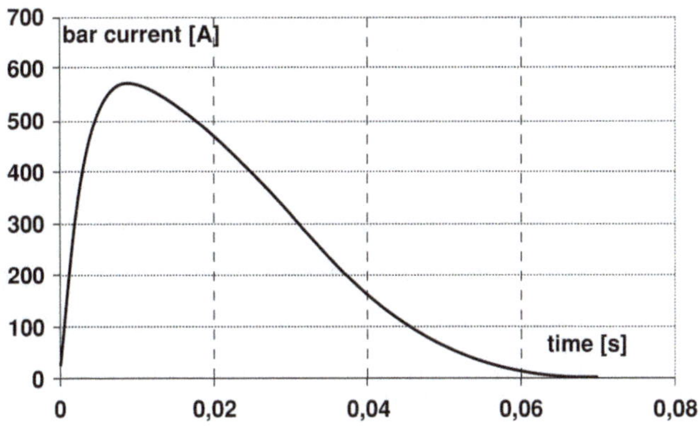

Figure 25. Rotor bar current vs. time for DC supply voltage 138 V.

Starting test with sinusoidal supplying voltages

The examination of the motor characteristics during starting is the most important method for comparison of the field-circuit dynamic model of the motor with the measurements. This test makes possible to compare all electromechanical quantities with the measurement. The solution of the mechanical transient equation due to moment of inertia value is not very sensible to torque higher frequency components. Therefore the speed vs. time curves, calculated using different methods, are very similar to the measured one. A similar situation can be observed for the current because of a significant value of magnetization current and the stator windings impedance. The most important is of course the torque characteristic vs. time (Figs. 26–28).

Conclusion

The presented paper has shown methods that can be used to verify the validity of the created field-circuit model for simulating transients occurring during the starting of the induction

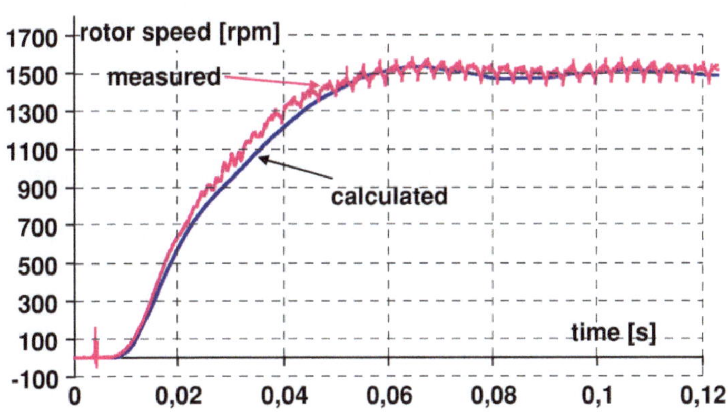

Figure 26. Rotor speed vs. time during starting.

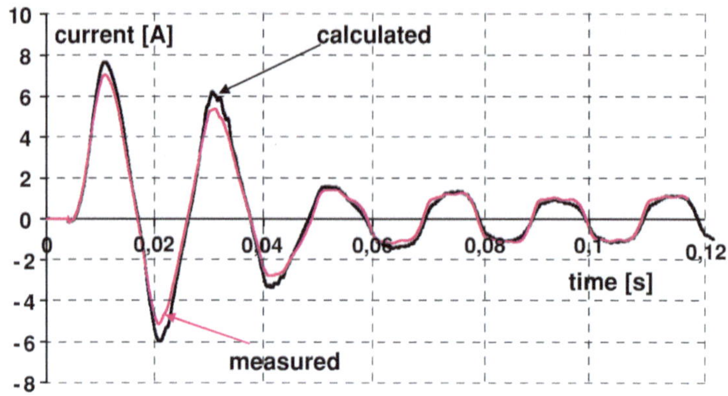

Figure 27. Current vs. time during starting.

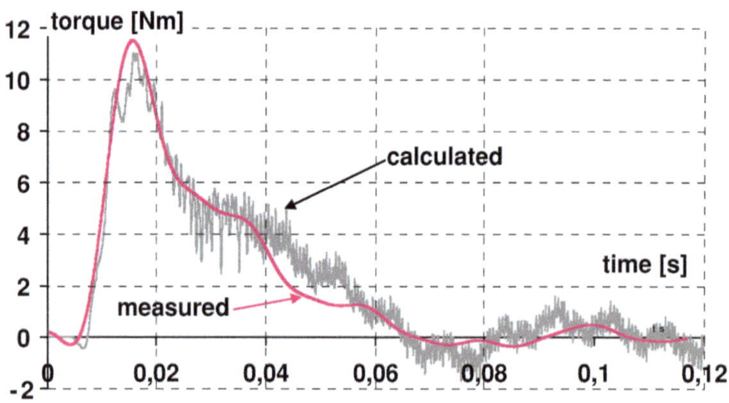

Figure 28. Torque vs. time during starting.

motor feed from an inverter. After verification this model can be successfully used for the analysis and optimization of the induction motor feed from an inverter.

References

[1] M. Dems, K. Komęza, "Circuit and Field-Circuit Analysis of Induction Motor with Power Controller Supply", International XIII Symposium Microdrives and Servomotors, MIS'2002, Krasiczyn, Poland, September 15–19, 2002, Tom II, Vol. 2, pp. 453–458.

[2] M. Dems, K. Komęza, Electromechanical transient processes of the induction motor with power controller supply, Electromotion, Vol. 10, No. 1, pp. 19–25, 2003.

[3] M. Dems, K. Komęza, "Simulations of Electromagnetic Field Distribution in an Induction Motor with Power Controller Supply", Proceedings of the XXII International Autumn Colloquium ASIS 2000.

[4] K. Komęza, M. Dems, "The Comparison of the Starting Characteristics of an Induction Motor for Frequency and Soft Starter Starting", Proceedings of the 8th Portuguese-Spanish Congress on Electrical Engineering, Portugal, July 3–5, 2003.

[5] M. Dems, K. Komęza, "A Comparison of Circuit and Field-Circuit Models of Electromechanical Transient Processes of the Induction Motor with Power Controller Supply", Proceedings COMPUMAG'2001, Lyon-Evian, France, July 2–5, 2001, pp. 206–207.
[6] PC OPERA-2D—version 10.5, Software for Electromagnetic Design from VECTOR FIELDS, 2005.
[7] S. Wiak, K. Komęza, M. Dems, "Electromagnetic Field and Parameters Modelling of Induction Motors by Means of FEM", Proceedings 32 Spring. International Conference MOSIS'98, Ostrava, Czech Republic, May 5–7, 1998, Vol. 3, pp. 275–281.
[8] M. Dems, K. Komęza, Influence of mathematical model simplifications on dynamic calculations of induction motors, Zeszyty Naukowe Politechniki Łódzkiej, Elektryka, wrzesień, Łódź, 2005.
[9] M. Dems, J. Zadrożny, J. Zadrożny Jr., "Comparison of Simulation Methods of Small Induction Motor Electromechanical Transients", International XII Symposium Micromachines and Servodrives, MIS'2000, Kamień Śląski, Poland, September, 2000.
[10] K. Komęza, M. Dems, S. Wiak, Analysis of the influence of the assumption of equivalent saturation on starting currents in induction motor, COMPEL Int. J. Comput. Math. Electr. Electron. Eng., Vol. 19, No. 2, pp. 463–468, 2000.

SECTION III
ELECTRICAL DRIVES APPLICATIONS

Introductory Remarks

The papers selected to first chapter are mainly focused on recently developed electrical drives applications with very important topics as: new motor structures, wind generators, use of advanced materials and new technologies.

The papers accepted for this chapter concerns the following leading problems:

- Steel-cored permanent magnet linear synchronous motor for large thrust force and high speed operation is designed, manufactured, and tested. The machine is analyzed by finite element method considering dynamic and static constraints. The designed model is optimized to reduce force ripples and to avoid magnetic saturation. Test machine is manufactured and the measured result of EMF constant shows good agreement with designed one. Thrust force characteristic shows good linearity and the measured maximum thrust force is over 15,000 N, the objective value. The measured maximum velocity is 3.98 m/s. The performances of the designed motor can guarantee the objective large thrust force and high speed.
- A high pole number, PM synchronous motor is presented, employing novel two-layer, special armature windings consisting of concentrated coils wound around the stator teeth. This kind of machine is characterized by excellent e.m.f. and torque waveform quality: it is well suited not only as an inverter driven motor, but also for mains feeding, self-starting, applications. In the paper, the main features of the machine are shown, together with some design, FEM and test results.
- An axial flux PM machine with field control capability for variable speed application is presented. To achieve such as control, surface mounted PM rotor pole configuration is shaped so that, a low reluctance path is included. In this way, controlling the armature reaction based on vector control allows us to command the airgap flux in a wide range. Magnetizing and demagnetizing effect can be reached with a low stator current requirement. In order to handle the rotor reluctance, an iron and PM sections are include. 3D FEA is carried out to confirm the viability of the proposed topology. Also a procedure to estimate the *dq* parameters for the topology is presented.
- An analysis of three topologies of iron-powder electrically magnetized synchronous machines by means of Finite Element Method. The first topology has the field winding placed in magnetically conducting end-plates, eliminating the need of slip-rings. In the second topology this is achieved by placing the winding above the outer rotor, and the third topology corresponds to the more conventional design with the field coils in the rotor. The results show that the first topology outputs 60% more torque than the other designs, although the three topologies present similar characteristics with regard to torque density.
- Performance of several motors where copper has been substituted for aluminum in the rotor squirrel cage is reported. Copper rotor motors die-cast in India for agri-pumping

S. Wiak, M. Dems, K. Komęza (eds.), Recent Developments of Electrical Drives, 291–293.

were dynamometer and field tested. Copper rotors resulted in higher electrical energy efficiency, slightly higher rotational speed, lower operating temperature, higher pumping rates and volume pumped per unit of input energy. SEW-Eurodrive motors with copper rotors are also described. A 1.1-kW motor with copper simply substituted and a 5.5-kW motor with redesigned rotor and stator are described. The copper rotor reduced losses in all major categories. Full-load efficiency was increased 6.7 and 3.1 percentage points, respectively. Finally, a study to minimize formation of large pores in die-cast rotors is summarized.

- Design and performance aspects of a brushless double fed cage induction generator as an economic and technical alternative to wind power generation are discussed. The above are focused on the main design criteria and on performance analysis to establish its behavior in load condition. The performance of a 15-kW prototype, comprising torque, current, efficiency, and power factor is compared to simulation results and to other types of machines as synchronous and wound rotor induction machines.
- The Permanent Magnet Induction Machine, a new wind generator concept, is considered to be a highly efficient, low maintenance solution for offshore wind turbines. Static and dynamic measurements have been performed with a test machine. Due to the inherent soft behavior of that machine type compared to normal synchronous machines, no dynamic excitation is found during operation that might endanger the stability of the system. Results of static measurements show high efficiency and little reactive power consumption.
- The study of the maximum conversion of the wind power for a wind diesel system with a battery storage using a current control. The maximum power points tracking have been achieved using a step down converter. This study are carried out taking into account the wind speed variations. The diesel generator is controlled using the power-speed characteristics. The results show that the control strategy ensures the maximum conversion of the wind power. The complete model is implemented in Matlab-Simulink environment.
- The comparative study for sinusoidal and trapezoidal waveforms in order to reduce the torque ripple and the power to grid fluctuation for large direct-drive PM wind generator. Trapezoidal waveform brings 28% higher power density but also two major drawbacks: necessity to vary the DC bus voltage and requirement for an additional filter on the DC bus.
- The equivalent thermal conductivity of insulating materials for a high voltage bar in slots of electrical machines is calculated using the finite element method. This allows the use of much coarser meshes with an equivalent thermal conductivity k_e, without accuracy loss in the hot spot temperature calculation. It is shown the dependency of k_e value on the equivalent mesh used. Some considerations are also presented on the heat flux finite element calculation.
- A simulation models of losses in Soft Magnetic Composites (SMC) components in new 3D-design solutions of electrical machine is defined. A method for the simulation of iron losses in SMC components is presented.
- The potential of the new technologies considered is evaluated through different examples of novel actuators, of centimetric or decimetric dimensioned actuators, which aim at meeting the increase of the performances or the expansion of required functionalities in the face of varied types of applications. An experimental study concerning friction drag reduction for a supersonic aircraft is briefly described. The aim is the control of turbulent

streaks with spanwise traveling wave. A piezoelectric demonstrator was designed for windtunnel testing at different configurations of frequency and wave-length.

- The electrical machine design considerations introduced by exploiting new magnetic material characteristics. The materials considered are amorphous alloy ribbons as well as neodymium alloy permanent magnets involving very low eddy current losses. Such advance materials enable electric machine operation at higher frequencies compared with the standard iron laminations used in the traditional magnetic circuit construction and provide better efficiently.
- A new approach for the study of the steady-state and transient behavior of three phase transformers. This approach based on magnetic equivalent circuit diagrams, takes into account the nonlinear B-H curve as well as zero-sequence flux. The nonlinear B-H curve is represented by a Fourier series, based on a set of measurement data. For the numerical simulations, two methods have been developed, by considering the total magnetic flux, respectively, the currents as state variables. Numerical results compared with test results and with FEM computations confirm the validity of the proposed approach.

III-1.1. DESIGN AND MANUFACTURING OF STEEL-CORED PERMANENT MAGNET LINEAR SYNCHRONOUS MOTOR FOR LARGE THRUST FORCE AND HIGH SPEED

Ho-Yong Choi[1], Sang-Yong Jung[2] and Hyun-Kyo Jung[1]

[1]Electromechanics Labratory, School of Electrical Engineering, Seoul National University, Korea
Shillim-Dong, Kwanak-Gu, Seoul, Korea

[2]Namyang R&D Center, Hyundai Motor Company, Hwasung-Si, Kyunggi-Do 445-706, Korea

plate@elecmech.snu.ac.kr

Abstract. Design characteristics of steel-cored PMLSM (Permanent Magnet Linear Synchronous Motor) are presented. Particularly, dynamic constraints resulted from repeated short-stroke travel are applied to the design criteria determining the machine specification. In addition, distinctive undesirable feature of detent force in steel-cored PMLSM and its notable minimizing methods based on manufacturing feasibility are considered. The designed machine is manufactured and tested for the verification.

Introduction

In these days, linear machine is widely used in industrial field because of its better characteristics, such as high acceleration and speed, large thrust force, and so on. To realize sufficiently large thrust force and power density, steel-cored permanent magnet linear synchronous motor is superior to air-cored motor. However, steel-cored motor has some demerits such as large normal force and force ripples caused by the high detent force. The normal force problem can be overcome by the LM (Linear-Motion) guide or the bearing system and the detent force reduction method should be considered during the machine design procedure. The magnetic saturation from large operation currents for the high speed operation can be a problem and proper steel-core design is also required [1,2].

Many linear applications require high acceleration and velocity with the short-stroke movement. In such applications, general design strategies considering the steady-state condition are not agreeable because linear motor operates mostly under the accelerating or decelerating circumstances on short travel displacements [3]. In addition, since the servo-capability responding to various motional profiles is recently regarded to be necessary, new plans considering dynamic constraints under the maximum input voltage and current should be made for better efficiency in machine sizing. Generally, capability of steel-cored

S. Wiak, M. Dems, K. Komęza (eds.), Recent Developments of Electrical Drives, 295–306.

PMLSM is defined by maximum input voltage and current, which expresses the maximum thrust force according to the specified mover velocity [4]. This capability, directly connected to a motor size, must be at least larger than the required motional profiles. Accordingly, dynamic constraints can be induced from a relation between such different dynamic capability and required motional profiles under the limited input voltage and current. These dynamic constraints express the admissible design range from which the design variables meeting the required trajectory can be obtained after all [5–7].

In this paper, steel-cored permanent magnet linear synchronous motor for large thrust force and high speed operation is designed and tested. The required maximum thrust force of the motor is 15,000 N and the maximum speed is 4 m/s. The continuous thrust force is 3,000 N under input voltage of 220 V and maximum peak current of 300 A. Finite element analysis is applied in the machine design procedure and some optimization method is used to minimize the detent force problem. The machine is manufactured and tested for verification of the designed model's validity.

Design of steel-cored PMLSM

Steel-cored PMLSM

Fig. 1 shows the moving-coil type steel-cored PMLSM with the magnetic combination of four poles and three coils which shows better operation in control. Self-bonded wires are much convenient to be attached to the core simply with a voltage of 5 to 6 V or thermal heating. Hall-sensors and incremental encoders are used for feedback control system, and Cable-Veyor, LM guide, and Shock-Absorber must be equipped. In addition, steel-cored

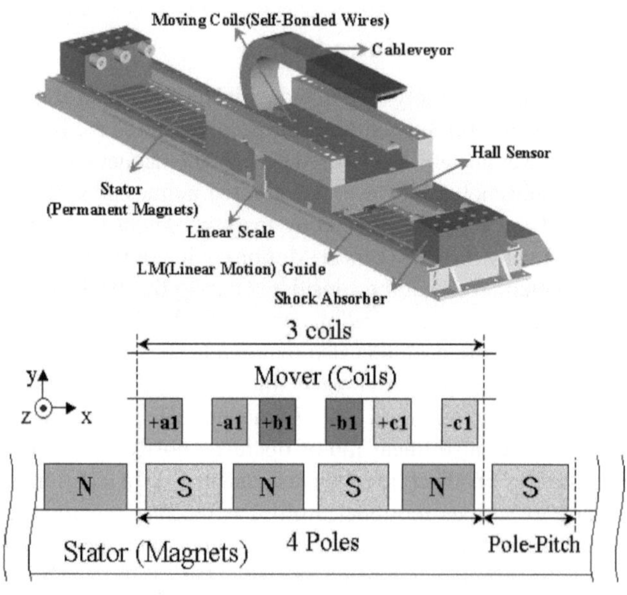

Figure 1. Steel-cored PMLSM (moving coils, four poles + three coils).

PMLSM has been much advanced from coreless one, and getting widely used mainly due to its large force productivity.

Capability and required motional profile

PMLSM with limited input voltage ($V_{\max}$) and current ($I_{\max}$) has dynamic capability as follows, which is induced with commands as $i_d = 0$ (force maximization).

$$F_{e,\max} = \frac{3}{2} K_e \min \left\{ \frac{C_1 + \sqrt{C_2 - C_3}}{R_s^2 + (\pi/\tau)^2 L_s^2 v^2}, I_{\max} \right\}, \qquad (1)$$

where,

$$C_1 \equiv -R_s \left\{ K_e v + \left(\frac{2L_s}{3K_e} \right) \left(m \frac{da}{dt} + Ba \right) \right\},$$

$$C_2 \equiv \left\{ \left(\frac{\pi}{\tau} \right)^2 L_s^2 v^2 + R_s^2 \right\} V_{\max}^2,$$

$$C_3 \equiv \left(\frac{\pi}{\tau} \right)^2 L_s^2 v^2 \left\{ K_e v + \left(\frac{2L_s}{3K_e} \right) \left(m \frac{da}{dt} + Ba \right) \right\}^2,$$

τ:Pole pitch [m], K_e: EMF constant [V/(m/sec)]
R_s:Resistance [Ω], L_s: Synchronous Inductance [H]

Equation (1) indicates the maximum thrust force at specified velocity under the maximum input voltage and current, and also includes the time-varying component, such as acceleration (a) and jerk ($J = da/dt$), available in dynamic analysis. Proposed dynamic capability has more meaning in linear machine than the conventional static capability under the acceleration and jerk set to be zero, which has been conventional approaches to the design process until now.

In Fig. 2, motional profiles of trapezoidal acceleration mode, most common in actual operation, and its relevant Force-Speed curve are shown. Force-Speed characteristics, which are obtained at each time interval, can be summarized as follows.

$$F_e(v) = \begin{cases} m\sqrt{\dfrac{2a_{\max}}{\Delta t_1}v} + Bv + F_l & (0 < v \le v_1) \\ ma_{\max} + Bv + F_l & (v_1 < v \le v_2) \\ m\sqrt{\dfrac{2a_{\max}(v_{\max} - v)}{\Delta t_1}} + Bv + F_l & (v_2 < v \le v_{\max}) \end{cases} \qquad (2)$$

where, $v_1 = (a_{\max}/2)\Delta t_1,\ v_2 = a_{\max}(\Delta t_1/2 + \Delta t_2)$

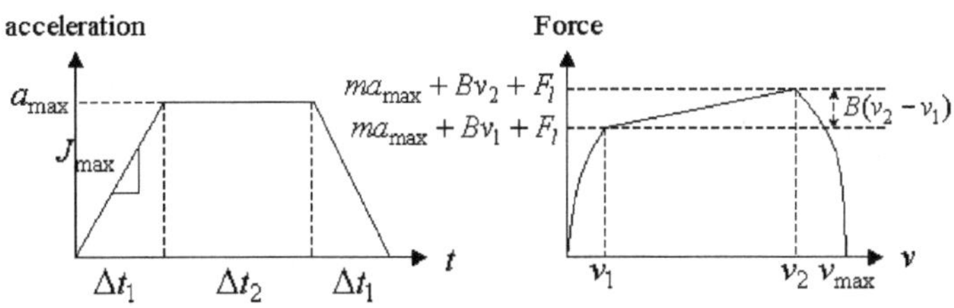

Figure 2. Motional profiles of trapezoidal acceleration and Force-Speed curve.

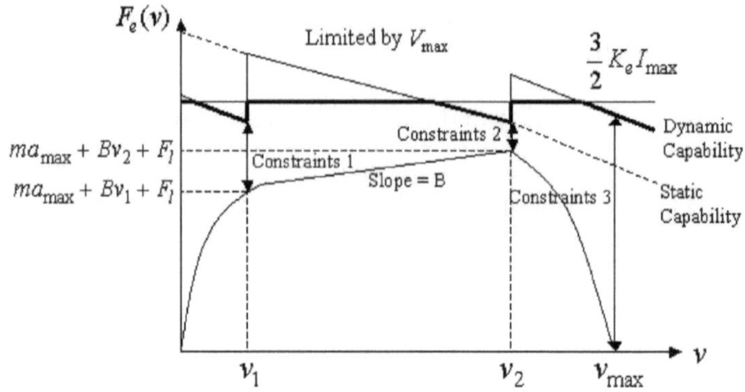

Figure 3. Dynamic constraints between dynamic capability and required motional profiles.

Design strategy using dynamic constraints

In principle, dynamic capability shown in (1) should be at least larger than Force-Speed relation of required trajectory shown in (2). This relation is shown in Fig. 3, where static and dynamic capability and the required motional trajectory are compared. Particularly between two forces from maximum voltage and current respectively, the smaller one could be the final dynamic capability, which is based on (1). Therefore, heavy-line in Fig. 3 indicates the final capability of PMLSM, which should be larger than motional profile, and this conclusion gives effective design criteria referred as *dynamic constraints*. Therefore, it is reasonable that only dynamic capability at the velocity of $v_1,\ v_2,\ v_{\max}$ should be larger than the required one, which is summarized as follows.

$$< \text{Constraint } 1; v = v_1,\ J = J_{\max},\ a = a_{\max} >$$

$$\frac{3}{2}K_e\frac{C_1 + \sqrt{C_2 - C_3}}{R_s^2 + (\pi/\tau)^2 L_s^2 v^2} >> ma_{\max} + Bv_1 + F_l \tag{3}$$

$$< \text{Constraint } 2; v = v_2,\ J = 0,\ a = a_{\max} >$$

$$\frac{3}{2}K_e \min\left\{\frac{C_1 + \sqrt{C_2 - C_3}}{R_s^2 + (\pi/\tau)^2 L_s^2 v^2},\ I_{\max}\right\} >> ma_{\max} + Bv_2 + F_l \tag{4}$$

$$< \text{Constraint } 3; v = v_{\max},\ J = -J_{\max},\ a = 0 >$$

$$\frac{3}{2}K_e\frac{C_1 + \sqrt{C_2 - C_3}}{R_s^2 + (\pi/\tau)^2 L_s^2 v^2} >> 0 \tag{5}$$

In (3), the dynamic constraints only from voltage limitation are considered, because the other constraints from maximum input current can be neglected due to the same kind of constraints in (4). In practical application, it is not fixed which one between dynamic capability and the force from maximum input current has larger value. Thereby, at velocity of v_2 in (4), both of them must be satisfied at the same time, where dynamic capability ($J = 0,\ a = a_{\max}$) and static capability ($J = 0,\ a = 0$) show little difference which can be verified through (1). In constraints 3, it is sufficient to judge whether motor has an ability to produce the force or not. However, constraints 3 can be replaced by other different constraint like $\partial F_{e,\max}/\partial v >> \partial F_e(v)/\partial v$ (at $v = v_2$) which means that, if the slope of dynamic capability is larger than that of required motional profile at $v = v_2$ (slope < 0), constraint 3 at $v = v_{\max}$ is satisfied by itself. However, this constraint is so strict that a lot of combination of design variables could fail to be selected even though they could survive through constraint 3. Therefore, it is reasonable to apply constraint 3 at design procedure, and then check the force margin in the interval of $v_2 < v < v_{\max}$ after work.

In addition to three basic constraints, such a relations as $v_{\max} = V_{\max}/K_e$ and another constraint, $C_2 >> C_3$, should be obeyed also in all of dynamic constraints.

Meanwhile, major difference between constraints from conventional static capability and proposed dynamic capability would be noticed at $v = v_1$ and $v = v_2$. Although the properly designed machine can satisfy constraints given by static capability, required motional profile cannot be realized due to the dynamic constraints, especially at $v = v_1$ (at $v = v_2$, there is little difference due to the zero jerk). Since discontinuous force change at $v = v_1$ and $v = v_2$ results purely from jerk and acceleration, high accelerating PMLSM used in short traveling displacements should be designed along the dynamic constraints.

Defined design parameters in (1) will be τ, K_e, R_s, L_s which are strongly regulated by dynamic constraints, and used as decision criteria to the combination of the design variables judging that the dynamic constraints are fully satisfied, i.e. designed machine can be driven successfully satisfying the required motional profile. Actually, sensitivity to the design parameter variance is most serious to τ and K_e relatively than R_s, L_s which are occasionally neglected in simplified design flow. Accordingly, in addition to the dynamic constraints, more generalized design consideration at the primary stage should be done focusing on the influence of τ and K_e, which makes entire design process performing more effectively.

Generalized design consideration and determination of design variables

In Fig. 4, the point where the maximum output power could be generated will be near $v = V_{\max}/K_e/2$ (half to the no-load velocity). Likewise, the maximum required mechanical power exists at $v = v_2$, therefore a design basis should be oriented as $v_2 \approx V_{\max}/K_e/2$ (K_{e2}

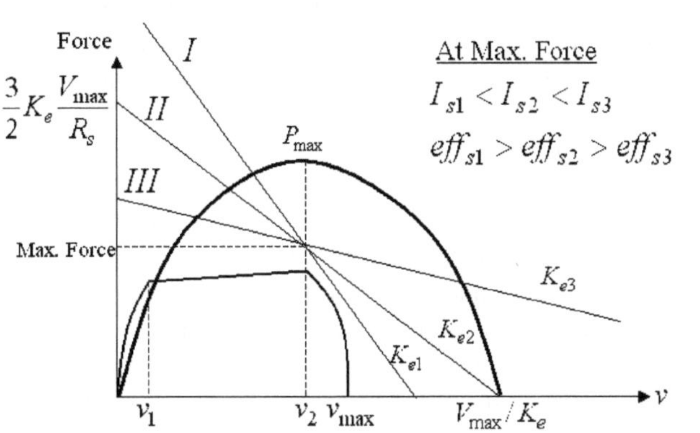

Figure 4. Generalized design schematic diagram ($K_{e1} > K_{e2} > K_{e3}$).

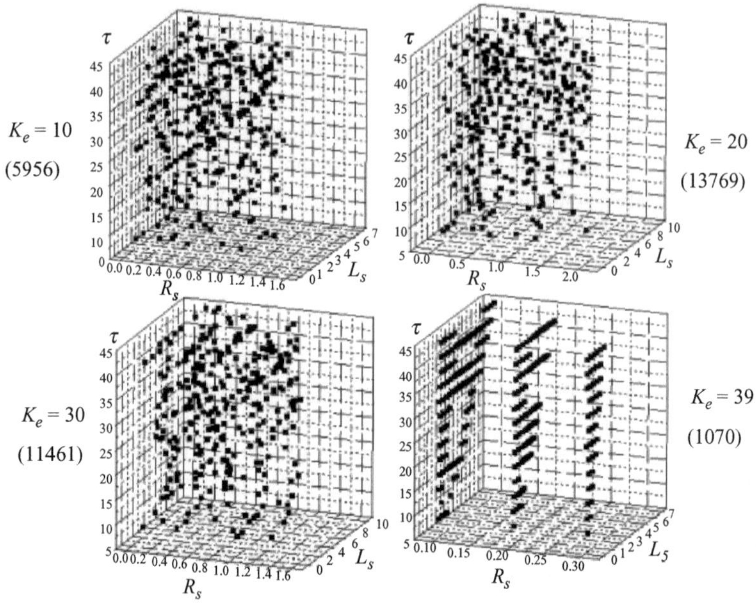

Figure 5. Admissible design combination vs. K_e (where $V_{max} = 160$ V, $I_{max} = 150$ A, $m = 37$ kg, $B = 100$ N/(m/s), $F_l = 50$ N, $a_{max} = 20$ m/s^2, $V_{max} = 4$ m/s, $J_{max} = 3{,}000$ m/s^3).

model in Fig. 4). However, if required input current ($I_s = F_{e,max}/(1.5K_e)$) is considered, K_{e1} model needs smaller current (I_{s1}) than K_{e2} model (I_{s2}), which could be also interpreted as better efficiency. In conclusion, EMF coefficient, K_e [V/(m/s)], should be designed at least in the interval as follows.

$$V_{max}/v_{max} \leq K_e \leq V_{max}/(2v_{max}) \tag{6}$$

Another sensitive design parameter, pole pitch (τ), should be defined from the magnetic combination (four poles and three coils) and the manufacturing feasibility. One module length τ_m corresponding to 4τ and $3\tau_c$ (where, τ_c is coil pitch) should be multiplied with 12, which has validated itself compared with the other combinations. Hence, its minimum and maximum size are strongly restricted by the manufacturing feasibility and the cost. Acceptable τ_m range for relatively larger power in continuous operation is approximately from 36 mm ($\tau = 9$ mm) to 180 mm ($\tau = 45$ mm).

In Fig. 5, distribution of design combination vs. K_e is shown, and if it is applied to Fig. 4, $K_e = 20$ corresponds to K_{e2} model manifesting the best design point from a viewpoint of size-effectiveness and usefulness in application. As K_e increases, better efficiency (lower input current) can be realized, and near the no-load speed ($K_e = 39$), dynamic constraints strongly restrict the design combinations in the speed range of $v_2 \leq v \leq v_{max}$. Particularly, the number of admissible design combination is maximum at the $K_e = 20$, which means the possibility to implement the machine successfully is highest at the best design point.

Design parameters (τ, K_e, R_s, L_s) are electrically and magnetically composed of the design variables expressing the machine dimension, hence the design process will be done by changing the design variables and checking the validity of sets of the design variables under

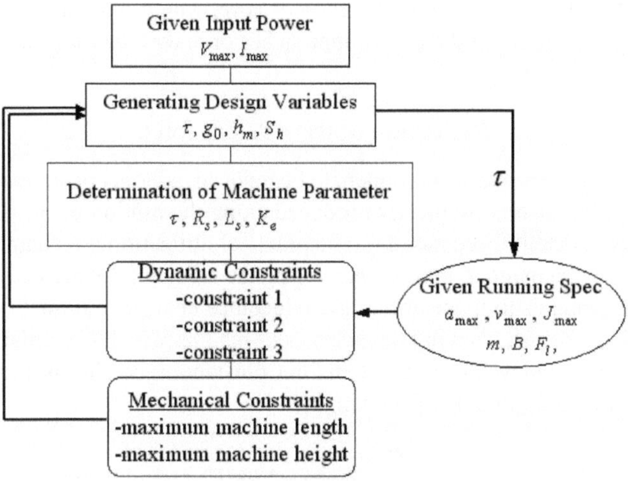

Figure 6. Optimal design flowchart.

the criteria proposed by dynamic constraints. Particularly, pole pitch (τ) will be sufficient to represent the moving-directional (longitudinal) design aspects, because coil pitch and one module-length are also determined accordingly. Then, the other variables including pole pitch can be summarized as air-gap length (g_0), height of magnet (h_m), height of slots (S_h) (or number of turns in slots), which are flexible to the normal direction. With the proposed design variables, the optimization method can be applied to the design process under the constraints such as dynamic constraints, the maximum mover length, and the maximum machine height, which are shown in Fig. 6 as a flowchart.

Detent force reduction

Theoretically, the detent force is the resultant one of two different reaction forces, i.e. the core detent force and the teeth detent force. The core detent force is the existing force between the permanent magnets and the primary core, which has a large period equal to the pole pitch. Whereas, the teeth detent force between the permanent magnets and the primary teeth has a relatively small period, the greatest-common-divider (GCD) of the pole pitch and the tooth pitch (τ_c, same with coil pitch)

Reduction of core detent force

Reduction of the core detent force can be done by giving the core a suitable length in order to cancel the core detent force at both end cores each other by adjusting the phase difference between the two core detent forces, and by reforming the edge of the core to minimize the reluctance variation. To begin with, making the geometric length such that the two forces at both end cores cancel each other can be effective, which could be realized by adjusting the electrical phase difference as follows [5].

$$\theta = (2k - 1)\pi, \tag{7}$$

where k is integer.

The other candidate is reforming the edge of core, which is induced from avoiding a rapid reluctance change when the mover approaches or leaves the magnets.

Reduction of teeth detent force

The teeth detent force, the main component to be reduced, not only occupies the total detent force up to 80%, but also is frequently produced along the motion track. Feasible ways to minimize the teeth detent force based on the practical utilization are chamfering the teeth edges and skewing the magnet. Firstly, chamfering the teeth edge, which is a similar idea to core chamfering, intends to make abrupt the reluctance changes minimal due to the sharp tooth edge. The other one, skewing the permanent magnet which is similar in principle to rotary machines, can remove the teeth detent force outstandingly. The optimized skew-angle should be determined through the following relation.

$$\text{Skew-angle} = \frac{GCD(\tau, \tau_c)}{2}\frac{1}{\tau}180 \text{ [Electrical degree]} \tag{8}$$

Equation (7) manifests the electrical 30° in four poles and three coils combination. However, this is so small one in a mechanical length. In case of $\tau = 45$ mm, mechanical skew-length correspondent to skew-angle (electrical 30°) is 7.5 mm.

Investigation on reduction result

Fig. 7 shows the reduction of the detent force.

The peak detent force of conventional model is about 300 N. But the peak value is reduced to 150 N after applying the chamfering and the skew, about 5% of the continuous thrust force. The detent force pattern has many harmonics because the width of magnet is very large and many teeth affect same magnet. In this case the effect of the skew is not so notable.

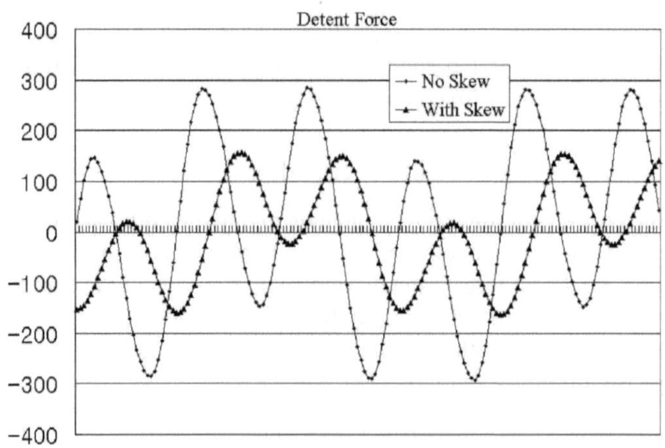

Figure 7. Detent force reduction by proposed methods.

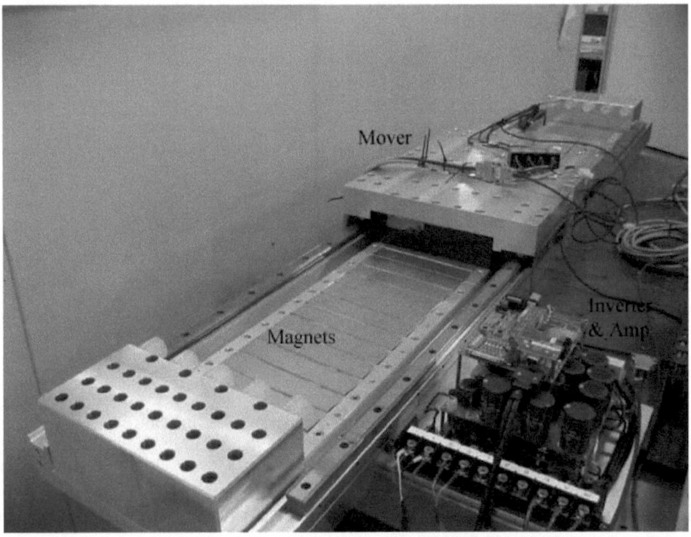

Figure 8. Manufactured PMLSM.

Design, manufacturing, and testing

The designed steel-cored PMLSM is manufactured and tested. The picture of the manufactured PMLSM is presented in Fig. 8 and specifications of the designed machine are listed in Table 1.

The magnetic flux distribution and air-gap flux density are shown in Figs. 9 and 10. In this model, very large input current is needed to get large thrust force, so that sufficient amount of iron core should be secured to avoid magnetic saturation.

The stroke of the linear motor is 1,000 mm and the maximum force/continuous force is 15,000 N/3,000 N. This motor can run up to 4 m/s under the input voltage of 220 V and the maximum current of 300 A.

Table 1. Design specification of sample steel-cored PMLSM

	Specification	Dimension
General (with water cooling)	Voltage/current	220 V/41 A
	Stack length	200 mm
	Magnet height	9 mm
	Magnet width	41 mm
Stator (NdFeB, 45 H)	Pole pitch	45 mm
	Slot width	22 mm
	Tooth height	30 mm
	Tooth width	38 mm
Mover (coil size = 1.2 Ø)	No of turns	90 per coil
	Coil connection	3 parallel
	Chamfering	10 × 6 mm

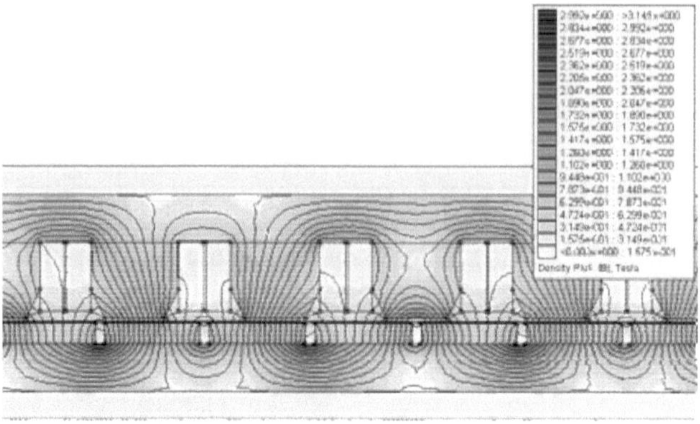

Figure 9. Magnetic flux density distribution.

The dynamic capability of the designed PMLSM is shown in Fig. 11 and the capability curve has force margin about 500 N.

By using the load cell, the thrust force is measured and the input current is measured with the current probe and the oscilloscope.

Fig. 12 shows the measured current-thrust force curve. The graph shows very good linear relation of the input current to the thrust force. The continuous thrust force is generated with the input current of 58 A and the maximum thrust force with the input current of 305 A is 15,890 N, which satisfies the objective output. Over 300 A region, the linearity of the curve is broken, because it is the highest available measuring value of the current probe.

The thrust force constant resulting from the measured curve is 54.81 [N/A] and EMF constant is 36.54 V/(m/s). The measured results have a good agreement with calculated thrust force constant 51.53 [N/A] and EMF constant is 34.35 V/(m/s).

The measured input current is shown in Fig. 13 when the motor is operated with the maximum speed. The pole pitch of the machine is 90 mm and the pitch of the measured

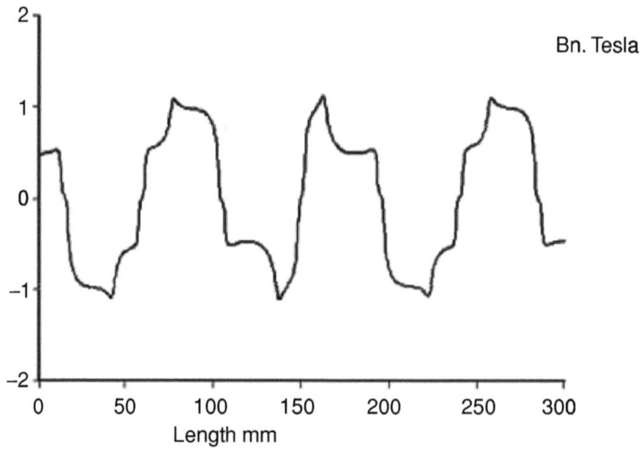

Figure 10. Air-gap flux density distribution.

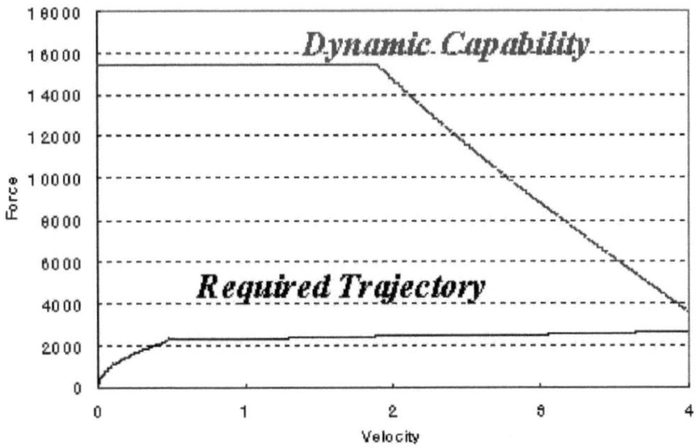

Figure 11. Running characteristics of designed motor.

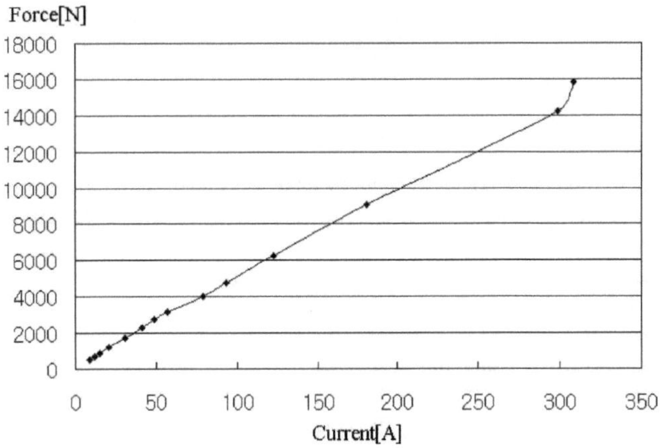

Figure 12. Current-thrust force curve.

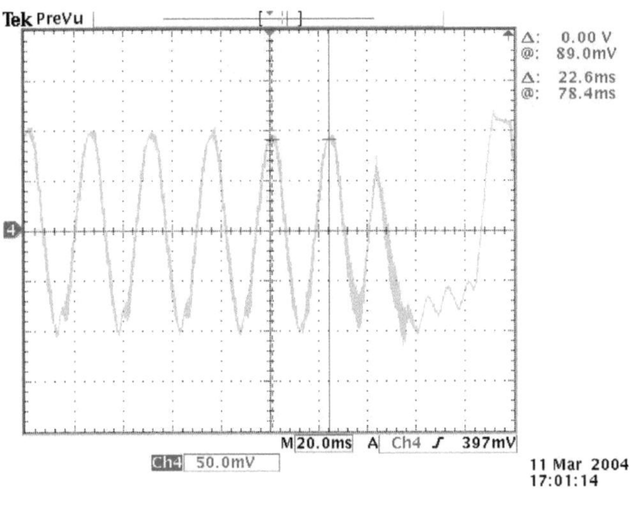

Figure 13. Measured input current.

current wave form is 22.6 ms. Therefore the moving speed can be calculated and the result is 3.98 m/s. Because the stroke is short, very large acceleration is needed to achieve the speed of 4 m/s. In addition, the power capability of the testing building is not sufficient, so that the resultant speed is not over 4 m/s. If long stroke or better power source is available, the machine can achieve the speed of 4 m/s.

Conclusion

In this paper, steel-cored permanent magnet linear synchronous motor for large thrust force and high speed operation is designed, manufactured, and tested. The machine is analyzed by finite element method considering dynamic and static constraints. The designed model is optimized to reduce force ripples and to avoid magnetic saturation.

Test machine is manufactured and the measured result of EMF constant shows good agreement with designed one. Thrust force characteristic shows good linearity and the measured maximum thrust force is over 15,000 N, the objective value. The measured maximum velocity is 3.98 m/s. The performances of the designed motor can guarantee the objective large thrust force and high speed.

References

[1] T. Sebastian, V. Gangla, Analysis of induced EMF waveforms and torque ripple in a brushless permanent magnet machine, IEEE Trans. Ind. Appl., Vol. 32, No. 1, pp. 195–200, 1996.

[2] T. Yoshimura, H.J. Kim, M. Watada, S. Torii, D. Ebihara, Analysis of the reduction of detent force in a permanent magnet linear synchronous motor, IEEE Trans. Magn., Vol. 31, No. 6, pp. 3728–3730, 1995.

[3] D.L. Trumpher, W.-J. Kim, M.E. Williams, Design and analysis framework for linear permanent-magnet machines, IEEE Trans. Ind. Appl., Vol. 32, No. 2, pp. 371–379, 1996.

[4] S.-Y. Jung, H.-K. Jung, J.-S. Chun, Performance evaluation of slotless permanent magnet linear synchronous motor energized by partially excited primary current, IEEE Trans. Magn., Vol. 28, No. 2, pp. 3757–3761, 2001.

[5] N. Bianchi, S. Bolognani, F. Tonel, "Design Criteria of a Tubular Linear IPM Motor", Proc. of IEMDC'03, 2001, pp. 1–7.

[6] S.-Y. Jung, S.-Y. Kwak, S.-K. Hong, C.-G. Lee, H.-K. Jung, "Design Consideration of Steel-Cored PMLSM for Short Reciprocating Travel Displacements", Proc. of IEMDC'03, Vol. 2, June 1–4, 2003, pp. 1061–1067.

[7] S.-Y. Jung, J.-K. Kim, H.-K. Jung, C.-G. Lee, S.-K. Hong, Size optimization of steel-cored PMLSM aimed for rapid and smooth driving on short reciprocating trajectory using auto-tuning niching genetic algorithm, IEEE Trans. Magn., Vol. 40, No. 2, pp. 750–753, 2004.

III-1.2. HIGH POLE NUMBER, PM SYNCHRONOUS MOTOR WITH CONCENTRATED COIL ARMATURE WINDINGS

Antonino Di Gerlando, Roberto Perini and Mario Ubaldini

Dipartimento di Elettrotecnica—Politecnico di Milano Piazza Leonardo da Vinci, 32-20133 Milano, Italy

antonino.digerlando@polimi.it, roberto.perini@polimi.it, mario.ubaldini@polimi.it

Abstract. A high pole number, PM synchronous motor is presented, employing novel two-layer, special armature windings consisting of concentrated coils wound around the stator teeth. This kind of machine is characterized by excellent e.m.f. and torque waveform quality: it is well suited not only as an inverter driven motor, but also for mains feeding, self-starting, applications. In the paper, the main features of the machine are shown, together with some design, FEM, and test results.

General features of the windings

In recent times, a large attention has grown toward the electrical machines equipped with concentrated coils, thanks to their great constructional and functional advantages [1–12]; nevertheless, a general approach to the concentrated winding theory seems not fully developed yet. In the proposed paper, a PM machine is considered, with two-layer, armature concentrated windings [13].

The features of this kind of machines are (see Figs. 1 and 2):

- uniformly distributed and equally shaped magnetic saliencies of the structures (stator teeth and rotor PMs);
- practical equality among tooth pitch τ_t and PM pitch τ_m (it can be $\tau_m < \tau_t$ or $\tau_m > \tau_t$, but $\tau_m \neq \tau_t$);
- series inverted connection of coils belonging to adjacent teeth of the same phase (controverse coils).

By adopting the representation of Fig. 1 (right) to specify the winding sense of each coil around its tooth, a typical three-phase, two-layer, winding appears as shown in Fig. 2.

Referring to Fig. 2, the following quantities and properties should be defined and considered:

- cycle: space period (periphery portion at which bounds the faced structures show the same mutual disposition);

S. Wiak, M. Dems, K. Komęza (eds.), Recent Developments of Electrical Drives, 307–320.

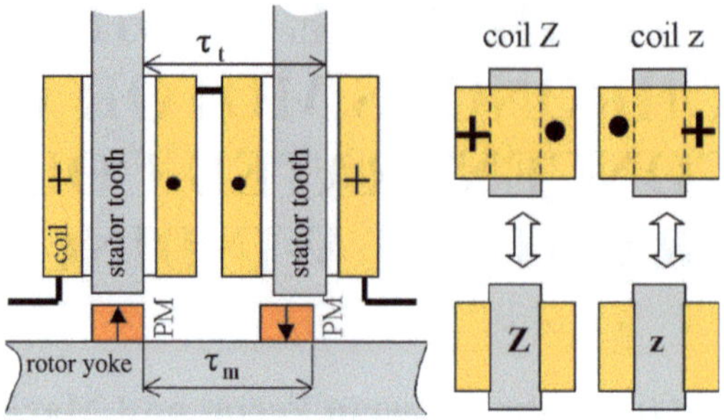

Figure 1. Left: basic structure of a PM synchronous machine, with tooth coil armature winding. Right: coil winding senses around teeth.

- cycle-phase: referring to a layer, portion of one cycle including adjacent coils belonging to the same phase; parent coil: in each layer, the first coil of every cycle-phase; its succession assignment defines the winding;
- the no. of teeth/cycle N_{tc} and the no. of coils/cycle N_{cc} must be multiple of the no. of phases N_{ph};
- links about no. of teeth/cycle-phase N_{tcph} and no. of coils/cycle-phase N_{ccph}: $N_{tc} = N_{ph}N_{tcph}$; $N_{cc} = N_{ph}N_{ccph}$;
- in case of controverse coils, the no. of coils/cycle-phase N_{ccph} coincides with the no. of teeth/cycle-phase N_{tcph};
- the optimal no. of PMs/cycle N_{mc} differs by one with respect to N_{tc}: $N_{mc} = N_{tc} \pm 1$ (→ highest winding factor);
- the optimal displacement among layers equals a no. of teeth N_{ts} nearest to $N_{ccph}/2$ (low harmonic distortion);
- the no. of cycles N_c equals the maximum no. of parallel paths "a" of each phase;
- the total no. of PMs $N_m = N_{mc}N_c$ of a rotating machine must be even; thus, if N_{mc} is even, the no. of cycles N_c can be any integer; if N_{mc} is odd, N_c must be even;
- the no. of coils/cycle-phase N_{ccph} can be any integer;

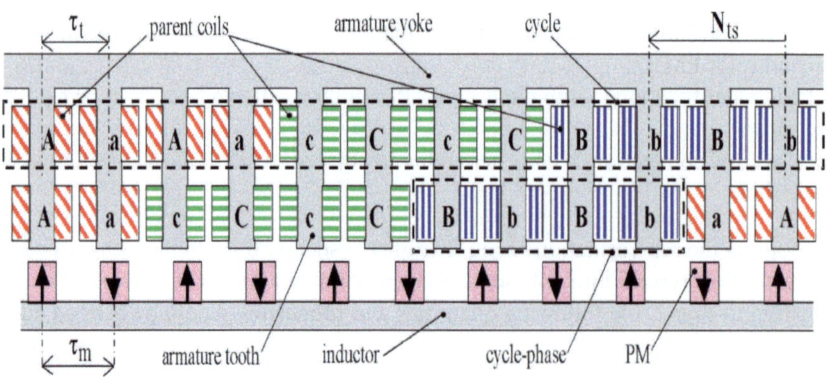

Figure 2. Double layer winding (two coils/tooth), with controverse tooth coils: $N_{tc} = 12$; $N_{cc} = 12$; $N_{ph} = 3$; $N_{tcph} = N_{ccph} = 4$; $N_{ts} = 2$.

- it can be shown that the winding factor k_w of a three-phase tooth coil machine (with two-layer windings) equals the product of a distribution factor k_d times a displacement factor k_s;
- for the phase winding e.m.f. of the jth order harmonic (j = 1, 3, 5, ...), we have:

$$k_{w_j} = k_{d_j} \cdot k_{s_j} \quad \text{with} \tag{1}$$

$$k_{d_j} = \frac{\sin(j \cdot \pi/6)}{N_{ccph} \cdot \sin[(j/N_{ccph}) \cdot \pi/6]}, \tag{2}$$

$$k_{s_j} = \cos(j \cdot (N_{sp}/N_{dcf}) \cdot \pi/6; \tag{3}$$

a traditional machine, with two-layer distributed windings, q slots/(pole-phase) and coil pitch shortening of c_a slots, exhibits a winding factor f_a equal to the product of a distribution factor f_d times a pitch factor f_p:

$$f_{a_j} = f_{d_j} \cdot f_{p_j}, \quad \text{with} \tag{4}$$

$$f_{d_j} = \frac{\sin(j \cdot \pi/6)}{q \cdot \sin[(j/q) \cdot \pi/6]}, \tag{5}$$

$$f_{p_j} = \cos(j \cdot (c_a/q) \cdot \pi/6); \tag{6}$$

these expressions and the previous ones are exactly corresponding each other, provided that we associate N_{ccph} with q and N_{ts} with c_a: the difference lies in the fact that, with a traditional machine, good quality performances (high winding factor and good e.m.f. waveform, no cogging, teeth harmonics, magnetic noise, and vibrations) can be obtained by adopting structures with $q \approx 5$–6, while a tooth coil machine (with the described features) exhibits similar performance quality with q values practically equal to 0.33: thus, machines with a given no. of poles can be realized with armature structures with a very low no. of slots;

- the other main advantages of these machines are:
  - the stator assembly is simplified: no skewing is required; only concentrated coils are used, that can be prepared separately (no endwindings overlapping; reduced copper mass; and armature losses);
  - the torque is high at low speed, allowing to eliminate any gears.

Table 1 shows some combinations of N_t and N_p (i = inferior; s = superior), for three-phase windings.

Design analysis of a basic prototype

In order to study the basic features of this kind of machine, we have decided to modify an existing induction motor, by re-winding its stator according to the previous theory and designing a new rotor, equipped with surface mounted PMs: of course, this choice has prevented from obtaining an optimized stator core, but, besides to easily provide a first test motor, it has also allowed to evaluate the suitability of existing laminations for the new machine. The main data of the used stator core are given in Table 2.

Table 1. Combinations of N_t and N_p (i = inferior; s = superior) of three-phase controverse windings, for some values of N_{ccph} and N_c ($N_{cmin} = 2$); S_{cph} = sequence of the parent coils within two cycles

N_{ccph}	N_{tc}	N_c	N_t	N_{pci}	N_{pi}	$S_{cph.i}$	N_{pcs}	N_{ps}	$S_{cph.s}$
2	6	2	12	5	10	AcBaCb	7	14	AbCaBc
3	9	3	27	8	24	ACBACB	10	30	ABCABC
4	12	2	24	11	22	AcBaCb	13	26	AbCaBc
5	15	3	45	14	42	ACBACB	16	48	ABCABC
6	18	2	36	17	34	AcBaCb	19	38	AbCaBc

About the rotor design, the available degrees of freedom are air-gap width and PM sizes and material: their choice is made by considering the operating point of the PM and the flux density B_t in the stator teeth. Considering the alignment condition between the PM axis and the tooth axis, from the analysis of the equivalent magnetic circuit concerning a zone extended to a tooth pitch, the no-load peak tooth flux φ_{t0} can be expressed as follows:

$$\varphi_{t0} = \varphi_r \cdot \eta_{PM} = (B_r \cdot b_m \cdot \ell) \cdot \frac{1}{1 + (1 + \varepsilon_\ell) \cdot \mu_{rPM} \cdot g/h_m}, \tag{7}$$

where $\varphi_r = B_r \times b_m \times \ell$ is the PM residual flux, η_{PM} the air-gap magnetization efficiency of the PM, ε_ℓ, μ_{rPM}, and h_m the PM leakage, the relative reversible permeability and the PM height respectively, g the air-gap width.

We adopted a NdFeB PM material (MPN40H: $B_r = 1.2$ T; $H_{cB} = 700$ kA/m at 80°C), choosing $N_c = 2$, $N_{tcph} = 6$, $N_m = 34$, $b_m = 10$ mm, central air-gap $g = 0.65$ mm: with these values, $h_m = 3$ mm is suited to gain an acceptable no-load magnetization (in fact, with $\varepsilon_\ell \approx 0.15$, it follows: $\eta_{PM} \approx 0.75$; $B_t = 1.32$ T; tooth flux $\varphi_{t0} = 0.761$ mWb); FEM simulations [14] confirmed (7) ($\varphi_{tanalytical} = 1.012 \times \varphi_{tFEM}$).

Fig. 3 shows the designed rotor during the construction process: the PMs are glued on the steel surface, inserted in suited slots for their correct and accurate positioning.

As the stator yoke, also the rotor yoke results definitely oversized (in fact, it was designed for a four pole motor).

Table 2. Main constructional data of the stator magnetic core used for the PM machine (obtained from an available standard induction machine lamination); main PM data

Stator internal diameter, D_i	140 mm
Stator external diameter, D_e	220 mm
Stator yoke width, h_y	19.5 mm
Lamination stack length, ι	85 mm
No. of stator teeth, N_t	36
No. of PMs, N_m	34
Slot opening width, b_a	2.7 mm
Slot opening height, h_a	0.55 mm
Tooth body width, b_t	6.7 mm
Tooth body height, h_t	20.00 mm
Tooth head width, b_e	9.5 mm
PM polar arc, α_m	0.77 pu

Figure 3. Picture of the PM rotor, during the assembling process: just some PMs are glued on the rotor surface; small slots (0.3 mm deep) allow a precise and reliable PM positioning, without appreciable increase of the flux leakage among adjacent PMs.

The complete cross section of the machine is represented in Fig. 4, that shows also the adopted winding disposition (in it, a layer displacement $N_{ts} = N_{ccph}/2 = 3$ has been adopted).

The FEM evaluated distribution [14] of the no-load flux density amplitude in the toothed zone (at half stator tooth height) is shown in Fig. 5; the following remarks are valid:

- the FEM peak value B_t confirms the analytical result;
- the peripheral amplitude distribution of $|B_{t0}|$ appears substantially sinusoidal, thanks to the gradual displacement among PMs and teeth within each cycle.

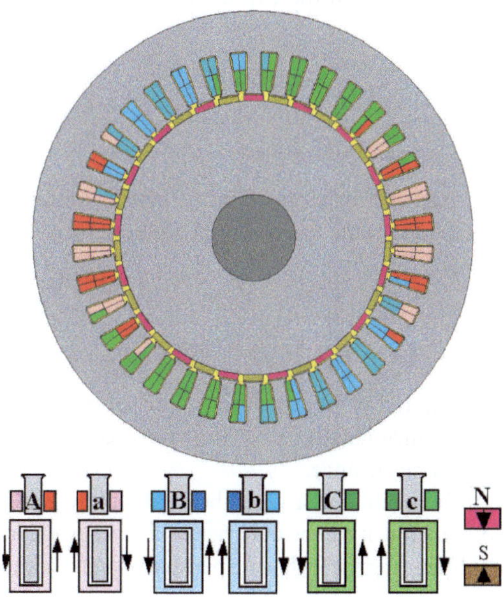

Figure 4. Top: magnetic structure and winding arrangement of the analyzed and constructed concentrated coil PM motor. Bottom: disposition conventions of coils and PMs.

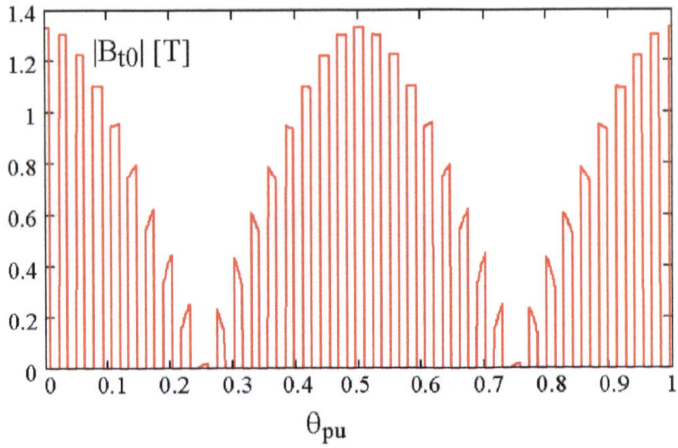

Figure 5. Peripheral amplitude distribution of the no-load flux density B_{t0} in the stator teeth (evaluated by FEM simulation, at half the tooth height) for the machine described in Table 2.

This sinusoidal distribution allows to express the r.m.s. no-load fundamental flux linkage Ψ_0 as follows:

$$\Psi_0 = (k_{w1} \cdot N_c \cdot 2 \cdot N_{tcph} \cdot \varphi_{t0}/\sqrt{2}) \cdot N_{tuc} = \Psi_{0_1} \cdot N_{tuc}, \qquad (8)$$

where the dependence on the no. of turns of each coil (N_{tuc}) is evidenced. In a two-layer winding, the no. of turns around each tooth N_{tut} is even: in fact, $N_{tut} = 2 \times N_{tuc}$ occurs.

The no-load flux linkage Ψ_0 can be evaluated also by FEM: some simulations have shown the accuracy of (8).

Of course, N_{tuc} is included also in the expressions of the equivalent resistance R and synchronous inductance L:

$$R = R_1 \cdot N_{tuc}^2 \qquad (9)$$

$$L = L_1 \cdot N_{tuc}^2. \qquad (10)$$

Ψ_{01}, R_1, and L_1 are the corresponding parameters of a phase winding consisting of one-turn series connected coils, being the same the coil total copper cross section:

$$R_1 = 2^2 \cdot N_{tcph} \cdot \left(N_c/a^2\right) \cdot \rho_{cu} \cdot [\ell_{tu}/(\alpha_{cu} \cdot (A_s/2))], \qquad (11)$$

$$L_1 = 2^2 \cdot \left(N_c/a^2\right) \cdot N_{tcph} \cdot \Lambda_e, \qquad (12)$$

with: a = no. of winding parallel paths, equal to N_c, or sub-multiple of it (here a = 1 has been chosen); ℓ_{tu} = average turn length; A_s = slot cross section; α_{cu} = slot filling factor; Λ_e = "per tooth" equivalent permeance.

While R_1 is simple to be evaluated, L_1 can be analytically evaluated only with some approximation; on the other hand, it can be obtained with energy calculations by a magnetostatic FEM simulation, substituting the PMs with passive objects, with the same permeability of the PMs.

For the machine of Table 2, Fig. 4, the values of Table 3 have been obtained.

Table 3. Calculated parameters of a PM motor with the data of Table 2, Fig. 4, equipped with "single turn per coil" windings

Flux linkage, Ψ_{01} (equation 8)	11.5 mWb_{rms}
Resistance, R_1 (equation 11)	8.03 mΩ
Inductance, L_1 (equation 12)	51.5 μH

The choice of N_{tuc} is a key design issue, greatly affecting the performances. In the following, just the Joule losses will be taken into account, neglecting the core $P_{\ell c}$ and mechanical losses $P_{\ell m}$, that can be considered separately. To evaluate the influence of N_{tuc}, the phasor diagram of Fig. 6 must be considered, analyzing the machine operation under sinusoidal feeding, at voltage V.

It is useful to define the quantities ρ_E and I_k as follows:

$$\rho_E = \frac{E}{V} = \frac{\omega \cdot \Psi_0}{V} = \frac{\omega \cdot \Psi_{0_1}}{V} \cdot N_{tuc} \tag{13}$$

$$I_k = \frac{V}{Z} = \frac{V}{\sqrt{R^2 + (X)^2}} = \frac{V}{N_{tuc}^2 \cdot \sqrt{R_1^2 + (\omega \cdot L_1)^2}} : \tag{14}$$

they represent the e.m.f./voltage ratio and the locked rotor current respectively, and depend on the number N_{tuc}.

The input current in loaded operation is given by:

$$I = I_k \cdot \sqrt{1 + \rho_E^2 - 2 \cdot \rho_E \cdot \cos(\delta)}, \tag{15}$$

where δ is the load angle (see Fig. 6).

Called $p = N_m$ the no. of poles, the torque T is given by:

$$T = 3 \cdot \Psi_0 \cdot (p/2) \cdot I_k \cdot [\cos(\varphi_z - \delta) - \rho_E \cdot \cos(\varphi_z)], \tag{16}$$

where

$$\varphi_z = \text{atan}(X / R) = \text{atan}(\omega \cdot L_1 / R_1) \tag{17}$$

is the characteristic angle of the motor internal impedance (independent on N_{tuc}) and δ the load angle (see Fig. 6).

From (16), the load angle δ in loaded operation follows:

$$\delta = \varphi_z - \text{acos}\{T / [3 \cdot \Psi_0 \cdot (p/2) \cdot I_k] + \rho_E \cdot \cos(\varphi_z)\}. \tag{18}$$

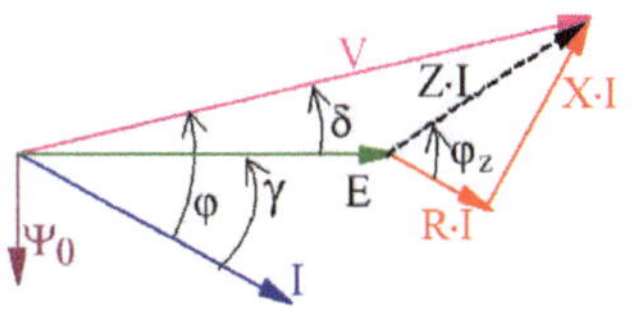

Figure 6. Phasor diagram for the analysis of the tooth coil synchronous motor, in sinusoidal feeding operation, at voltage V.

Moreover, (16) shows that the max. torque T_{max} (pull-out torque) occurs for the static stability limit angle δ_{max}:

$$\delta_{max} = \varphi_z, \tag{19}$$

$$T_{max} = 3 \cdot \Psi_0 \cdot (p/2) \cdot I_k \cdot [1 - \rho_E \cdot \cos(\varphi_z)]. \tag{20}$$

Imposing the condition $T = 0$ in (18) leads to evaluate the no-load angle δ_0 and the corresponding no-load current I_0:

$$\delta_0 = \varphi_z - \mathrm{acos}(\rho_E \cdot \cos(\varphi_z)), \tag{21}$$

$$I_0 = I_k \cdot \sqrt{1 + \rho_E^2 - 2 \cdot \rho_E \cdot \cos(\delta_0)}. \tag{22}$$

Assuming a suited value of the rated current density S_n, the rated current I_n can be expressed as follows:

$$I_n = S_n \cdot [(\alpha_{cu} \cdot A_s)/(4 \cdot N_{tuc})] \tag{23}$$

(in our motor, thermal status suggested: $S_n = 6.5\ \mathrm{A/mm^2}$). Substituting (23) in (15) gives the rated load angle:

$$\delta_n = \mathrm{acos}\left(\left(1 + \rho_E^2 - (I_n/I_k)^2\right)/(2.\rho E)\right), \tag{24}$$

and inserting (24) in (16) gives the rated torque T_n.

The reactive power absorbed by the motor is expressed by:

$$Q = 3 \cdot V \cdot I_k \cdot [\sin(\varphi_z) - \rho_E \cdot \sin(\varphi_z + \delta)]; \tag{25}$$

while the ideal input power P_i equals ($P_{\ell c}$, $P_{\ell m}$ neglected):

$$P_i = T \cdot \Omega + 3 \cdot R \cdot I^2. \tag{26}$$

From (25) and (26), the power factor:

$$\cos\varphi = 1 \Big/ \sqrt{1 + (Q/P_i)^2}. \tag{27}$$

is a function of ρ_E and N_{tuc}, by (9), (15), and (16).

As concerns the transient model, the differential equations in terms of Park vectors are as follows:

$$\begin{cases} \dfrac{d\theta}{dt} = \Omega \\ L \cdot \dfrac{d\vec{i}_P}{dt} = \vec{v}_P - R \cdot \vec{i}_P - j \cdot \dfrac{p}{2} \cdot \Omega \cdot \sqrt{3} \cdot \Psi_0 \cdot e^{j \cdot \theta \cdot p/2} \\ J_{tot} \cdot \dfrac{d\Omega}{dt} = \dfrac{p}{2} \cdot \sqrt{3} \cdot \Psi_0 \cdot \boldsymbol{Im}\left(\vec{i}_P \cdot e^{-j \cdot \theta \cdot p/2}\right) - T_{load} \end{cases} ; \tag{28}$$

θ is the mechanical angle between PM and phase "a" axes; $J_{tot} = J_{rot} + J_{load}$ the total inertia, T_{load} the load torque.

In the following, the diagrams in Figs. 7–12 will show the effect of N_{tuc} changes on the previously defined quantities: all the curves refer to steady state operation under sinusoidal feeding ($V = 380$ Vrms, $f = 50$ Hz).

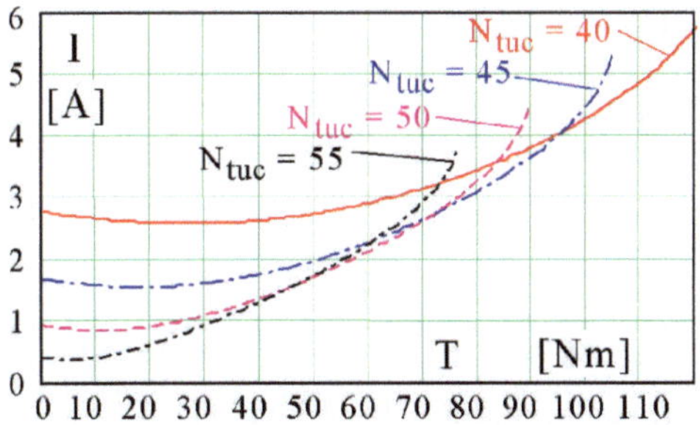

Figure 7. Input current I of the motor of Table 2 and Fig. 4, as a function of the torque T, in sinusoidal operation under V = 380 Vrms, f = 50 Hz, for different values of the no. of turns/coil N_{tuc}.

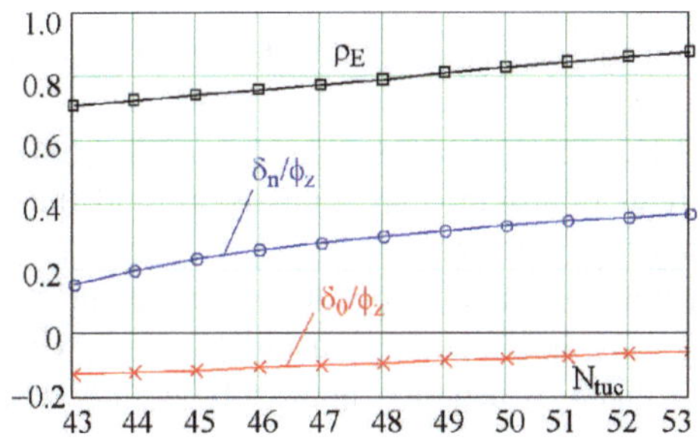

Figure 8. Ratio ρ_E as a function of N_{tuc}, together with the curves of the ratios δ_0/φ_z and δ_n/φ_z (see equations (13), (21), and (24)), in sinusoidal operation under V = 380 Vrms, f = 50 Hz, for different values of the no. of turns/coil N_{tuc}.

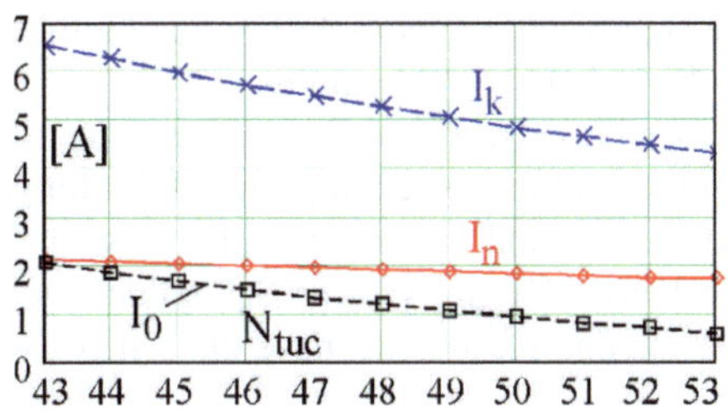

Figure 9. Locked rotor (I_k), rated (I_n), and no-load (I_0) input currents of the motor of Table 2 and Fig. 4, as a function of the no. of turns/coil N_{tuc} (sinusoidal feeding: V = 380 Vrms, f = 50 Hz).

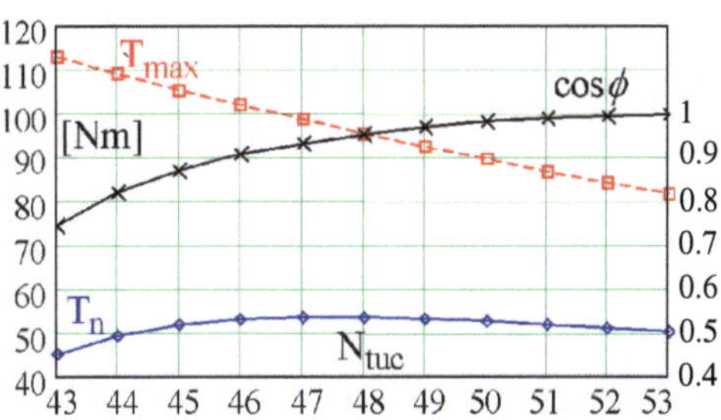

Figure 10. Power factor ($\cos\varphi$), rated (T_n) and maximum torque (T_{max}) of the motor of Table 2 and Fig. 4, as a function of the no. of turns/coil N_{tuc} (sinusoidal feeding: V = 380 Vrms, f = 50 Hz).

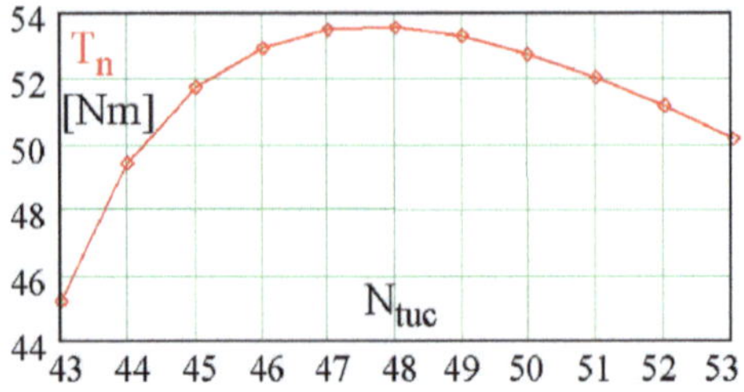

Figure 11. Rated torque (T_n) of the motor of Table 2 and Fig. 4, as a function of N_{tuc} (sinusoidal feeding: V = 380 Vrms, f = 50 Hz).

Fig. 7 shows the current-torque characteristics, for some N_{tuc} values, traced by (15) and (16), for $\delta_0 \leq \delta \leq \delta_{max} = \varphi_z$.

The adoption of high N_{tuc} values ($N_{tuc} \to 61$, corresponding to $\rho_E \to 1$) allows to reduce the no-load current, but reduces also the maximum torque and, thus, the motor overloading capability and the self-starting performances.

Fig. 8 shows ρ_E as a function of N_{tuc}, together with the curves of the ratios δ_0/φ_z and δ_n/φ_z (see equations (13), (21), and (24)), in sinusoidal feeding with V = 380 Vrms, f = 50 Hz: it is worth to observe that δ_0 is negative, approaching unity when ρ_E approaches unity too ($E \to V$).

Fig. 9 confirms the remark concerning the no-load current I_0 as a function of N_{tuc}, also showing the change of the rated current I_n and of the locked rotor current I_k.

Fig. 10 illustrates the decrease of the power factor cosφ when lowering N_{tuc}, while the maximum torque shows a significant increase. As the rated torque, it shows an almost flat maximum around $N_{tuc} = 48$, as better visible in Fig. 11.

On the other hand, a correlative property is shown in Fig. 12, showing that the ratio among the Joule losses and the output power has a minimum for $N_{tuc} = 48$.

As regards losses, rated torque and power factor, the best choice would be $N_{tuc} = 48$; considering also the importance of T_{max}, a lower N_{tuc} value can allow better overloading and self-starting features: for this reason, we have chosen $N_{tuc} = 46$ (→wire diameter: 0.63 mm).

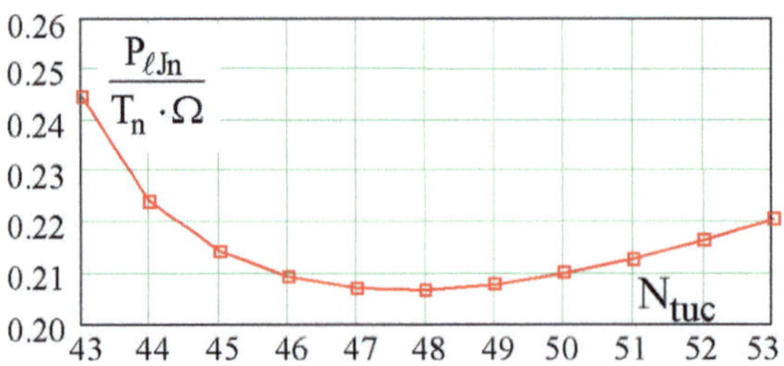

Figure 12. Ratio between stator Joule losses and output power of the motor of Table 2 and Fig. 4, as a function of N_{tuc}, in sinusoidal feeding (V = 380 Vrms, f = 50 Hz).

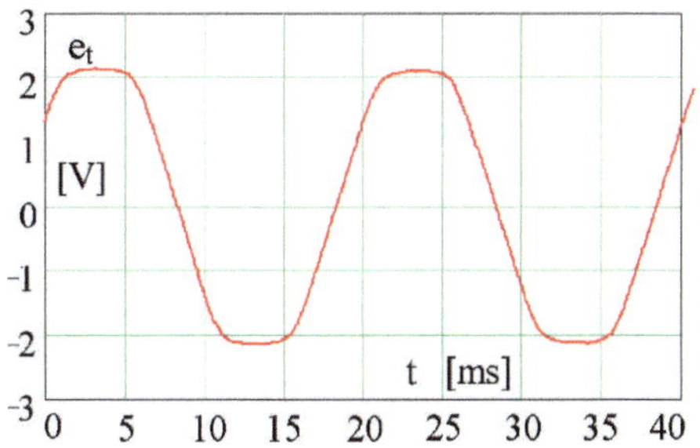

Figure 13. Measured waveform of the no-load e.m.f. at the terminals of a probe coil of $N_p = 10$ turns, disposed around one stator tooth: the typical trapezoidal shape can be observed.

Simulation and experimental results

Several simulations and experimental tests have been performed on a constructed prototype based on the previous data, in order to validate the design and operation models and to verify the achievable performance levels.

Fig. 13 shows the measured waveform of a "tooth" e.m.f., i.e. the no-load e.m.f. at the terminals of a probe coil of $N_p = 10$ turns, disposed around one stator tooth: even if a certain distortion can be observed, the amplitude estimable from (8) is fairly confirmed.

Fig. 14 shows the measured waveform of the no-load phase-to-neutral e.m.f. e_{ph}: the amplitude evaluated by (8) is confirmed; moreover, it is evident the great shape improvement compared with the tooth e.m.f.

It is particularly noticeable the absolute absence of slotting effects, in spite of the very low no. of slots/(pole-phase). The phase-to-neutral e.m.f. is almost sinusoidal: in fact, the harmonic analysis e_{ph} has evidenced limited harmonics, except for an appreciable, even if low, third harmonic e.m.f.; but, as known, this component is cancelled in the line-to-line voltage, while the actual lowest order harmonics (fifth, seventh order) are reduced by the layer displacement (see (3)).

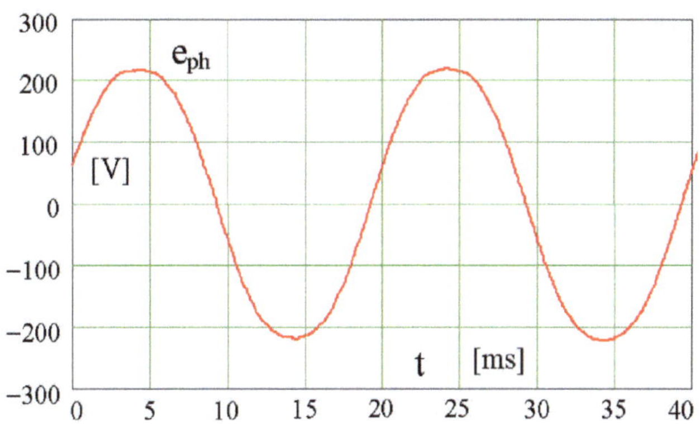

Figure 14. No-load phase-to-neutral measured e.m.f., for the constructed motor (data of Table 2, Fig. 4, $N_{tuc} = 46$ turns/coil)

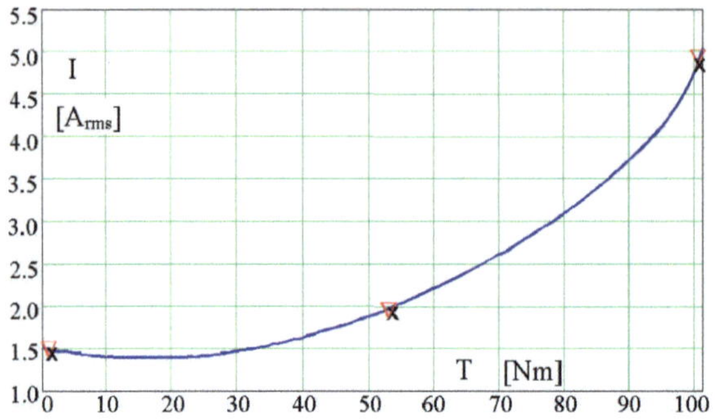

Figure 15. Simulated and test results of the motor of Table 2 and Fig. 4, with $N_{tuc} = 46$, in sinusoidal feeding (380 Vrms, 50 Hz); x axis: torque (from no-load to T_{max}); y axis: input current; ∇: analytical simulation, by (15), (16), $\delta_0 \geq \delta \geq \delta_{max} = \varphi_z$: ∇: experimental result; x: FEM simulation result [14].

Another important effect connected to the PM winding arrangements adopted in this kind of machine is the very low level of cogging: by manually handling the rotor of the unfed motor, we have verified no appreciable cogging torque, as confirmed also by FEM simulations [14].

Fig. 15 shows simulated and test results of the input current in loaded operation with sinusoidal feeding ($V = 380$ Vrms, $f = 50$ Hz), with the torque ranging from zero to T_{max}: the analytical result (see also Fig. 7) is confirmed both by measurements and FEM simulation, for no-load, rated torque and pull-out torque conditions.

The rated operation has been verified also by a thermal test (Fig. 16), that indicated acceptable temperature levels.

Figs. 17 and 18 report some simulations, performed by integrating equation (28), aimed to show the motor dynamic behavior, evidencing its self-starting capabilities, under mains sinusoidal supply, in loaded conditions.

Considering that the rotor inertia equals $J_r = 0.023$ kgm^2, we have considered to drive a load with the same inertia ($\rightarrow J_{tot} = 0.046$ kgm^2); several simulations have been performed, with different mains voltage phase conditions.

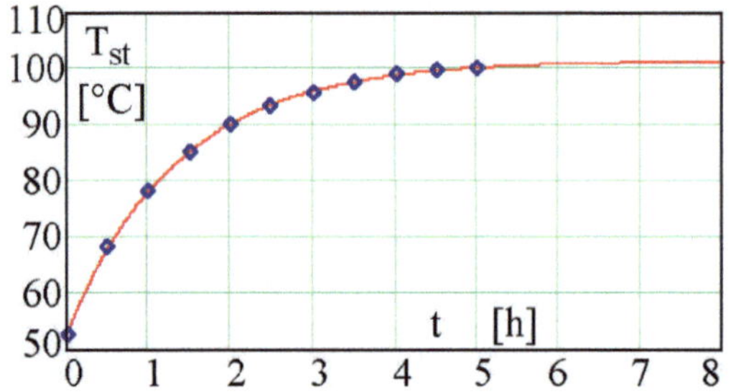

Figure 16. Experimental thermal test of the motor of Table 2, Fig. 4, $N_{tuc} = 46$, running with sinusoidal feeding (380 Vrms, 50 Hz), with rated torque ($T_n = 53$ Nm); the points are the temperatures measured by a thermocouple put in contact with the endwindings (ambient temperature: $T_a = 25.5$°C).

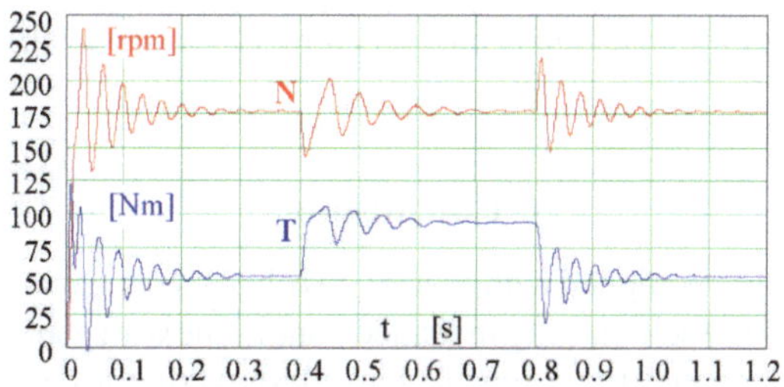

Figure 17. Simulated transient of the motor of Table 2, Fig. 4, $N_{tuc} = 46$, with sinusoidal feeding ($V = 380$ Vrms, $f = 50$ Hz): synchronization from zero speed with rated torque ($T_n = 53$ Nm); total inertia: 0.046 kgm^2; response to torque steps of DT = 40 Nm.

Fig. 17 refers to a synchronization from zero speed with rated torque ($T_n = 53$ Nm), followed by two opposite torque steps of $\Delta T = 40$ Nm: the response appears stable and acceptable, both at starting and after load variations.

Fig. 18 shows another starting transient under the same conditions of Fig. 17, except for the initial values of the supply voltages (in opposition to the previous one): the starting transient has the same duration as before (roughly 0.4 s), but torque and speed show different instantaneous values, even significantly negative. At $t = 0.5$ s, a torque ramp is applied, up to the pull-out torque, that occurs exactly at the analytically estimated torque value ($T_{max} = 101$ Nm), with the consequent loss of synchronization.

Corresponding results have been obtained also by FEM transient simulations: these simulations gave the additional information of the absence of torque ripple: this result, confirming the absence of cogging of the unfed machine, appears particularly interesting, also considering that no skewing have been applied between teeth and PMs.

Experimental starting tests in loaded conditions demonstrated the correctness of the simulations, with a satisfying behavior, both at starting and during steady state operation:

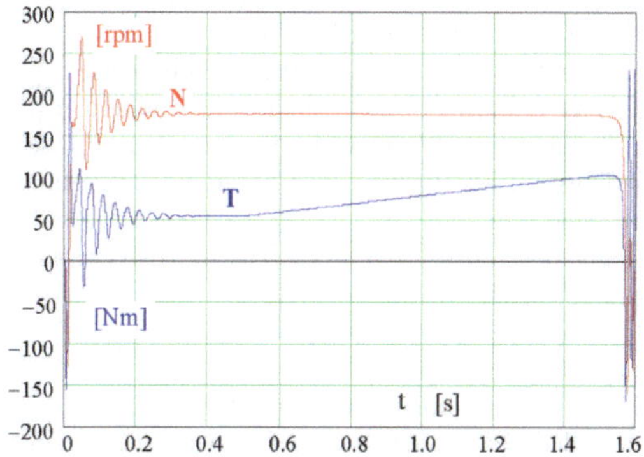

Figure 18. Simulated electromechanical transient of the motor of Table 2, Fig. 4, $N_{tuc} = 46$, with sinusoidal feeding ($V = 380$ Vrms, $f = 50$ Hz): synchronization from zero speed with rated torque, with initial voltages in opposition to those in Fig. 17; application of a torque ramp, up to the pull-out torque ($T_{max} = 101$ Nm).

the sinusoidal nature of the machine is confirmed by the practical absence of noise in any operating condition.

Conclusion

A PM synchronous motor equipped with special, two-layer, concentrated coil windings have been described, capable of self-starting in loaded conditions with mains supply: the winding structure have been illustrated, together with some design criteria, developing useful figures of merits for the best choice of the main constructional parameters.

Several simulations by analytical and FEM models have demonstrated the interesting performances of the machine, confirmed also by corresponding experimental tests.

The activity will be intensively continued, both as regards the optimization of the motor, and concerning the application of the developed winding theory to different machine configurations.

References

[1] E. Spooner, A.C. Williamson: "Direct coupled, PM generators for wind turbine applications", IEE Proc. on Electr. Power Appl., Vol. 143, No. 1, pp. 1–8, January 1996.
[2] E. Spooner, A.C. Williamson, G. Catto: "Modular design of PM generators for wind turbines", ibidem, Vol. 143, No. 5, pp. 388–395, September 1996.
[3] E. Spooner, A.C. Williamson: UK Patent 2278738: "Modular Electromagnetic Machine".
[4] P. Lampola: "Electromagnetic Design of an Unconventional Directly Driven PM Wind Generator", Proceedings ICEM'98, Istanbul, Turkey, pp. 1705–1710, 1998.
[5] M. Lukaniszyn, M. Jagiela, R. Wrobel: "Influence of Magnetic Circuit Modifications on the Torque of a Disc Motor with Co-axial Flux in the Stator", Proceedings ICEM'02, Brugge, Belgium, paper No. 069, 2002.
[6] A. Muetze, A. Jack, B. Mecrow: "Alternate Designs of Low Cost Brushless DC Motors using Soft Magnetic Composites", ibidem, paper No. 237.
[7] Th. Koch, A. Binder: "PM Machines with Fractional Slot Winding for Electric Traction", ibidem, paper No. 369.
[8] S. Tounsi, F. Gillon, S. Brisset, P. Brochet, R. Neji: "Design of an axial flux brushless DC motor for electric vehicle", ibidem, paper No. 581.
[9] W.R. Canders, F. Laube, H. Mosebach: "PM Excited Poly-phase Synchronous Machines with Single-Phase Segments. Featuring Simple Tooth Coils", ibidem, paper No. 610.
[10] F. Magnussen, C. Sadurangani: "Winding Factors and Joule Losses of PM Machines with Concentrated Windings", IEEE-IEMDC '03 Conference Proceedings, Madison, Wisconsin, USA, pp. 333–339, June 1–4, 2003.
[11] N. Bianchi, S. Bolognani, F. Luise: "Analysis and Design of a Brushless Motor for High Speed Operation", ibidem, pp. 44–51.
[12] N. Bianchi, S. Bolognani, P. Frare: "Design Criteria of High Efficiency SPM Synchronous Motors", ibidem, pp. 1042–1048.
[13] A. Di Gerlando, M. Ubaldini: Italian Patent Application MI2002A 001186, "Synchronous Electrical Machine with Concentrated Coils", May 31st, 2002; International PCT Patent pending.
[14] Maxwell 2D and 3D FEM codes, Vol. 10, Ansoft Corporation, Pittsburgh, PA, USA, November 2003.

III-1.3. AXIAL FLUX SURFACE MOUNTED PM MACHINE WITH FIELD WEAKENING CAPABILITY

J.A. Tapia, D. Gonzalez, R.R. Wallace and M.A. Valenzuela

Electrical Engineering Department, University of Concepcion, Casilla 160-C, Correo 3 Concepcion, Chile

juantapia@udec.cl, degonzalez@udec.cl, rwallace@udec.cl, anivalenz@udec.cl

Abstract. In this paper an axial flux PM machine with field control capability for variable speed application is presented. To achieve such as control, surface mounted PM rotor-pole configuration is shaped so that, a low reluctance path is included. In this way, controlling the armature reaction based on vector control allows us to command the airgap flux in a wide range. Magnetizing and demagnetizing effect can be reached with a low stator current requirement. In order to handle the rotor reluctance, an iron and PM sections are include. 3D-FEA is carried out to confirm the viability of the proposed topology. Also a procedure to estimate the *d*-*q* parameters for the topology is presented

Introduction

Permanent magnets (PM) machines have gained great popularity due to higher power density and efficiency compared to conventional electromagnet excitation [1]. In fact, Modern PM based on *NdFeB* allows us to mounted directly on the rotor surface provide high airgap flux density [2], with no field losses, reduced volume, and lower requirement for the machine manufactured. However, for variable speed applications, PM machines offer difficulties because of the fix excitation provided for the magnets. Induce voltage increases linearly with frequency reducing the speed range over rated speed [3]. Controlling the airgap flux is the main issue for PM machines.

Axial flux machines offer several advantages when high torque and power density [4,5] compared with radial flux topologies. Sandwich configuration allows us to stack several rotors and stators in a single shaft with direct control over the airgap length. These features are required especially in traction and power generation.

In this paper, an axial flux surface mounted PM (AFPM) machine configuration with field weakening capability is proposed. The topology allows us to control the airgap flux with a reduced *d-axis* current requirement for operation over the rated speed. To perform such as control, machine design considers a modification of the rotor-pole magnetic configuration including an iron-pole piece. In this manner, a negative *d-axis* flux component can easily be generated by the armature reaction, so that the total airgap flux is reduced (or increased) accordingly [6]. As a result, magnets are not submitted to any significant demagnetizing field and control action is made over iron mostly.

S. Wiak, M. Dems, K. Komęza (eds.), Recent Developments of Electrical Drives, 321–334.

3D finite element analysis (3D-FEA) has been carried out with commercial software FLUX3D provided by MAGSOFT Co. These analyses demonstrate that airgap flux can be commanded with an appropriate armature reaction control. As a result speed range can be increased without significant requirement of the stator *d-axis* current. In addition a procedure to estimate the *d-q* parameters based on armature reaction waveform analysis is used. Geometry and iron to magnet ratio define the machine reactances.

In the following sections description of the proposed machine, FEA for no-load and load conditions, and parameters procedure calculation are shown.

Axial flux machine topology

Description

The AFPM machine topology proposed is shown in Fig. 1. This machine is composed of two rotors and one central stator. Rotors are north-north (NN) PM surface mounted type containing the excitation poles. Each of these poles is assembled by two parts: PM and iron piece. PM section is a magnet part axially magnetized which provides excitation to the machine. On the other hand an Iron section which offers an easy path for the stator current armature reaction. Due to the short airgap length, the total flux per pole can be considered as two components: one associated to the magnets (high reluctance), the other associated to iron (low reluctance).

The stator contains two set of three-thase AC windings (one in front of each rotor) allocated in radial slots. Stator iron yoke completes the rotor-stator magnetic circuit, so that each side can be considered an independent magnetic circuit. Control can be performed separately in each side of the machine.

From the construction point of view, stator and rotor yoke, and stator teeth are made using iron lamination, which is compacted by epoxy glue and enrolled as a spiral. In that

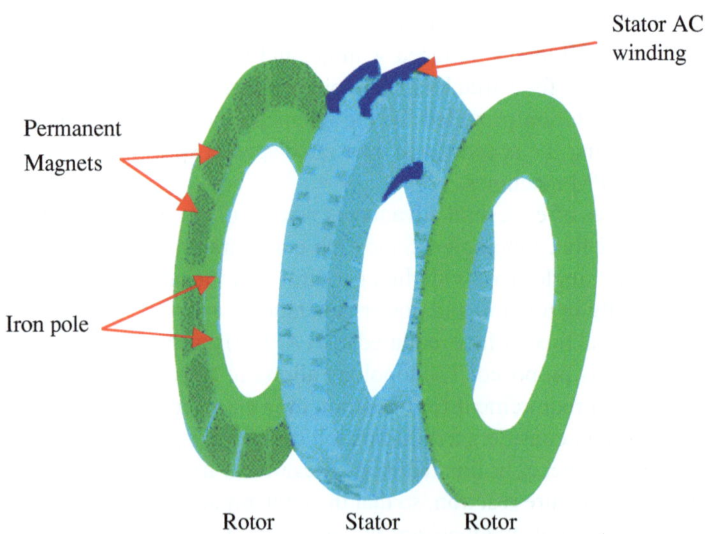

Figure 1. Axial flux surface mounted PM machine for field weakening application.

way, radial flux is minimized, avoiding zigzag leakage flux and axial paths are allowed to conduct flux.

Operation

Below rated speed, the machine is controlled using maximum torque per ampere (MTA) trajectory [3]. *dq-Axis* currents are calculated according to the machine parameters and operating condition to obtain maximum torque with rated armature current.

Over the rated speed, the voltage and current inverters constrain obligate an appropriated current control. In fact, linear variation of the *back-emf* with the speed makes this internal voltage increase above the rated value. The operating condition became critical due to fix and uncontrollable PM magnetization. Therefore, in order to reduce the total airgap flux, armature reaction is utilized to demagnetize the machine by controlling of the stator current. Phasor diagram and flux distribution for the synchronous PM motor is depicted in Fig. 2. Armature reaction flux, Φ_{ad}, is divided it in two components. As is observed, negative *d-axis* current introduces a flux component, Φ_d, which neutralizes to the PM flux. In this manner adequate stator current can be used to control the total flux on the machine. For the circuit point of view, *d-axis* voltage drop ($jX_d I_d$) compensates the increment in the *back-emf* (E_f). Observe that both voltages have linear dependence of the frequency (speed). However, in regular PM machine, the amount of *d-axis* current required to perform such a control, is extremely high due to the large PM reluctance. Using this approach elevated copper losses are generated and PM demagnetization risk reduce its application.

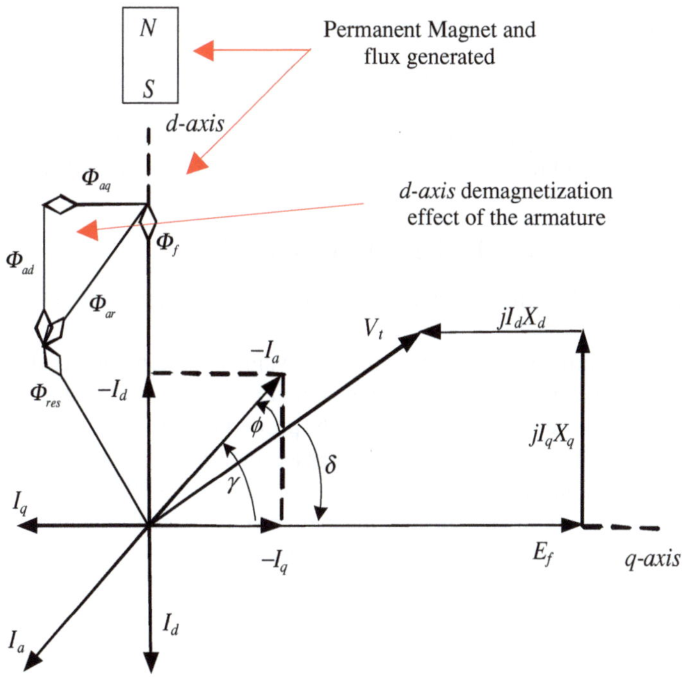

Figure 2. Phasor diagram and flux relationship for the salient pole synchronous machine.

The proposed machine topology, total rotor-pole reluctance is modified using a small iron-pole section. In this manner, *d-axis* flux has two components: one for the magnets and the other for the iron. Because of the low iron reluctance, low stator current is required to perform the airgap flux control.

Adequate selection of the PM to iron ratio allows us to adjust machine parameter so that the speed range can be extended for the machine over the rated speed [6]. The optimum operation condition requires at high speed both *back-emf* and *d-axis* reactance voltage drop increase in the same proportion, so that their variation are balanced [7]. As a result, stable operation is achieved.

Features

In addition to the natural axial flux machines advantages, the machine topology presented shows several others, in comparison with the regular PM machine such as:

- Wide range of airgap flux control to reduce or increase its value, this is made with low requirement of *d-axis* current.
- This particular configuration allows us to control the level of excitation of the machine without any demagnetization risk for the permanent magnet.
- Airgap flux control allows to increase and to improve the power capability at high-speed range of the drive-motor configuration.

Drawbacks

However, this configuration has some problems that can be summarized as follow:

- Lower power density respect to regular PM machine due to the reduction on the amount of magnet.
- gsymmetrical flux density distribution over the stator teeth introduces additional saturation over the stator and rotor yoke. This is because of their trapezoidal shape.

Finite element analysis

In order to determine the effectiveness of the proposed configuration a 3D-FEA is carried out.

Rotor and stator domain and 3D-mesh used to evaluate the topology is shown in Fig. 3. Stator winding representation is depicted. One detail has to be incorporated in the model.

Lamination core for the magnetic circuit has the property to carry flux mainly in tangential and axial directions. However, due to the interlamination airgap radial flux is reduced considerably. To take into account this effect in the model, additional radial airgaps are introduced in the 3D-FE model. In this manner, flux is forced to flow in the ordinary directions given by the iron permeability and lamination.

No-load operation

For no-load operation, the only excitation present on the machine is provided for the magnets. Flux density distribution for this operation is depicted in Fig. 4. As expected, there is

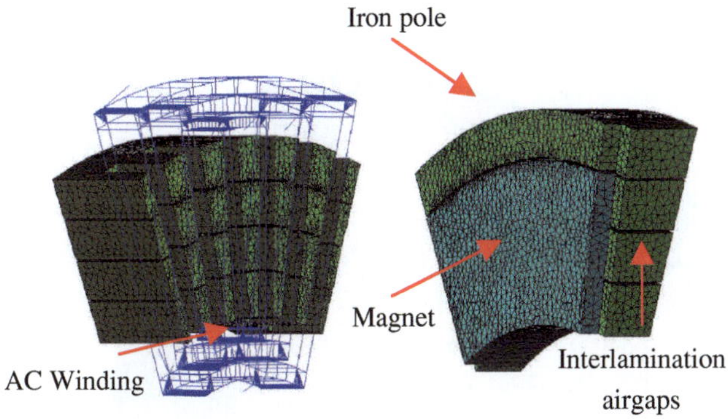

Figure 3. 3D-mesh for one pole of the AFPM machine. (a) Stator with armature winding (b) Rotor.

magnetic activity mostly over the magnet area. Due to the no stator current, armature reaction does not apply flux over the iron section of the airgap. As a result, flux density is negligible in this area. Total flux crossing airgap correspond that impressed by the PMs.

On-load operation

Under vector control strategy, stator current can be positioned in any location over the airgap respect to the PM flux. According to the required demagnetization effect, current angle (γ in Fig. 2) is calculated so that necessary *d-axis* flux is generated to counteract magnet flux. Combined flux density distribution for maximum *d-axis* demagnetization effect ($I_d = 1$ pu) and magnet excitation is presented in Fig. 5. For the operating condition indicated, 3D-FEA establishes that armature reaction acts mostly over the iron section and flux density over the magnet has almost no variation respect to the *no-load* condition. Iron rotor pole offers a very low axial reluctance for the armature reaction respect to the magnet. As a result, flux imposes by the stator current increase consequently and direction over iron section of the airgap is opposite respect to the magnet flux. Total airgap flux is the difference between these two fluxes. Magnetic effort over the magnets is reduced substantially, with minimum demagnetization risk and low current.

To evaluate the airgap flux as a function of the *d-axis* current, the airgap region is divided in two sections: one considering the area in front of the magnet and the other in front of the iron section. Each component of the airgap flux is plotted in Fig. 6.

Flux associated to the magnet has minimum variation as the stator current increases. At the same time, iron section flux increases with the current. As a result, total airgap flux is varied according to the *d-axis* current.

Iron-magnet rotor pole

Iron to magnet ratio over the rotor pole has an important impact in the flux control capacity in this topology. While the iron section is reduced, saturation diminishes the armature reaction effect over the total airgap flux, as depicted in Fig. 7. As the iron section increases, higher variation of the flux is encountered. In addition a linear dependence respect to the demagnetizing current is presented. However magnets size makes that power density

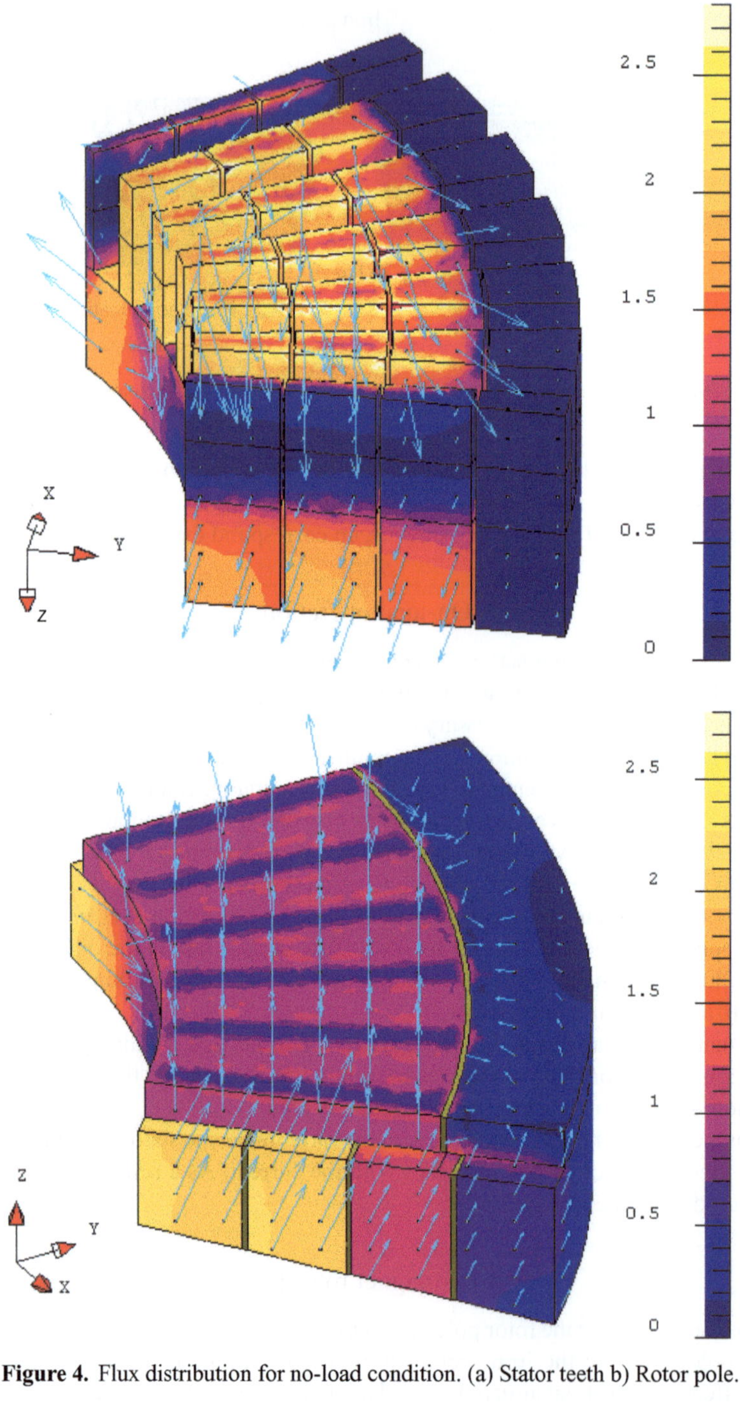

Figure 4. Flux distribution for no-load condition. (a) Stator teeth b) Rotor pole.

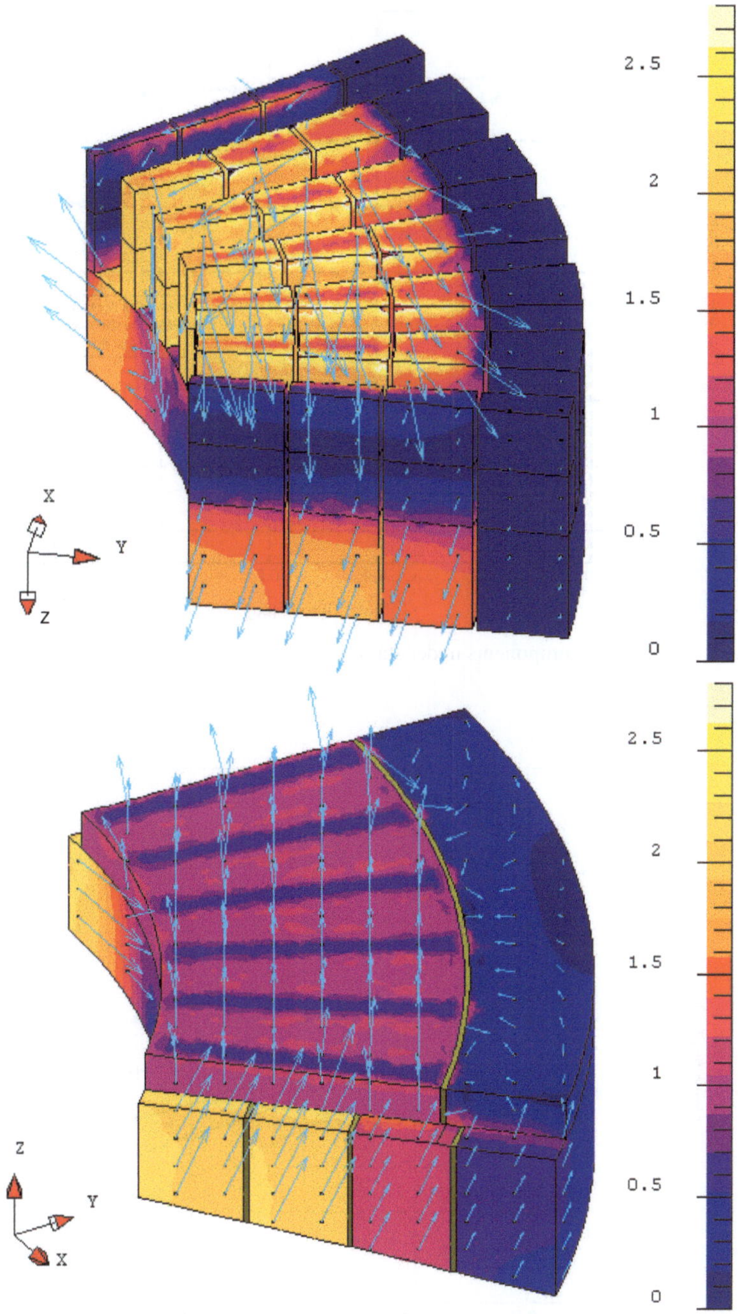

Figure 5. Flux distribution for maximum demagnetization condition. (a) Stator teeth (b) Rotor pole.

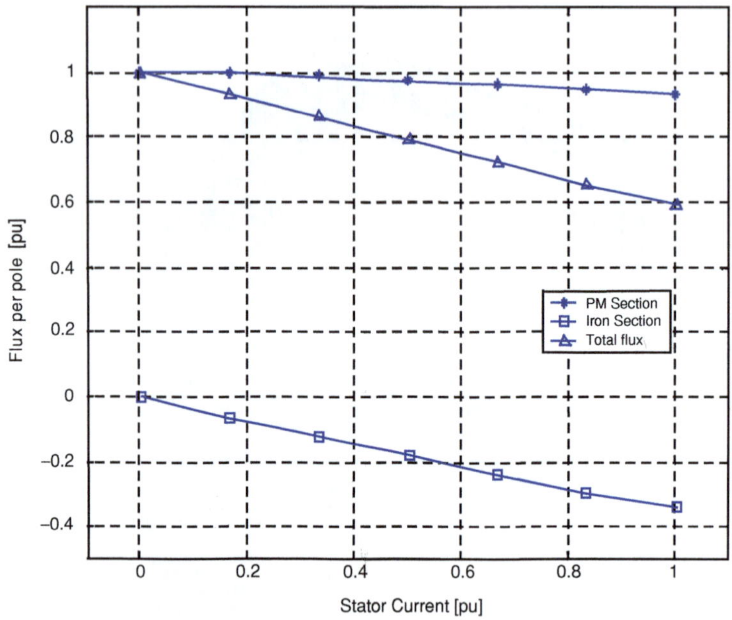

Figure 6. Airgap flux components under stator *d-axis* current variation. Iron-PM ratio 0.45.

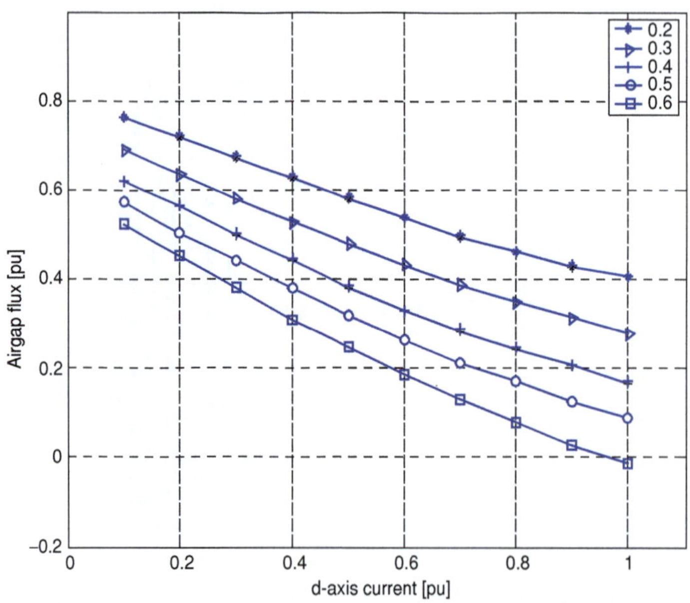

Figure 7. Total airgap flux as a function of the stator *d-axis* current for different iron/magnet ratio.

diminishes in the same proportion. Proper selection of this ratio must be selected in order to accomplish field weakening capability and power delivered for the machine.

Reactance parameters

Neglecting saturation, synchronous machine parameters can be estimated from the fundamental flux density distribution which is by the MMF of the armature reaction. Based on the two-reaction theory, this armature reaction is solved for each of the characteristic axis. In this manner, it is possible to express reactance parameters in term of the *form factors* [8].

Form factor

The *dq-axis* mutual reactances can be calculated in terms of the form factor of the stator field (armature reaction factors) k_{fd} and k_{fq} as:

$$X_d = k_{fd} X_a \tag{1}$$

$$X_q = k_{fq} X_a \tag{2}$$

This approach considers the amplitude reduction of the *d-axis* and *q-axis* fundamental harmonic of the armature reaction field due to airgap non-uniformity. This is caused by the presence of the PM pole and the air space between the poles. They make it possible to express each component of the armature reaction scaled respect to the magnetic field created in a cylindrical rotor synchronous machine. The armature reaction peak values of the fundamental harmonics are calculated using Fourier series coefficients expressions for the fundamental.

d-Axis form factor

Where X_a is the mutual reactance for the cylindrical rotor synchronous machines. These form factors are calculated based on harmonic distribution of the flux density over the airgap as:

$$k_{fd} = \frac{B_{d1}}{B_{ad}} \quad \text{and} \quad k_{fq} = \frac{B_{aq1}}{B_{dq}} \tag{3}$$

The rotor of the AFPM machine combines the structure of a surface mounted PM machine with the structure of the reluctance machine for the reactance calculation purposes. The *d-axis* flux's path is composed by the salient PM pole and the iron-pole component. Rotor configuration and *d-axis* armature reaction field is shown in Fig. 8. Magnet sections represent a large airgap because their permeability (close to unity). According to (3) *d-axis* form factor for the magnet section is

$$k_{fd\_PM} = 1 \tag{4}$$

For the iron-pole section, according to Fig. 8, low reluctance of the iron makes a large penetration of the armature reaction into the rotor and very small for the air section between poles. Position of the *d-axis* armature reaction MMF respect to the rotor is presented in

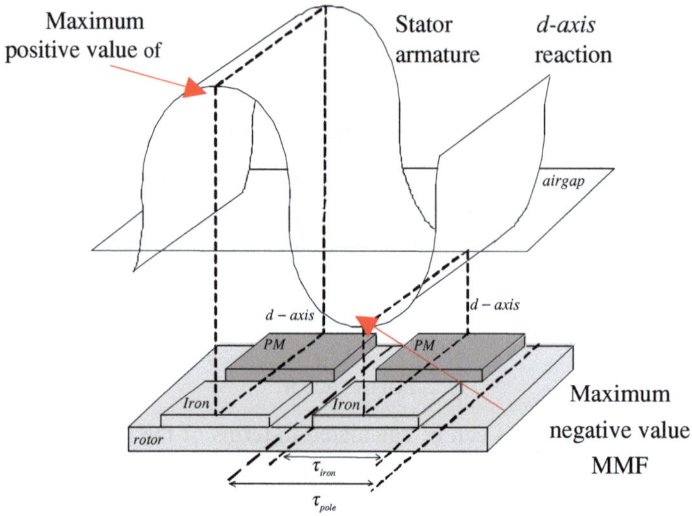

Figure 8. Rotor reluctance for the *d-axis* armature reaction force.

Fig. 9. If saturation is ignored, the MMF generates a linear variation of the flux density over the iron pole. Neglecting fringing effects, the flux density waveform, $B_{ad1}\cos\alpha_{iron}$, appears as is shown in this figure. The ratio α_{iron} is defined as

$$\alpha_{iron} = \frac{\tau_{iron}}{\tau_{pole}} \tag{5}$$

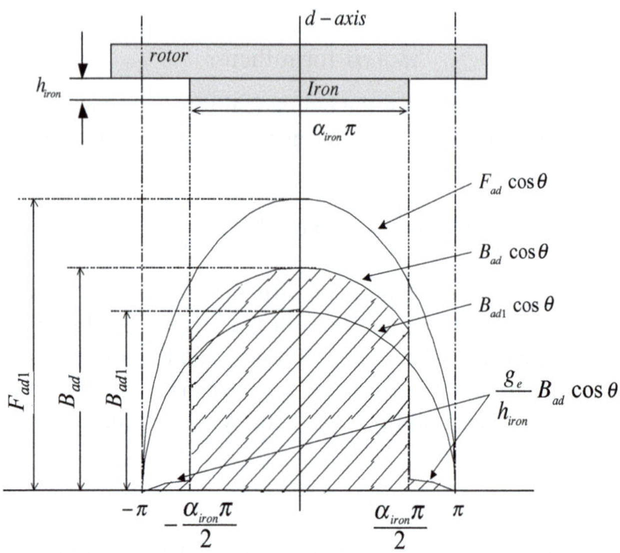

Figure 9. The *d-axis* armature reaction MMF, flux density, and fundamental harmonic of the field reaction.

Based on this procedure the *form factor* for the iron is calculated as:

$$k_{fd\_iron} = \frac{g_e}{h_{iron}}\left((1-\alpha_{iron}) - \frac{\sin\alpha_{iron}\pi}{\pi}\right) + \left(\alpha_{iron} + \frac{\sin\alpha_{iron}\pi}{\pi}\right) \tag{6}$$

The ratio g_e/h_{iron} in (6), considers the effect of the larger airgap between poles. Due to the double component of the airgap flux, total *d-axis* form factor, k_{df}, is the combination of (4) and (6) affected by the proportion of the total airgap occupied by each section. This is:

$$k_{fd} = \left(\frac{A_{iron}}{A_{pole}}\right) k_{fd\_iron} + \left(\frac{A_{pm}}{A_{pole}}\right) k_{fd\_PM} \tag{7}$$

However, the reluctance ratio between the PM and iron pole has to be considered to obtain the correct *form factor* value. In fact, definitions of these two factors are based on their own relation between the fundamental to maximum flux density in the airgap (3). To normalize this expression the PM height, h_{pm} (which is the same as the iron) to the airgap length ratio is introduced. Using the result given in (7), the *d-axis* form factor for the AFPM machine is

$$k_{fd} = \left(\frac{A_{iron}}{A_{pole}}\right) k_{fd\_iron} + \left(\frac{A_{pm}}{A_{pole}}\right) \frac{g_e}{h_{pm}} \tag{8}$$

This equation establishes that the form factor is a function of the iron section parameter (α_{iron}), the PM height to airgap length ratio and the iron and magnet section. For actual surface PM machines, g_e/h_{pm} ratio is numerically much larger than k_{fd_iron} (which is less than 1). As a result, the *d-axis* form factor can be approximated as:

$$k_{fd} \approx \left(\frac{A_{iron}}{A_{pole}}\right) k_{fd\_iron} \tag{9}$$

Equation (9) implies that the *d-axis* reactance (1) is mainly defined by the iron section parameters of the rotor pole. The high reluctance of the PM is reduced by the parallel path with very low reluctance given by the iron. This reduces the saliency of the machine, but also reduces the amount of Ampere-Turn required to demagnetize the machine for field weakening application.

q-Axis form factor

The *q-axis* armature reaction waveform force is directed over the interpolar section of the rotor, as depicted in Fig. 10. Due to the high reluctance of the magnet, the *q-axis* presents an axial symmetry, composed by two section: magnet and iron. From the point of view of the stator demagnetization force, the rotor acts in different manner in both sections as is studied previously. A simplified diagram for the flux density distribution over the iron section is shown in Fig. 11. According to the distribution, *q-axis* form factor for each section are calculated as: for the magnet section

$$k_{fd\_PM} = 1 \tag{10}$$

Similarly as the previous section, for the magnet section

$$k_{fq\_iron} = \frac{g_e}{h_{iron}}\left((1-\alpha_{iron}) + \frac{\sin\alpha_{iron}\pi}{\pi}\right) + \left(\alpha_{iron} - \frac{\sin\alpha_{iron}\pi}{\pi}\right) \tag{11}$$

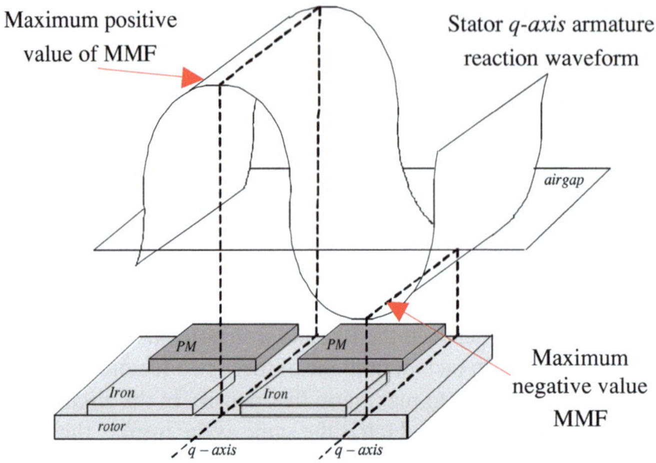

Figure 10. Rotor reluctance for the *q-axis* armature reaction force.

Therefore, in conjunction iron and magnet sections define the total form factor. Following the same procedure for the *d-axis*, the approximate value for the *q-axis form factor*

$$k_{fq} \approx \left(\frac{A_{iron}}{A_{pole}}\right) k_{fq\_iron} \tag{12}$$

As the iron section over the rotor pole increases, *dq-axis* form factor vary as is depicted in Fig. 12.

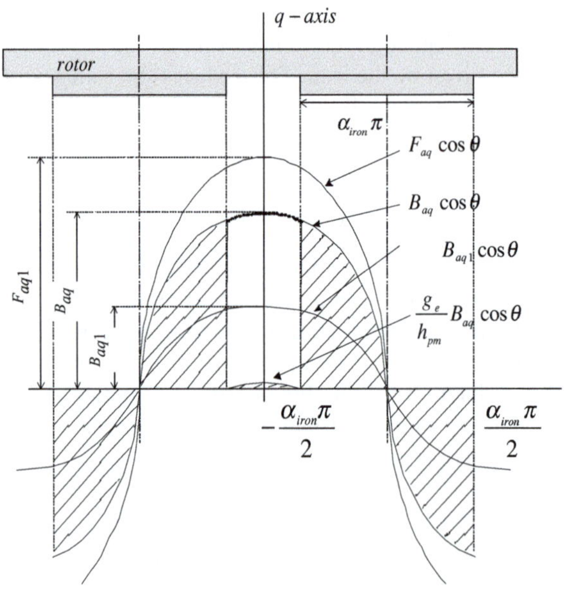

Figure 11. The *q-axis* armature reaction MMF, flux density, and fundamental harmonic of the field reaction.

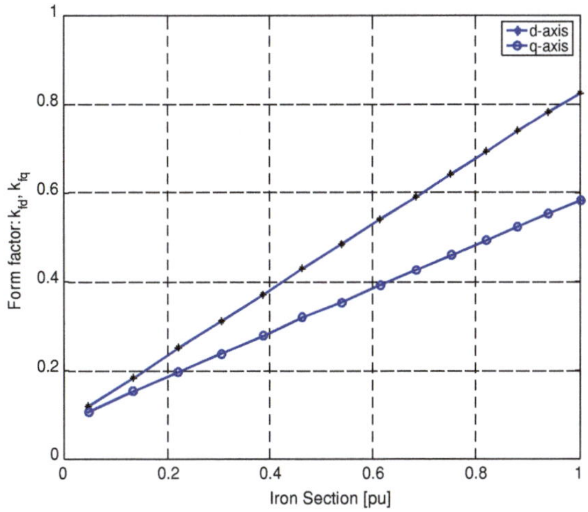

Figure 12. Form factor for the AFPM machine according to the amount of iron section. $A_{iron} = 0.83$, $g_e = 0.5$ mm, $h_{iron} = 5$ mm.

Machine saliency changes as the iron section increase. In fact, for very low portion of iron, to rotor pole is mostly cover by the magnet. For this range, *form factors* tend to be similar, like is found in cylindrical rotor synchronous machine. However, large amount iron increases the *d-axis* inductance, due to the low reluctance. *q-Axis* form factor experiment a lower variation that *d-axis* form factor, because of the interpolar space, which, for the surface mounted PM machine is air. As a result, different values for the *dq-axis* reactances are found if the iron to magnet is varied. Stable operation over the rated speed is commanded by the machine parameters and the relation between them.

From the previous analysis it is clear that flux control over the AFPM machine, can be visualized as an interaction between PM and armature reaction flux or through the reactances values. In both cases 3D-FEA and analytic approach, airgap flux control to keep the *back-emf* at 1 pu value, is achieved by adequate selection of the iron to magnet ratio.

Conclusions

In this paper an axial flux surface mounted PM machine topology with field weakening capability has been presented. The rotor-pole configuration is composed by PM and iron section which provides, in conjunction, a low *d-axis* reluctance for an easy airgap flux control using the armature reaction. As a result speed range can be increased for variable speed applications. In this structure, low *d-axis* stator current is required to control airgap flux with minimum risk for PM demagnetization. 3D-FEA demonstrates that airgap flux can be controlled essentially using demagnetizing effect of the armature reaction over the iron section. Iron to magnet ratio is key factor that determine the maximum reduction of the airgap flux.

Analytic parameter estimation based on fundamental armature reaction establishes that proper value for the reactances is defined by this ratio. Optimal parameter selection can be achieved so that power capability is extended over the rated speed.

A 3 kW prototype using this configuration and proper control strategy are part of the next steps of this investigation.

Acknowledgment

The authors wish to thank the financial support given by the Chilean Research Council (Fondecyt), through the Project # 1030329.

References

[1] F. Profumo, Z. Zheng, A. Tenconi, Axial flux machines drives: a new viable solution for electric cars, IEEE Tran. Ind. Electron., Vol. 44, No. 1, pp. 39–45, 1997.

[2] N. Bianchi, S. Bolognani, Design techniques for reducing the cogging torque in surface-mounted PM motors, IEEE Trans, Ind. Appl., Vol. 38, 2002.

[3] T.M. Jahns, Motion control with permanent magnet AC machine, Proc. IEEE, Vol. 82, No. 8, pp. 1241–1252, 1994.

[4] S. Huang, M. Aydin, T.A. Lipo. "Comparison of (Non-Slotted and Slotted) Surface Mounted PM Motors and Axial Flux Motors for Submarine Ship Drives", Third Naval Symposium on Electric Machines, Philadelphia, December 2000.

[5] B.J. Chalmers, W. Wu, E. Spooner. An axial-flux permanent-magnet generator for a gearless wind energy system, IEEE Trans. Energy Convers., Vol. 14, No. 2, 1999.

[6] R.F. Schiferl, T.A. Lipo, Power capability of salient pole permanent magnet synchronous motors in variable speed drive applications, IEEE Trans. Ind. Appl., Vol. 26 No. 1, pp. 115–123, 1990.

[7] J.A. Tapia, Development of the Consequent Pole Permanent Magnet Machine, Ph.D. thesis, University of Wisconsin-Madison, February 2002.

[8] M. Kostenko, L. Piotrovsky, Electrical Machines, Moscow: MIR Publishers, 1972, pp. 191–202.

III-1.4. COMPARISON BETWEEN THREE IRON-POWDER TOPOLOGIES OF ELECTRICALLY MAGNETIZED SYNCHRONOUS MACHINES

David Martínez-Muñoz, Avo Reinap and Mats Alaküla
Department of Industrial Electrical Engineering and Automation, Lund University, IEA/LTH, Box 118, 221 00 Lund, Sweden
david.martinez@iea.lth.se, avo.reinap@iea.lth.se, mats.alakula@iea.lth.se

Abstract. The finite element method has been used to analyze three topologies of iron-powder electrically magnetized synchronous machines. The first topology has the field winding placed in magnetically conducting end-plates, eliminating the need of slip-rings. In the second topology this is achieved by placing the winding above the outer rotor, and the third topology corresponds to the more conventional design with the field coils in the rotor. The results show that the first topology outputs 60% more torque than the other designs, although the three topologies present similar characteristics with regard to torque density.

Introduction

Soft magnetic composite (SMC) materials consist of iron powder and have isotropic properties, which is particularly useful in machines with 3D flux flow, such as claw-pole machines. SMC materials are more conveniently used in machines with separate excitation, since the demands on the permeability of the magnetic material is lower compared to machines without separate field provision. A comparison between an electrically magnetized claw-pole machine without slip-rings and a permanent magnet machine was presented in [1]. In this paper the claw-pole machine has been compared to two other topologies that use electrical magnetization. One topology has an outer claw-pole rotor, and the slip-rings are removed by placing the field coil in a yoke above the rotor. The other topology corresponds to a more conventional design, with the field coils in the rotor accessed through slip-rings. All the machines have a similar arrangement for the armature coils, which form a concentrated winding. The properties of the machines have been calculated using the finite element method (FEM). The iron losses at load and no load have also been calculated in FEM, implementing a detailed formulation that includes alternating and rotational effects. Thermal models have been set up for the machines, and the current loading has been adjusted for the same temperature rise in the windings.

S. Wiak, M. Dems, K. Komęza (eds.), Recent Developments of Electrical Drives, 335–347.

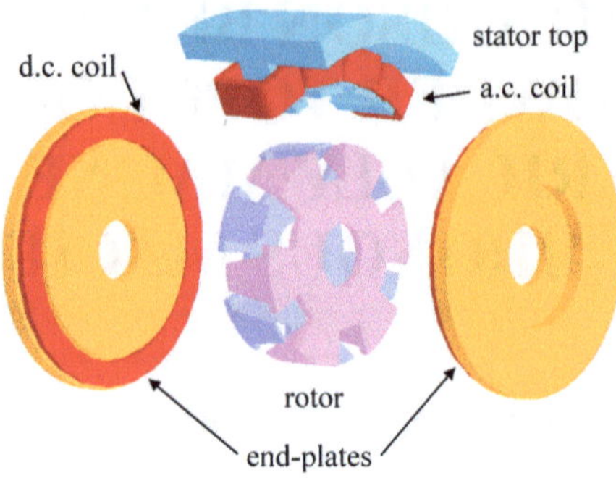

Figure 1. Design 1.

Machine topologies

The claw-pole machine presented in [1] is shown in Fig. 1, and it will be referred to as *Design 1*. The novelty of the machine lies in that the slip-rings are removed by placing the field coils in magnetically conducting end-plates attached to both sides of the stator. The iron plates close the magnetic circuit between the stator and the claw-pole rotor. The field coils are wound around the salient part of these plates, remaining therefore stationary. The stator coils are wound around a single tooth, forming a three phase concentrated winding. This has advantages from the manufacturing point of view since the coils can be pressed in order to increase the filling factor [2]. A simplified plot of the machine with the flux flow is shown in Fig. 2. The magnetizing flux flows from one of the rotor claw-poles toward

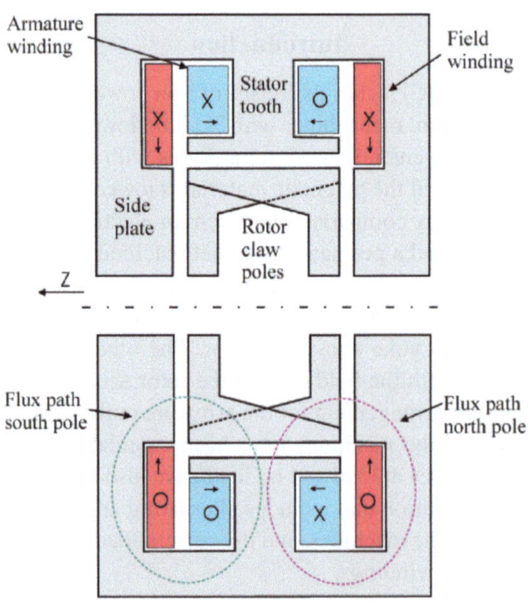

Figure 2. Axial plot of Design 1.

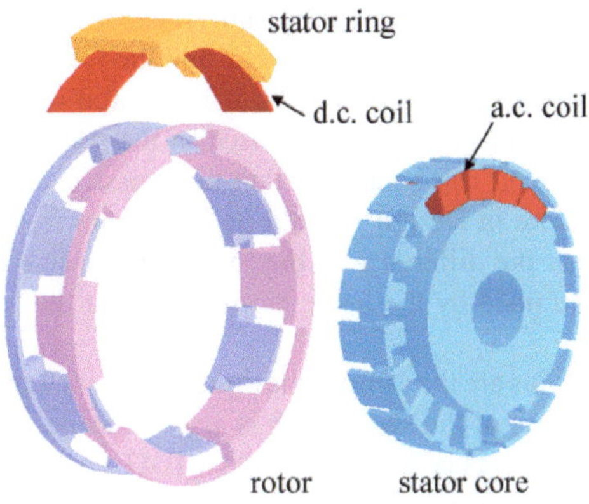

Figure 3. Design 2.

the stator and then back to the same claw-pole through the corresponding iron plate, as indicated in the figure. The flux crosses two airgaps in a complete loop, the radial between the rotor claw-pole and the stator teeth, and the axial between the iron plate and the rotor claw-pole. This is an example of a machine where 3D flux flow is required, and the iron powder used in the simulations for all the designs in this paper is SOMALOY500 [3].

The second topology is shown in Fig. 3 and it will be referred to as *Design 2*. This design is an outer rotor variant of Design 1, and the claw-pole rotor is now sandwiched between the stator core and the stator ring, where the field coil is placed. There are two radial airgaps, between the stator core and the rotor, and between the rotor and the stator ring. With this topology the slip-rings are also removed, but the mechanical coupling from the rotor to the shaft is more complicated. A mould could be used to fit both sides of the rotor into one piece, which is then joined to the shaft using a set of radial bars on one side of the machine. On the other side the stator ring and the stator core are coupled together mechanically. Finally, the third topology is shown in Fig. 4 and it will be referred to as *Design 3*. This configuration resembles the conventional design of an electrically magnetized synchronous machine, where the field coils in the rotor are accessed through slip-rings.

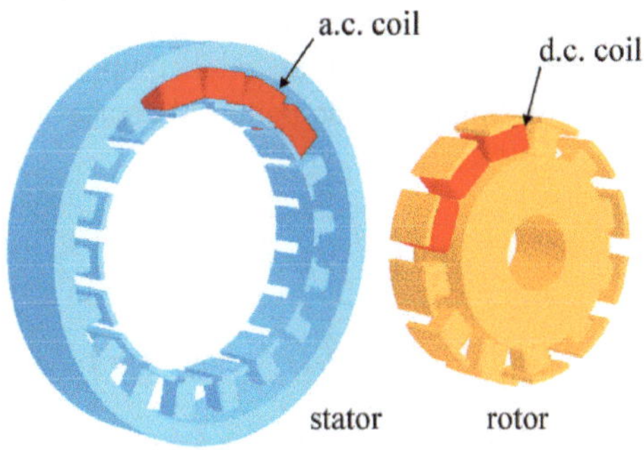

Figure 4. Design 3.

Design 3 is a radial machine that could be entirely built from iron laminations. However, using iron powder allows both the stator yoke and the teeth tips to be extended above the teeth body, which leaves more space for the winding and reduces the amount of copper [2]. For the same reason iron powder was used in the stator core in Design 2, where the flux flow is only radial. It can be argued that the massive yokes in the stator core in Design 2 and the rotor core in Design 3 could be built from iron laminations. However, it was tested in FEM that the difference in the torque response was negligible, since the linking flux is still constrained by the lower permeability of the iron powder used in the teeth.

The inner and outer radius are the same in the three designs, 19 and 100 mm respectively. The axial length of the body of the stator teeth is also the same, as well as the tip length. The total length of the machines is 76 mm in Design 1, 43 mm in Design 2, and 50 mm in Design 3. Design 1 was optimized using the magnetic-equivalent-circuit model described in [4]. Designs 2 and 3 were optimized directly using FEM, and adjusting the dimensions obtained for the stator top and the claw-pole rotor in Design 1. The number of turns in the a.c. windings was calculated for a d.c. link voltage in the converter of 310 V, giving a peak phase voltage of 179 V, and for a nominal speed of 1500 rpm. The voltage for the d.c. winding was provided by a 12 V source.

Iron losses

The iron losses in a rotating electrical machine consist of an alternating and a rotating component [5–7], and can be expressed as in (1). For pure alternation and rotation the trajectory of the flux density loci describes a line and a circle respectively. But in general, alternating and rotating effects interact yielding an elliptical trajectory, and B_{major} and B_{minor} represent the major and minor axis of the ellipse. Their ratio $R_B = B_{minor}/B_{major}$ determines the contribution of the alternating and rotating components to the total core losses. When R_B is 0 or 1 the losses are purely alternating or rotational respectively.

$$P_{core}^{ellipse} = R_B \cdot P_{core}^{circle} + (1 - R_B)^2 \cdot P_{core}^{line} \tag{1}$$

In this paper, R_B has been calculated as in (2), where the minor and major axis have been selected from the minimum and maximum value between the modulus of the radial an axial components together and the tangential component, respectively. Basically, in the regions in Designs 1 and 2 where the flux flows in the radial direction, the axial component can be neglected and vice versa. In Design 3 the axial component is negligible. When the rotor rotates, the change in the magnetization pattern increases the tangential component in the three designs.

$$R_B = \frac{\min\left(\sqrt{B_{rad}^2 + B_{ax}^2},\, B_{\tan}\right)}{\max\left(\sqrt{B_{rad}^2 + B_{ax}^2},\, B_{\tan}\right)} \tag{2}$$

The specific alternating and rotational components in (1) were calculated according to the procedure presented in [7], using equations (3)–(6), where is the peak modulus of the

Table 1. Loss coefficients for SOMALOY500 [7]

Coefficient	Value
C_{ha}	0.1402
C_{ea}	1.233×10^{-5}
C_{aa}	3.645×10^{-4}
h	1.548
C	2.303×10^{-4}
C_{ar}	0
a_1	6.814
a_2	1.054
a_3	1.445
B_s	2.134 T

flux density at each element and f the frequency. The loss coefficients are summarized in Table 1.

$$P_{core}^{line} = C_{ha} \cdot f \cdot \hat{B}^h + C_{ea}(f \cdot \widehat{B})^2 + C_{aa}(f \cdot \widehat{B})^{1.5} \tag{3}$$

$$P_{core}^{circle} = P_{hr} + C_{er}(f \cdot \widehat{B})^2 + C_{ar}(f \cdot \widehat{B})^{1.5} \tag{4}$$

$$P_{hr} = f \cdot a_1 \left[\frac{1/s}{(a_2 + 1/s)^2 + a_3^2} - \frac{1/(2-s)}{(a_2 + 1/(2-s))^2 + a_3^2} \right] \tag{5}$$

$$s = 1 - \frac{\widehat{B}}{B_s}\sqrt{1 - \frac{1}{a_2^2 + a_3^2}} \tag{6}$$

This procedure has been implemented in the finite element (FE) model for each machine, which was set up using a commercial package, OPERA 3D. The fields were calculated at 24 rotor positions, comprising one electrical cycle. A table was created at each position with the Cartesian components of the flux density at each element. These tables were processed in MATLAB, where they were transformed into cylindrical coordinates and their FFT was calculated up to the 11th harmonic. The value of R_B was also calculated for each element and harmonic, and the results were stored in tables. The tables were imported into the FE postprocessor, where (1) was implemented for each element and harmonic, and the losses were calculated performing a volume integral and multiplying by the density of the material. The total loss at each element was approximated simply by adding the fundamental and all its harmonic components. Finally, the losses from the elements corresponding to the same region in the thermal model were added together. It should be noted that P_{hr} in (5) becomes negative for values of $\widehat{B} > B_s$. Although the total flux density in some local heavily saturated part of the machines passed this limit, it was observed that this condition was never satisfied for the fundamental or the harmonic components on their own.

The distribution of the alternating and rotating losses in the stator of the machines for the fundamental component is shown in Fig. 5. In the three machines, alternating losses are concentrated in the body of the teeth, while the losses in the tips are dominated by the rotational component. In the back core, rotational losses appear around the regions where the teeth are connected to the core, while alternating losses are more important in the regions between the teeth. In Design 3 the flux density is almost zero in most regions of the stator core, especially those close to the shaft, which is the reason for the unity ratio in these

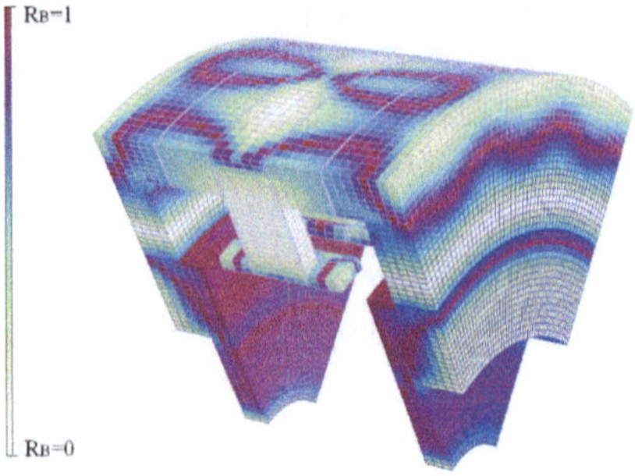

(a) Design 1

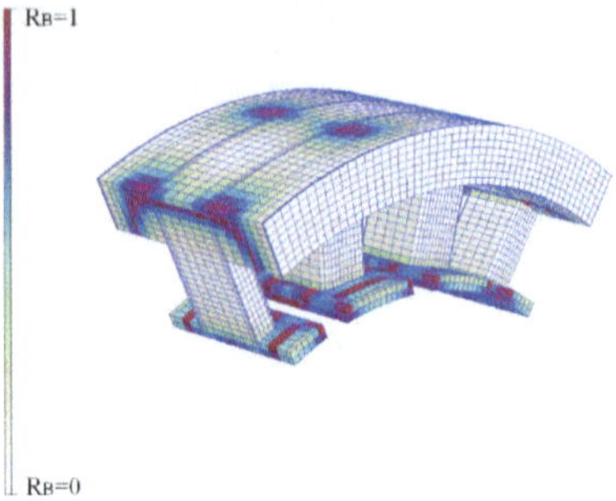

(b) Design 2

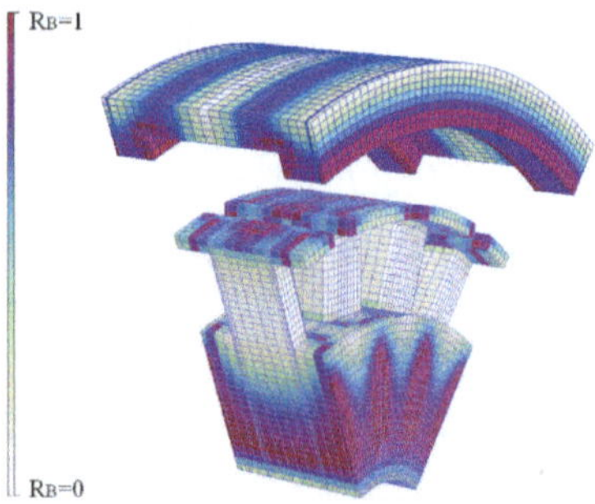

(c) Design 3

Figure 5. Value of RB in the stator for the fundamental component at no load. (a) Design 1. (b) Design 2. (c) Design 3.

elements. In Design 1, the rotational losses are clearly dominating in the regions of the end-plates close to the rotor, as it is also the case in the stator ring in Design 3.

A summary of the losses in the machines including the harmonics is presented in Figs. 6 and 7 at no load and load respectively.

Above each plot the total loss is referred to as "Tot," and the losses in the stator and the rotor are referred to as "St" and "Rt" respectively. Since Design 1 has higher magnetic loading than the other machines, it also presents the highest losses. This is due to the higher current loading given by the better cooling, as it will be shown in the next section. The ampere-turns per a.c. coil is 1049, 691, and 1057 A for Designs 1, 2, and 3 respectively. The ampere-turns for the two d.c. coils together in Design 1 is 2202 A, for the d.c. coil in Design 3 it is 1433 A, and for the d.c. coil around one pole in Design 2 it is 457 A. The copper filling factor for the pressed windings in all the coils is 75%. It can be observed that when the machine is loaded, the interaction between the fields produced by the armature and the field windings sinks the fundamental component of the magnetic loading in the stator. The losses in the stator due to this component are reduced by around 30% in Design 1 and 2, and 50% in Design 3. At the same time, the field interaction gives rise to new harmonics in the rotor, which now appear in the whole spectrum. The rotor losses at load are between 2.5 and 6.5 times higher than at no load, depending on the machine. The total losses at load

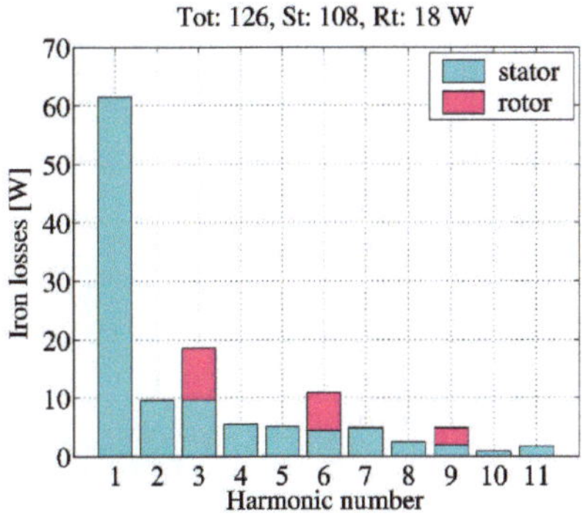

(a) Design 1

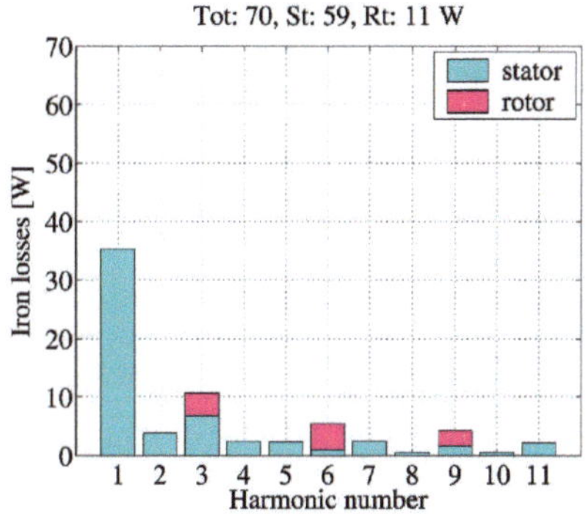

(b) Design 2

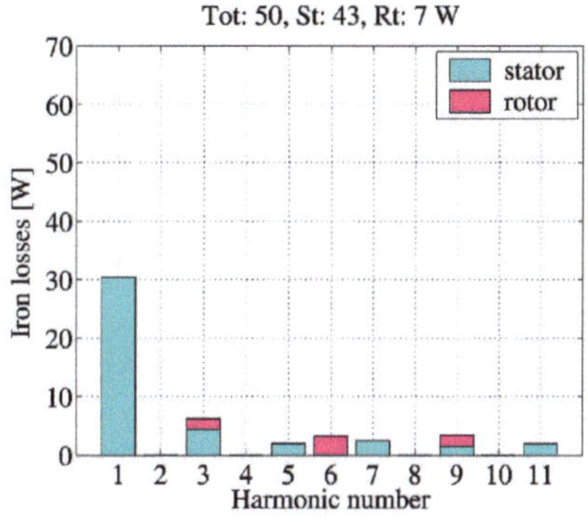

(c) Design 3

Figure 6. Iron losses at no load. (a) Design 1. (b) Design 2. (c) Design 3.

are increased by 10%, 50%, and 60% for Design 1, 2, and 3 respectively, compared to the no load case.

Thermal model

A simplified thermal model was implemented for each machine in order to assess the increase of temperature in the windings, which will limit the current loading. The maximum temperature rise allowed was 100°C above an ambient temperature of 40°C. Water-cooling will be used by default. The total water flow through the machine was limited to 1.2 l/min with a temperature of 30°C. Half of the coolant flow is used to cool the armature winding, and the other half the field winding. It was assumed that the coolant would flow through a duct of exactly the same shape as the cooled surface and a thickness of 3 mm.

The heat conduction inside the coils was modeled by calculating an equivalent thermal conductivity for round conductors λ_r using (7) [8], where λ_i is the conductivity of the copper, d_1 is the diameter of the conductor, d_2 is the diameter of the conductor and the coating, and δ_i is the shortest distance between the surface of two conductors, which was

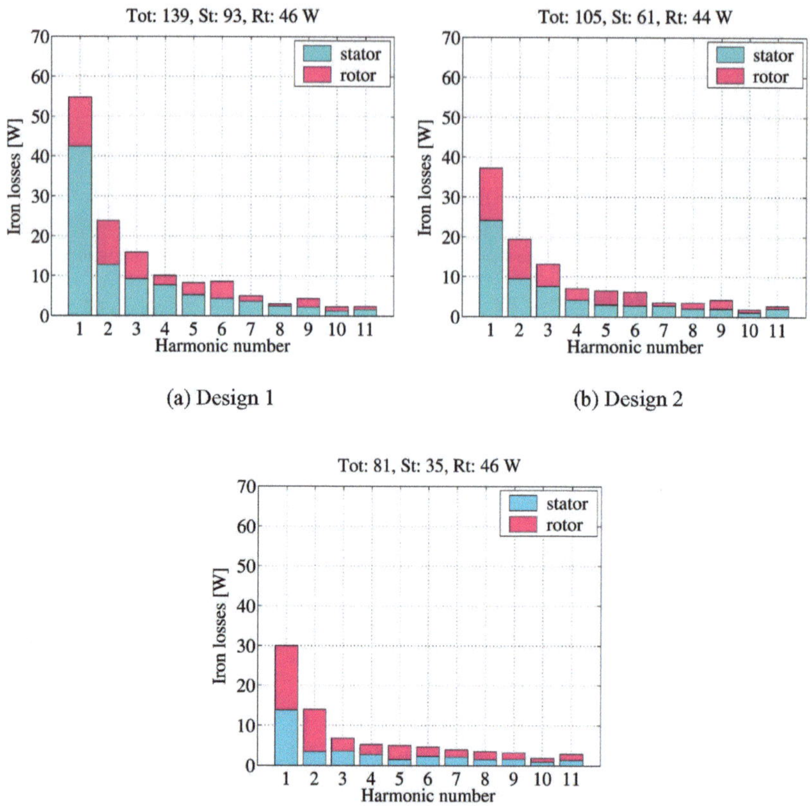

(a) Design 1

(b) Design 2

(c) Design 3

Figure 7. Iron losses at load. (a) Design 1. (b) Design 2. (c) Design 3.

approximated as two times the thickness of the coating. This thickness was selected as 7% the diameter of the conductor, and the coating material was bonding epoxy.

$$\lambda_r = \lambda_i \left(\frac{d_1}{\delta_i} + \frac{\delta_i}{d_2} \right) \tag{7}$$

In [2] it was stated that the thermal resistance of the pressed windings was reduced by 46%, so the thermal conductivity calculated from (7) was increased by this factor. The heat is transferred from the surface of the coil to the iron through a 0.5 mm kapton wall insulation, both for the a.c. and d.c. windings. The thermal conductivity of the iron powder was taken as 13 W/mK. It was assumed that only the surface of the active parts of the machine was used for cooling. The convection factors from these surfaces were calculated from the known formulas for simple geometries given in the basic heat transfer theory [9,10].

Design 1

From the thermal point of view, Design 1 allows a very good cooling of the copper losses from the field winding, given the considerable dissipating area from the sides of the

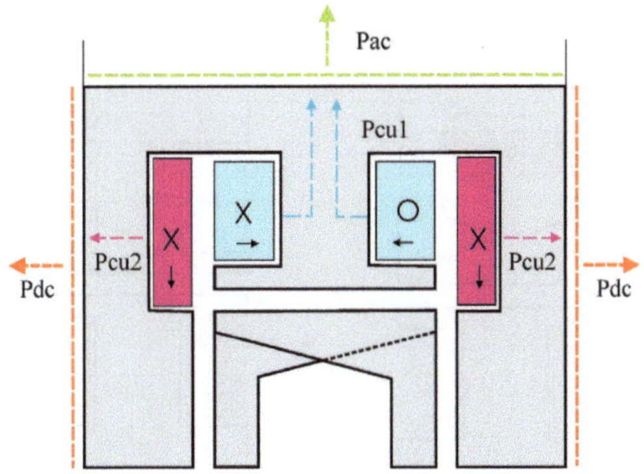

Figure 8. Thermal model for Design 1.

end-plates. In general, two paths were defined for the dissipation of the losses in the machine, and they are shown in Fig. 8.

The copper losses of the a.c. winding ("Pcu1") and the iron losses of the stator teeth and the stator yoke were referred to as "Pac." These losses were dissipated in the radial direction through the outer cylindrical surface of the yoke. The whole surface area of the end-plates was used to cool the copper losses of the d.c. winding ("Pcu2") and the iron losses of the end-plates, and they were referred to as "Pdc."

It was assumed that 20% of the copper losses in the a.c. winding was transferred directly to the core through the top of the coil and that the other 80% was transferred through the teeth, following the path shown in Fig. 8. This ratio is kept constant for the three designs. The iron losses from the FE model at load were grouped into the macro-elements in the thermal model, namely the tooth tip, the tooth body, the stator core, the end-plates, the rotor claw-poles, and the rotor sides attached to these claw-poles. The losses from the rotor claw-poles were added to "Pac," whereas the losses from the rotor sides were added to "Pdc."

The convection factor for "Pac" was calculated from the formulation for forced convection in a cylinder in cross flow, and the value obtained was 242 W/m^{2}K. For "Pdc," the convection factor was approximated from the formulation for forced convection on a flat plate without energy dissipation. The length of the plate was approximated as half the circumferential length at the average radius between the inner and outer radius of the end-plate. The equivalent area is half the area of the end-plate including the axial surface, and the water flow is one quarter of the total flow for the field winding. The convection factor obtained was 195 W/m^{2}K.

Design 2

The thermal model for Design 2 is shown in Fig. 9(a). Only the sides of the stator core can be used to cool the a.c. copper losses, since the shaft passes through the center of the core. The rotor is not completely enclosed as in Design 1 and therefore it was assumed that the rotor losses were dissipated directly through the airflow caused by the rotor rotation. The area available to cool the field winding is considerably reduced compared to

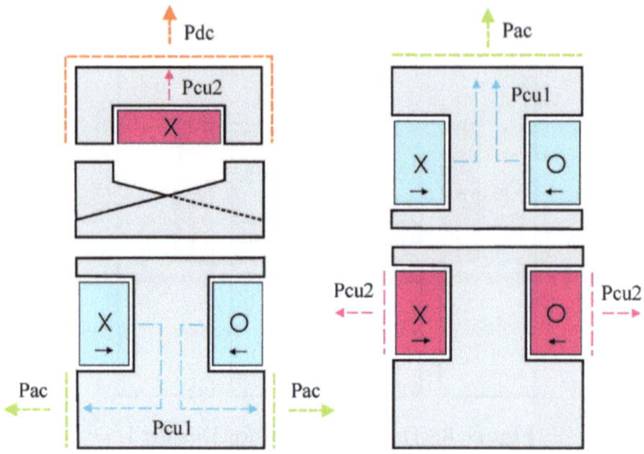

Figure 9. Thermal model for Design 2 (a) and Design 3 (b).

Design 1, and this will constrain the field current. The iron losses in the stator core were grouped in a similar way as in Design 1, whereas the iron losses in the stator ring were separated into those in the core above the d.c. coil and on its sides. The convection factor for "Pac" was calculated using a similar formulation as for "Pdc" in Design 1, obtaining 426 W/m^2K. This formulation was also used for "Pdc" along the lateral sides of the stator ring, and a convection factor of 236 W/m^2K was calculated. The factor for the core above the coil was calculated using a similar formulation as for "Pac" in Design 1, obtaining 426 W/m^2K.

Design 3

The thermal model for Design 3 is shown in Fig. 9(b). In this case the d.c. coils are mounted on the rotor and they must be air-cooled. It was assumed that a separate fan would be used to provide an airflow of 10 m/s along the lateral surfaces of each coil. The water volumetric flow in the stator is the same as in the stators in the previous designs and therefore a smaller pump could be purchased, which in turn will also compensate for the additional cost of the air-cooling system. The iron losses in the stator were grouped as in Design 1. The iron losses in the rotor were assumed to be dissipated directly to the air, thus not contributing to heat the coils. No losses were transferred between the stator and the rotor through the airgap.

The convection factor for "Pac" was calculated as in Design 1, obtaining 347 W/m^2K. The increase in this factor is mainly due to the higher coolant speed of flow. This is a consequence of the smaller duct cross-sectional area for a constant volumetric flow since the machine is shorter. The convection factor for "Pcu2" was calculated again using the formulation for forced convection on a flat plate without energy dissipation. It was assumed that the cooling airflow was divided into two axial paths along both sides of each coil. The length of each path is equal to the axial length of the coil plus half its length in the circumferential direction at the front and at the back. The convection factor calculated was 53 W/m^2K.

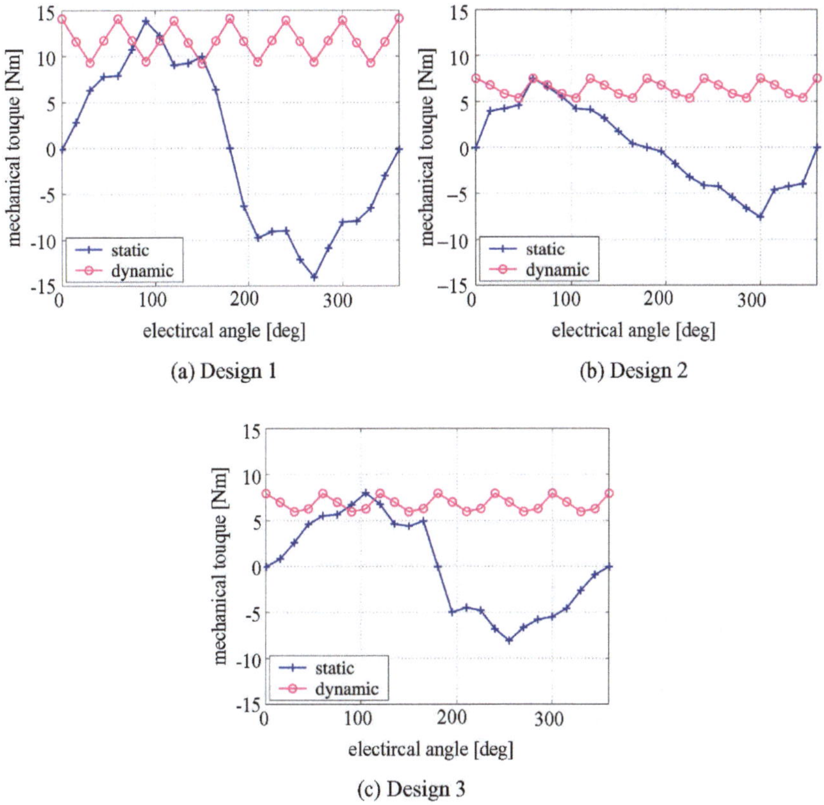

Figure 10. Static and dynamic torque response. (a) Design 1. (b) Design 2. (c) Design 3.

FEM results

The torque response was calculated for the three models, and it is shown in Fig. 10. The static characteristic is obtained simply by maintaining constant the armature current and rotating the rotor along one electrical cycle. The dynamic at each position. This response gives information about the level of the torque ripple in the machine. A summary of the properties of the machines is shown in Table 2.

It can be observed that Design 1 presents a much higher torque than the other two designs, which is due to the higher current loading. However, its total weight is also considerably higher, mainly due to the extra weight from the end-plates. In fact, the ratio of torque per weight in Design 1 is the same as in the other two designs. With regard to torque per volume of the active parts, Design 1 and Design 3 present similar performance, which is 18% better than in Design 2. The same applies for the efficiency, which is around 8% better in Design 1 and 3 compared to Design 2. The percentage of torque ripple is measured as the ratio of the ripple with respect to the maximum torque at thermal limit. This value is highest for Design 1, reaching around one third of the peak torque. Finally, the rotor inertia is almost double in Design 2 compared to Design 1 and 3, and this is due to the higher diameter of the outer rotor.

Table 2. Summary of the properties of the machines

Component	Design 1	Design 2	Design 3
Temperature rise (°C)	100	100	100
Peak torque (Nm)	13.9	7.5	8.0
Mass iron (kg)	9.7	5.5	4.7
Mass copper (kg)	2.8	1.5	2.5
Total weight (kg)	12.5	7.0	7.2
Volume (l)	2.4	1.6	1.4
Torque/weight (Nm/kg)	1.1	1.1	1.1
Torque/volume (Nm/l)	5.8	4.7	5.7
Rotor inertia (10^{-3} kg m^2)	4.0	7.9	3.9
Torque ripple (%)	34	29	25
Efficiency (%)	77	68	76

Conclusions

The comparison between the three topologies has been carried out for the same maximum temperature rise in the windings. The iron losses have been calculated implementing in the finite element model an advanced formulation taking into consideration alternating and rotating losses. The better cooling capability in Design 1, given by the higher cooling surface, implies that this design presents a peak torque around 60% higher than in the other two designs. However, Design 1 is also the heaviest due to the additional end-plates, and the torque per kilo is actually the same in the three designs. The efficiency however is around 8% lower in Design 2 compared to the other two designs. Overall, it was observed that Design 1 and 3 present similar characteristics, including torque per volume, being the main advantage of Design 3 its lower torque ripple, while the most attractive feature of Design 1 is that slip-rings are avoided.

References

[1] D. Martínez-Muñoz, M. Alaküla, "Comparison Between a Novel Claw-Pole Electrically Magnetized Synchronous Machine Without Slip-Rings and a Permanent Magnet Machine", IEEE International Electrical Machines and Drives Conference, Madison, WI, USA, June 1–4, 2003, pp. 1351–1356.

[2] A.G. Jack, B.C. Mecrow, P.G. Dickinson, D. Stephenson, J.S. Burdess, N. Fawcett, J.T. Evans, Permanent-magnet machines with powdered iron cores and prepressed windings, IEEE Trans. Ind. Appl., Vol. 36, No. 4, pp. 1077–1084, 2000.

[3] A.B. Höganäs, SOMALOY™ 500, SMC 97-1, AB Ruter Press, Sweden, 1997.

[4] D. Martínez-Muñoz, M. Alaküla, "A MEC Network Method Based on the BH Curve Linearisation: Study of a Claw-Pole Machine", International Conference on Electrical Machines, ICEM'04 conf. proc., Cracow, Poland, September 5–8, 2004, p. 6.

[5] J.G. Zhu, V.S. Ramsden, Improved formulations for rotational core losses in rotating electrical machines, IEEE Trans. Magn., Vol. 34, No. 4, pp. 2234–2242, 1998.

[6] L. Ma, M. Sanada, S. Morimoto, Y. Takeda, Prediction of iron loss in rotating machines with rotational loss included, IEEE Trans. Magn., Vol. 39, No. 4, pp. 2036–2041, 2003.

[7] Y. Guo, J.G. Zhu, J.J. Zhong, W. Wu, Core losses in claw pole permanent magnet machines with soft magnetic composite stators, IEEE Trans. Magn., Vol. 39, No. 5, pp. 3199–3201, 2003.

[8] A. Arkkio, "Thermal Analysis of High-Speed Electrical Machines", Postgraduate Seminar on Electromechanics, Laboratory of Electromechanics, Helsinki University of Technology, Finland, May 2002.
[9] J.P. Holman, Heat Transfer, 7th edition, McGraw-Hill, London, UK, 1992.
[10] H.Y. Wong, Handbook of Essential Formulae and Data on Heat Transfer for Engineers, Longman, New York, USA, 1977.

III-1.5. RECENT ADVANCES IN DEVELOPMENT OF THE DIE-CAST COPPER ROTOR MOTOR

E.F. Brush Jr.[1], D.T. Peters[2], J.G. Cowie[2], M. Doppelbauer[3] and R. Kimmich[3]

[1]BBF Associates, 68 Gun Club Lane, Weston, MA 02493, USA
ebrush@comcast.com
[2]Copper Development Association Inc., 260 Madison Avenue, New York, NY 10016, USA
dtpeters@adelphia.net, jcowie@cda.copper.org
[3]SEW Eurodrive GmbH & Co KG, Ernst-Blickle Str. 42, D-76646 Bruchsal, Germany
rainer.kimmich@sew-eurodrive.de, martin.doppelbauer@sew-eurodrive.de

Abstract. Performance of several motors where copper has been substituted for aluminum in the rotor squirrel cage is reported. Copper rotor motors die cast in India for agri-pumping were dynamometer and field tested. Copper rotors resulted in higher electrical energy efficiency, slightly higher rotational speed, lower operating temperature, and higher pumping rates and volume pumped per unit of input energy. SEW-Eurodrive motors with copper rotors are also described. A 1.1 kW motor with copper simply substituted and a 5.5 kW motor with redesigned rotor and stator are described. The copper rotor reduced losses in all major categories. Full-load efficiency was increased 6.7 and 3.1 percentage points, respectively. Finally, a study to minimize formation of large pores in die-cast rotors is summarized.

Introduction

At ICEM 2002, we reported on the performance of motors with die-cast copper rotors. Rotor I^2R losses were reduced by 29% to 40% and motor total losses were reduced by 11% to 19% resulting in increased motor efficiencies of no less than 1.5 percentage points. In this paper, motor test data for another group of motors where copper has been directly substituted for the aluminum in the rotor are reported. These copper rotors were die cast in India and motors built and tested by several motor manufacturers there. In the two years since the last ICEM conference, important advances have been made in designing and optimizing the rotor and the entire motor to properly utilize the higher electrical conductivity of copper. Ongoing work reported by Kirtley in this conference is showing the importance of conductor bar shape to accommodate the high electrical conductivity of copper to achieve high starting torque and to further reduce stray load losses. SEW-Eurodrive in Germany has made notable advances in design and performance of a series of motors in drives now commercially available. This paper describes the design approach and test results for 1.1 and 5.5 kW motors optimized for the copper rotor.

S. Wiak, M. Dems, K. Komęza (eds.), Recent Developments of Electrical Drives, 349–359.

Development work on die casting a high melting point metal such as copper has been reported elsewhere [1,2]. Substantial progress in understanding and managing the porosity problem characteristic of high pressure die casting has also been made. The results are applicable to die casting in general and apply to die casting of the rotor in aluminum as well as copper. This work is summarized here. Together with development of the heated nickel-base alloy die system to achieve economically attractive die life, this work is significant to the ability to manufacture the copper rotor.

Motor tests in India

A project to test the suitability of the copper rotor technology upgrade for motors used for water pumping in agriculture in India was carried out by a cluster of motor and pump manufacturers at Coimbatore, Tamil Nadu. Copper rotors were cast by a small Indian die-casting firm for all the tests. Rotor laminations designed for aluminum were used in this direct substitution evaluation. Motors were built and tested by six motor manufacturers. Field test of motors fitted to pumps pumping water for agricultural use and one test of a motor driving a doffing machine in a textile plant were then conducted.

Results for two of the two-pole motors are shown in Tables 1 and 2 and two four-pole motors in Tables 3 and 4. All of these motors are 415 V, 50 Hz, three-phase. As expected with a higher conductivity rotor material, the speed is increased slightly, the slip is reduced, and the efficiency is increased. Starting (locked rotor) torque is also reduced somewhat when copper is substituted for aluminum in laminations with slots designed for aluminum as shown in Table 5. Copper rotors generally result in reduced motor operating temperatures compared to the aluminum counterpart. This is true in these examples except in the 2-Hp (1.5 kW) motor where the copper rotor was cast with no cooling fins and the aluminum counterpart had fins. Even without fins, the motor with the copper rotor ran only about 3°C warmer than the cooled aluminum rotor motor. The temperature rise data in Table 5 were obtained by the winding resistance method. Temperature rise by direct measurement of the core temperature showed the same trends but the temperatures measured were 20°C to as much as 40°C lower.

In addition to the locked rotor torque values reported, the Indian manufacturers of the four-pole motors of Tables 3 and 4 also reported the pull out (breakdown) torque values. Here the copper rotors showed improved torque in these particular motors. The 3-Hp (2.2 kW) motor with a copper rotor had a pull out torque of 408.8% of the rated torque compared to 340.5% for the same motor with an aluminum rotor. Similarly, the 5-Hp (3.7 kW) copper

Table 1. Test results for 2-Hp (1.5 kW), 415-V, two-pole, three-phase, 50-Hz motor, copper rotor compared to aluminum

Rotor material	Load (%)	Input power (W)	Speed (rpm)	Eff. (%)
Copper	100	1,824	2,949	82.54
Aluminum	100	1,856	2,926	81.14
Copper	75	1,440	2,955	79.19
Aluminum	75	1,456	2,940	77.80

Table 2. Test results for 5-Hp (3.7 kW), 415-V, two-pole, three-phase, 50-Hz motor, copper rotor compared to aluminum

Rotor material	Load (%)	Input power (W)	Speed (rpm)	Eff. (%)
Copper	100	4,256	2,947	87.09
Aluminum	100	4,496	2,925	83.99
Copper	75	3,232	2,960	85.99
Aluminum	75	3,408	2,935	82.19

Table 3. Test results for 3-Hp (2.2 kW), 415-V, four-pole, three-phase, 50-Hz motor, copper rotor compared to aluminum

Rotor material	Load (%)	Input power (W)	Speed (rpm)	Eff. (%)
Copper	100	2,600	1,451	85.88
Aluminum	100	2,660	1,411	83.55
Copper	75	1,960	1,465	84.15
Aluminum	75	2,040	1,433	82.82

Table 4. Test results for 5-Hp (3.7 kW), 415-V, four-pole, three-phase, 50-Hz motor, copper rotor compared to aluminum

Rotor material	Load (%)	Input power (W)	Speed (rpm)	Eff. (%)
Copper	100	4,344	1,469	85.97
Aluminum	100	4,544	1,429	83.01
Copper	75	3,280	1,473	85.54
Aluminum	75	3,400	1,443	82.56

Table 5. Locked rotor torque and temperature rise measurements for two- and four-pole motors, copper rotor compared to aluminum

Hp	Poles	Rotor material	Locked rotor torque (% of rated torque)	Temp. rise (°C)
2	2	Cu	406.2	39.6[1,2]
		Al	442.2	36.8[1,2]
5	2	Cu	174.0	66.7[2]
		Al	260.9	80.1[2]
3	4	Cu	242.4	57.7
		Al	268.4	68.8
5	4	Cu	168.4	61.8
		Al	205.2	68.9

[1]No cooling fins on this die-cast copper rotor.
[2]Measured at reduced voltage of 353 V.

Table 6. Field test results of motors of tables 1 and 2 fitted to pumps for agricultural application

Motor/rotor material	Input power, kW	V	Discharge rate, l/s	Energy, kWh	PerkWh
2 Hp					
Cu	2.413	462	2.43	0.551	3,630
Al	2.171	446	2.01	0.599	3,333
5 Hp					
Cu	3.86	389	12.93	0.824	12,114
Al	3.77	377	11.71	0.894	11,003

rotor motor had a measured pull out torque of 350.7% of rated torque while the aluminum version measured 294.2%.

Field testing of three of the motors described above are summarized here. The two-pole 2-Hp (1.5 kW) and 5-Hp (3.7 kW) motors of Tables 1 and 2 were applied to pumping water for agricultural use. Voltages at many locations in India vary substantially over time.

Table 6 shows that the field voltages were both higher and lower than the nomonal 415 V. But the tests comparing the pumping performance of motors with copper and aluminum rotors were decisive in terms of pumping time to fill the tanks and energy consumed in pumping a liter of water. The 2-Hp (1.5 kW) motor-pump combination was tested filling a 2,000 l tank. The tank was brought near to the top in 823 s with the copper rotor motor, 170 s faster than with the aluminum rotor motor, a result of the higher rotational speed of the copper motor. But importantly, less total energy was consumed even at the higher pumping rate by the copper motor and the volume of water pumped per kWh was 8.9% higher. The larger motor was tested filling a 5,000 l tank. Filling time was reduced by 82 s with the copper motor, i.e., 772 s vs. 854 s. The volume of water pumped per unit of energy was increased by 10.1% by using copper in the rotor.

It should be noted that the increased speed of a low slip copper rotor motor can be a problem in pump and fan applications when higher flow rates are not desired. Energy can be wasted with the aluminum to copper rotor substitution because the power increases with the cube of the rotational speed to produce the increased flow rate. In the examples above, the higher flow rate was actually a benefit. If this was not the case, adjustment of the gear or drive belt ratio could be done to keep the flow rate constant.

The four-pole 5-Hp (3.7 kW) motor of Table 4 was tested in the doffing operation in a textile plant. At this plant, the available voltage at the time of the tests was about 345 V. The hourly rate of energy consumption decreased from 1.95 to 1.68 kWh comparing the aluminum rotor to the copper. This translates to an annual energy savings of 2365 kWh. Power costs are generally high in India and are $0.109 per kWh at the location of this textile plant. Annual electricity cost would be reduced by $265.00. The initial cost of the copper rotor version of the motor was $167.08 which was 10.35% higher than the aluminum rotor motor. The payback period for the extra investment in the high efficiency copper rotor motor is only 22 days.

These results generated by motor manufacturers in India on motors where die-cast copper has been simply substituted for the aluminum in the rotor with no design modifications support the conclusions for similar material-substituted motors previously reported [3,4]. In this latest contribution, the several loss components have not been measured and therefore there is less information for the designer in attempting to optimize a motor for the high electrical conductivity of copper. The field tests support the supposition that improved performance and energy savings will result from using improved motors and that the payback period for the more expensive motor is very short. Having demonstrated that the copper rotor consistently results in increased electrical energy efficiency and lower operating temperature and with solutions to the manufacturing and die-casting tool life problems in hand, the industry is moving on to the problem of design of the motor as a whole and the rotor slots in particular for copper's high conductivity.

Sew-Eurodrive experience

SEW-Eurodrive has been active in an extended effort to design the motor to optimally use copper in the rotor. In April 2003, this company announced the availability of a range of EFF1 motors. Motors to 50 Hp (37 kW) are now available.The higher efficiency had been obtained in large part by employing electrical grade copper in the rotor although stator lamination and winding designs were also modified. These modifications succeeded in raising efficiency over the entire load spectrum while at the same time maintain torque at critical points on the torque-load curve including starting torque. This section presents the major design considerations and results of motor performance tests by IEEE standard 112B for 1.1 and 5.5 kW motors at both 50 and 60 Hz.

Table 7 presents efficiency data for 1.1 and 5.5 kW SEW aluminum and copper rotor motors. Comparison of these motors is especially interesting because two different design concepts have been employed for the 1.1 and the 5.5 kW copper motors. The 1.1 kW motors essentially have the same layout of stator and rotor laminations. Aluminum rotor bars have simply been replaced by die-cast copper but the lamination material is of a higher grade. In contrast the high efficiency DVE132S4 (5.5 kW) has a completely new lamination and winding design.

The data in Table 7 shows that the copper rotor leads to a significant increase in efficiency while maintaining the outer motor dimensions standard for aluminum—regardless of design.

The advantages of design modifications will became clearer when starting behavior is discussed below.

Table 7. Full-load efficiencies according to IEEE 112-B for high efficiency motors DTE/DVE-series and standard efficiency motors DT/DV-series

	50 Hz	60 Hz
Copper rotor motors		
DTE90S4 – 1.1 kW	82.8%	84.1%
DVE132S4 – 5.5 kW	88.1%	89.7%
Aluminum motors		
DT90S4 – 1.1 kW	75.7%	77.4%
DV132S4 – 5.5 kW	84.8%	86.6%

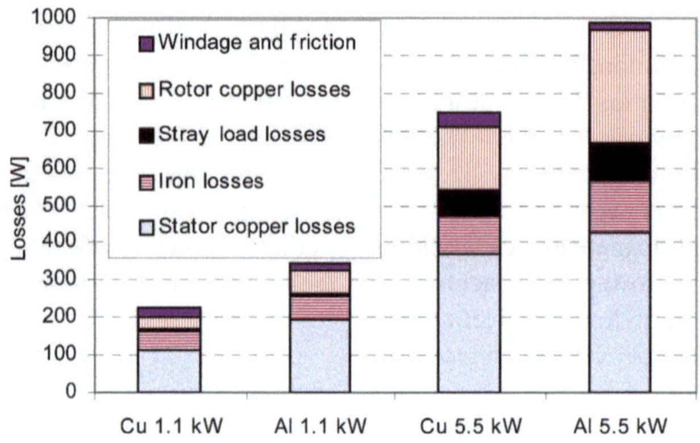

Figure 1. Loss distribution at 50 Hz.

In order to evaluate the efficiency contribution of the copper rotor, Fig. 1 shows the loss distribution for both motor ratings at 50 Hz. Fig. 2 contains the same data for 60 Hz operation.

The graphs clearly show that the main effects arise from reduced rotor losses. Especially for the 1.1 kW motor at 50 Hz operation, a decrease of more than 50% in rotor copper losses was observed. Because of the diagram scaling, the effect for 1.1 kW at 60 Hz does not show clearly, but indeed a reduction of rotor losses from 39 to 27 W was observed which is a drop of more than 30%. Since lower losses also lead to decreased operating temperatures, stator copper losses are also reduced.

A loss component which becomes more and more important with increasing power ratings are stray load losses (SLL). In Fig. 3 these losses for the motors of this study are compared. Generally one observes that copper motors have lower SLL than their aluminum counterparts except for the 1.1-kW/60-Hz measurement where the SLL per unit input power is 0.57% for aluminum and 0.7% for copper. This might be due to a poor correlation in SLL

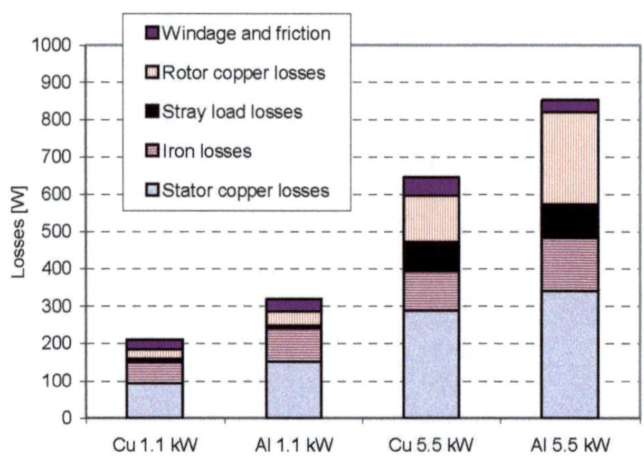

Figure 2. Loss distribution at 60 Hz.

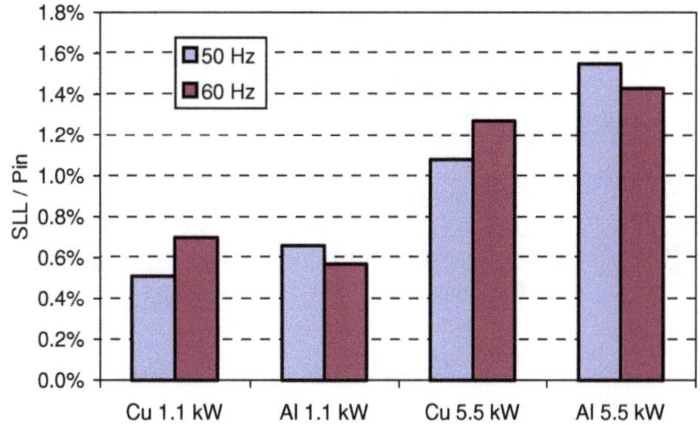

Figure 3. Stray load loss per input power for 50 and 60 Hz.

estimation. In the aluminum case a correlation coefficient of 0.95 was calculated whereas all other measurements exhibit a coefficient of about 0.98 to 0.99.

In industrial applications, it is quite common that drives do not run at full load at all times. As a consequence full load efficiencies are not the "one and only" and rather partial load efficiencies must also be taken into account. For that reason Fig. 4 shows the dependence of efficiency on output power.

It can be stated that even in the partial load regime the efficiency of the copper rotor motors stays above the corresponding standard efficiency aluminum motors. On the other hand, the efficiency drop for output powers greater than 100% is smaller than it is for aluminum motors. This is due to the lower temperature rise of the high efficiency motor and therefore these motors have more thermal reserves which support good overload capabilities.

If aluminum bars are simply substituted by copper bars (the 1.1 kW motors for example, as mentioned above) the breakdown slip s_k becomes lower since $s_k \sim R_2$. Focusing on starting conditions, this approach leads to decreased starting torque and higher starting current. In Fig. 5, torque-speed and current-speed curves for both 1.1-kW motors are compared. The

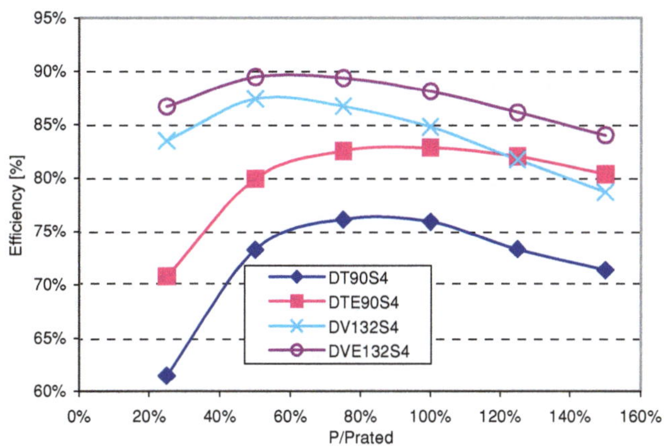

Figure 4. Efficiency dependence on output power (50 Hz only).

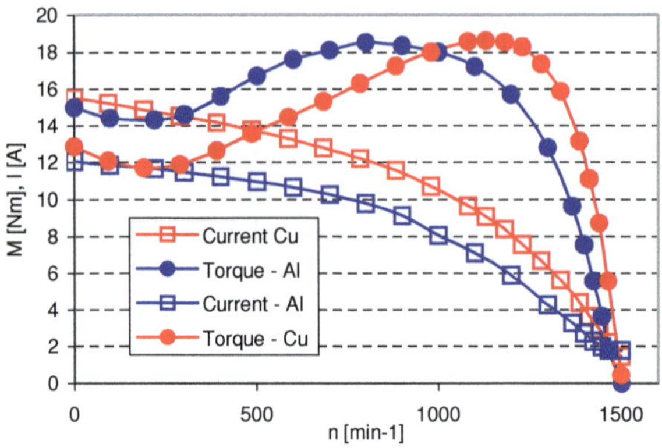

Figure 5. Torque-speed and current-speed curves for 1.1 kW motors. Standard efficiency aluminum motor (blue); copper high efficiency motor (red).

starting torque of the copper motor is 15% below that of the aluminum motor but well above two times rated torque. On the other hand, starting current is increased by about 30%. But the absolute numbers are still controllable and far from being critical. For that reason only minor design changes had been necessary for 1.1 kW motors.

The situation is different for motors of higher power rating where starting currents become more and more critical. Therefore a completely new lamination design was developed for all SEW high efficiency motors above 3 kW. The curves in Fig. 6 display the results for the 5.5-kW motor.

Again the R_2 effect with lower breakdown slip and a steeper torque curves is obvious. But comparing the starting conditions, currents are nearly of the same magnitude, despite the lower rotor bar resistance. On the other hand the starting torque is approximately 20% lower but this was indeed a desired effect, since lower, but sufficient starting torque is beneficial for gear box life. Sufficient, starting torque has a positive effect on gear box lifetime.

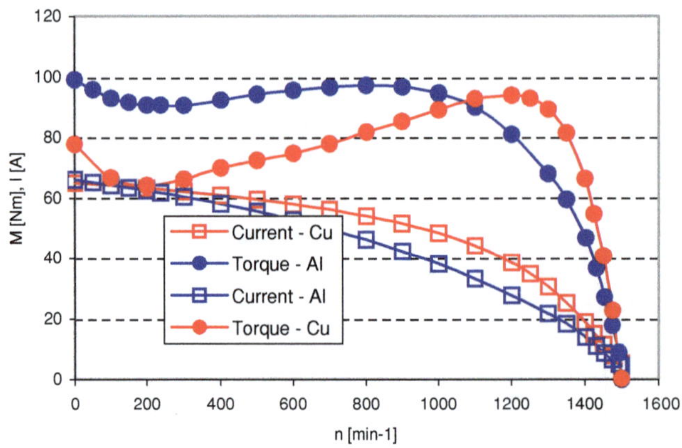

Figure 6. Torque-speed and current-speed curves for 5.5 kW motors. Standard efficiency aluminum motor (blue); copper high efficiency motor (red).

Porosity control in copper die castings

Copper is a very fluid metal and, apart from its high melting temperature, is readily die cast. High pressure die casting is the most economical process to form the squirrel cage of the induction motor rotor, but the high entrance rate of liquid metal through the gates generally results in some trapped air porosity in the casting. Despite the potential for porosity, rotors tested by motor manufacturers were remarkably easy to balance and stray load losses were reduced compared to the aluminum rotor. Both factors seemed to indicate the absence of large pores in the copper cage. Some larger rotors cast later were found to be difficult to balance and sectioning of the end rings revealed large pores. Porosity was as much as 25% in some castings and 8 to 10% in others. These findings prompted an investigation of the origins of the porosity and means to eliminate formation of large pores. This work is fully described elsewhere [5] and is summarized here.

Flow 3D software using computational fluid dynamics methods was used to simulate metal flow into the cavity. These were analyzed to identify shot speed—time profiles that would cause large pores in the end rings or conductor bars and profiles that would eliminate large pores in favor of uniformally dispersed small pores.

Simulation of the shot profile used in casting many rotors successfully predicted the large end ring pores. This baseline shot profile used to die cast many rotors extended the initial slow plunger speed so that about 10% of the gate end ring was filled before transition to the fast shot speed and completion of fill. Sections of end rings typical of this baseline shot profile are shown in Fig. 7. It is noteworthy that the large porosity was always confined to the end rings. Copper rotors machined to expose the conductor bars revealed only pin hole porosity in the conductor bars as shown in Fig. 8. The significant result from the model simulations was the discovery that slow prefill of the die cavity beyond the gates of 40% to as much as 55% was predicted to be a strategy to consistently eliminate large trapped air pores in the end rings.

Experimental runs to test the prediction of the modeling were then conducted. The shot profile was varied so that the speed transition occurred below the gates about half way up the runner and at prefills of 33% and 55%.

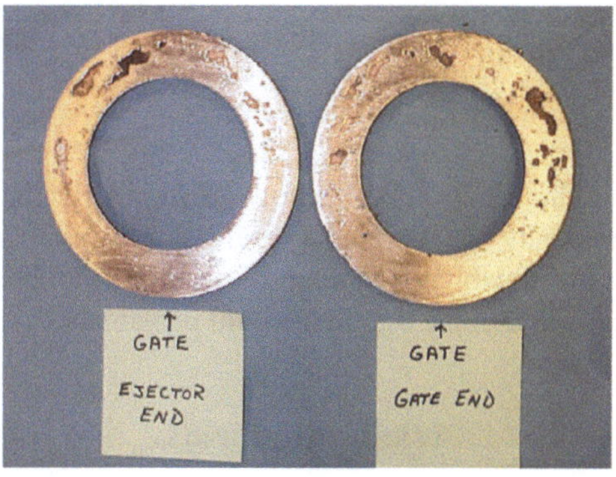

Figure 7. Photographs of sectioned end rings from copper rotors typical of baseline die-casting conditions.

Figure 8. Photograph of die-cast copper rotor turned on the OD to expose the conductor bars. Trapped air bubbles are not seen in the bars but are clearly visible in the end ring.

Figure 9. Photographs of sectioned end rings with 55% prefill. Ejector end ring on left; gate end ring on right.

Results are shown in the sawed cross sections of Fig. 9 for the 55% prefill. Porosity was seen to decrease markedly with increasing prefill compared to acceleration before the metal reaches the gate. Presumably the amount of prefill cannot be increased indefinitely. Additional experiments to determine the limit would be valuable.

Acknowledgments

The pilot project for producing and laboratory and field testing of copper rotor motors in the motor and pump manufacturing cluster at Coimbatore, Tamil, India was initiated by the International Copper Promotion Council, India and was in part supported by the International Copper Association Ltd. (ICA) and in part by a grant from the Small Scale Industries Development Bank of India/Technology Bureau of Small Industries fund of Nextant under the USAID Eco Project. The Copper Development Association Inc. (CDA) provided technical support. ICA and CDA provided the funding and technical support for the ongoing development of the copper motor rotor including the die-casting porosity studies summarized in this paper. The contributions of W.G. Walkington to the 3D modeling and of S.P. Midson to the die-casting trials are gratefully acknowledged.

References

[1] D.T. Peters, J.G. Cowie, E.F. Brush Jr., S.P. Midson, "Advances in Pressure Die Casting of Electrical Grade Copper", Amer. Foundry Society Congress Paper No. 02-002, Kansas City, MO, 2002.

[2] D.T. Peters, J.G. Cowie, E.F. Brush Jr., S.P. Midson, "Use of High Temperature Die Materials and Hot Dies for High Pressure Die Casting Pure Copper and Copper Alloys", Trans. of the North Amer. Die Casting Assoc. Congress, Rosemont, IL, 2002.

[3] J.G. Cowie, D.T. Peters, D.T. Brender, "Die-Cast Copper Rotors for Improved Motor Performance", Conference Record of the 49th IEEE-IAS Pulp and Paper Conference, Charleston, SC, June 2003.

[4] E.F. Brush Jr., J.G. Cowie, D.T. Peters, D.J. Van Son, "Die-Cast Copper Motor Rotors: Motor Test Results, Copper Compared to Aluminum", Trans. of the Third International Conference on Energy Efficiency in Motor Driven Systems (EEMODS), Treviso, Italy, September 2002, pp. 136–143.

[5] D.T. Peters, S.P. Midson, W.G. Walkington, E.F. Brush Jr., J.G. Cowie, "Porosity Control in Copper Rotor Die Castings", Trans. of the North Amer. Die Casting Assoc. Congress, Indianapolis, IN, 2003.

III-2.1. PERFORMANCE ANALYSIS OF A DOUBLY FED TWIN STATOR CAGE INDUCTION GENERATOR

F. Rüncos[1], R. Carlson[2], N. Sadowski[2] and P. Kuo-Peng[2]

[1]WEG MÁQUINAS, C.P. 3000, 89250-900, Jaraguá do Sul-SC, Brazil

fredemar@weg.com.br

[2]GRUCAD-UFSC, C.P. 476, 88040-900, Florianópolis-SC, Brazil

rcarlson@grucad.ufsc.br, nelson@grucad.ufsc.br, patrick@grucad.ufsc.br

Abstract. This paper analyzes design and performance aspects of a brushless double fed cage induction generator as an economic and technical alternative to wind power generation. It focuses on the main design criteria and on performance analysis to establish its behavior in load condition. The performance of a 15 kW prototype, comprising torque, current, efficiency, and power factor, is compared to simulation results and to other types of machines as synchronous and wound rotor induction machines. Vibration analyses are performed and experimental results are shown.

Introduction

The increasing interest in wind power generation directs the study and development of several alternatives of gearless electrical generators that operate at variable speeds. One of these alternatives consists of a Doubly Fed Twin stator Squirrel Cage three-phase Induction Generator (DFTSCIG), as shown in Fig. 1, because when it is doubly supplied its performance presents certain features of practical interest. By using an appropriate drive, it is possible to control the induction machine to operate as generator working above the synchronous speed as well as under the synchronous speed. This is especially convenient when a variable speed prime mover is used, as is the case of wind turbines. This machine has been studied up to these days only in small power ranges, not allowing conclusions about its ability to properly operate at larger power range systems as required in a modern Wind Power Station [1,2,5]. Thus, to evaluate its capabilities it is important to use pertinent analytical models to aid in the machine design and to have a better insight on its peculiar characteristics mainly in what concerns the rotor cage [3,4]. To evaluate its capabilities it is important to make a performance analysis in order to verify its behavior in different load conditions. This paper focuses firstly on machine operation and main design aspects and secondly on steady-state and dynamic analytical models that enable the efficient prediction of the DFTSCIG performance. Experimental results are presented and discussed. An analysis of the DFTSCIG vibration behavior is presented and discussed when compared to experimental results.

S. Wiak, M. Dems, K. Komęza (eds.), Recent Developments of Electrical Drives, 361–373.

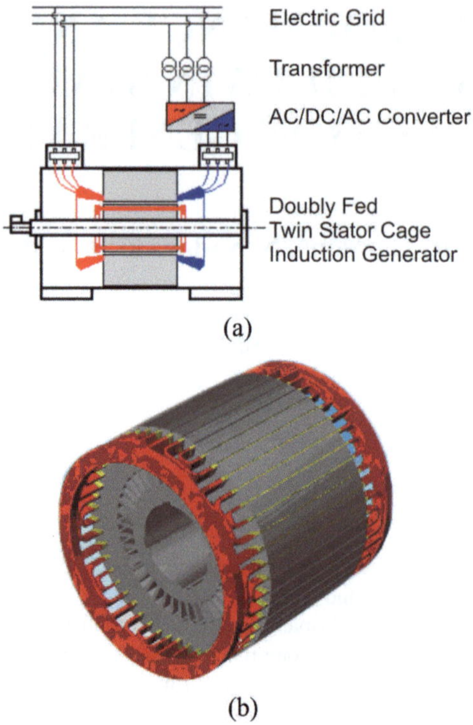

Figure 1. (a) Grid connection of the DFTSCIG. (b) Nested rotor cage.

Principles of operation

The DFTSCIG is an induction machine with a main three-phase winding with $2p_p$ poles directly connected to the electric grid, and a three-phase auxiliary winding with $2p_a$ poles connected to the electrical grid through a vector-controlled converter (Fig. 1). The electrical connections shown in Fig. 1 allow the control of the torque, the speed, and the power factor of the main winding by the converters connected to the auxiliary winding.

The special cage, shown in Fig. 1(b), is designed with inner loops to reduce the harmonic content of the flux in the air gap [3,4]. The advantage of this system is the fact that it is compact and brushless. The operation of this machine depends greatly on the rotor construction with that special cage [6,7].

The fundamental of the air-gap induction wave generated by the main winding induces a current density in the cage, with a frequency f_g calculated by:

$$f_g = f_p - p_p f_m \tag{1}$$

where f_p is the main winding frequency and f_m is the shaft mechanical frequency, both in Hertz.

In the auxiliary winding is induced a current density with a negative phase sequence with a frequency f_a in Hertz, given by

$$f_a = -\left\lfloor f_p - (p_p + p_a) f_m \right\rfloor \tag{2}$$

Therefore, the mechanical frequency of the shaft of the machine in hertz can be calculated by:

$$f_m = \frac{f_p + f_a}{p_p + p_a} \tag{3}$$

This equation shows that it is possible to control the speed of the DFTSCIG by changing the frequency of the imposed voltage on the auxiliary winding [2].

The frequency converter connected to the auxiliary winding, as shown in Fig. 1, not only imposes the frequency, but also controls the amplitude and phase of the voltage applied to the auxiliary winding, allowing the complete control of the DFTSCIG.

When the frequency induced in the auxiliary winding f_a is null, the machine is running at its natural synchronous speed f_{sn} [6,7].

Design criteria

Physically the DFTSCIG consists of two three-phase windings sharing the same stator magnetic core. To avoid the magnetic coupling between these windings, the number of poles of the main winding $2p_p$ and of the auxiliary winding $2p_a$ must have a Maximum Common Divisor which divides the two numbers of poles giving as a result an odd number for one of them and an even number for the other.

To avoid also the unbalanced electromagnetic pull, the difference between the two numbers of pole pairs must obey the relation [3,4]:

$$\left|p_p - p_a\right| > 1 \tag{4}$$

The main winding generates a set of induction harmonic waves in the air-gap of the machine with the following numbers of pole pairs:

$$\nu_p = p_p \left(1 + M_p \frac{g_p}{c_p}\right) \tag{5}$$

where M_p is the number of the phase belts per pole pair of the main winding; c_p is the fractional part of the main winding, and $g_p = 0; \pm1; \pm2; \pm3; \pm4 \ldots$ assumes integer values from $-\infty$ to $+\infty$.

The auxiliary winding is able to generate air-gap harmonic induction waves with the following number of pole pairs:

$$\nu_a = p_a \left(1 + M_a \frac{g_a}{c_a}\right) \tag{6}$$

where M_a is the number of the phase belts per pole pair of the auxiliary winding; c_a is the fractional part of the auxiliary winding and $g_a = 0; \pm1; \pm2; \pm3; \pm4 \ldots$ assumes integer values from $-\infty$ to $+\infty$.

To guarantee the magnetic uncoupling between the main and the auxiliary windings, ν_a and ν_p must obey the relation:

$$\nu_p \neq \nu_a \tag{7}$$

A good performance is obtained when the N_{pg} rotor bars produce N_{pg} poles, which couple the main and auxiliary windings producing additive torques.

The cage is capable to generate induction waves in the air gap with ν_g pole pairs given by:

$$\nu_g = \nu_p + g_g N_{pg} \tag{8}$$

where $g_g = 0; \pm 1; \pm 2; \pm 3; \pm 4 \ldots$ assumes integer values from $-\infty$ to $+\infty$.

Taking these considerations in account, according to [6,7], the number of bars N_{pg} of the cage can be calculated by:

$$N_{pg} = p_p + p_a \tag{9}$$

Equation (9) gives us a rule of how to choose the number of rotor cage bars. To minimize the harmonic content each pole of the cage may be constructed not only with one bar but with several loops, as shown in Fig. 1(b).

Analytical modeling

Dynamic model

The analytical dynamic model is obtained by transforming the equations written in machine variables into equations written in an arbitrary reference frame [6]. The stator circuit is considered fixed to the stationary reference θ_{p1} and all machine variables (rotor and auxiliary winding parameters) are referred to the main stator winding [6].

The stator circuits of the auxiliary winding are physically fixed to the stator (stationary), but in order to consider the cascade effect in our dynamic model, we are forced to admit that their axes rotate with an angular speed ω_{a1} electric rad/s that represents the angular speed of the stator circuits of the auxiliary winding and is given by:

$$\omega_{a1} = \left(p_p + p_a\right)\omega_m \tag{10}$$

In (10), ω_m represents the mechanical angular speed of the rotor; p_p and p_a are the number of poles of the main winding and the auxiliary winding, respectively. By transforming the equation system to the arbitrary reference frame, we obtain the following set of equations:

$$\begin{bmatrix} [u_{pqdo1}] \\ [0] \\ [u'_{aqdo1}] \end{bmatrix} = \begin{bmatrix} [R_{p1}] & [0] & [0] \\ [0] & \{[R'_{p2}]+[R'_{a2}]\} & [0] \\ [0] & [0] & [R'_{a1}] \end{bmatrix} \begin{bmatrix} [i_{pqdo1}] \\ [i'_{pqdo2}] \\ [i'_{aqdo1}] \end{bmatrix} + \begin{bmatrix} [\omega_{qdo}] & [0] & [0] \\ [0] & [\omega_{qdo} - \omega_{p2}] & [0] \\ [0] & [0] & [\omega_{qdo} - \omega_{a1}] \end{bmatrix} \begin{bmatrix} [\lambda_{pdq1}] \\ [\lambda'_{pdq2}] \\ [\lambda'_{adq1}] \end{bmatrix} + \frac{d}{dt} \begin{bmatrix} [\lambda_{pqdo1}] \\ [\lambda'_{pqdo2}] \\ [\lambda'_{qdoq1}] \end{bmatrix} \tag{11}$$

The system of differential equations in (11) is solved by the fourth order Runge-Kutta method and, as a result, the dynamic behavior of the machine is obtained.

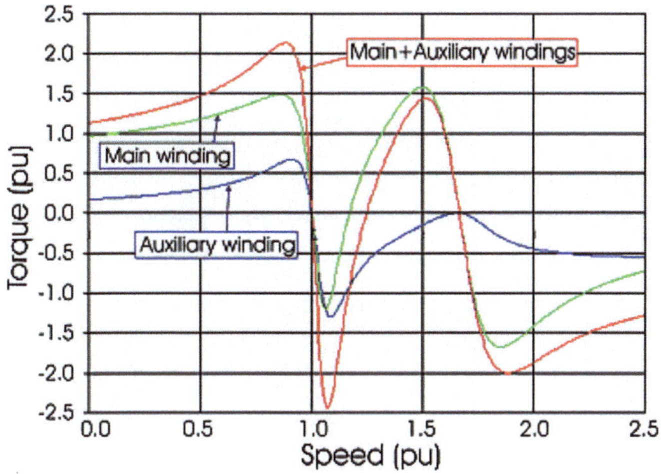

Figure 2. Steady-state torque-speed curves.

Steady-state model

The steady-state behavior is obtained using the equivalent circuit of the machine considering the plus connection of the two stator windings [6]. With this model it is possible to analyze the machine operating at steady state both as motor and as generator for any load condition with inductive and capacitive power factor.

Fig. 2 displays the curves of torque of the DFTSCIG, in steady state, obtained by the equivalent circuit model with the auxiliary winding short-circuited. This figure shows the torque developed by the auxiliary winding (8 poles), the torque of the main winding (12 poles), and the total torque which is, the sum of the main and auxiliary windings torque, proving the desired additive behavior of the torque. In the point of 1 pu rotating speed, the three torque pass by zero indicating that the machine is in its natural synchronous speed.

At 1.667 pu rotating speed, again the three torque pass by zero. At this point we have the synchronous rotation of the main stator winding.

In Fig. 2 it is also possible to observe that in the speed interval from 0 to 1 pu the machine behaves as motor because the torque is positive. From 1 to 1.667 pu speed the machine behaves first as generator (negative torque) until the torque of the main stator winding becomes positive again. Then the total torque also becomes positive and the machine behaves again as a motor. For speeds above 1.667 pu the three torques are negative again, and the machine works as a generator one more time.

This demonstrates that the DFTSCIG can work perfectly as motor or as generator, when controlled by the static converter, as shown in Fig. 1, in a speed range of ±30% around the 1 pu natural synchronous rotation.

This machine can be controlled through an external action over the auxiliary winding as was commented earlier. As the power electronics control is not yet available for the prototype, external resistances were connected to the auxiliary winding terminals to show that how it affects the torque vs. speed characteristics. Fig. 3 shows a set of total steady-state torque curves over 2.5 pu speed range with five external resistance steps.

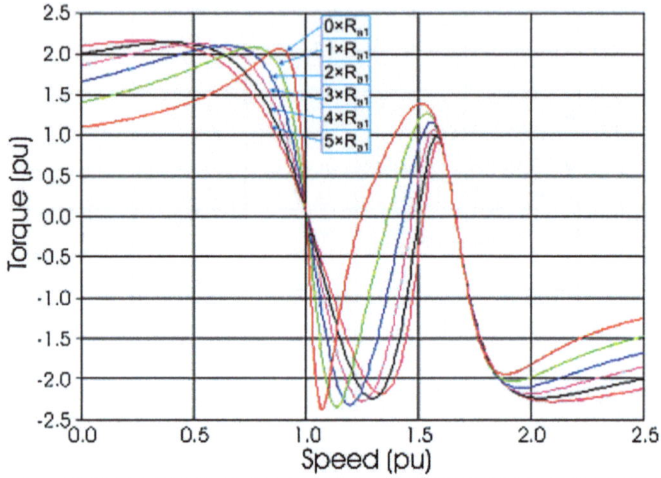

Figure 3. Total steady-state torque-speed curves with external auxiliary winding resistances.

Experimental tests

The prototype here analyzed is a 15 kW—440 V/760 V—60 Hz, 12 poles for the main winding and 8 poles for the auxiliary winding. The main stator winding is considered Y-connected and fed directly by three-phase balanced voltage sources 760 V_{RMS} – 60 Hz. The auxiliary winding is Y-connected, with its external terminals short-circuited or connected to external resistances.

The dynamic test was performed by applying a negative torque (motor torque) to the DFTSCIG shaft, this imposed torque being enough to drag it up to approximately two and half times the DFTSCIG natural synchronous speed. A Rotary Torque Sensor (RTS) was inserted between the DFTSCIG shaft and the dynamometer and its signals have been recorded directly by an analog plotter.

Comparison between simulation and experimental results

Fig. 4 show the transient torque obtained in the simulation of the DFTSCIG acceleration process using the dynamic model, with the time scale in seconds. We can identify in these dynamic results the instants the rotor passes through the natural synchronous speed ($t \cong 1.08$ s) and the main winding synchronous speed ($t \cong 1.45$ s). These characteristic points of the DFTSCIG operation were already identified in Fig. 2 for the steady-state regime.

Tables 1 and 2 show the machine performance when operating as a motor and as a generator, with the auxiliary winding terminals short-circuited.

Comparing the experimental results with the analytical simulation, we observe that they present a good agreement.

The analysis of the performance data presented in Tables 1 and 2, with the machine operating with the auxiliary winding short-circuited, makes very clear that the main issue of the DFTSCIG concerns its power factor. The low value of the power factor is a direct consequence of the low number of rotor cage bars (cage pole number). The nested loops of

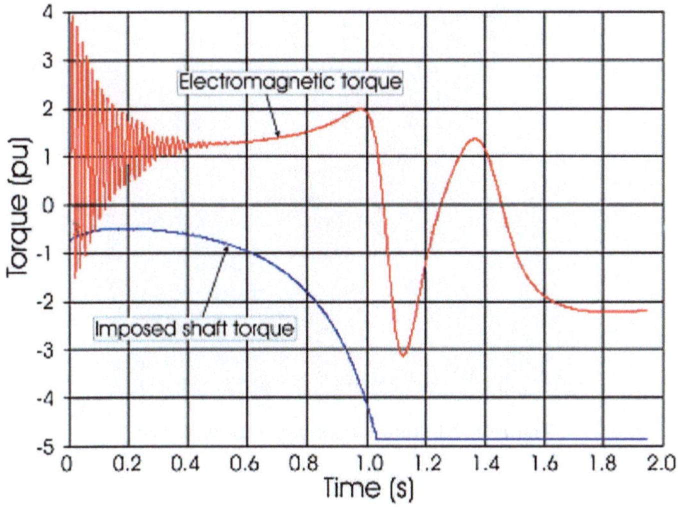

Figure 4. Dynamic torque simulation result.

the rotor are intended to reduce this effect. The number of bars given by (9) is a necessary condition to the operation of the DFTSCIG. One way to improve the power factor is to substitute the rotor cage by a wounded cage with multiturn windings, as shown in Fig. 5 [8].

Figs. 6–9 show the main winding experimental current and torque vs. speed curves.

Table 1. DFTSBIG performance data—100% loaded

	Motor		Generator	
	Analytical	Test	Analytical	Test
Speed (rpm)	351.3	355.1	370.8	366.3
Torque (Nm)	433.6	403.0	609.0	578.0
I_{p1} (A_{RMS})	43.1	48.1	51.5	59.4
Power factor	0.37	0.33	0.22	0.20
Efficiency (%)	74.1	74.2	63.4	67.6

Table 2. DFTSBIG performance data—75% loaded

	Motor		Generator	
	Analytical	Test	Analytical	Test
Speed (rpm)	354.3	356.6	368.3	365.9
Torque (Nm)	308.7	300.0	474.8	500.0
I_{p1} (A_{RMS})	41.1	46.4	47.5	55.7
Power factor	0.31	0.27	0.18	0.16
Efficient (%)	69.1	68.9	61.4	58.7

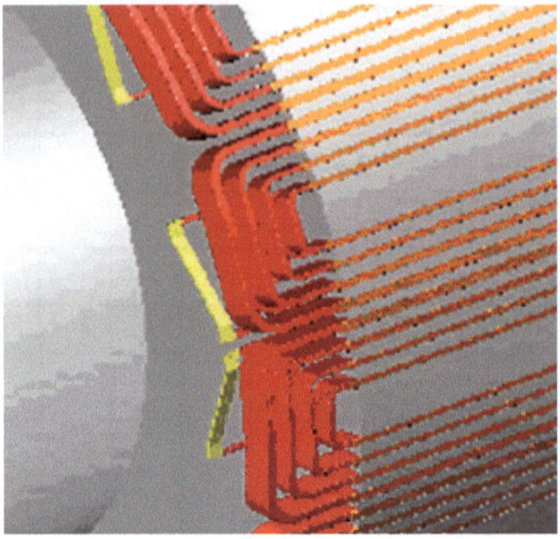

Figure 5. Multiturn wounded rotor windings.

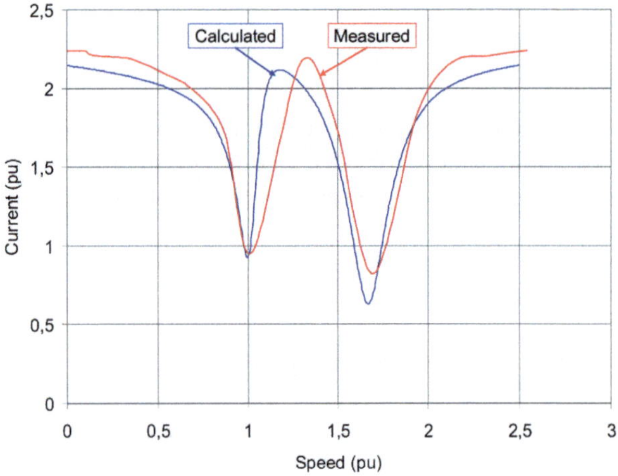

Figure 6. Total steady-state torque-speed curves without external auxiliary winding resistances.

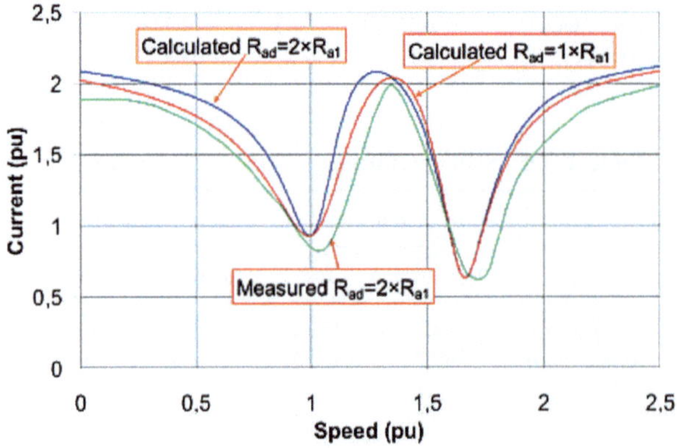

Figure 7. Main winding current vs. speed characteristics with external resistances connected to the auxiliary winding terminals.

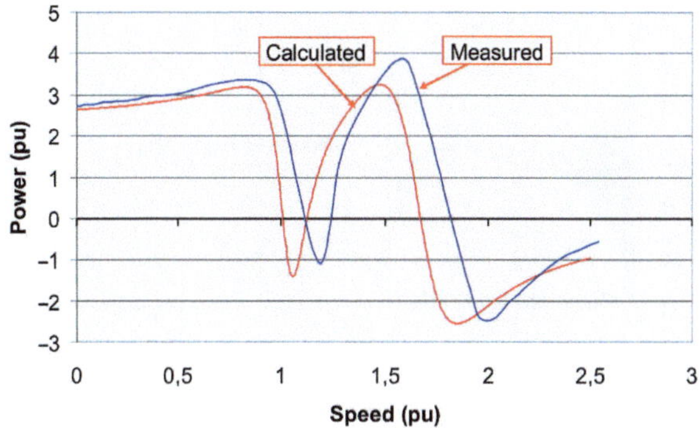

Figure 8. Main winding power vs. speed characteristics (Rad = 0 × Ra1).

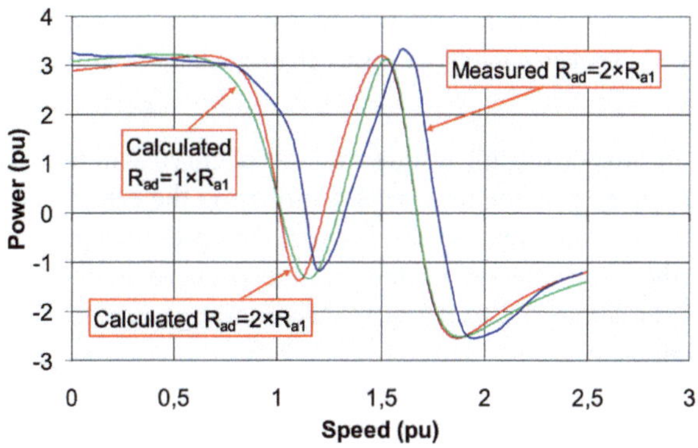

Figure 9. Main winding power vs. speed characteristics (Rad = 2 × Ra1).

Analyzing Figs. 6–9, we can see that the experimental curves are shifted to the right direction when compared with the analytical steady-state curves. This is due to the dynamic measurement method. This displacement is not observable in the Figs. 10 and 11 where the comparisons are made with the analytical dynamic curves.

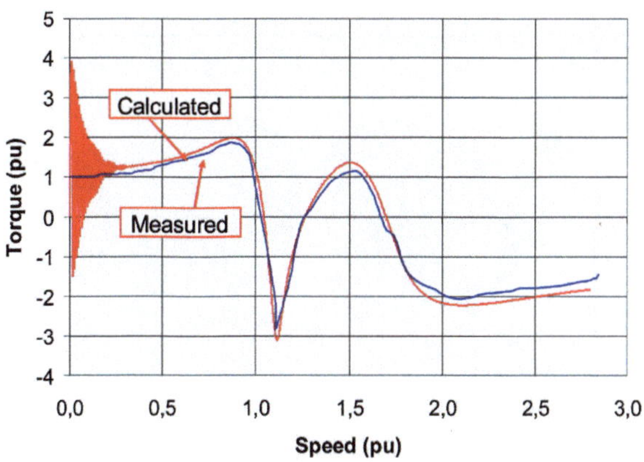

Figure 10. Dynamic torque vs. speed characteristics (Rad = 0 × Ra1).

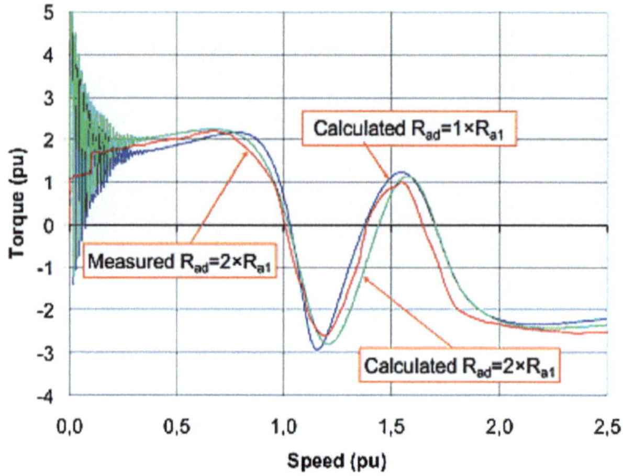

Figure 11. Dynamic torque vs. speed characteristics (Rad = 2 × Ra1).

Fig. 10 shows the experimental torque vs. speed curve without external resistances connected to the auxiliary winding terminals. Comparing the experimental torque × speed curves to the analytical simulation we observe that they present a good agreement.

Fig. 11 shows the experimental torque vs. speed curve with an external resistance (Rext = 2.00 × Ra1) connected to the auxiliary winding terminals, in comparison with analytical simulation curves.

Vibration analysis

For the vibration analysis of the DFTSCIG it is necessary to determine the induction waves generated in the air gap [7]. With these induction waves it is possible to calculate the mechanical forced vibrations due to the electromagnetic excitation of electrical machines regarding to structural vibration. This type of problem can accurately solved using the *Modal Superposition Method* [7].

The vibration measurements where performed on the four points indicated in Fig. 12.

Figure 12. Vibration measurement points.

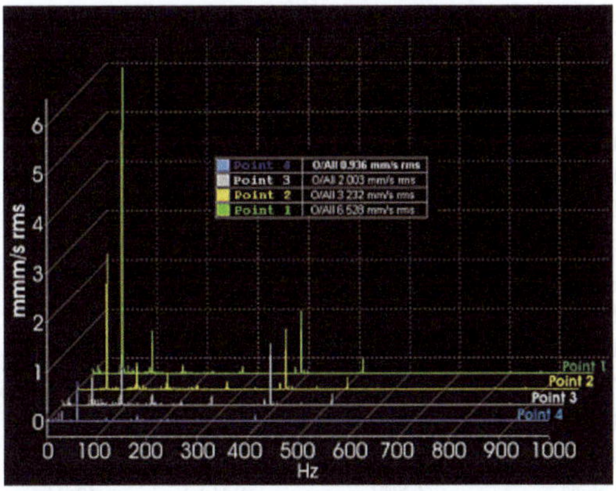

Figure 13. Vibration velocities spectrum measured in RMS.

Analyzing the measurement results shown in Fig. 13, we see that there are the main vibration frequencies are 60.625, 120, and 415.625 Hz with RMS amplitudes of 6.217, 0.863, and 1.258 mm/s, respectively. The frequency 60.625 Hz, whose vibration mode is r = 2 produces the maximum vibration amplitude measured at point 1.

The peaks at 60.625 and 120.0 Hz are a response to an electromagnetic forced excitation [7]. However, the peak at 415.625 Hz is a response to a mechanical excitation force (70 × 5.9375 Hz) and not an *electromagnetic force* produced by the machine currents.

Analyzing the entire measured vibration spectrum, the maximum vibration occurs at 60.625 Hz with excitation mode r = 2. This excitation wave force is produced by the fundamental induction harmonic wave $p_p = 6$ of 60 Hz frequency, generated by the main winding and the fundamental induction harmonic wave $p_a = -4$ of 0.625 Hz frequency generated by the auxiliary winding. This excitation force is dangerous for the structural vibration behavior because it depends of the load condition and must be considered in the machine design.

The rigidity and the damping factor of the lamination core is strongly dependent of the vibration mode, meaning that for low vibration mode the DFTSCIG can present severe structural vibration making the machine not operational. To avoid low vibration modes, the difference between the pole pair number of the main and auxiliary winding must be large. On the other hand, the lamination core must be designed with enough rigidity to support the electromagnetic excitation in acceptable level.

Comparative analysis

The prototype here presented is not a large machine, but based on the analysis performed above it is possible to believe that the DFTSCIG may be a solution for large wind power generators. Table 3 shows a comparison of this machine with other electrical machines, like Induction Squirrel Cage Machine (ISCM), Induction Wound Rotor Machine (IWRM), and Salient multi Pole Synchronous Machine (SPSM). Those machines were designed using classic industry methods. In large wind power stations one important issue is the multistage gearbox, which is expensive and presents maintenance problems. Low speed machines like

Table 3. Performance comparison

kW Volt	rpm	Machine	D_{e1} (mm)	L_{fe1} (mm)	*Eff* (%)	*PF* (%)	Cost (%)
15	360	DFTSCIG	480	420	75.0	31	100
		IWRM	770	200	83.5	66	400
440		ISCM	530	200	81.7	58	120
		SPSM	620	130	75.1	80	500
100	180	DFTSCIG	980	350	77.3	30	100
		IWRM	1,250	350	79.5	60	178
440		ISCM	980	350	85.1	59	100
		SPSM	1,300	300	81.8	80	225
1,000	120	DFTSCIG	3,250	300	93.0	32	100
		IWRM	3,250	250	93.2	54	110
690		ISCM	2,900	400	93.5	54	96
		SPSM	3,250	400	90.0	80	190
2,000	120	DFTSCIG	3,250	400	93.7	33	100
		IWRM	3,250	350	94.0	55	120
690		ISCM	2,900	575	93.5	55	00
		SPSM	250	00	9.9	0	80

those in Table 3 can work with one stage planetary gear witch is cheaper and requires less maintenance. All machines in Table 3 where designed for the same output and speed condition.

In this table D_{e1} is the outer diameter of the lamination core, L_{fe1} is the lamination core length, *Eff* is the efficiency, and *PF* is the power factor. The last column represents the relative manufacturer cost of these machines taking the DFTSCIG cost as the basis. Of all these machines, only the prototype of DFTSCIG 15 kW—440/760V was manufactured and tested. The other machines in the Table 3 are designed in real condition but not manufactured.

The small size machines, 15 and 100 kW, the IWRM and SPSM need a larger diameter to accommodate the rotor slots and rotor poles, respectively. The largest ones, 1,000 and 2,000 kW, present similar diameters, but the lamination core length for the IWRM is smaller indicating a higher torque density. The efficiency of all large size machines, 1,000 and 2,000 kW, are similar.

Comparing the costs on the last column, one can see that the DFTSCIG and the ISCM have very similar cost and much less than the synchronous machines SPSM.

As commented before, the power factor of the DFTSCIG is the smallest one. This low power factor can be improved through the static converters connected to the auxiliary winding. However, carrying this reactive power increases the converters ratings, but even so we believe the DFTSCIG could be cheapest solution when compared with the SPSM. The advantage of the DFTSCIG is the fact that it is a brushless solution and needs lower rating static converters because it is connected to the auxiliary winding which carries only about 40% of the machine main power.

Conclusion

A discussion about operational characteristics and design criteria of a DFTSCIG machine was presented. The simulation models have been validated by experimental results obtained from a prototype. The theoretical analysis has presented good results when compared with the experimental ones and also the performance analysis, in what concerns the torque, was quite satisfactory confirming the initial statements that the DFTSCIG may be an advantageous alternative to generate electric energy from wind power.

The main difficulty encountered during the design phase is related to the high harmonic content generated in the air-gap flux due to the small number of rotor cage bars. These harmonics create high leakage reactances. The loops introduced in the rotor cage help minimize this harmonic content, but showed not enough for the prototype. One solution that is being analyzed is to substitute the rotor cage by the by multiturn windings, as was shown in Fig. 5 This new rotor circuit will be analyzed using a finite elements model that can represent this multiturn winding in what concerns its physical connections and its distribution giving the real current distribution waveform and amplitude [6]. After this analysis and if this solution shows effective for improving the machine power factor, a new prototype will be manufactured.

Another important aspect is that windmill speed is quite low and then the DFTSCIG should have a large number of poles in the main winding. However, this increases the leakage reactances of the generator affecting its power factor and efficiency.

The results presented in this paper, although very important for the machine performance comprehension and development of design tools, are still preliminary. Continuous efforts will be engaged to improve the DFTSCIG performance.

Acknowledgment

Authors wish to thank WEG S.A. for the construction of the prototype and for the use of the testing facilities.

References

[1] Y. Liao, "Design of a Brushless Doubly-Fed Induction Motor for Adjustable Speed Drive Applications", Thirty-First IEEE/IAS Annual Meeting, San Diego, USA, pp. 850–855, 1996.

[2] R. Li, R. Spée, A.K. Wallace, G.C. Alexander, Synchronous drive performance of brushless doubly-fed motors, Trans. Ind. Appl., Vol. 30, No. 4, 1994.

[3] S. Williamson, A.C. Ferreira, A.K. Wallace, Generalized theory of brushless doubly-fed machine. Part 1: Analysis, IEE Proc. Elect. Power Appl., Vol. 144, No. 2, 1997.

[4] S. Williamson, A.C. Ferreira, A.K. Wallace, Generalized theory of brushless doubly-fed machine. Part 2: Model verification and performance, IEE Proc. Elect. Power Appl., Vol. 144, No. 2, 1997.

[5] A.R.W. Broadway, L. Burbridge, Self cascaded machine: A low-speed motor or high-frequency brushless alternator, Proc. IEE, Vol. 117, No. 7, 1970.

[6] F. Rüncos, R. Carlson, A.M. Oliveira, P. Kuo-Peng, N. Sadowski, "Performance Analysis of a Brushless Double Fed Cage Induction Generator", Nordic Wind Power Conference, Göteborg, Sweden, 2004.

[7] R. Carlson, F. Rüncos, A.M. Oliveira, P. Kuo-Peng, N. Sadowski, C.G.C. Neves, "Vibration Analysis of a Doubly-Feed Twin Stator Cage Induction Generator", Symposium on Power Electronics, Electrical Drives, Automation and Motion–Speedam, Italy, 2004.

[8] C. Fräger, Neuartige Kaskadenmaschine für Burstenlose Drehzahlstell-Antriebe mit geringen Strom-richteraufwand, Dusseldorf, VDI-Verlag Gmbh., 1995 (in German).

III-2.2. STATIC AND DYNAMIC MEASUREMENTS OF A PERMANENT MAGNET INDUCTION GENERATOR: TEST RESULTS OF A NEW WIND GENERATOR CONCEPT

Gabriele Gail, Thomas Hartkopf, Eckehard Tröster, Michael Höffling, Michael Henschel and Henning Schneider

Department of Renewable Energies, Institute for Electrical Energy Conversion, University of Technology Darmstadt, Landgraf-Georg-Straße 4, 64283 Darmstadt, Germany

ggail@ew.tu-darmstadt.de, thomas.hartkopf@ew.tu_darmstadt.de, etroester@ew.tu_darmstadt.de

Abstract. The Permanent Magnet Induction Machine, a new wind generator concept, is considered to be a highly efficient, low maintenance solution for offshore wind turbines. Static and dynamic measurements have been performed with a test machine. Due to the inherent soft behavior of that machine type compared to normal synchronous machines, no dynamic excitation is found during operation that might endanger the stability of the system. Results of static measurements show high efficiency and little reactive power consumption.

Introduction

The Permanent Magnet Induction Machine (PMIM) is an internally excited induction machine which combines the advantages of a permanent excited synchronous machine (PMSM) and an induction machine (IM). In 1992 Low and Schofield [1] first investigated this machine concept for wind power application for the following reasons: Working like an induction machine, the PMIM provides a soft grid connection so **no converter** is necessary. Additionally, through the use of a freely rotating magnet rotor in the air gap (Fig. 2) the reactive power demand is reduced so that the PMIM can be built to operate at low speeds like synchronous generators, and **no gear** needs to be installed.

According to the WindEnergy Study 2004 [2], steady growth in the wind energy market is forecasted in the next decade, and the industry aims to achieve an installed capacity of 150,000 MW wind power world wide by 2012.

That means the wind energy industry has an increasing demand for reliable solutions especially for offshore wind turbines. Ongoing projects, e.g. the Horns Rev Windpark in Denmark with drastic technical problems as mentioned in [3], show how important it is to have low maintenance concepts at less-accessible sites like those offshore. This means

S. Wiak, M. Dems, K. Komęza (eds.), Recent Developments of Electrical Drives, 375–384.

that the PMIM without gear and its oil together with a simple grid connection, without frequency converter, represents the ideal generator concept for future wind turbines.

The idea to use the PMIM as a wind generator concept was followed up by Epskamp et al. [4] and Hagenkort et al. [5] in recent years. Results of numeric and analytical calculations, together with dynamic simulation of a 2 MW generator, showed the PMIM to be a suitable and high efficiency solution for wind power applications. As mentioned in [4], the next step in this process was to validate the mathematical models and calculations with tests. Thus a PMIM test machine (Fig. 2) has been used to produce measurement data and gain first practical experience with that machine type. Due to a very different design of the test machine, comparisons with the planned 2 MW generator are avoided. However, the results can give a first impression of what the PMIM's behavior will be and on which aspects later designs should be focused. The PMIM's applicability as a wind generator will especially depend on how stable the system operates after dynamic excitations. Results of static and dynamic measurements are presented and discussed here.

New generator concept: the PMIM

The basic idea of this generator concept is a low maintenance solution for offshore wind turbines. To begin with the PMIM is considered to be an induction generator. That means we have an asynchronous rotating rotor with slip which leads to a soft behavior towards the grid during load changes in the wind. Therefore the error-prone and expensive frequency converter can be avoided. To avoid the need for gears the generator must be built with a large diameter and hence with a high number of poles. Since the large diameter and the high number of poles cause a large air gap/pole pitch ratio, a very high magnetizing current arises. Thus the PMIM has a second rotor mounted with permanent magnets which rotates freely in a widened air gap (Fig. 2). This rotor provides an additional flux in the air gap with its permanent magnets so that the demand for magnetizing current and hence for reactive power can be minimized.

The equivalent circuit of the PMIM in Fig. 1 reflects this phenomenon.

The large air gap/pole pitch ratio causes a small main reactance X_m compared to large leakage reactances $X_{\sigma s}$ and $X'_{\sigma r}$. This evokes a high magnetizing current, which is now limited by the internal voltage V_p which is induced by the permanent magnets. A basic theoretical analysis of the steady-state equivalent circuit is carried out in [6].

To understand the machine's dynamic behavior more clearly one can describe the interaction between stator, magnet rotor, and asynchronous rotor with a "model of four machines". Each interaction between the generator parts can be explained with a normal electrical

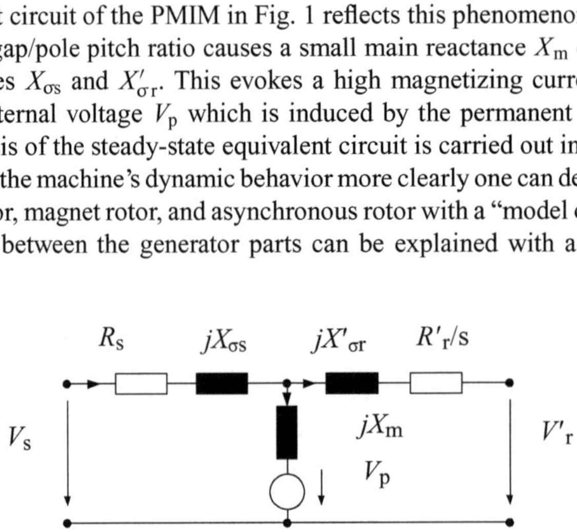

Figure 1. Equivalent circuit of the PMIM.

machine. The aluminum cylinder on which the magnets are fixed serves here as a damper circuit. The interaction can be seen as follows:

- Stator to Inner Rotor as **Asynchronous Machine**
- Stator to Magnet Rotor as **Synchronous Machine**
- Magnet Rotor to Inner Rotor as **Asynchronous Machine**
- Stator to Damper Circuit as **Asynchronous Machine**

However, the overall concept has two apparent drawbacks: Because no frequency converter is used, the wind turbine will be a fixed speed system. Nevertheless, from the results of Hoffman [7] one can draw the conclusion that highly sophisticated wind turbine systems like pitch-controlled variable speed power plants do not lead to a significant gain in energy capture at sites with high wind speed and low turbulence as is the case offshore. Another drawback is definitely the mechanical assembly. The bearing of now two rotors of very large size has to be managed and each component has to be designed to withstand the mechanical stresses.

The final 2 MW wind turbine generator might be designed with the following specifications [5]:

- Power factor $cos\varphi = 0.9$
- Stator Voltage $V_s = 840$ V
- Slip $s = 2.5\%$
- Turbine Speed $n = 33.3$ rpm
- Diameter $d = 5.2$ m
- Pole number $2p = 180$
- Permanent magnets NdFe

Test machine and test set up

Initially the design of the PMIM test machine is very different to the above specifications for financial and practical reasons. Fig. 2 shows the basic assembly of the test machine.

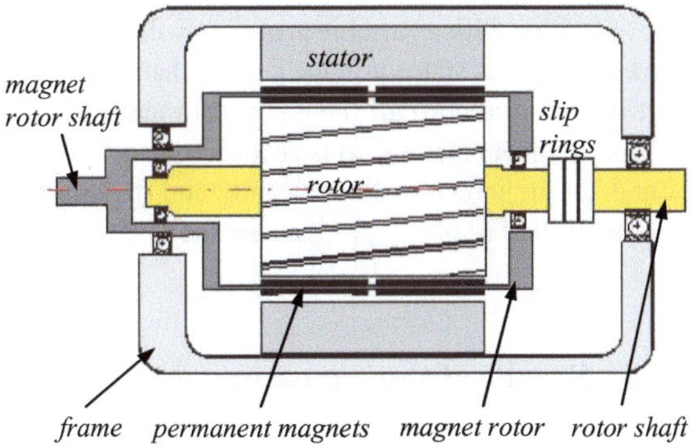

Figure 2. Cross section of the test machine assembly.

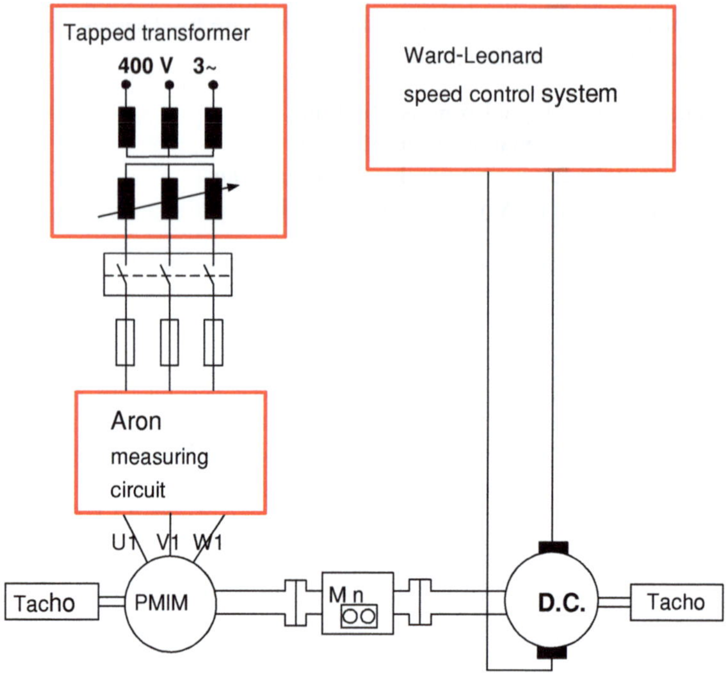

Figure 3. Test set up for the PMIM.

A standard 15 kW stator of a four pole asynchronous machine is combined with a 3 kW slip ring rotor, although the final solution will be realized with a squirrel cage rotor due to its simplicity and robustness. However, the slip ring rotor is advantageous for laboratory use, as rotor currents can be measured via the slip rings. This assembly provides a sufficiently large air gap to fit the freely rotating magnet rotor between the stator and rotor. The magnet rotor consists of an aluminum cylinder mounted with ferrites in four poles.

Regarding the results note that this test machine is not an optimized generator but can provide useful information about the machine concept.

Fig. 3 shows the test set up with which the PMIM test machine is examined. The PMIM is coupled with the driving engine, a D.C. machine of 103 kW. A Ward Leonard speed control system supplies the D.C. machine with a variable armature voltage. The PMIM is connected with a tapped transformer, so that its stator voltage can be changed for tests up to 400 V. An Aron measuring circuit is used to record all the electrical measurements and the transient data of the rotor speed and magnet rotor speed is recorded via a DMCplus amplifier.

With the experimental set up shown in Fig. 3, static and dynamic measurements were performed to figure out the PMIM's special properties. The first results are presented in the following paragraphs

Results of static measurements

Although the test machine is rather different to the planned design, it does show the desired effects of the generator.

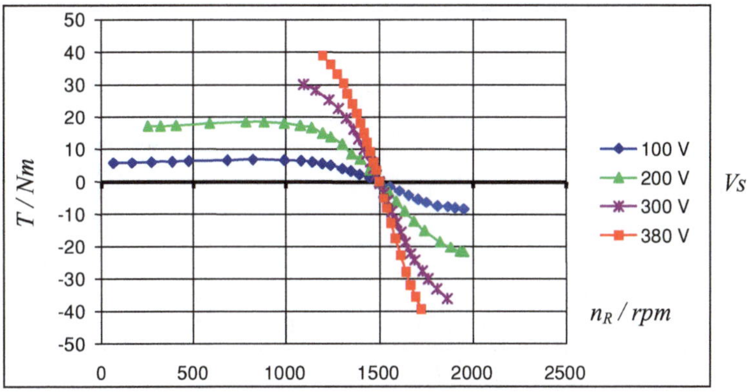

Figure 4. Torque T vs. rotor speed n_R (stator voltage V_s as parameter).

Fig. 4 shows the static torque curve vs. the rotational speed. A maximum stator current of approximately 40 A and a maximum speed of 2000 rpm were limiting factors so that there was no risk of damage to the test machine during operation. The load angle increased to maximum 20° so that the magnet rotor was stable during the test. The torque curves are plotted for four different stator voltages V_s between 100 and 380 V. Although the curves do not show a distinct breakdown torque they point out the PMIM's soft behavior like that of induction machines.

In order to measure the internal voltage of the permanent magnets V_p, the no-load test was performed as it is done for synchronous machines. At synchronous speed the permanent magnets induce a voltage V_p of 100 V which can be measured at the stator terminals. The influence of the internal voltage V_p becomes clear with the help of the next two figures.

The plot in Fig. 5 shows the current locus diagram of the PMIM at different voltages. The current locus represents the position of the current vector in the imaginary area.

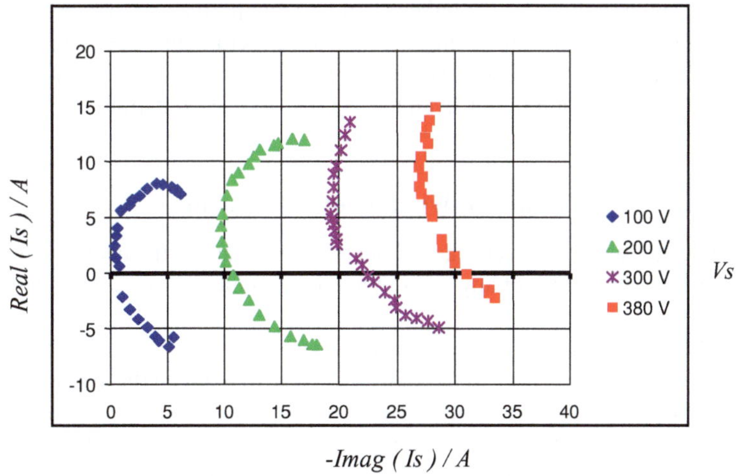

Figure 5. Stator current locus diagram (stator voltage V_s as parameter).

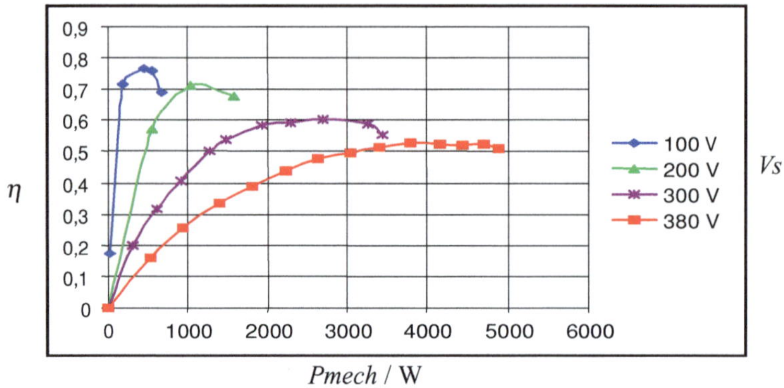

Figure 6. Efficiency η vs. load (stator voltage V_s as parameter).

Here the shape of the current locus deviates from the circle, as it was predicted in [5]. One can further see that if the internal voltage V_p is in the range of the stator voltage V_s, as it is for the blue curve, the magnetizing current is small and the current locus is displaced to the left. If the stator voltage increases the permanent magnets have less influence and the current locus diagram moves to the right towards higher reactive currents.

Fig. 6 plots now the efficiency vs. the load, once again for four different stator voltages. The highest efficiency is found at a stator voltage of 100 V where the internal voltage V_p is as big as V_s and no reactive power consumption is necessary. According to [5] the PMIM has a high efficiency at partial load, which is very suitable for wind turbines, which operate most often at partial load. But with a smaller ratio of V_p/V_s the internal voltage has less impact and the efficiency curves resemble more a normal induction machine with a large air gap.

So we can draw the conclusion that the PMIM reaches its desired characteristics of small reactive power consumption together with high efficiency when the internal voltage V_p is in the range of the stator voltage V_s. With the use of stronger permanent magnets, the internal voltage V_p can match higher stator voltages. We can further assume that with the help of a tapped voltage transformer the demand of reactive power and hence the power factor *cos* φ can be changed within certain limits, even towards capacitive excitation.

Results of dynamic measurements

Firstly, dynamic measurements were carried out to ensure that the magnet rotor in particular causes no dynamic oscillations or any other dynamic excitations that might endanger stable operation.

Figs. 7 and 8 show the result of the following test.

Initially the asynchronous rotor and the magnet rotor rotate with the synchronous speed of 1500 rpm. The peak value of the stator voltage is 100 V as expected and the stator current is ideally zero. At around 700 ms the stator terminals are connected to the grid. In Fig. 7 stator voltage V_s and stator current I_s are plotted vs. the time. Fig. 8 shows the magnet rotor speed n_{MR}.

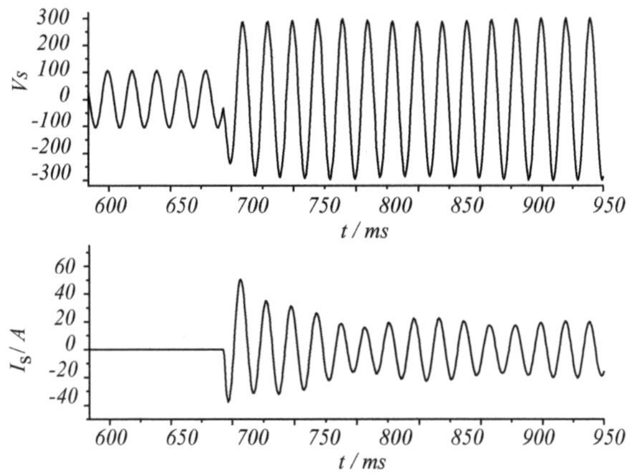

Figure 7. Stator voltage V_s and stator current I_s vs. time after switching on the grid voltage.

At 700 ms the stator voltage V_S changes to the value of the grid voltage with a peak value of here 300 V. In the plot of the stator current we can recognize decaying electric transients until around 800 ms. Later, we find oscillations caused by the transient condition of the magnet rotor.

The next plot in Fig. 8 shows the magnet rotor speed over a longer period up to 2000 ms.

The magnet rotor speed oscillates sinusoidal around synchronous speed which is caused by an oscillation of the magnet rotor around the static load angle at no load. Because of the aluminum cylinder which can be regarded as a damper circuit, the oscillations have exponential decaying amplitude until the system is stable again. If the PMIM will work as a wind generator this test represents a starting procedure for the wind turbine.

Finally the results of the acceleration process from standstill are presented, after the PMIM is directly connected to the grid. Fig. 9 compares the rotor speed and the magnet rotor speed. Because of the higher inertia the asynchronous rotor (blue line) accelerates slower than the magnet rotor. The magnet rotor, like a synchronous machine, can only perform an asynchronous run up with the help of its damper circuit. Because the stator field rotates much faster than the field of the permanent magnets during acceleration the magnet rotor sees an oscillating torque. In the plot we can find these oscillations transferred to the magnet rotor speed.

The smaller the slip of the magnet rotor becomes and thus the smaller the frequency of the oscillating torque, the stronger is its influence. Thus the amplitude of the speed oscillations

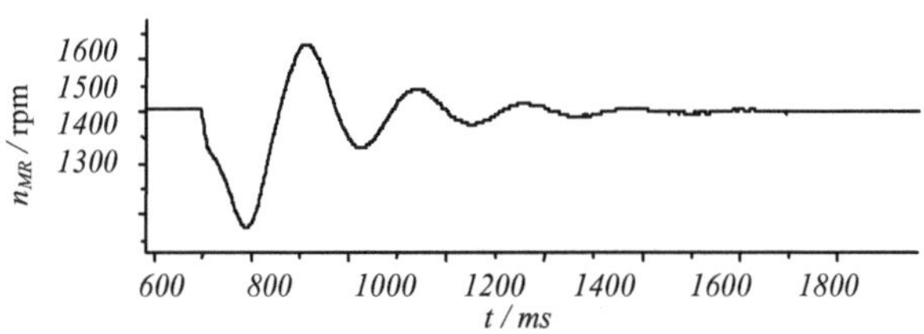

Figure 8. Magnet rotor speed $n_{\rm MR}$ vs. time after switching on the grid voltage.

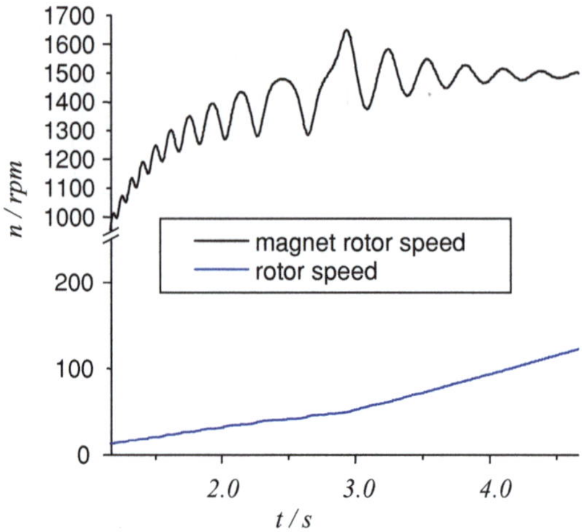

Figure 9. Rotor speed and magnet rotor speed vs. time during acceleration from standstill.

becomes larger. Before the magnet rotor reaches synchronous speed, the oscillations have another characteristic after the point of synchronization at around 2.75 s. Here the time span of higher speed is longer than the time span of lower speed in each oscillation period, because the magnet rotor field is lagging. After synchronous speed is reached, we find again the transient sinusoidal oscillations explained in Fig. 8. Once the magnet rotor reaches synchronous speed, its permanent magnets support the field in the air gap. From that point on the inner rotor can accelerate faster, as can be seen in the Figure.

The point of synchronization and the impact of the magnet rotor oscillations on stator and rotor current can be seen in detail in Fig. 10.

In Fig. 10 the magnet rotor speed is plotted again at the moment of synchronization. The magnet rotor reaches synchronous speed now at 2200 ms due to a different starting point. Additionally the stator current I_s and the rotor current I_r are plotted vs. the time.

During acceleration the stator current is high because the magnet rotor does not yet support the magnetic field in the air gap. The rotor current shows the opposite behavior because before synchronization the magnet rotor field weakens the stator field, but after synchronization they support each other. If the resulting field is small the stator current must be high to build up the magnetic field, but the weak field induces only a small rotor current. After synchronization the field is higher so that the magnetizing current is reduced but a high rotor current can be induced.

We can further regard superposed oscillations in the currents caused by the magnet rotor oscillations.

The measurements presented here can give a first impression of the dynamic characteristics of the PMIM during operation. Sudden load changes could not be performed with this test set up because of the large driving engine. But the demonstrated test results give an impression of what the PMIM will do during load changes. Compared to the synchronous machine the magnet rotor shows no different behavior during acceleration or after grid connection. If the operating point changes the magnet rotor reacts like a synchronous machine: the magnet wheel will oscillate around the new load angle and reach stable operation again.

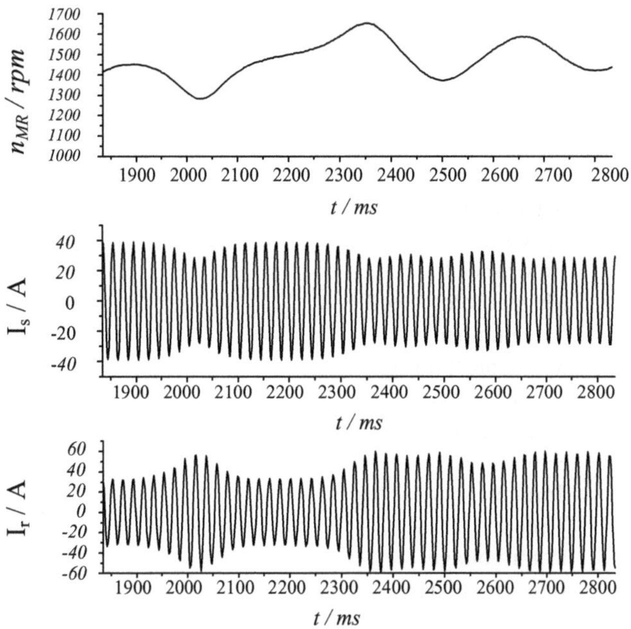

Figure 10. Magnet rotor speed n_{MR}, stator current I_{s}, rotor current I_{r} vs. time in the synchronization process.

The stability is even better because the load affects the asynchronous rotor first and is only indirectly coupled with the magnet rotor.

Conclusion

The PMIM represents a new wind generator concept for offshore wind power applications. It combines the advantages of PMSM and IM so that no gear and no converter are necessary. Working like an induction machine, the PMIM provides a soft grid connection together with stable operation. To achieve gearless operation at low speeds, a second permanent magnet rotor supports the magnetic flux so that the demand in reactive power can be minimized. After preliminary calculations a test machine was introduced to give first measurement results. Although the test machine is rather different to the planned design, it provides useful data to evaluate the PMIM's behavior. Initial static and dynamic measurements show accordance with the predicted and desired properties. The characteristics of that concept are good efficiency at partial load together with small reactive power consumption. However, these desired effects only appear if the internal voltage of the permanent magnets V_{p} is in the range of the stator voltage V_{s}. The idea is to control the demand of reactive power by changing the stator voltage with the help of a tapped transformer. Dynamic measurements do not show any significant drawbacks of the system that may be caused by dangerous oscillations. The dynamics of the magnet rotor are the same as for conventional PMSMs. All the PMIM's qualities lead to a low maintenance and reliable solution for fixed speed offshore wind turbines

Nomenclature

List of symbols

Symbol	*Qunatity*
d	Diameter
I_s	Stator current
I_r	Rotor current
n	Speed
n_R	Rotor speed
n_{MR}	Magnet rotor speed
$2p$	Pole number
P_{mech}	Mechanical power
R'_r	Rotor resistant
R_s	Stator resistant
s	Slip
T	Torque
t	Time
V'_r	Rotor voltage
V_p	Internal voltage
V_s	Stator voltage
$X_{\sigma s}$	Stator leakage reactance
$X'_{\sigma r}$	Rotor leakage reactance
X_m	Main reactance
$cos\ \varphi$	Power factor
η	Efficiency

References

[1] W.F. Low, N. Schofield, "Design of a Permanent Magnet Excited Induction Generator", Proc. ICEM 1992, Manchester University, 1992, Vol. 3, pp. 1077–1081.

[2] Wind Energie 2004, Short Version of the Findings of the WindEnergy Study 2004, Hamburg Messe und Congress GmbH, March 2004, http://www.hamburgmesse.de/Scripte/allgemein_Info/Bestellung_DEWIStudie/Studie_WindEnergy_en.htm.

[3] Horns Rev: Gondeln müssen runter, article in neue energie, magazin no. 6, pp. 78–79, June 2004, Bundesverband WindEnergie, Osnabrück.

[4] T. Epskamp, B. Hagenkort, T. Hartkopf, S. Jöckel, "No Gearing No Converter—Assessing the Idea of Highly Reliable Permanent-Magnet Induction Generators", Proceedings of EWEC 1999, Nice, France, 1999, pp. 813–816.

[5] B. Hagenkort, T. Hartkopf, A. Binder, S. Jöckel, "Modelling a Direct Drive Permanent Magnet Induction Machine", Proc. ICEM 2000, Helsinki University of Technology, 2000, Vol. 3, pp. 1495–1499.

[6] E. Tröster, T. Hartkopf, H. Schneider, G. Gail, M. Henschel, "Analysis of the Equivalent Circuit Diagram of a Permanent Magnet Induction Machine", ICEM 2004, Cracow, 2004.

[7] R. Hoffmann, "A Comparison of Control Concepts for Wind Turbines in Terms of Energy Capture", PhD Thesis, D17 Darmstädter Dissertation, 2002.

III-2.3. MAXIMUM WIND POWER CONTROL USING TORQUE CHARACTERISTIC IN A WIND DIESEL SYSTEM WITH BATTERY STORAGE

M. El Mokadem[1], C. Nichita[1], B. Dakyo[1] and W. Koczara[2]

[1] Groupe de Recherche en Electrotechnique et Automatique du Havre, University of Le Havre, 25, rue Philippe Lebon, 76058 Le Havre Cedex, France

mostafa.elmokadem@univ-lehavre.fr, nichita@univ-lehavre.fr, dakyo@univ-lehavre.fr

[2] Institute of Control and Industrial Electronics, Technical University of Warsaw, 75 Koszykowa, 00-662 Warszawa, Poland

koczara@isep.pw.edu.pl

Abstract. The purpose of our work is to study the maximum conversion of the wind power for a wind diesel system with a battery storage using a current control. The maximum power points tracking have been achieved using a step down converter. This study was developed taking into account the wind speed variations. The diesel generator is controlled using the power-speed characteristics. The results show that the control strategy ensures the maximum conversion of the wind power. The complete model is implemented in Matlab-Simulink environment.

Introduction

Actually the most autonomous feeding systems of electricity, in remote areas, are the diesel generators or hybrid wind diesel systems or wind-photovoltaic-diesel. The diesel generator is used to provide the necessary power to the costumers for insufficient wind periods. The wind generator is used in this case to save the maximum of fuel by the diesel generator when the wind power is abundant (ecological criterion). The random characteristic of the wind power constitutes a considerable technical problem for the integration of the wind generators in such systems. This imposes to develop control intelligent structures for the subsystems: diesel generator, wind generator, accumulators (battery, flywheel), and load (energy criterion). In order to develop a coherent approach of control, we study the optimization of the quality of the energy produced in remote area by the wind diesel hybrid system (stability of voltage and frequency). Increasing the life time of the equipment by the efficiency of the wind energy conversion and by the control diesel engine means to save the maximum of fuel.

The main goal of our approach is to study the connection of a hybrid wind diesel system to a DC variable load with battery storage. The wind diesel hybrid power systems are required to provide a maximum power under stochastic wind. But, the integration of wind

S. Wiak, M. Dems, K. Komęza (eds.), Recent Developments of Electrical Drives, 385–396.

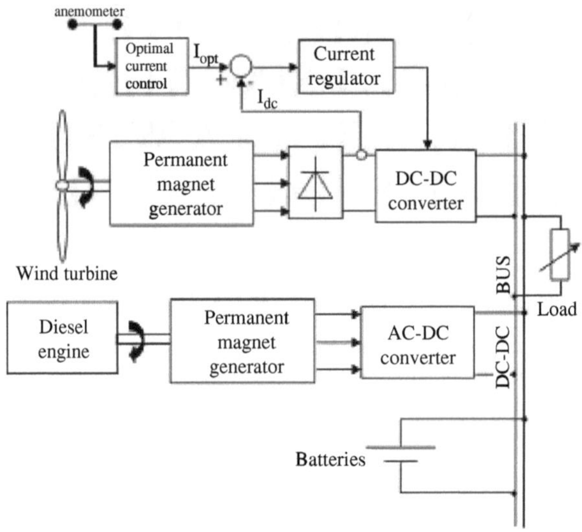

Figure 1. Wind diesel system with battery storage.

turbines into electric power systems generates some problems, which is the rejection of power fluctuations at the output of wind turbine generator. When the grid is large, these fluctuations have a little effect of the quality of the global delivered energy. But, with weak autonomous networks, the power fluctuations could have a marked effect, which must be instantaneously eliminated [2,3].

When the wind resource is sufficient, the diesel unit is shooting down to slow motion for saving the fuel. When wind resource is not abundant, the diesel is started at full load regime; its control is developed according to the power required by the main load. The excess of energy is dissipated by the dump load. Also, when it is necessary, the batteries take over to supply the load [1].

The proposed structure of our system is based on the following elements (Fig. 1): a permanent magnet synchronous wind generator which feeds an AC-DC converter, a diesel generator unit with permanent magnet synchronous generator feeding an AC-DC converter, a bank of batteries, a variable passive load, and a dump load.

Wind speed model

To take into account the random behavior of the wind power, we have modeled the wind speed.

Studies were already carried out to simulate numerically the wind speed which is considered as a random process. This process can be assumed to two components [4]:

- The slower component, which describes the slow evolution of the wind on a defined time horizon.
- The turbulence component, considered as a nonstationary, is assumed to be dependent on the lower component.

One of the well known method used for the modeling of the wind is the wind spectral characteristic of Van Der Hoven. In this model, the turbulence component is considered as a stationary random process where fluctuations magnitude does not depend on the wind mean value. Wind speed is then obtained by means of direct discretization of the power spectral characteristic S_{vv}.

The task is achieved as follows:

- Discretization of the pulsation w_i.
- Calculation of the areas between the $S_{vv}(w_i)$ curve and pulsation, which correspond to the consecutive discrete values of the pulsation.

$$S_i = \frac{1}{2}\left[S_{vv}(w_i) + S_{vv}(w_{i+1})\right](w_{i+1} - w_i) \tag{1}$$

- Determination of the magnitude A_i of each spectral component characterized by the discrete pulsation w_i

$$A_i = \frac{2}{\pi}\sqrt{S_i} \tag{2}$$

- Calculation of the wind speed v(t) which is the sum of the harmonics characterized by the magnitudes A_i, the pulsation w_i, and the phase φ_i generated in a way random.

In order to provide more relevant wind speed related to an actual site, it is necessary to consider nonstationary turbulence component as follows:

$$v(t) = v_l(t) + v_t(t) \tag{3}$$

where

$$v_l(t) = \frac{2}{\pi}\sum_{i=0}^{N_l} A_i \cos(w_i t + \varphi_i) \tag{4}$$

And

$$v_t(t) = \frac{2}{\pi}\sum_{N_t}^{N} A_i \cos(w_i t + \varphi_i) \tag{5}$$

N_l: Samples for the slow component $v_l(t)$;
$N - N_t$: Samples for the component of turbulence $v_t(t)$.

The amplitude of the turbulence component is adjusted by a coefficient K which increase with v_l and then modified by a filter which has time constant τ_F [4]. These quantities depend on the direct component v_l.

$$K = \frac{\alpha_1 v_l}{\beta_1 + v_l} \tag{6}$$

$$\tau_F = \tau_0 - a_1 v_l \tag{7}$$

α_1, $\beta_1 \tau_0$, and a_1 are constants.

In Fig. 2 we present the result of the wind speed using the method mentioned above. The speed of wind v(t) is generated with a sampling period $T_e = 1$ s, on a temporal horizon of half hour [4,5].

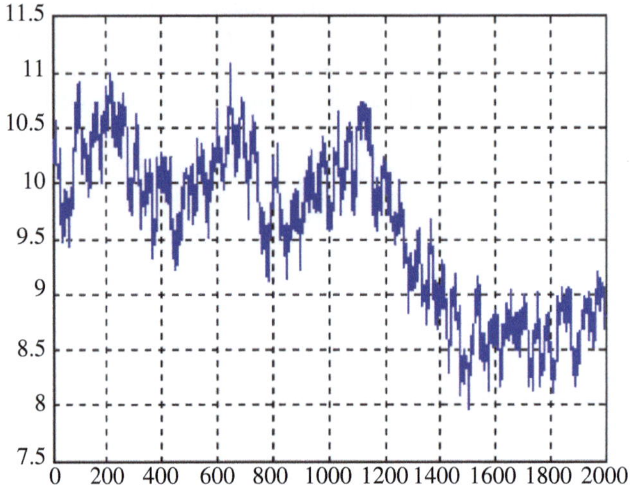

Figure 2. Wind speed profile used in the simulation.

Model of the wind turbine

We have considered that the blades are rigidly attached to the wind turbine; consequently the pitch angle of the blades is constant. The wind generator is connected with the DC common coupling point (Fig. 1) [5,6]; the AC-DC Converter unit is composed by a six pulse rectifier and DC-DC buck converter. The characteristics modeling have been made by a six-order polynomial regression. The power coefficient characteristic C_p is a function of tip-speed-ratio λ and in this case is given by:

$$C_p(\lambda) = \sum_{i=0}^{n} a_i \lambda^i \tag{8}$$

$$\lambda = \frac{R\Omega}{v} \tag{9}$$

where
R radius of the rotor;
Ω mechanical angular velocity of the rotor;
v wind speed.

The a_i parameters ($i = 0 \ldots 6$) are determined by a Matlab computing program [7].
The output power of the wind turbine is calculated from the following equation:

$$P_t = \frac{1}{2} C_p(\lambda) A v^3 \tag{10}$$

Where ρ is air density in kg/m^3 and A is the frontal area of the wind turbine in m^2.
The torque developed by the wind turbine is expressed by [8–10]:

$$T_t = \frac{P_t}{\Omega} = \frac{1}{2} \rho A R v^2 C_\Gamma(\lambda) \tag{11}$$

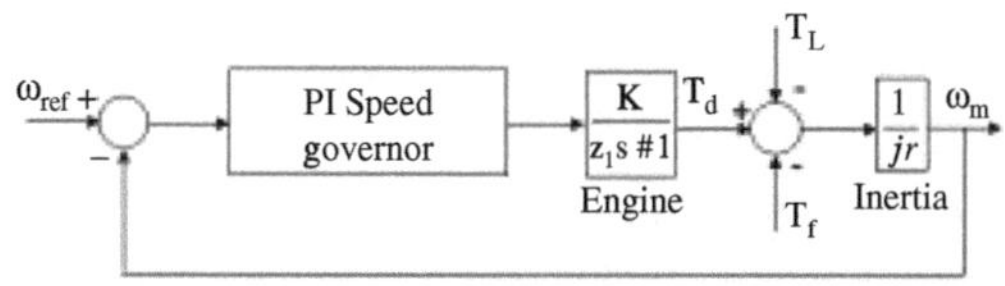

Figure 3. Scheme of diesel engine and governor.

Where,

$$C_\Gamma(\lambda) = \frac{C_p(\lambda)}{\lambda}$$

is the torque coefficient

Diesel engine and governor modeling

A diesel generator is a device which converts fuel into mechanical energy in an engine and subsequently converts mechanical energy to electrical energy in a generator or alternator. Speed regulation and controls are necessary to maintain useful power of the generator. Governors occur in two basic configurations, these being mechanical or electronic [13]. The mechanical governor is most often utilized on installations under 500 kW and where shared loads fluctuate by ±5–10%. The electronic governor is used where frequency stability is very important or in automatic parallel operation. Loads are generally managed within 5%.

The diesel engine is a non-linear system. It presents dead-times, delays, non-linear behaviors, making difficult its control.

A simplified general functional diagram for a diesel engine and the respective speed regulator system is presented in Fig. 3. The model has three blocks: the speed governor, the fuel flow, and the combustion process. The speed governor determines the power (torque) output of the diesel engine. Its dynamic behavior can be approximate by a first-order model, with a time constant τ_1. The fuel flow block is a gain that adjusts the relationship between the torque and fuel consumption [13,14].

T_L is the load torque, T_f represents the friction and mean effective pressure torques, and J is the total inertia

Model of the permanent magnet synchronous generator (PMSG), rectifier, and DC-DC buck converter

In our study, we consider that the wind generator and the diesel generator drive both a permanent magnet synchronous generator. The three-phase output of the PMSG is rectified with a full wave diode bridge rectifier, filtered to remove significant ripple voltage components, and fed a DC-DC buck converter. For an ideal (unloaded and loss-less) PMSG, the line to line voltage is given as [11–13]:

$$V_{ll}(t) = K_v \omega_e \sin(\omega_e t) \tag{12}$$

Where K_v is the voltage constant in V/(rad/s) and ω_e is the electrical frequency.

The electrical frequency is related to the mechanical speed ω_m by

$$\omega_e = \omega_m p \tag{13}$$

where p is the number of pair poles of the PMSG.

Neglecting commutation delays, the DC rectifier voltage V_{dc} is reduced from $\frac{3}{\pi}\omega_e L_s I_{dc}$ value:

$$V_{dc} = \frac{3\sqrt{2}}{\pi} V_{ll} - \frac{3}{\pi}\omega_e L_s I_{dc} \tag{14}$$

Where V_{ll} is in RMS volts, I_{dc} is the average rectified PMSG current and L_s is the stator inductance.

Assuming negligible loss, the electrical power output (equal to mechanical power input) of the PMSG as a function of I_{dc} or V_{dc} is given as:

$$P_{dc} = V_{dc} I_{dc} = K_e \omega_m I_{dc} - K_x \omega_m I_{dc}^2 = V_{dc}\left(\frac{K_e}{K_x}\right) - \frac{V_{dc}^2}{K_x \omega_m} \tag{15}$$

where

$$K_e = \frac{3pK_v}{\pi} \tag{16}$$

$$K_x = \frac{3pL_s}{\pi} \tag{17}$$

The mechanical shaft torque (loss-less operation) can be found as:

$$T_m = \frac{P_m}{\omega_m} = \frac{P_{dc}}{\omega_m} = K_e I_{dc} - K_x I_{dc}^2 = V_{dc}\left(\frac{K_e}{K_x \omega_m}\right) - \frac{V_{dc}^2}{K_x \omega_m^2} \tag{18}$$

The average output voltage of the DC-DC buck converter is given by:

$$V_s = \alpha V_{dc} \tag{19}$$

Assuming negligible loss, the electrical power input equal to the electrical power output of the DC-DC buck converter, the average output current of the DC-DC buck converter is given by:

$$V_{dc} I_{dc} = V_s I_s \tag{20}$$

$$I_{dc} = \alpha I_s \tag{21}$$

I_s represents the contribution output current of the wind generator or the diesel generator.

Modeling of the battery

The model assumes that: (a) the electromotive force voltage of the battery increases with charging current and state of charge (C_{soc}) and (b) the electromotive force voltage decreases with discharging current and state of charge [14].

$$E_{bat} = (E_{CC} - E_{CD})C_{SOC} + E_{CD} \tag{22}$$

E_{CC} and E_{CD} design the electromotive force when the battery is completely charged and discharged respectively.

The battery is represented by a voltage source in series with a resistance and capacitance. The internal resistance, R_{bat}, is assumed to be constant and the internal voltage, E_{bat}, varies with state of charge. The internal terminal voltage, V_s, in discharge and charge operations, is given respectively by [14]:

$$V_s = E_{bat} - I_{bat}R_{bat} \tag{23}$$

$$V_s = E_{bat} + I_{bat}R_{bat} \tag{24}$$

I_{bat} is the battery current. In steady state, the terminal voltage of the capacitance is negligible.

In our case the battery current will have two different expressions:

$$I_{discharge} = \frac{E_{bat}}{2R_{bat}} - \frac{1}{2}\sqrt{\frac{E_{bat}^2}{R_{bat}^2} - \frac{4P_{out}}{R_{bat}}} \tag{25}$$

$$I_{charge} = -\frac{E_{bat}}{2R_{bat}} + \frac{1}{2}\sqrt{\frac{E_{bat}^2}{R_{bat}^2} + \frac{4P_{out}}{R_{bat}}} \tag{26}$$

P_{out} is the power delivered or received by the battery.

The state of charge of the battery may be calculated by:

$$C_{SOC\_discharge} = \frac{\eta_{disch}}{C_{cap}{}^{*}3600} \int_{t=0}^{t} I_{discharge}\, dt \tag{27}$$

$$C_{SOC\_charge} = \frac{\eta_{ch}}{C_{cap}{}^{*}3600} \int_{t=0}^{t} I_{charge}\, dt \tag{28}$$

C_{cap} is the capacity of the battery in Ampere-hours and (η_{disch}, η_{ch}) are efficiency factors of discharge and charge operations respectively.

The C_{soc} can have a value between 0% and 100%. The 0% corresponds to a fully discharged state and 100% correspond to a fully charged state.

Control strategy

Unfortunately, the wind energy is not completely predictable and it fluctuates rapidly. As a result it is difficult to balance the system. For efficient capture of wind power, turbine torque or turbine speed should be controlled to follow the optimal tip-speed-ratio (TSR).

Our study is based on the current control of the wind generator (Fig. 1). Assuming that the permanent magnet generator torque is proportional to the machine current, the control structure allows the torque and rotational speed to be controlled. The reference current of the wind generator rectified current is calculated for steady state points where the turbine torque and the generator torques are equals.

$$J\frac{d\omega}{dt} = T_t - T_g \tag{29}$$

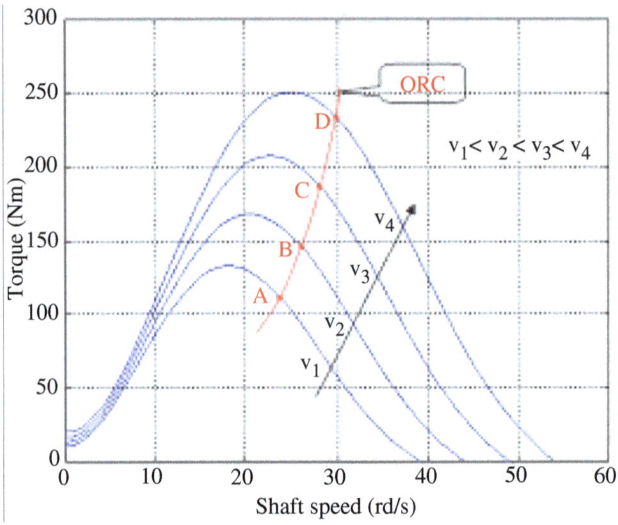

Figure 4. Wind turbine torque characteristics.

J is the inertia in kgm^2.

$$T_g = KI \tag{30}$$

K is a constant, it depends on generator characteristics.

$$I = \frac{T_t}{K} \tag{31}$$

If the turbine torque changes when the wind speed increases, the system will be able to accelerate more quickly to the next steady state which corresponds to the maximum power points tracking (MPPT) (Fig. 4). We can see that the optimal operating points are different for every wind speed v_i.

Consequently, the wind maximum power transfer is ensured by the operations points following the curve controlled with MPPT unit (Fig. 5). In our approach, the MPPT function is realized by a step down converter.

In order to control the diesel generator, we have considered two possibilities. For the first case, we assume that the diesel generator operates at constant power and constant speed. In this case, the diesel generator is started on when the terminal voltage of batteries falls bellow a minimum value E_{batmin} and the wind power is not sufficient to supply the load. The diesel engine should be started to re-charge the battery and supply the load. In the contrary case, if the terminal voltage of batteries exceeds a maximum value E_{batmax} and the wind power is sufficient to supply the load, the diesel generator is slow down.

The second possibility: the diesel generator is controlled using the power-speed characteristics. A power sensor detects the load power and produces reference correction speed that is compared to actual speed signal. A speed controller provides signal for adjustment of the fuel injection unit, feeding the engine prime mover [16].

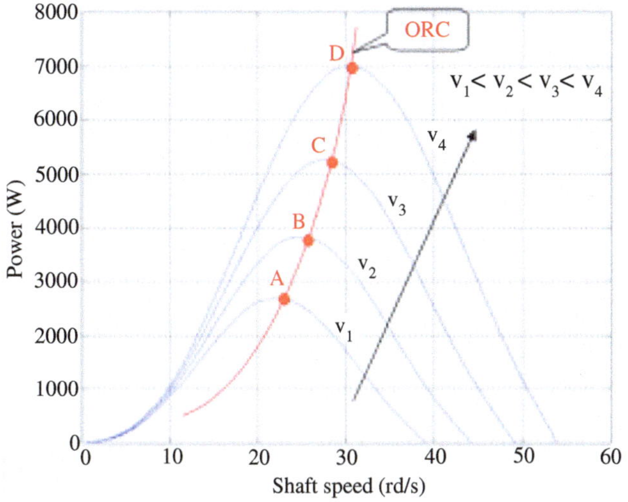

Figure 5. Wind turbine power characteristics.

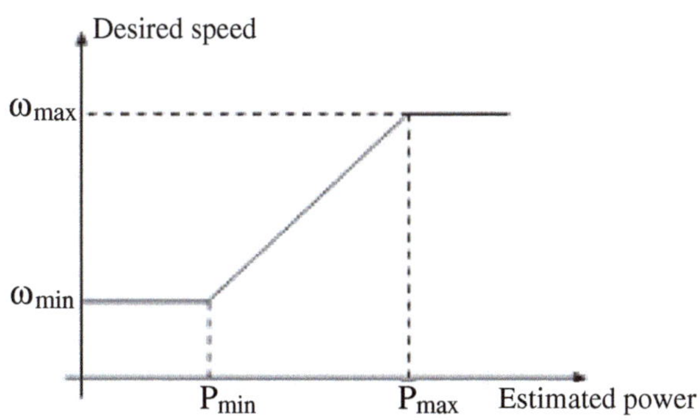

Figure 6. Speed vs. power characteristics.

Actual speed is adjusted according to the power required by the load in steady state operation (Fig. 6). The speed is relatively low when power demand is not important. When the load power increases, the diesel governor controls the speed evolution according to the linear law designed for this purpose (Fig. 6).

Simulation results

The complete model of the system has been implemented on Matlab-Simulink environment. In our study, we have used the wind speed profile depicted in Fig. 2. According to Figs. 7 and 8, we can see that the wind turbine operates at its most efficient operating points for different values of wind speed.

Fig. 9 presents the output current of the wind generator. We can conclude that the current follows well the reference generated according to the MPPT law.

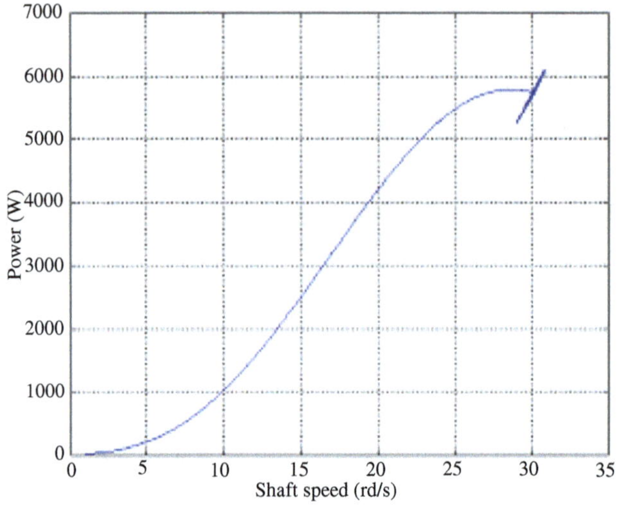

Figure 7. Simulation of the maximum regime characteristics (MPPT) for a given wind turbine.

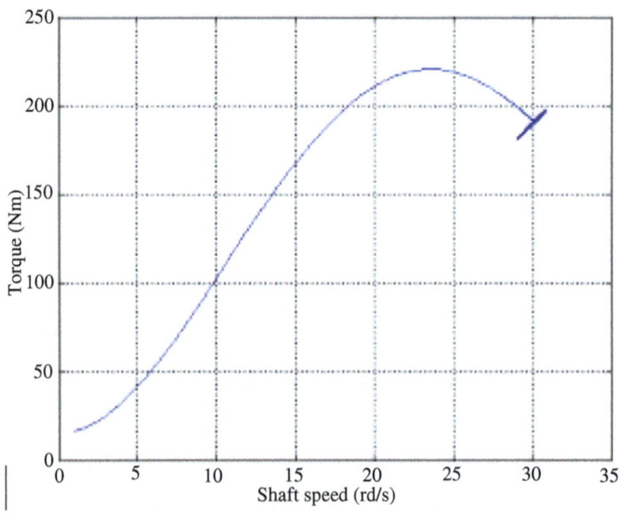

Figure 8. Simulation of the torque characteristics for a given wind turbine.

Fig. 10 presents the diesel generator speed and its reference. This reference is a function of the power required by the load and the wind fluctuations. Because of the diesel engine inertia, the diesel generator speed cannot follow the dynamics of this reference.

Conclusion

The purpose of our work is to study and develop a maximum wind power control using torque characteristic for a wind diesel system with battery storage.

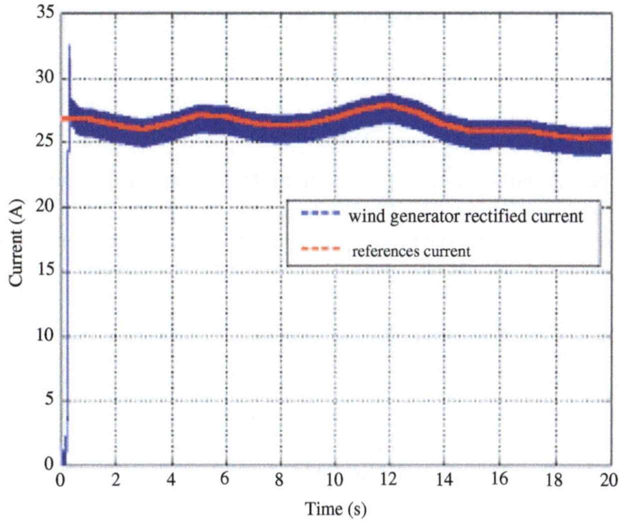

Figure 9. Wind generator rectified current of the wind generator and its reference.

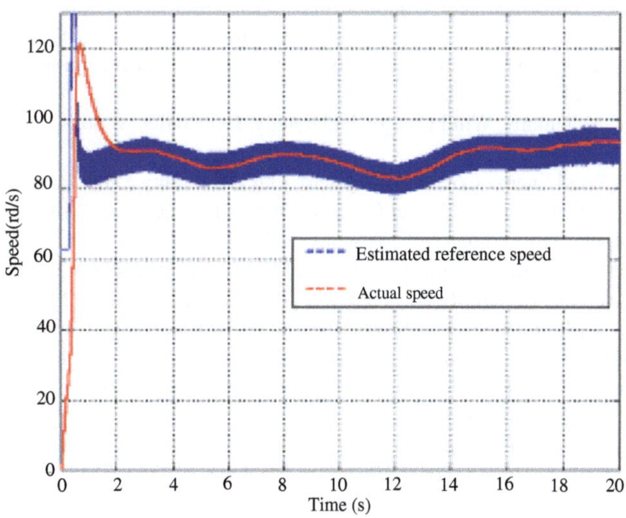

Figure 10. Diesel generator speed and its reference.

Our study is based on the current control of the wind generator. We have assumed that the permanent magnet generator torque is proportional to the machine current; the control structure allows the torque and rotational speed to be controlled.

Moreover, the diesel generator power contribution is a function of the wind power and the load variations. When wind resource is not abundant, the diesel is started on to supply the load, the excess of energy could be dissipated by the dump load. Also, when it is necessary, the batteries take over to supply the load.

We have point out that this control strategy, based on the maximum power point tracking, could ensure the maximum conversion of the wind power.

References

[1] C.V. Nayar, S.J. Philips, W.L. James, T.L. Pryor, D. Remer, Novel wind/diesel/battery hybrid energy system, Solar Energy, Vol. 51, No. 1, pp. 65–78, 1993.
[2] K.B. Saulnier, R. Reid, "Modélisation, simulation et régulation d'un réseau Eolien/Diesel autonome", Rapport IREQ4340, Institut de Recherche de l'Hydro-Quebec Varennes, P.Q., 1989, Québec.
[3] R.B. Chedid, S.H. Karaki, C. El-Chamali, Adaptive fuzzy control for wind-diesel weak power systems, IEEE Trans. Energy Convers., Vol. 15, No. 1, pp. 71–78, 2000.
[4] C. Nichita, D. Luca, B. Dakyo, E. Ceanga, Large band simulation of the wind speed for real time wind turbine simulators, IEEE Trans. Energy Convers., Vol. 17, No. 4, pp. 523–530, 2002.
[5] M. El Mokadem, N. Nichita, G. Barakat, B. Dakyo, "Control Strategy for Stand Alone Wind-Diesel Hybrid System Using a Wind Speed Model", Electrimacs Proceeding CD-ROM, Montreal, 2002.
[6] M. El Mokadem, "Structure d'un conditioneur depuissance pour un système éolien-diesel", JCGE'03 Proceedings CD-ROM, Saint-Nazaire, June 2003.
[7] C. Nichita, "Etude et développement de structures et lois de commande numériques pour la simulation en temps réel d'actionneurs. Application à la réalisation d'un simulateur d'aérogénérateur de 3 kW", Thèse de Doctorat, Université du Havre, 1995.
[8] D. Le Gourieres, Energie éolienne, théorie, conception et calcul pratique des installations, Eyrolles, Paris, France, 1982.
[9] J.F. Walker, N. Jenkins, Wind Energy Technology, John Wiley & Sons, Inc., Chichester, UK, 1997.
[10] L.L. Freris, Wind Energy Conversion System, England: Prentice Hall International Ltd., 1990.
[11] M.J. Ryan, R.D. Lorenz, "A Novel Controls-Oriented Model of a PM Generator with Diode Bridge Output", Proceedings EPE, Trondheim, 1997, pp. 1.324–1.329.
[12] M.N. Eskander, Neural network controller for a permanent magnet generator applied in a wind energy conversion system, Renewable Energy, Vol. 26, pp. 463–477, 2002.
[13] R. Hunter, G. Elliot, Wind-Diesel Systems, New York: Cambridge University Press, 1994.
[14] G.S. Stavrakakis, G.N. Kariniotakis, A general simulation algorithm for the accurate assessment of isolated diesel-wind turbines systems interaction, IEEE Trans. Energy Convers., Vol. 10, No. 3, pp. 577–590, 1995.
[15] M.J. Hoeijmakers, "The (In)Stability of a Synchronous Machine with Diode Rectifier", Proceedings of the International Electrical Machines Conference, UK, September 1992, pp. 83–87.
[16] W. Koczara, J. Leonarski, R. Dziuba, "Variable Speed Three Phase Power Generation Set", EPE_2001, Graz, Austria, August 2001.

III-2.4. STUDY OF CURRENT AND ELECTROMOTIVE FORCE WAVEFORMS IN ORDER TO IMPROVE THE PERFORMANCE OF LARGE PM SYNCHRONOUS WIND GENERATOR

D. Vizireanu[1], S. Brisset[1], P. Brochet[1], Y. Milet[2] and D. Laloy[2]

[1]L2EP, Ecole Centrale de Lille, Cité Scientifique, BP 48, 59651 Villeneuve d'Ascq Cedex, France
darius.vizireanu@ec-lille.fr, stephane.brisset@ec-lille.fr, pascal.brochet@ec-lille.fr
[2]Framatome ANP, 27 rue de l'Industrie, BP 189, 59573 Jeumont Cedex, France
daniel.laloy@framatome-anp.com, yves.milet@framatome-anp.com.

Abstract. The paper presents a comparison between sinusoidal and trapezoidal waveforms in order to reduce the torque ripple and the power to grid fluctuation for large direct-drive PM wind generator. Trapezoidal waveform brings 28% higher power density but also two major drawbacks: necessity to vary the DC bus voltage and requirement for an additional filter on the DC bus.

Introduction

During last decades, an important development of permanent magnet machines domain has been observed, due to the improvement of the permanent magnet characteristics and the occurrence of new power electronic components. The magnets have allowed to eliminate the excitation and the slip rings, and consequently to increase the power of the machines. The new power converters using IGBT or IGCT technologies allow supplying the machines with different waveform voltages and different frequencies, depending on the application.

In the present, permanent magnet synchronous machines are used in large power applications as high torque low speed systems for wind energy generators. For these kinds of systems, an important parameter is the electromagnetic torque, and the interest is to minimize torque oscillations which cause lower mechanical stability, audible noise, and accelerated aging of the machine due to vibrations.

In this moment, the efforts are concentrated to increase the power of PM synchronous generators. But a special attention should be paid to the conception of the power converters. The power electronic devices have a certain limit, and special architectures are used to increase the voltage and current capability. Multi-level structures are used to obtain higher voltage capability. To obtain higher rated current, a solution is to do parallel connection of several converters, which corresponds to an increase of the number of the legs, or to an increase of the phase number of the machine. A resulting advantage is the possibility to obtain a certain modularity, which allows facilities for the fabrication process, transportation, and maintenance.

S. Wiak, M. Dems, K. Komęza (eds.), Recent Developments of Electrical Drives, 397–413.

The goal of this paper is to study the influence of the electromotive force (e.m.f.) and current waveforms over the electromagnetic torque and to search an optimum topology of the PM synchronous machine (the shape of the magnet and the winding) and associated converter in order to obtain minimum torque pulsation and highest efficiency.

System description

The system that will be studied is a direct-drive wind generator (Fig. 1).

The machine topology is an axial-flux machine with two rotor discs and one inner stator with teeth (Fig. 2). Refs. [1,2] suggest that this architecture has higher power density than the radial-flux PM machine. The converter used is a back-to-back converter, which consists in two PWM converters, a rectifier, and an inverter (Fig. 3).

The capacitor from the intermediate circuit is an advantage of this topology, allowing a separate control for both converters and the possibility to compensate asymmetries that appears on both sides. The rectifier control strategy realizes a vector control of the generator,

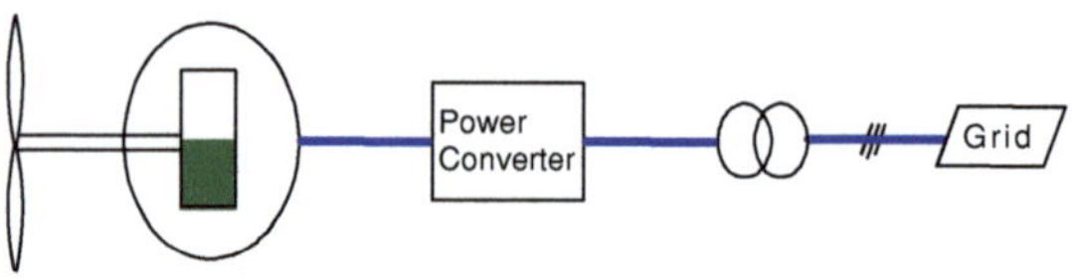

Figure 1. Direct-drive wind turbine system with PM synchronous generator.

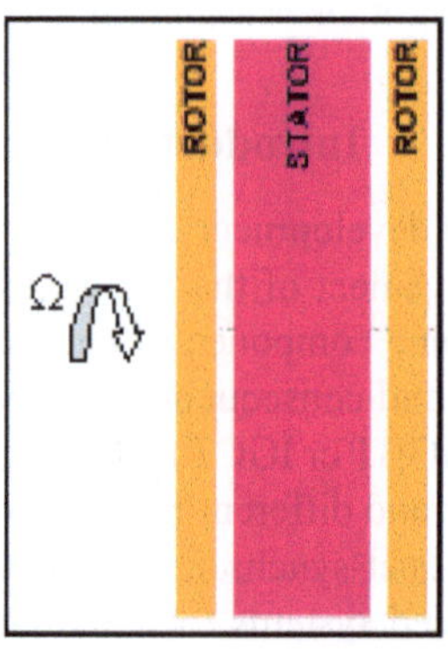

Figure 2. Generator's topology.

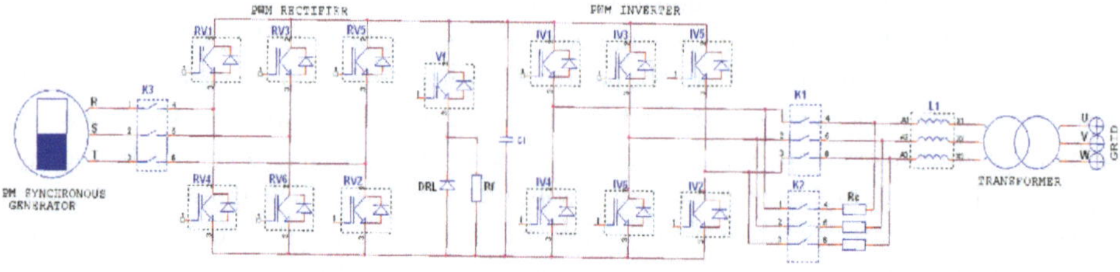

Figure 3. The back-to-back PWM-VSI power converter.

while the inverter controls the energy transfer to the grid. The chopper connected between the converters allows the control of the DC bus voltage and current during breaking regime of the generator. The energy is dissipated over a breaking resistor.

The two converters are decoupled at the level of the DC bus. That will permit to reduce the studied system: from the shaft of the generator until the DC bus. To avoid over voltages and protect the transistors, the control of the inverter imposed a constant DC voltage. If the DC voltage is maintained constant, the DC current waveform will give an indication about the power transfer. Reducing harmonic content of the DC bus current will allow reducing the size of the DC bus filter and the harmonic filter at the output of the converter. As mentioned before, the goal is to reduce torque oscillations, but also to observe the influence over the quality of the DC bus current. At constant speed, low level of DC bus current harmonics means reduced power fluctuation at the output.

Analytical approach

In this part, the influences of e.m.f. and current waveforms over the electromagnetic torque are analytically studied, even if the waveforms are not practically feasible.

Sinusoidal waveform

For a three-phase PM synchronous machine, without damping, the electromagnetic torque has the following expression:

$$T_{elmg} = \frac{P_{elmg}}{\Omega} = \frac{1}{\Omega} \cdot \sum_{i=1}^{3} e_i \cdot i_i \tag{1}$$

where Ω is the mechanical speed, e_i is the e.m.f. corresponding to phase i, andi_i is the phase current.

Using a FFT for the e.m.f,

$$\begin{aligned} e_1 &= \sum_{k=0}^{\infty} E_{2k+1} \cdot \cos[(2k+1)\theta] \\ e_2 &= \sum_{k=0}^{\infty} E_{2k+1} \cdot \cos\left[(2k+1)\left(\theta - \frac{2\pi}{3}\right)\right] \\ e_3 &= \sum_{k=0}^{\infty} E_{2k+1} \cdot \cos\left[(2k+1)\left(\theta - \frac{4\pi}{3}\right)\right] \end{aligned} \tag{2}$$

where $\theta = \omega \cdot t$. Imposing sinusoidal currents in phase with the e.m.f,

$$\begin{aligned} i_1 &= I \cdot \cos[(2k+1)\theta] \\ i_2 &= I \cdot \cos\left[(2k+1)\left(\theta - \frac{2\pi}{3}\right)\right] \\ i_3 &= I \cdot \cos\left[(2k+1)\left(\theta - \frac{4\pi}{3}\right)\right] \end{aligned} \tag{3}$$

The electromagnetic torque could be written as:

$$T_{elmg} = \frac{3E_1 \cdot I}{2 \cdot \Omega} + \frac{3 \cdot I}{2 \cdot \Omega} \cdot \sum_{i=1}^{3} [(E_{6k-1} + E_{6k+1}) \cos(6k\theta)] \tag{4}$$

It is easy to observe that the electromagnetic torque contains only sixth or multiple by six harmonics. The torque harmonics are proportional with the current amplitude. If $(6k - 1)$ and $(6k + 1)$ harmonics have opposite phases, the effect will be to reduce the torque oscillation. Voltage harmonics can be reduced using different winding techniques, but it is impossible to completely eliminate them. But controlling the machine's phase currents, torque ripple minimization can be realized by injecting current harmonics.

When the currents contain odd harmonics, their expressions are:

$$\begin{aligned} i_1 &= \sum_{k=0}^{\infty} I_{2k+1} \cdot \cos[(2k+1)\theta] \\ i_2 &= \sum_{k=0}^{\infty} I_{2k+1} \cdot \cos\left[(2k+1)\left(\theta - \frac{2\pi}{3}\right)\right] \\ i_3 &= \sum_{k=0}^{\infty} I_{2k+1} \cdot \cos\left[(2k+1)\left(\theta - \frac{4\pi}{3}\right)\right] \end{aligned} \tag{5}$$

Then, the electromagnetic torque becomes:

$$\begin{aligned} T_{elmg} &= \frac{3}{2 \cdot \Omega}(E_1 I_1 + E_3 I_3 + E_5 I_5 + E_7 I_7) \\ &+ \frac{3}{2 \cdot \Omega}(E_1 I_5 + E_1 I_7 + E_3 I_3 + E_5 I_1 + E_7 I_1) \cdot \cos(6k\theta) \\ &+ \frac{3}{2 \cdot \Omega}(E_5 I_7 + E_7 I_5) \cdot \cos(12k\theta) + \cdots \end{aligned} \tag{6}$$

It is possible to reduce the $6k$ torque harmonics by injecting current harmonics as suggested in (6). The DC component of the torque can be increased using odd current harmonics in phase with the same order e.m.f. harmonics. Equation (6) shows the interest of trapezoidal waveforms to increase the mean value of the torque.

Trapezoidal waveforms

The interest to study PM machines with trapezoidal waveforms of the e.m.f. emerges from the necessity to increase the power density. For the same structure of a machine, it is possible to increase the effective value of the e.m.f. using a full-pitch winding. The result will be also an increase of the harmonic content of the e.m.f.

The ideal e.m.f. waveform is an 120 electrical degrees trapezoidal form, while for the current the ideal shape is an 120 electrical degrees rectangular one (Fig. 4) [3,4]. The electromagnetic torque, in this case, can be expressed as:

$$T_{elmg} = \frac{2 \cdot E_{(tr)} \cdot I_{(tr)}}{\Omega} \tag{7}$$

where $E_{(tr)}$, $I_{(tr)}$ are the peak values of the e.m.f., respectively the current. But in reality, the shape of the e.m.f. is not perfectly trapezoidal, and the current has not a perfect squared waveform.

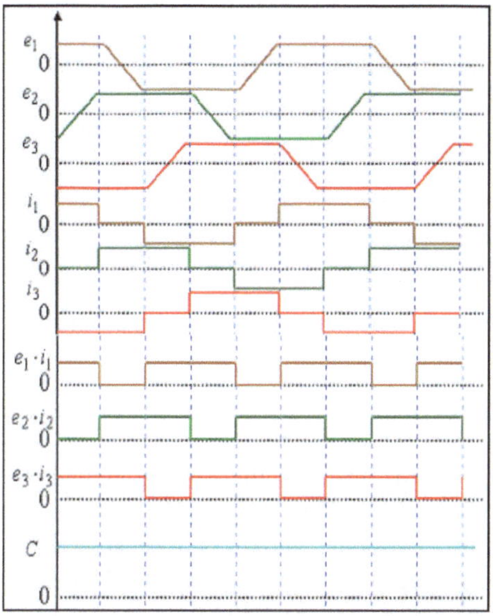

Figure 4. E.m.f., currents, electromagnetic power, and torque for a three-phase trapezoidal waveform system.

For the e.m.f., the width of the interval corresponding to the peak value depends on the width of the magnet. It could vary between 100° and 150°. Deviation from the ideal forms of e.m.f. and current implies torque ripple.

Finite element models

Due to the fact that the ideal waveforms are not achieved, numerical models are elaborated to calculate the harmonic content of the current and e.m.f.

If the shape of the magnet is defined it is possible to use a finite element model to calculate the air-gap magnetic field. High power disk PM synchronous generators have large diameters, and a flat representation of the machine is used to represent a symmetry period. Finite elements models are elaborated, for the sine-wave and trapezoidal square-wave machines. The models are built for an average value of the diameter. The magnetic characteristics are introduced for each region of the machine: the stator, the rotor, the magnets, and the air-gap. Due to the symmetry of the machine's geometry, only half-stator and one rotor are designed. The magnetic periodicity is 9. For each model, boundary conditions are imposed: the tangential component of the magnetic induction is equal to zero for the line delimiting the two stators and the normal component of the magnetic induction is equal to zero for the inferior limit of the rotor.

Sinusoidal waveforms

For the sinus machine, the slot pitch is 3.6 slots/pole, the width of the magnet reported to the pole pitch is 0.6, and a 5/6 short-pitch winding is used. The flux for each coil is determined,

Figure 5. Flux lines for nominal charge conditions.

and consequently the phase e.m.f. Fig. 5 presents a zoom of the geometry over three poles and the flux lines for nominal charge conditions.

For the sinusoidal waveforms machine, the e.m.f. obtained for no-load conditions are presented in Fig. 6. A FFT analysis (Fig. 7) shows the time harmonics spectrum: 4.3% of third harmonic, 0.7% of fifth harmonic, and 0.1% of seventh harmonic.

For a constant rotational speed, and imposing sinusoidal currents in phase with the e.m.f, the electromagnetic torque computed for a 750 kW generator has the waveform presented in the Fig. 8. The blue line is the result done by the finite elements software, while the red

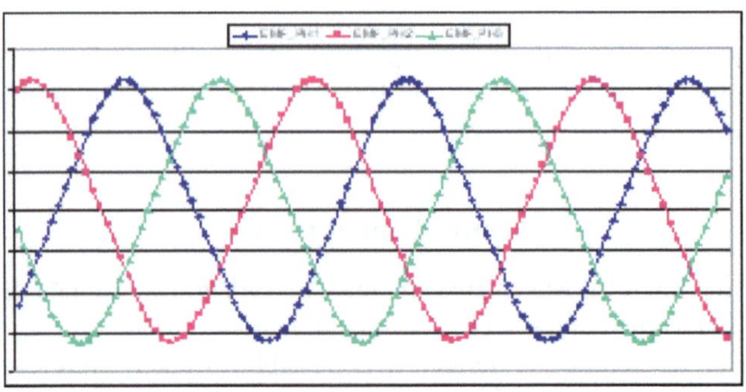

Figure 6. The e.m.f. for no-load conditions.

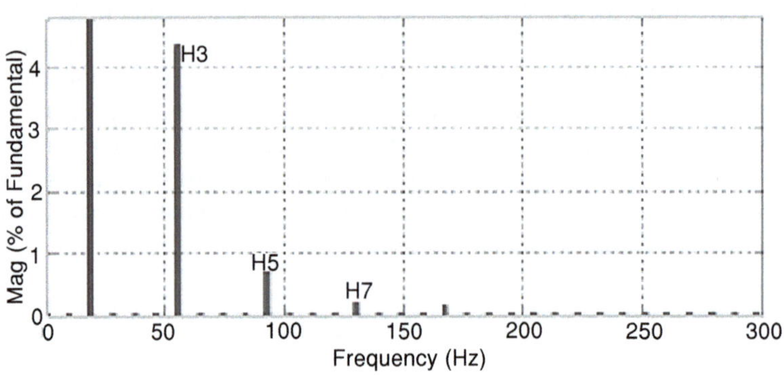

Figure 7. E.m.f. time harmonics.

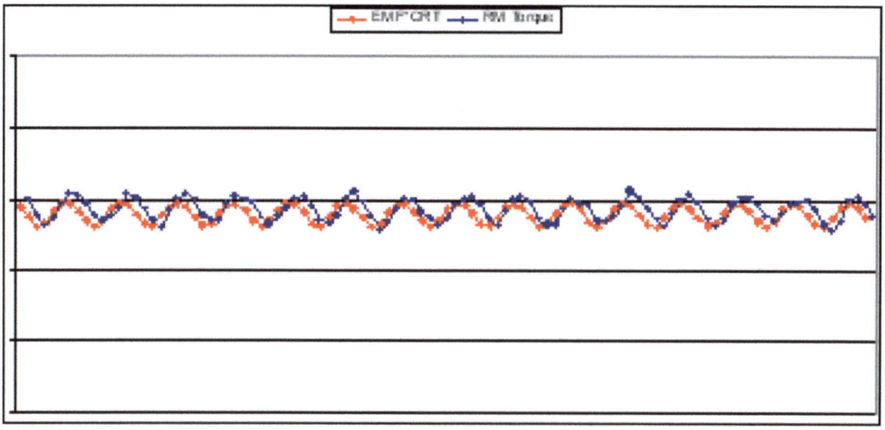

Figure 8. Electromagnetic torque.

line represents the torque calculated as the sum of products between phase e.m.f. and phase currents divided by the rotational speed. The curves are almost identically, and it confirms that the armature reaction has a low influence due to the large air-gap. Due to the fifth and seventh harmonics of the e.m.f. the torque presents a 0.5% 6th harmonic pulsation.

A comparison is done between the simulation model and a real system (Table 1). The compared parameters are: the air-gap flux density, the per-unit RMS and harmonics values of the e.m.f., the synchronous reactance, and the per-unit electromagnetic torque.

Trapezoidal waveforms

For the trapezoidal machine, the dimensions of the slots height, the inner and outer diameters are maintained. The goal is to impose the same core and copper volumes for both machines. This allows to impose the same level of losses and to compare the performances for the same level of heating. To obtain a trapezoidal e.m.f. waveform, the slot pitch in this case is 3 slots/pole and the winding is a full-pitch one.

Table 1. Comparison between simulation and experimental results

Compared parameters	Simulation	Measure
Air-gap flux density (T)	0.77	0.77
FEM (pu)	0.98	1
Third Harmonic (%)	4.3	3.7
Fifth Harmonic (%)	0.7	0.7
Seventh Harmonic (%)	0.1	0.1
Phase current (u.r.)	1	1
Current THD (%)	1	1
Synchronous reactance (%)	0.52	0.52
Electromagnetic torque (pu)	0.98	1

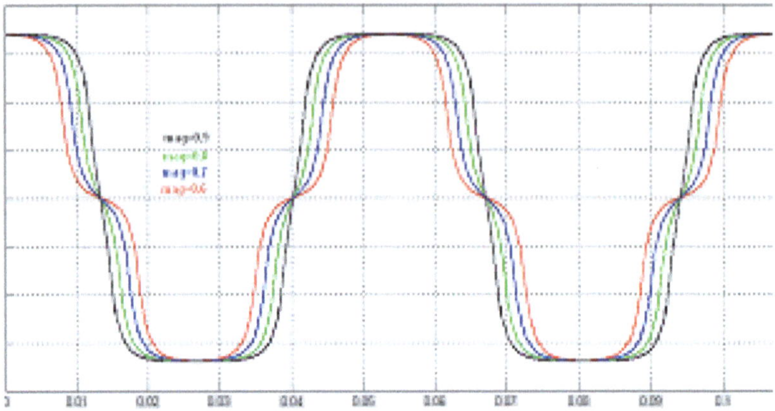

Figure 9. Trapezoidal e.m.f. for different magnet width.

The width of flat zone of the e.m.f. depends on the magnet's width. An e.m.f. close to the ideal trapezoidal waveform is obtained for a width of the magnet equal to 90% of the pole pitch (Fig. 9). Larger magnets induce higher leakage flux between magnets.

The number of coils decreases by 20%, but the width is 20% larger, compared to the sinusoidal machine. Larger slots are used to obtained same volume of copper for trapezoidal waveform machine as for the sinusoidal waveform machine.

For both systems, the same DC bus voltage is imposed. To avoid oscillations of the current during commutation of phases, it can be analytically proved that the peak value of the line voltage at the output of the generator should be half of the DC bus voltage. Therefore, the peak value of the e.m.f. for trapezoidal waveform is approximately half of the peak value for sinusoidal waveform. The number of turns/coil is decreased to obtain the proper peak value of the phase e.m.f. The three-phase no-load e.m.f. calculated for nominal speed and a magnet width equal to 90% of the pole pitch are presented in Fig. 10. The width of the e.m.f.'s flat zone is 120 electrical degrees and significant third harmonic due do the large coils and magnets can be noticed.

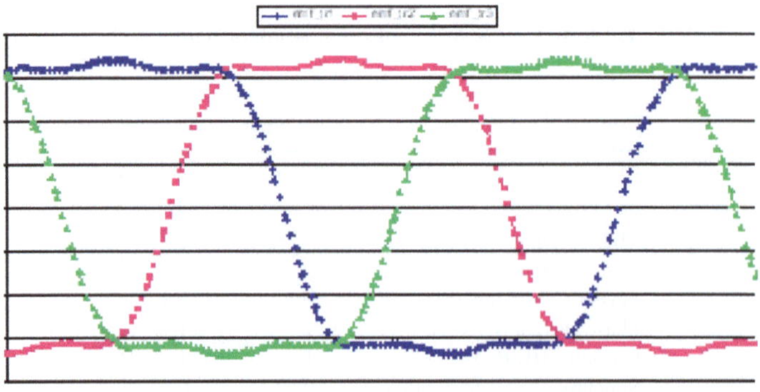

Figure 10. Trapezoidal e.m.f. for no-load conditions.

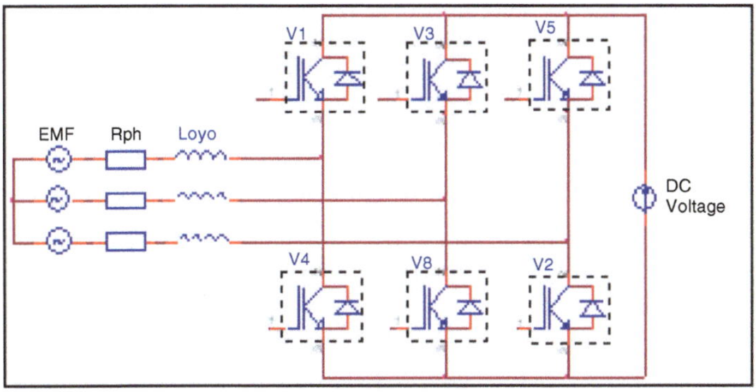

Figure 11. Matlab-Simulink model of the machine and converter.

Simulation model

A simulation model is used to study the behavior of the wind generator system, from the generator's shaft until the DC bus. The model uses the e.m.f. waveforms computed with the finite element model. Other parameters of the machine are also introduced to simulate the system: stator resistance, stator reactance, rotor position. The air-gap linkage inductance, the slots leakage inductance, and the mutual inductances are computed with the finite element model. Due to large air-gap and large slots, these inductances are linear, practically. The command strategy of the rectifier depends on e.m.f. waveform, imposing rectangular or sinusoidal currents, with or without harmonics injection. For the rectifier, the model allows also to introduce the parameters of each component (diode, transistor, snubber) which permits to estimate the converter losses.

The DC bus allows the decoupling of the converters, and the simulation model will be simplified. In this way the group inverter-transformer-grid is replaced by a DC voltage source connected in the DC bus circuit (Fig. 11). The control system imposes the generator speed depending on the wind speed, and using a speed-torque characteristic, a reference torque is generated, which consists in imposing a reference for the quadrature component of the current, *Iq*. The direct component, *Id*, is imposed zero. Finally, using an inverse Park transformation, the command voltages are generated (Fig. 12). Therefore, for both types of machine, the currents are imposed in phase with the e.m.f. (Fig. 13). For the trapezoidal waveform machine, the currents are controlled using hysteresis regulators.

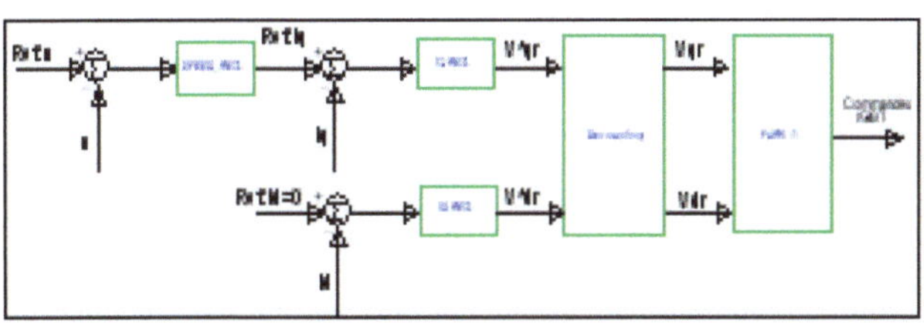

Figure 12. Control block for sinusoidal waveform machine.

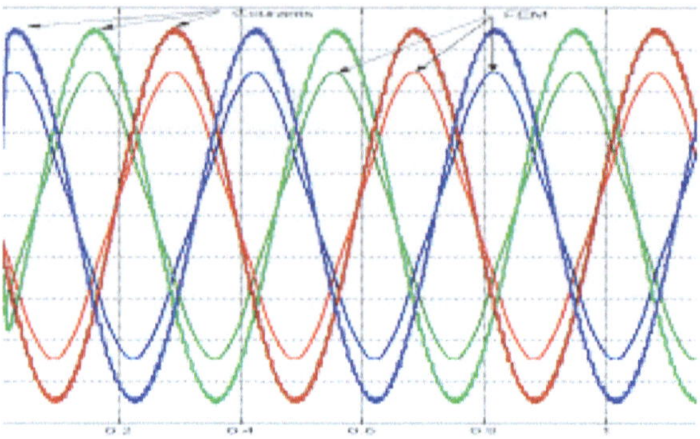

Figure 13. E.m.f. and currents for sinusoidal machine.

Sinusoidal waveforms

In the case of sinusoidal waveforms, the e.m.f. contains fifth and seventh order harmonics, which induce torque oscillations if the current is purely sinusoidal. It is possible to inject current harmonics, as suggested by [2], in order to minimize the torque ripple, but these current harmonics influence also the harmonic content of the DC bus current. An FFT analysis of the electromagnetic torque (Fig. 14) shows a small 6th and 12th harmonics, which confirms the results obtained with the finite element model. Compared to the results presented in Fig. 8, high frequency harmonics are associated with the PWM frequency.

For sinusoidal waveforms, the DC bus current obtained has the shape with an envelope done by the three-phase currents (Fig. 15). The harmonic content has very small low-frequency harmonics, and taking into consideration that the DC bus voltage is kept constant, the power transferred through the DC bus has low fluctuations.

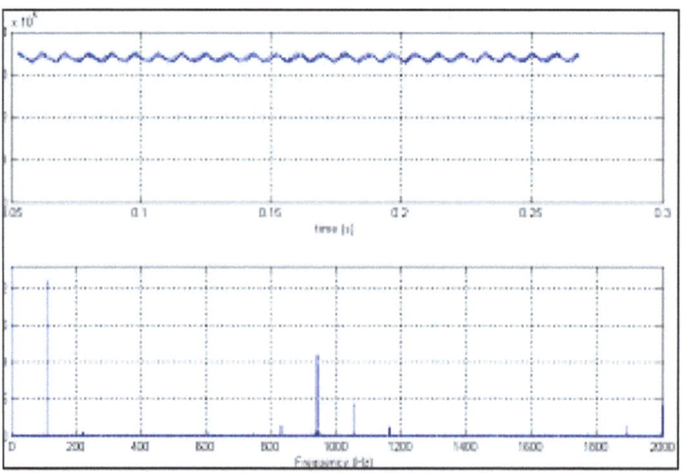

Figure 14. Electromagnetic torque for sinusoidal currents.

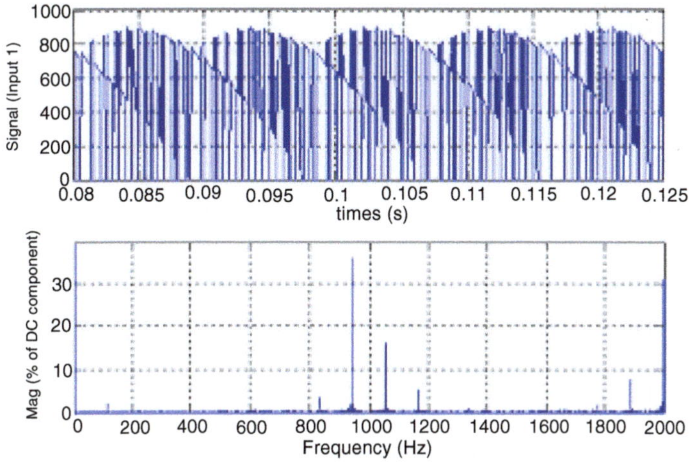

Figure 15. DC bus current for sinusoidal phase currents.

Trapezoidal waveforms

For trapezoidal e.m.f. waveforms, the imposed currents have square-waveforms, in phase with the e.m.f. (Fig. 16). In this case problems are encountered during the commutation between two phases. The current in the phase not under commutation has a distortion which depends on the DC bus voltage, the rotational speed, the phase resistance, and the phase inductance. Proper values of these parameters lead to a minimum torque ripple (Fig. 17).

In this case the DC bus current presents intervals where it drops to zero (Fig. 18), which means that energy transfer to the grid will have high oscillations. These oscillations have low frequencies (6*k* order) and depend on the rotational speed. Therefore, an additional filter on the DC bus is required to reduce them.

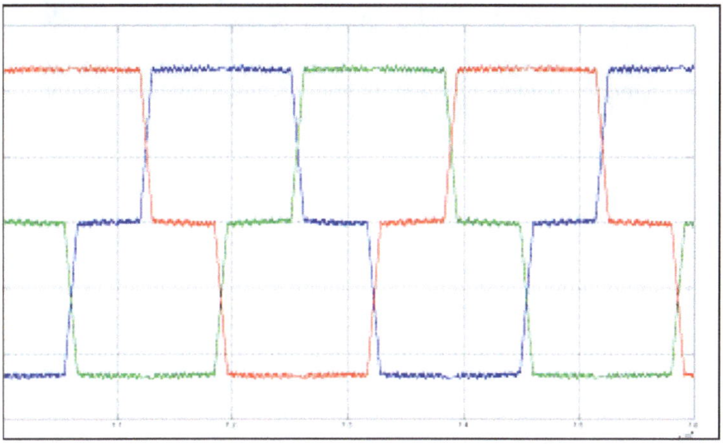

Figure 16. Currents for trapezoidal waveform machine.

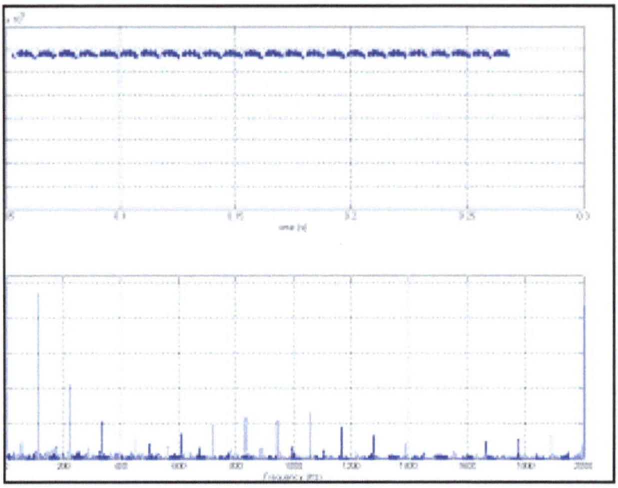

Figure 17. Electromagnetic torque for trapezoidal phase currents.

The number of turns/coil and the slots area decrease significantly for a trapezoidal waveform machine, compared to a sinusoidal waveform machine. Then, the phase resistance and inductance decrease too. The phase resistance for the trapezoidal waveform machine will be:

$$R_{(tr)} = R_{(\sin)} \frac{N_{(tr)}}{N_{(\sin)}} \cdot \frac{S_{(tr)}}{S_{(\sin)}} \tag{8}$$

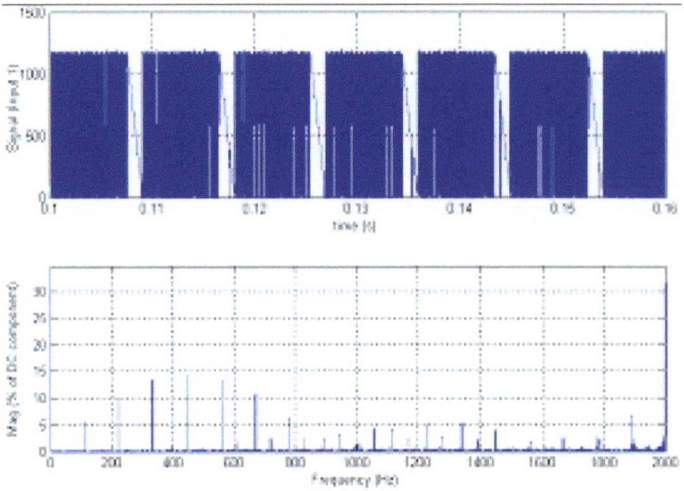

Figure 18. DC bus current for trapezoidal phase currents.

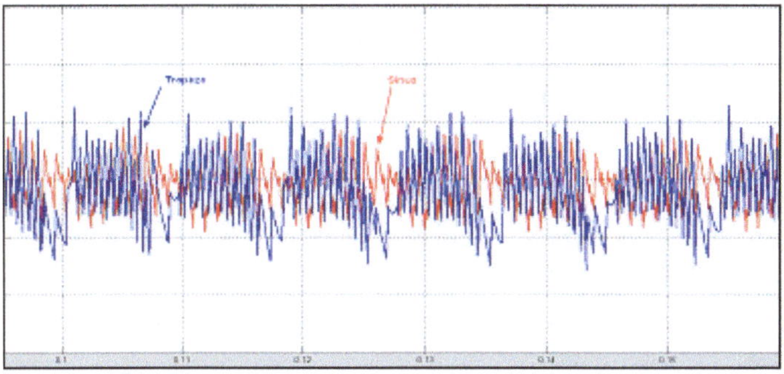

Figure 19. Comparison between electromagnetic torque of the sinusoidal and trapezoidal waveforms machines.

where N is the number of turns/coil and S is the number of slots of the studied machines. In order to have the same warming of stator winding with the same stator core volume for both machines, equal copper losses are imposed for both machines:

$$P_{J(\sin)} = 3 \cdot R_{(\sin)} I_{rms}^2 \tag{9}$$

$$P_{J(tr)} = 2 \cdot R_{(tr)} I_{(tr)}^2 \tag{10}$$

Comparing with the sinusoidal waveform machine, a comparable level of torque ripple is obtained for a magnet width equal to 90% of the pole pitch (Fig. 19). The trapezoidal waveform machine has 150% of the magnets volume but only 48% of the copper volume of the sinusoidal waveform machine. In this case, the copper losses of trapezoidal waveform machine are lower.

The electromagnetic torque fluctuation obtained for the trapezoidal waveforms machine depends on the width of the magnets (Fig. 20). Larger magnets give an electromagnetic torque with minimum oscillation. Decreasing the width of the magnets generates higher torque oscillations and lower average torque value (Fig. 20).

In Table 2, the electromagnetic torque obtained by a trapezoidal waveform machine and different magnets width is compared to the torque obtained by a sinusoidal waveform

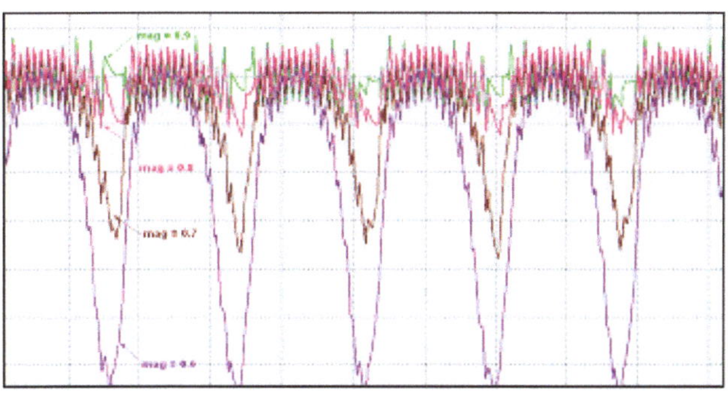

Figure 20. Electromagnetic torque for different magnet's width.

Table 2. Comparison of electromagnetic torque characteristics for different magnet width

Magnet	Torque (pu)	H6 (%)	H12 (%)	H18 (%)	H24 (%)
0.6	1.10	16.7	7.00	3.10	1.32
0.7	1.20	6.15	2.98	1.80	1.15
0.8	1.25	2.27	1.03	0.63	0.34
0.9	1.28	0.985	0.52	0.36	0.10

machine with the same copper volume. If the copper volume is the same for both machines, the rated current of the trapezoidal waveform machine can be increased to obtain the same amount of losses as the sinusoidal machine. As the stator core volume is not changed, the stator cooling area is the same such as the warming of coils.

The power-to-weight ratio of trapezoidal waveform machines is 28% higher than sinusoidal machine but the magnets weight is 50% higher (Table 2).

Conclusions

A study concerning the influence of the e.m.f. and currents waveform over the torque ripple and over the DC bus current harmonics has been done. Detailed analytical and numerical approaches were elaborated for sinusoidal and trapezoidal waveforms. Machines models have been elaborated using finite element models, and the system machine-converter has been simulated. The interest is to study the possibility of increasing the power density of permanent magnet synchronous machine without altering the quality criteria imposed for wind generator systems: low torque oscillation and low DC bus current harmonics.

Simulations show that increasing the magnets width, higher power density is obtained for trapezoidal waveforms machine. But problems are also encountered due to the fact that the DC bus voltage should be adapted to the rotational speed to avoid phase current distortions and torque ripple. Another disadvantage is that the DC bus current has high low-frequency harmonics, compared to the sinusoidal waveform machine. That means that additional active and/or passive components should be added, increasing the price of the converter.

A solution to decrease the harmonics of the DC bus current is to increase the phase number. Even for sinusoidal waveform machine, poly-phased machine seems a good solution to reduce the torque ripple, to minimize the DC bus current oscillation, and to overcome the technological limits of the power electronic components.

Acknowledgment

The work presented in this paper was done within ARCHIMED project of "Centre National de Recherche Technologique en Génie Electrique", with the support of ERDF, French government, and Région Nord—Pas de Calais.

References

[1] M.R. Dubois, "Review of Electromechanical Conversion in Wind Turbines", Report EPP00.R03, TU Delft, ITS Faculty, The Netherlands, April 2000.

[2] M.R. Dubois, H. Polinder, J.A. Ferreira, "Comparison of Generator Topologies for Direct-Drive Wind Turbines", Proceedings of the Nordic Countries Power and Industrial Electronics Conference (NORPIE), Denmark, 2000, p. 22–26.

[3] J.F. Gieras, M. Wing, Permanent Magnet Motor Technology, 2nd edition, Revised and Expanded Marcel Dekker Inc., New York.

[4] J.R. Henderson Jr., T.J.E. Miller, Design of Brushless Permanent-Magnet Motors, Oxford: Magna Physics Publishing and Calderon Press, 1994.

III-3.1. EQUIVALENT THERMAL CONDUCTIVITY OF INSULATING MATERIALS FOR HIGH VOLTAGE BARS IN SLOTS OF ELECTRICAL MACHINES

P.G. Pereirinha[1,2] and Carlos Lemos Antunes[1]

[1]ISR-Lab. CAD/CAE, University of Coimbra, 3030-290 Coimbra, Portugal
ppereiri@isr.uc.pt
[2]Inst. Sup. Engenharia de Coimbra. Rua Pedro Nunes, 3030-199 Coimbra, Portugal
lemos.antunes@deec.uc.pt

Abstract. The equivalent thermal conductivity of insulating materials for a high voltage bar in slots of electrical machines is calculated using the finite element method. This allows the use of much coarser meshes with an equivalent thermal conductivity k_e, without accuracy loss in the hot spot temperature calculation. It is shown the dependency of k_e value on the equivalent mesh used. Some considerations are also presented on the heat flux finite element calculation.

Introduction

One of the major thermal problems in electrical machines is the steady-state hot spot temperatures in the windings, which are responsible for its thermal aging and degradation. So it is important to correctly determine those hot spots in the thermal design of the machine.

In the thermal finite element (FE) modeling of electrical machines, all the different materials to be crossed by the heat flux should be considered, namely the windings insulations. For a bundle of conductors a simple explicit formula for the thermal conductivity [1] and a statistical approach for the temperature calculation [2] were presented elsewhere. Another study was presented [3] for a voice coil loudspeaker motor in which the real coil was replaced by an equivalent bulk coil.

A different problem arises in electrical machine slots with high voltage bars containing several different narrow insulation materials. Despite the amazing development of the computer and FE software capabilities it is still a problem to model the different materials when analyzing the thermal problem for a full machine due to the different dimensions involved and the consequent huge number of nodes required for the mesh discretization.

The aim of this paper is to present a method to replace the several insulation layers by a single layer with equivalent thermal conductivity allowing the use of much less number of FE elements without significant accuracy loss, namely in the bar maximum temperatures.

S. Wiak, M. Dems, K. Komęza (eds.), Recent Developments of Electrical Drives, 413–422.

Formulation

The classic heat diffusion model is [4,5]

$$\rho c \frac{dT}{dt} + \nabla(-k\nabla T) = q \tag{1}$$

where ρ is the density [kg/m^3], c the specific heat capacity (also called specific heat) [J/kgK], T the temperature [K], ∇ the nabla differential operator, k the thermal conductivity [W/(Km)], and q the thermal sources [W/m^3].

The boundary conditions depend on the problem type. The heat flow due to conduction is given by the Fourier's law,

$$\vec{F} = -k\nabla T \tag{2}$$

where $\vec{F}$ is the heat flux vector [W/m^2], and the heat flux ϕ_h [W] crossing a surface, closed or not, is then given by

$$\phi_h = \int \vec{F}\hat{n}\, dS \tag{3}$$

where $\hat{n}$ is the unit outer normal vector to the boundary. In a Cartesian coordinate system (3) becomes

$$\phi_h = \int_{yz} F_x\, dydz + \int_{xz} F_y\, dxdz + \int_{xy} F_z\, dxdz \tag{4}$$

where F_x, F_y, and F_z are the components of the heat flux density vector $\vec{F}$ in the x, y, and z directions respectively.

The case of the steady-state heat transfer problem, is described by the following partial differential equation [4,5]:

$$\nabla(-k\nabla T) = q \tag{5}$$

where the thermal sources q for the present problem are only the Joule losses in the bar copper which can be given by

$$q = \rho_0 J^2 \tag{6}$$

where ρ_0 is the electric resistivity [Ωm] at a reference temperature T_0 [K] and J the current density [A/m^2], or by

$$q = \rho_0(1 + \alpha(T - T_0))J^2 \tag{7}$$

where α is the linear expansion coefficient [K^{-1}], if it is necessary to consider the resistivity variation with the temperature rise [6].

The thermal problem was solved using first order FE thermal processor [6] of our finite element package CADdyMAG.

Case study and methodology

As the case study it was considered the bar presented in Fig. 1(a), of a three phase high voltage synchronous generator (1 MVA, 6.3 kV) driven by a hydraulic turbine of 250

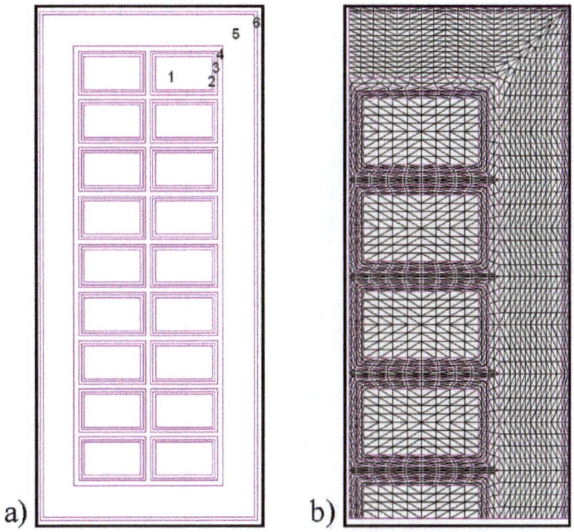

Figure 1. (a) Bar with 9 × 2 wires; (b) original FE mesh (2,035 nodes/3,920 elements) for 1/4 of the bar in a slot.

rpm. Each side of the winding (bar) consists of 9 × 2 copper wires ("1" in Fig. 1a) each one individually insulated with 0.2 mm layers of paper and cotton ("2" and "3" in Fig. 1a) packed together with a 0.4 mm "bitumen" layer and a final 2.5 mm layer of molded micanite ("4" and "5" in the same figure). Finally a 0.25 mm impregnation resin layer between the bar and the slot was also considered ("6" in Fig. 1a). As in each slot there are two bars with an additional spacer between them and supposing the analysis of half machine with 48 slots, this would lead to nearly 400,000 nodes only to model the stator slots in a 2D problem.

The idea is then to replace the different insulator materials by only one bulk material with a considered equivalent thermal conductivity k and significantly lower number of elements and nodes of the corresponding finite element mesh.

Due to symmetry it can be analyzed only 1/4 of the bar, and it was used for the original bar a mesh with 2,055 nodes and 3,920 first order finite elements, as shown in Fig. 1(b).

To the nominal current of 91.64 A, flowing in two wires in parallel, corresponds a current density $J = 3.916$ A/mm^2. In the original bar the insulators thermal conductivities vary from 0.17 to 0.7 W/(Km), and to solve the original bar problem it was considered a boundary condition corresponding to a temperature in the slot border $T_{ref} = 373.15$ K (100°C). It was used (5) for the thermal sources. The steady-state thermal conduction problem was solved using first order FE thermal processor of our FE package CADdyMAG [6]. The temperature results obtained for the original bar are presented in Fig. 2(a) and in Fig. 2(b) the heat flux density vector distribution can be seen.

The hot spot temperature rise for the bar

$$\Delta T_{ref} = T_{max} - T_{ref} \tag{8}$$

was $\Delta T_{ref} = 10.103$ K.

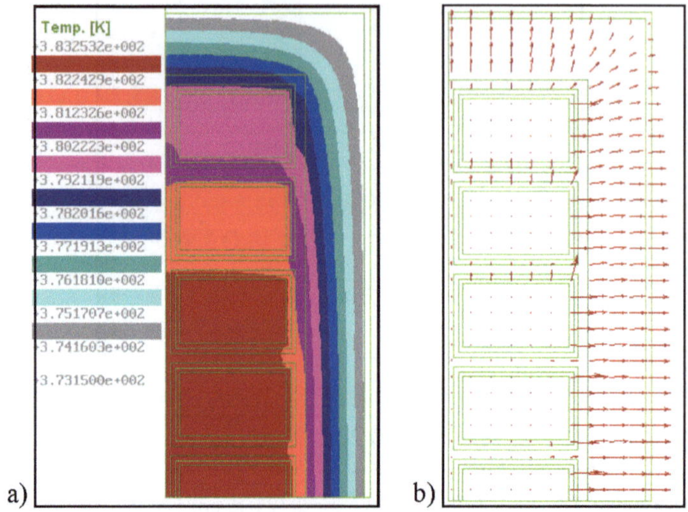

Figure 2. Original bar temperature; and (b) heat flux density vector F distribution.

To easily calculate the heat flux leaving the bar one can see that, for a 2D problem in the xy plane, like the presented in Figs. 1–5, F_z in (4) is zero. So (4) can be simplified to

$$\frac{\phi_h}{l} = \int_y F_x\,dy + \int_x F_y\,dx = -\int_y k\frac{\partial T}{\partial x}\,dy - \int_x k\frac{\partial T}{\partial y}\,dx \tag{9}$$

where l is the bar length in the z direction.

The thermal flux ϕ_h/l [W/m] leaving the cable crossing lines 1, 2, or 3 (red lines in Fig. 3), as the integration path only crosses one material with constant thermal conductivity,

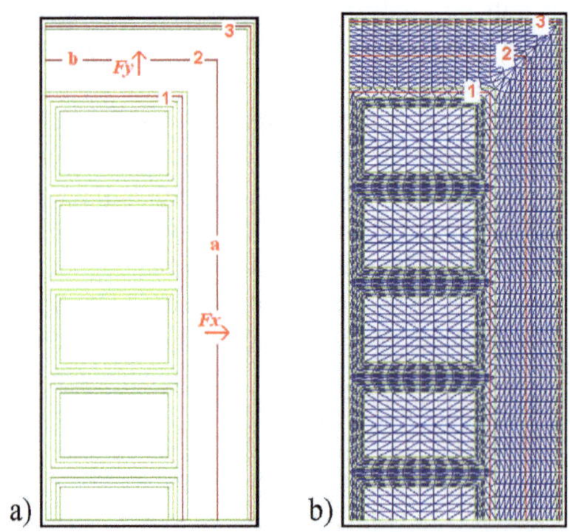

Figure 3. (a) Integration lines for the bar; (b) relative positions of lines 1 to 3 to the mesh elements.

is given by

$$\frac{\phi_h}{l} = \int_{y=a} F_x \, dy + \int_{x=b} F_y \, dx = k\left(-\int_{y=a} \frac{\partial T}{\partial x} dy - \int_{x=a} \frac{\partial T}{\partial y} dx\right) \tag{10}$$

As the lines chosen for the integral calculation are parallel to the Cartesian axes, it is interesting to note that for the vertical part of the integration lines (line "a" in Fig. 3a) only F_x is to be considered and for the horizontal part of the integration lines (line "b" in Fig. 3a) only F_y is to be considered.

As the lines chosen for the integral calculation are parallel to the Cartesian axes, it is interesting to note that for the vertical part of the integration lines (line "a" in Fig. 3a) only F_x is to be considered and for the horizontal part of the integration lines (line "b" in Fig. 3a) only F_y is to be considered.

The thermal sources q' [W/m] for 1/4 of the bar are $q' = 13.91942$ W/m. The heat flux crossing lines 1, 2, and 3 in Fig. 3 was calculated by using (10) for the original bar and it was confirmed that it is equal to the thermal sources q' (with an error of 0.839%, 0.256%, and −0,083%, respectively). So it was checked that (5) is verified and the finite element solution is validated.

As a methodology we have chosen to replace the original bar by one composed by copper, where the current losses are produced, surrounded by a bulk insulation material. Two different approaches were considered: first, replace the original bar by an equivalent one with the same copper distribution ("Equal," Fig. 4) and second, replace it by a bulk bar with all the copper area concentrated ("Conc.," Fig. 5) in only one bigger wire.

The idea of keeping the same copper area as in the original bar is to use the same current density J values for both the thermal and magnetic problems, although other possibilities may be considered. For both cases, the five different insulating materials were replaced by only one equivalent material. Two different meshes for each case were also considered: fine meshes ("Fine") and coarse meshes ("Coarse"). For the models with the same copper

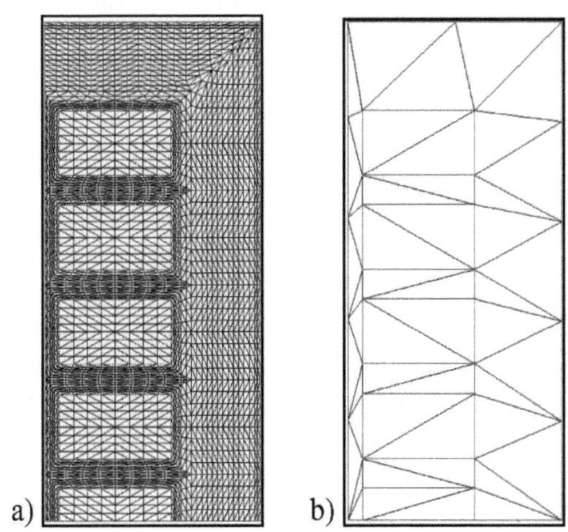

Figure 4. Models with same copper distribution, "Equal": (a) "Fine" mesh (2,035 nodes/3,920 elements); (b) "Coarse" mesh (33 nodes/49 elements).

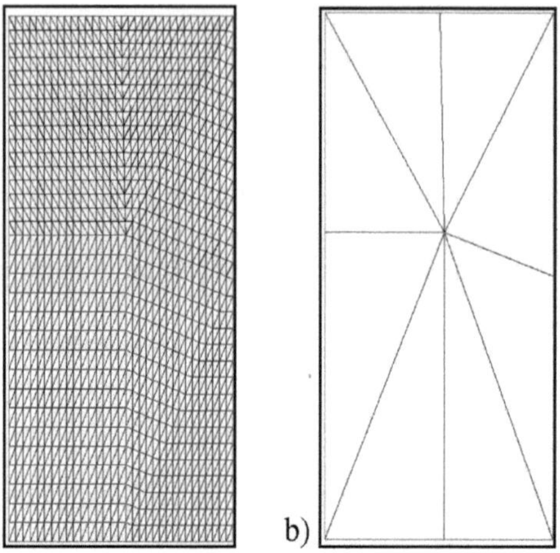

Figure 5. Models with different copper distribution, concentrated "Conc.": (a) "Fine" mesh (1,089 nodes/2,048 elements); (b) "Coarse" mesh (nine nodes/eight elements).

distribution, "Equal," the fine mesh (Fig. 4a) has the same geometry as the original bar mesh (Fig. 1b), and the coarse mesh (Fig. 4b) has the minimum number of nodes and elements required for the analysis. For the models with concentrated copper, "Conc.," the fine mesh (Fig. 5a) has about half of the nodes of the original bar (Fig. 1b), and the coarse mesh (Fig. 5b) has also the minimum number of nodes and elements required for the analysis.

As in the proposed methodology the heat does not have to cross the whole bar from side to side (what would probably lead to the consideration of two different thermal conductivities, one for the horizontal and another for the vertical directions of the bar), but instead it is generated inside the bar, the equivalent thermal conductivity was considered isotropic.

The issue here is how to calculate the value of the equivalent thermal conductivity. To calculate this, the steady-state thermal problem for the four cases mentioned before (concentrated and equal copper distribution, for both a very coarse and a fine mesh) were solved with thermal conductivities k ranging from 0.2 to 0.6 W/(Km) and the hot spot temperature rise were calculated (Fig. 6). An equivalent thermal conductivity k_e was calculated as will be seen in more detail in the next section "Results and Validation."

Results and validation

The steady-state thermal problem (5) for the four equivalent meshes in Figs. 4 and 5 was solved for several thermal conductivities ranging from 0.2 to 0.6 W/(Km), as in the bar the real insulators thermal conductivities vary from 0.17 to 0.7 W/(Km). The hot spot temperature raise ΔT is given by

$$\Delta T = T_{max} - T_{ref} \tag{11}$$

where T_{max} is the maximum temperature for each mesh, and is presented in Fig. 6 with thick color lines.

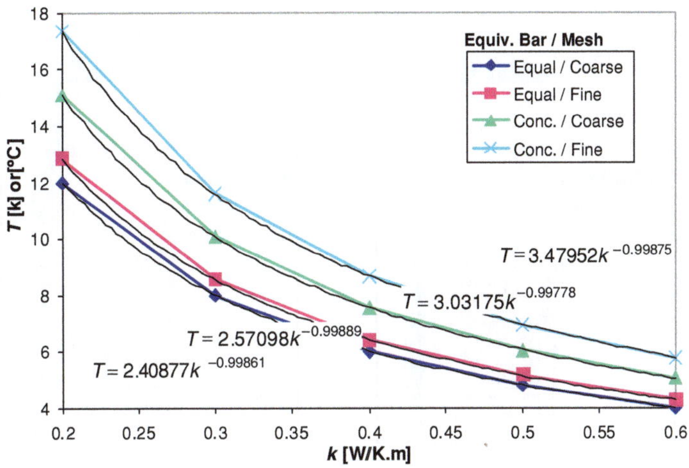

Figure 6. Hot spot temperature rise as a function of thermal conductivity for the different meshes.

It is seen that the curves presented can be perfectly fitted by power functions (black thin lines) in the form

$$\Delta T = ak^{-b} \tag{12}$$

where "*a*" and "*b*" are coefficients given in Fig. 6 using the "trendline" function of Microsoft Excel©.

To obtain the same hot spot rise of the original bar and mesh, ΔT_{ref}, the equivalent thermal conductivity k_e for each particular FE mesh (as the temperature solution will depend on the mesh) can then be very easily calculated by

$$k_e = \left(\frac{\Delta T_{ref}}{a}\right)^{-1/b} \tag{13}$$

The values of the "*a*" and "*b*" coefficients as well as the resulting k_e are presented in Table 1.

At this point, it should be mentioned that for the thermal sources, it should not be used the more accurate expression (7) which accommodates the resistivity variation with the temperature. Indeed, although (7) is the expression that should be used to solve the global thermal problem, for the particular case presented in this communication, it is mandatory to use (6). The reason is that if (7) is used the thermal sources will not remain constant and

Table 1. Coefficients and equivalent conductivities k_e

Bar/mesh	"*a*" coefficient	"*b*" coefficient	k_e, W/(Km)
Equal/coarse	2.40877	0.99861	0.237946
Equal/fine	2.57098	0.99889	0.25409
Conc./coarse	3.03175	0.99778	0.299282
Conc./fine	3.47952	0.99875	0.343945

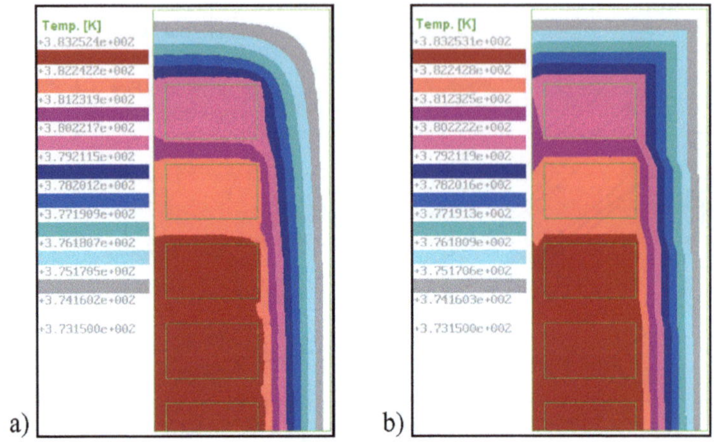

Figure 7. Temperature distribution for equal copper distribution, "Equal": (a) Fine mesh (2,035 nodes/3,920 elements), $k_e = 0.254090$; (b) Coarse mesh (33 nodes/49 elements), $k_e = 0.237946$.

consequently the results of the hot spot temperature rise ΔT would not be only a function of the thermal conductivity (as are those presented in Fig. 6) but also of the current density and of the boundary condition T_{ref}.

Using the equivalent k_e calculated by (13) and presented in Table 1, the thermal conduction problem was solved and the results for the fine and coarse meshes are presented in Figs. 7 and 8, for the "Equal" and "Conc." bars respectively.

Comparing these results with the reference ones in Fig. 2(a) it can be seen that the reference hot spot temperature, $T_{ref} = 383.253$ K ($\Delta T_{ref} = 10.103$ K), is obtained with the very small errors presented in Table 2.

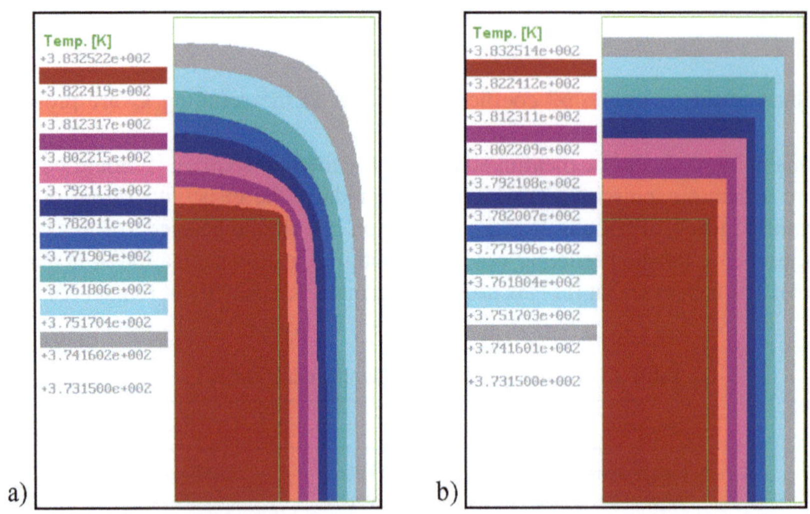

Figure 8. Temperature distribution for concentrated copper distribution, "Conc.": (a) Fine mesh (1,089 nodes/2,048 elements), $k_e = 0.343945$; (b) Coarse mesh (9 nodes/8 elements), $k_e = 0.299282$.

Table 2. Errors in hot spot temperature

Bar/mesh	Nodes/elements	k_e, W/(Km)	ΔT(K)	Error (%)
Equal/coarse	33/49	0.237946	10.1031	0.0010
Equal/fine	2,035/3,920	0.254090	10.1024	−0.0059
Conc./coarse	9/8	0.299282	10.1014	−0.0158
Conc./fine	1,089/2,048	0.343945	10.1020	−0.0099

Table 3. Errors in hot spot temperature for slightly less coarse meshes

Bar/mesh	Nodes/elements	k_e, W/(Km)	ΔT(K)	Error (%)
Equal/coarse2	35/53	0.241042	10.1031	0.0010
Conc./coarse2	10/10	0.319620	10.1017	−0.0129

It can also be seen that for the equivalent concentrated bar "Conc." the temperature distribution as well as the average copper temperature has some differences to the original bar. However, for the "Equal" mesh (Figs. 4b and 7b), which has 33 nodes, i.e. 62 times less nodes than the original one (Figs. 1b and 2a, 2,035 nodes), a very similar temperature distribution is obtained.

The temperature distribution in the coarse meshes can be further improved by simply adding two or one more nodes in the line from the upper right corner of the copper to the upper right corner of the FE mesh in Figs. 4(b) and 5(b), respectively, as presented in Fig. 9, where the thermal solution is plotted along with the new meshes "Coarse2." The k_e for these two new meshes are presented in Table 3, with the resulting hot spot temperature rise and the correspondent errors.

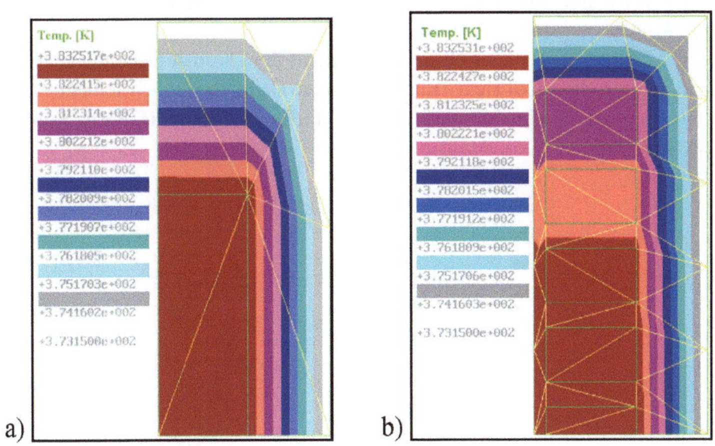

Figure 9. Temperature distribution for concentrated and equal copper distribution, with slightly less coarse meshes "Coarse 2": (a) "Conc." (10 nodes/10 elements), $k_e = 0.319620$; (b) "Equal" (35 nodes /53 elements), $k_e = 0.241042$.

It was also numerically verified that, as expected, the thermal equivalent conductivity k_e for each mesh is independent of the thermal source densities and boundary temperature. Indeed the problem was solved for two other different values of current densities (two and four times more) and two different values of boundary temperatures, which lead to the same results as those obtained by the original bar with the same conditions.

Conclusions

The presented method seems to be able to calculate very accurately the global equivalent thermal conductivity of any bar with several thin insulation materials and not trivial geometries with only conduction heat transfer.

It has been shown that the equivalent thermal conductivity depends on the mesh used.

The method performs a very good fitting of the hot spot temperature in the considered multilayer insulated bar in electrical machine slot with much less computational costs than modeling all the insulation materials.

The numerical thermal solution was checked by verifying that the total heat flux through the bar boundary was equal to the thermal sources applied.

References

[1] E. Matagne, "Macroscopic Thermal Conductivity of a Bundle of Conductors", Conference on Modelling and Simulation of Electrical Machine and Static Converters (IMACS TC1'90), Nancy, France, September 1990, pp. 189–193.

[2] E. Chauveau, E. Zaim, D. Trichet, J. Fouladgar, A statistical approach of temperature calculation in electrical machines, IEEE Trans. Magn., Vol. 36, pp. 1826–1829, 2000.

[3] M. Dodd, "The Application of FEM to the Analysis of Loudspeaker Motor Thermal Behavior", 112th Audio Engineering Society Convention (AES 112th Convention), München, Germany, May 2002.

[4] J. Holman, Heat Transfer, 6th edition, New York: McGraw-Hill Book Company, 1986.

[5] J.C. Coulomb, J.C. Sabonnadiere, CAO en Électrotechnique, Paris, France: Hermes Publishing, 1985, pp. 41–43.

[6] J. Pinto, C.L. Antunes, A.P. Coimbra, "Influence of the Thermal Dependency of the Windings Resistivity in the Solution of Heat Transfer Problem Using the Finite Element Approach", Proc. of the International Conference on Electrical Machines (ICEM94), Paris, France, September 1994, pp. 448–451.

III-3.2. LOSS CALCULATIONS FOR SOFT MAGNETIC COMPOSITES

Göran Nord[1], Lars-Olov Pennander[1] and Alan Jack[2]

[1]Höganäs AB, SE-263 83 Höganäs, Sweden,
Goran.Nord@hoganas.com, Lars-olov.pennander@hoganas.com
[2]University of Newcastle upon Tyne, Newcastle upon Tyne, NE1 7RU England,
alan.jack@ncl.ac.uk

Abstract. This paper focuses on iron loss measurements and simulations in perspective to evaluate a loss model that can be used for SMC materials in FEA.

Introduction

Soft Magnetic Composites (SMCs) are today a viable alternative to steel laminations in a range of electromagnetic applications, such as machines, sensors, and fast switching solenoids. SMC components are efficiently manufactured using the well established powder compaction process. The isotropic nature of the SMCs combined with the unique shaping possibilities opens up for new 3D-design solutions. If carefully implemented advantages such as better performance, reduced size, and weight, less number of parts at low cost can be obtained.

In Fig. 1 an example of an SMC component is displayed and in Fig. 2 the stator assembly is displayed with one SMC part missing.

This paper focuses on simulation models of losses in SMC components. A method for the simulation of iron losses in SMC components is presented. The approach is different from what is described in [1].

Iron losses

During the design process for a machine it is essential to predict the iron losses. This also applies when using SMC in the soft magnetic parts. The total iron losses are basically divided into hysteresis, eddy current, and anomalous losses according to the general equation:

$$P_t = P_h + P_e + P_a \tag{1}$$

Simulation tools such as FEA software have built-in models for the simulation of iron losses. These models are developed and adapted mainly for lamination steel materials and are normally not directly applicable on SMCs without modifications.

S. Wiak, M. Dems, K. Komęza (eds.), Recent Developments of Electrical Drives, 423–433.

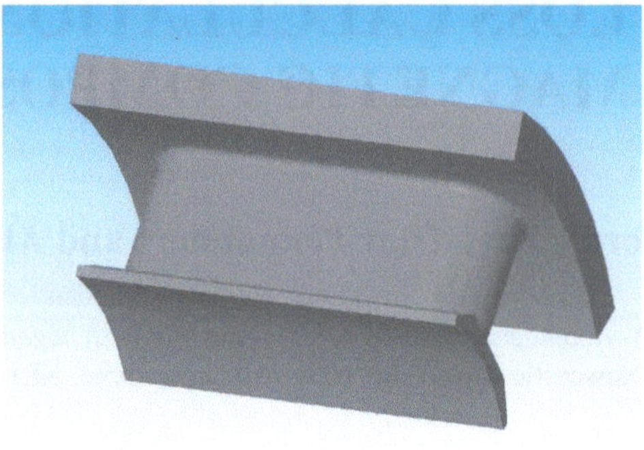

Figure 1. SMC stator component. Design Alan Jack, Newcastle University.

When using laminations the size factor is the lamination thickness that is equal through the whole motor. In a motor using SMC material the size factor is different depending on the components geometry and has to be calculated for each design. The components cross-section geometry will influence the value of the eddy current losses, the question is how much and in what way.

Material loss data in the case of laminations emanates from the Epstein test that mainly reflects the results in the plane of a single sheet. Therefore the simulated results are usually adjusted based on experience using established design factors. The SMC material must be considered from a different perspective due to its homogeneous isotropic magnetic properties.

Simulation of losses using FEA for laminations are often done for a single sheet in 2D and then are the results multiplied by the total number of sheets of the actual core. The geometrical distribution of losses is made by the calculated B-field. SMC components can theoretically be seen as single lamina structures with variable cross-sections/thickness and thereby cannot be treated as a modular system like the laminated structure. For SMCs it is though possible to separately calculate the eddy current losses by FEA for a homogeneous isotropic core with known electrical conductivity. A limited Steinmetz model can be used covering the basic hysteresis loss.

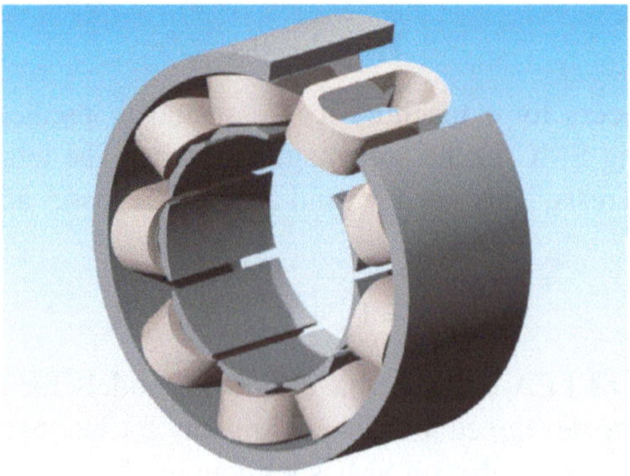

Figure 2. SMC stator. One SMC tooth missing. Design Alan Jack, Newcastle University.

Figure 3. Ring test samples.

Experimental measurements

Twelve slugs of different heights were compacted from a SMC powder material. The slugs were heat-treated using a non-optimized procedure with an estimated time and temperature resulting in slightly reduced material performance compared to parts in production. The electrical conductivities and BH-curves were therefore fluctuating somewhat between the rings. From the slugs 12 ring samples of different diameters and heights were manufactured by wire erosion, Fig. 3. The choice of ring dimensions were determined with a view to maintain the recommended ratio of inner to outer diameter of not less than 0.82 according to standard IEC60404-6 in order to limit non-uniform flux distribution.

To more easily be able to recognize the connection between eddy current losses and component size, the rings were made in three different logical series according to Fig. 4 using ring size 1 as the basic size. There were two rings of size 1A and B.

Two separate windings were put on each of the rings. Inner winding was used as a sense winding and outer winding to apply the AC field, Fig. 5. The number of coil turns

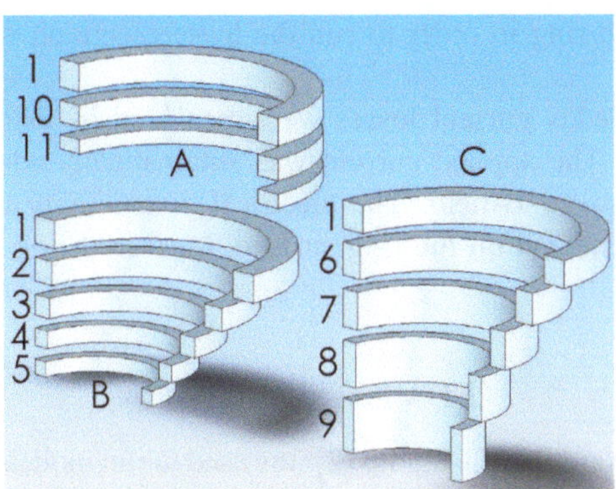

Figure 4. Ring test samples A, B, and C series. Series A: decreasing rectangular cross-section; Series B: decreasing square cross-section, and Series C: increasing rectangular cross-section—constant cross-section area.

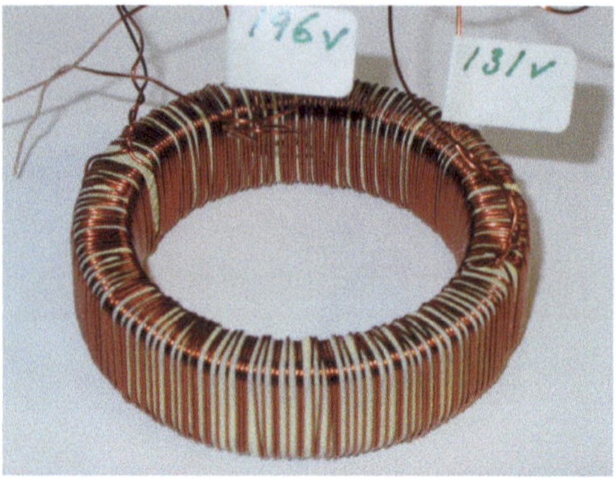

Figure 5. Wounded test ring.

Table 1. Ring sample data

Ring no.	ID (Mm)	OD (mm)	Th (mm)	Weight (g)	Density (kg/m^3)	Conductivity (S/m)
1A	94.08	115.24	10.62	269.92	7,301.8	14,451
1B	94.06	115.20	10.57	268.15	7,298.7	18,373
2	80.43	98.37	9.00	166.05	7,320.2	19,697
3	68.72	84.11	7.72	104.42	7,319.9	22,693
4	58.79	71.91	6.62	65.44	7,340.6	27,370
5	50.22	61.45	5.73	41.34	7,325.8	24,970
6	80.41	98.40	12.34	227.83	7,307.5	23,794
7	68.77	84.10	14.41	194.61	7,338.9	24,840
8	58.76	71.89	16.80	166.07	7,333.8	26,049
9	50.26	61.45	19.76	142.67	7,356.4	25,872
10	94.05	115.19	7.94	201.19	7,300.2	19,109
11	93.98	115.16	5.38	136.04	7,266.8	17,201

was optimized for each ring in order to run the hysteresisgraph within its measurement range.

The hysteresis and eddy current losses were measured by using a Brockhaus MPG-100D hysteresisgraph. The applied current was automatically adapted to ensure a pure single frequency sinusoidal B-field. The electrical conductivity of the rings was measured by a four-point method on the ring surface. Measured ring data is summarized in Table 1.

FEA loss simulations

Commercial FEA software JMAG-Studio [2] was used to run models of all the ring samples according to Table 1, with applied AC currents, measured electrical conductivities, and measured BH-curves. The FEA model was made very close to the measurement conditions. Model size was reduced by symmetries. In Fig. 6 the FEA model for ring sample number

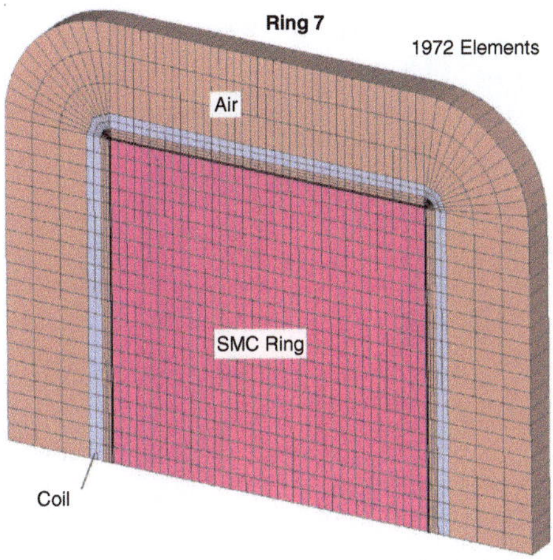

Figure 6. FEA model ring sample no. 7.

7 can be seen. Only a sector of one degree and the upper half of the ring were modeled. The boundary conditions were applied on the symmetry planes and on the outside of the air region. The current was set to give a flux density of 1 T in the cross-section.

It can be seen in Fig. 7 that the flux density is almost uniform over the cross-section.

In Fig. 8 simulated eddy currents together with the flux density are shown in a ring cross-section.

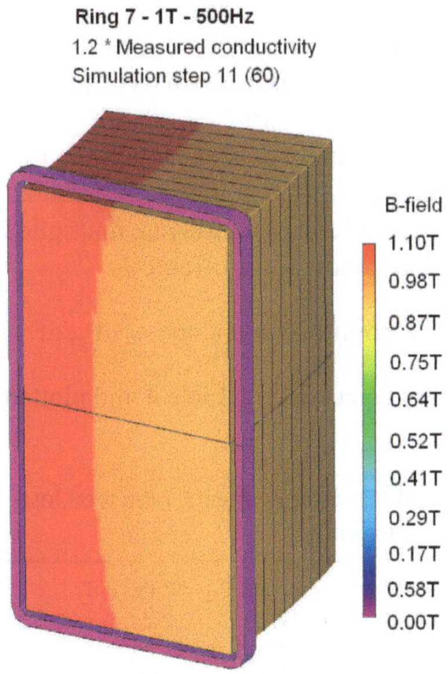

Figure 7. Simulation of B-field distribution in the ring cross-section. Part of ring (11°) and coil (1°).

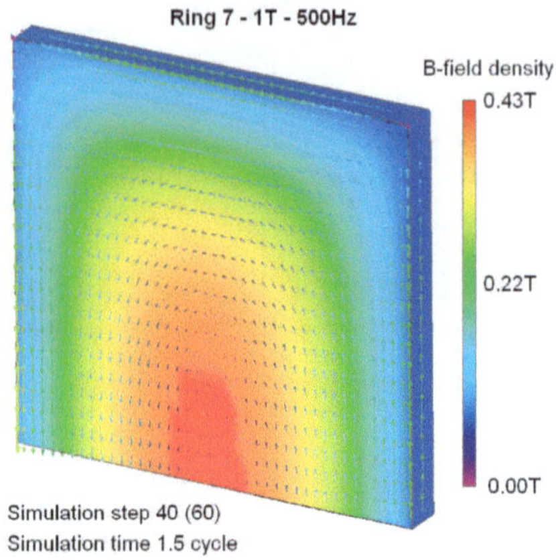

Figure 8. Simulation of B-field distribution and eddy currents in the ring cross-section.

Results

The separation of hysteresis and eddy current losses from the measured total loss is demonstrated mainly on ring sample 1A. The hysteresis loss was assumed to be of first order in frequency and eddy current losses of second order in frequency. In Table 2, with the frequency in the first column, measured flux density, the total measured losses, and the ratio total loss/frequency are shown.

The ratios are plotted in Fig. 9 and it can be seen that the ratio is a straight line. The value where the line is cutting the y-axle is the hysteresis loss at 0 Hz and for ring 1A this value is 0.1194 J/kg. By subtracting the measured total loss by this hysteresis loss value a separation can be made between hysteresis losses and eddy current losses for each frequency, Table 2.

In Table 3 measured and calculated AC-slopes with the measured conductivity (1.0 × conductivity) are shown. The mean value of the ratio measured slope/simulated AC-slope was calculated to 1.2. The measured conductivities are multiplied with 1.2 to create modified measured conductivities (1.2 × conductivity), Table 3.

$$\text{Definition: AC-slope} = dP_e/df$$

The AC-slopes for ring 1A are calculated in Table 4 and plotted in Fig. 10.

Table 2. Ring 1A total loss measurements, ratio total loss/frequency, and extracted eddy current losses

Frequency F (Hz)	Flux density B (T)	Total loss P_t (W/kg)	Ratio P_t/f (J/kg)	Eddy current loss $P_e = P_t - 0.1194$ (W/kg)
99.99	1.0	13.1910	0.1319	0.0125
249.97	1.0	38.3170	0.1533	0.0339
500.00	1.0	93.8360	0.1877	0.0683
750.00	1.0	165.7000	0.2209	0.1015

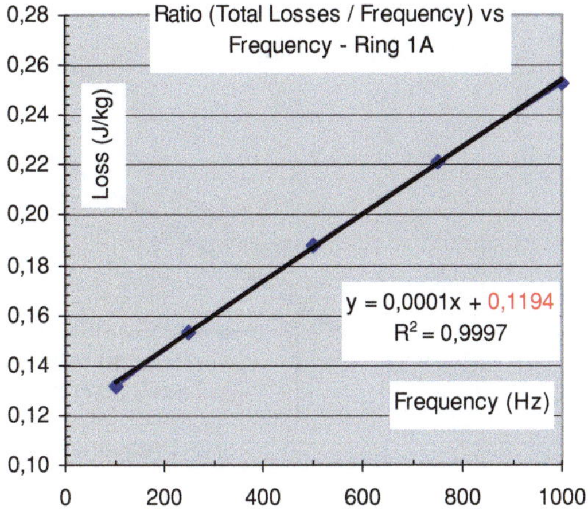

Figure 9. Ring 1A—measured ratio (total loss/frequency) vs. frequency—AC-slope and hysteresis value at 0 Hz.

Table 3. Measured and simulated AC-slopes for rings 1–11

Ring no.	Measured slope 1.0 × conductivity	FEA slope 1.0 × conductivity	Ratio M/FEA	Difference %	FEA slope 1.2 × conductivity
1A	1.345E–04	1.19E–04	1.131	11.6	1.384E–04
1B	1.725E–04	1.45E–04	1.188	15.8	1.707E–04
2	1.424E–04	1.14E–04	1.245	19.6	1.328E–04
3	1.287E–04	9.85E–05	1.306	23.4	1.150E–04
4	1.088E–04	8.71E–05	1.249	19.9	1.021E–04
5	8.920E–05	6.06E–05	1.471	32.0	7.171E–05
6	1.945E–04	1.67E–04	1.163	14.0	1.893E–04
7	1.715E–04	1.56E–04	1.102	9.2	1.776E–04
8	1.562E–04	1.40E–04	1.119	10.6	1.615E–04
9	1.338E–04	1.14E–04	1.169	14.5	1.338E–04
10	1.298E–04	1.14E–04	1.136	11.9	1.329E–04
11	9.095E–05	6.41E–05	1.419	29.5	7.585E–05
		Mean value	**1.2**	17.7	
		Standard deviation	0.120		
		Standard deviation	9.8%		

Table 4. Ring 1A—measured and calculated AC-slopes

Frequency (Hz)	JMAG 14,451 S/m	Measured 14,451 S/m	JMAG 17,341.2 S/m
100	0.013	0.013	0.016
1,000	0.120	0.134	0.140
AC-slope	0.0001188	0.0001345	0.0001384

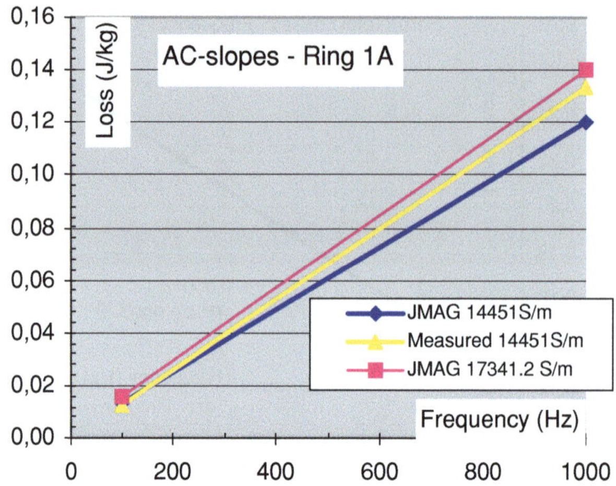

Figure 10. Ring 1A—AC-slopes.

It can be seen that the AC-slope calculated from FEA simulations using the modified conductivity is very close to the AC-slope calculated from measurements.

In Fig. 11 the measured and simulated AC-slopes are shown for ring samples 1A, 1B, and 2–5. It can be seen that the simulated results with a modified measured conductivity are quite close to the measured ones. For ring samples 3 and 5, though the correspondence is lower especially for the smallest one, ring 5. Tests with different numbers of finite elements came out to have very little impact on the results and could not explain why the FEA simulation of eddy current losses seemed to be more difficult with smaller ring samples.

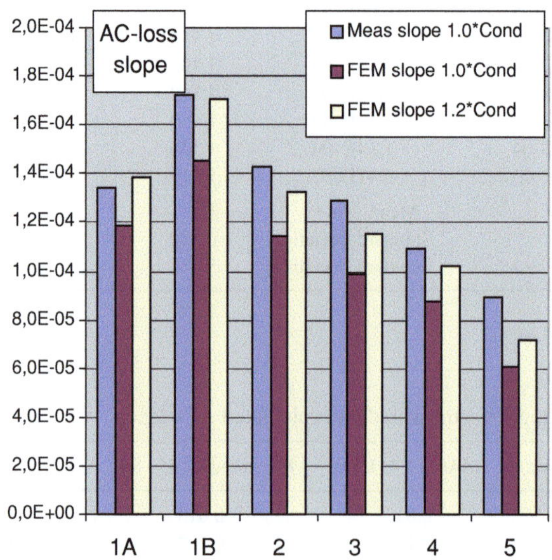

Figure 11. AC-slopes—ring samples 1A, 1B, and 2–5.

Table 5. Hysteresis losses—mean values measured for all ring samples with standard deviation and calculated losses using the calibrated formula

B (T)	Mean measurement (J/kg)	Standard deviation	$P_h/f = K_h \times B^a$ J/kg
0	0.0000	0.0000	0.0000
0.50	0.0354	0.0003	0.0361
1.00	0.1193	0.0018	0.1190
1.50	0.2390	0.0013	0.2390

An important conclusion from Fig. 11 is that the component size has big influence on the AC losses. This fact has to be considered when a new component is designed for an application.

K_h and α factors

The mean values of measured hysteresis losses for the 12 ring samples at three different flux densities are summarized in Table 5. The standard deviation between the different ring samples and the corresponding calculated hysteresis loss can also bee seen in Table 5.

The hysteresis loss calculations were using formula (2) with $K_h = 0.119$, αg 1.72, and $B =$ magnetic flux density (T).

$$P_h/f(J/kg) = K_h \cdot B^{\alpha}. \tag{2}$$

K_h and α were created by curve fitting. How well the values are corresponding to measurements is shown in Fig. 12. The formula is a part of the built-in iron loss calculation model that can be found in most FEA packages. How the K_h factor is modified depends on the FEA software in use, due to different implementations. When using JMAG-Studio [2] the K_h has to be multiplied with the mass density for each ring sample.

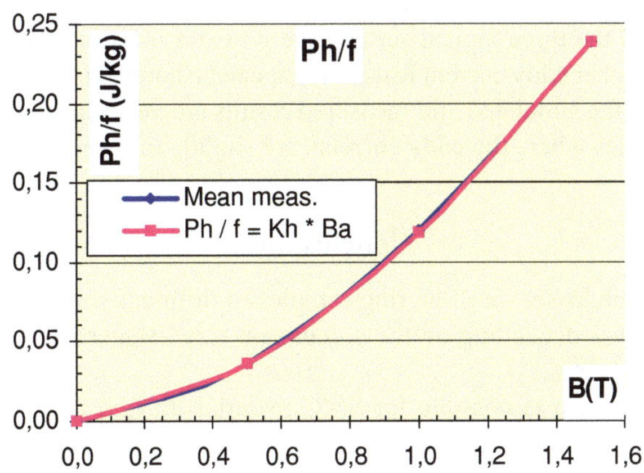

Figure 12. Measured and calculated curve for P_h/f. The curves are almost completely overlapping.

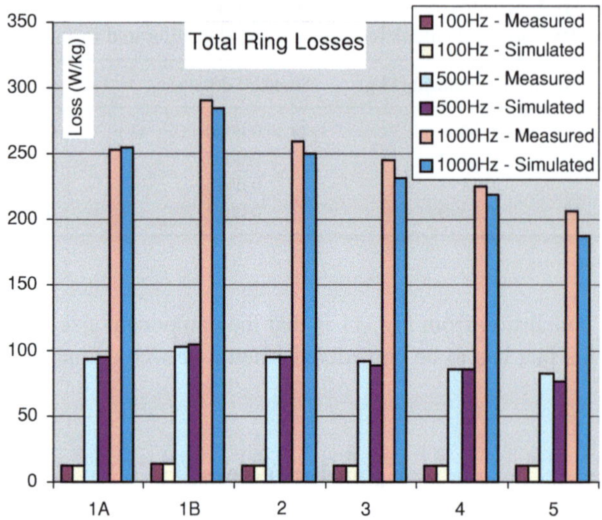

Figure 13. Ring samples 1A, 1B, and 2–5. Measured and simulated total loss at different frequencies using a modified conductivity and established factors for the hysteresis loss.

The correspondence between measured and calculated values is good. From Table 5 it can also be seen that the values of hysteresis loss are very close between the different ring samples.

Total loss comparison

Simulations of total losses were carried out using the modified measured conductivity to simulate the eddy currents. Hysteresis losses were calculated from simulated B-field and the built-in iron loss model, using the established values of K_h and α. The measured and simulated losses for ring samples 1A, 1B, and 2-5 are summarized in Fig. 13. It can be seen that the simulated values corresponds quite well with measurements.

An evaluation of the three logical series gave no extra information except that bigger components have higher eddy current losses. What can be noticed is that regardless the ring or components size the simulated and measured results are very close to each other except the small ring samples where the eddy currents are slightly underestimated.

Conclusions

Measurements of iron losses on SMC ring samples of different sizes showed that the total loss per mass unit was depending on the component size. Bigger SMC components have higher eddy current losses.

It is also shown that it is possible for SMC materials to find factors for eddy current and hysteresis losses resulting in good agreement between FEA simulations and measurements.

References

[1] P. Jansson, A. Jack, "Magnetic Assessment of SMC materials". Twenty first Conference on Properties and Applications of Magnetic Materials, Chicago, May 13–15, 2000.

[2] JMAG-Studio from the Japan Research Institute Ltd. Engineering Technology Division 16, Kudan Building 2F, 1-5-3, Kudanminami, Chiyoda-ku, Tokyo 102-0074, Japan.

III-3.3. ELECTROACTIVE MATERIALS: TOWARD NOVEL ACTUATION CONCEPTS

B. Nogarede, Jean-François Rouchon and Alexis Renotte
Electrodynamics–EM3 Research group, Laboratoire d'Electrotechnique et d'Electronique Industrielle INPT-ENSEEIHT, UMR-CNRS n°5828, 2 rue Charles Camichel, 31071 Toulouse, France
Bertrand.Nogarede@leei.enseeiht.fr, Jean-Francois.Rouchon@leei.enseeiht.fr, Alexis.Renotte@leei.enseeiht.fr

Abstract. After a brief recapitulation of diverse physical processes which can be used in electromechanical energy conversion, the present article proposes a survey of the modern stakes of electrodynamics in the range of centimetric or decimetric dimensioned actuators. The potential of the new technologies considered is evaluated through different examples of novel actuators which aim at meeting the increase of the performances or the expansion of required functionalities in the face of varied types of applications. An experimental study concerning friction drag reduction for a supersonic aircraft is briefly dealt with at the end of the article. The aim is the control of turbulent streaks with spanwise traveling wave. A piezoelectric demonstrator was designed for wind tunnel testing at different configurations of frequency and wave-length.

Introduction

The modern applications of electromechanics is characterized by a more and more intensive integration of actuator and sensor functions within mechanisms allying high mass performances and advanced functionalities. Recently accomplished progress in the field of electroactive materials (piezoelectric, electrostrictive ceramics, magnetostrictive alloys, shape memory alloys) reveal a very promising field of innovation aiming at ending up in high mass performance devices with a high functional integration level besides [1]. The present article proposes a non-exhaustive survey of new electrodynamic device stakes in the face of diverse types of emerging applications.

A wide variety of exploitable effects

In the centimetric dimensional field considered, the design of electroactive actuators constitutes a real technological brake. Let it be underlined that this potential essentially hangs on the possibility of generating high specific efforts (driving constraints to the order of 40 MPa in the PZT piezoelectric ceramics or to the order of 100 MPa in shape memory alloys) in a reduce bulk (the energy conversion is operated in the very volume of the materials). Of course, the amplitude of the elementary displacements produced by the transducers remains limited (relative strain to the order of 1,000 parts per million for multi-layer piezoelectric

S. Wiak, M. Dems, K. Komęza (eds.), Recent Developments of Electrical Drives, 435–442.

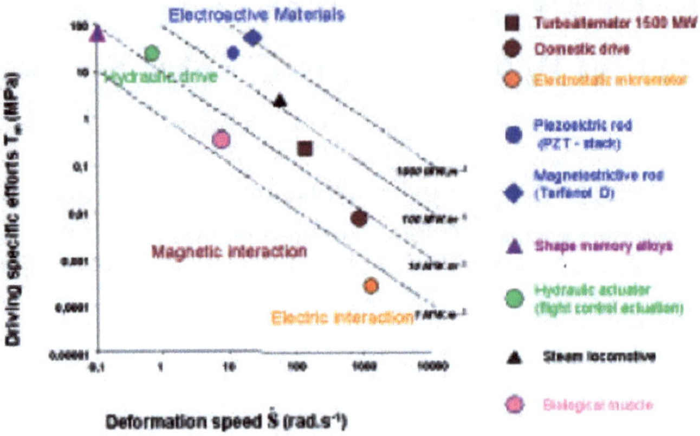

Figure 1. Comparison of the various electromechanical effects in terms of specific energy [3].

ceramics and of 8,000 ppm in the case of shape memory alloys [2]). Nevertheless, the possibility of cumulating these microdisplacements in space (e.g., by using the flexion of a beam), or in time (thanks to the transmission of a high frequency vibratory movement by intermittent contact) gives rise to remarkable performances in terms of specific power.

Thus, as illustrated in Fig. 1, the comparison of the intrinsical performances associated with the main exploitable processes clearly shows the potential represented by the use of these novel materials.

Toward an intensive functional integration

Modern electromechanical applications, notably in the field of servo-control or devices for electric assistance, are often characterized by the need to integrate the functions of a mechanical speed reducer, eventually associated with a brake, so as to reduce the global bulk of the device. One can try to integrate the driving and reducing functions into one and the same structure thanks to the exploitation of new actuation concepts [4].

The mechanism in Fig. 2, which corresponds to the active part of a motorized hand prosthesis, with two degrees of mobility (wrist bending, grasping) clearly brings out the

Figure 2. Hand prosthesis mechanism prototype driven by two rotating mode piezomotors [5].

Figure 3. Quasistatic operating piezoelectric actuator with its electronic supply.

potential represented by piezoelectric motor technology for this type of use [5]: two driving functions can be lodged instead of the classical mechanism with a sole degree of freedom worked by a direct current micromotor (for globally reduce bulk) whereas unobtrusive noise, the precision of positioning and the off-supplied locking which characterize piezoelectric drives (here two rotating mode motors), are all advantages to consider.

High torque–low speed actuator for direct drive

At the same time the approach which leads to a better integration of speed reducing functions in electric drives, the emergence of new needs to satisfy specifications of the high-effort-at-low-speed type, under more and more severe mass and bulk constraints (notably in space and aeronautic fields) favors the development of actuator using a direct drive of the mechanical load (absence of reduction stages in the cinematic chain).

Fig. 3 illustrates these possibilities through a new design of rotating piezoactuator which operates in quasistatic mode (clamp stator deformed by multi-layer PZT ceramics) intended for the direct drive of an aeronautic fuel gate and dimensioned to develop a maximum torque of 10 Nm (maximum speed to the order of 5 rpm), for a total mass of 1.5 kg.

Actuator combining several degrees of freedom

The new functionalities resulting from the simultaneous management of several degrees of freedom within a same mechanism (robotized microsurgery, micropositioning for microelectronics or near field microscopics [6]) also give the research scientist a particularly rich field of investigation. Opposed to the approach which aims at combining the different

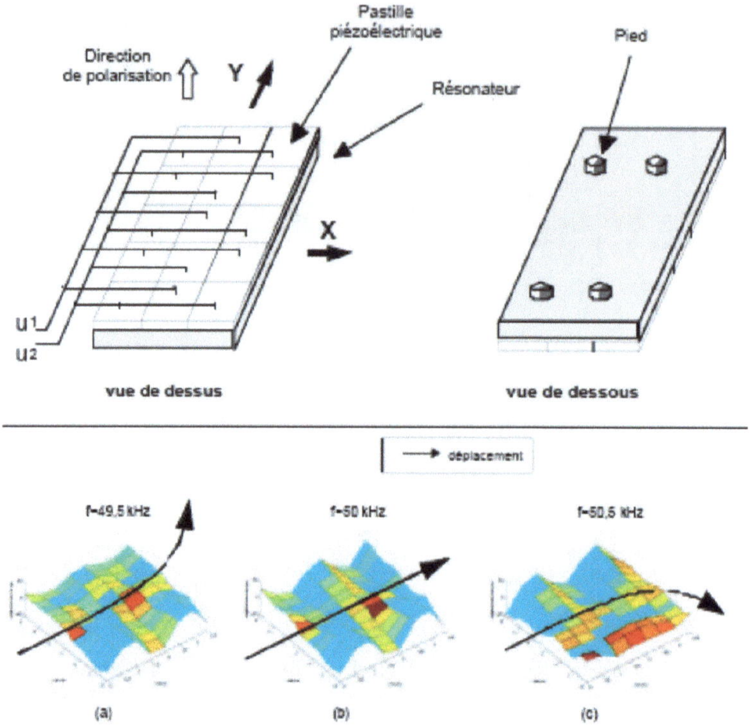

Figure 4. Piezoelectric planar translator–straight line driving (b), left turn (a), and right turn (c) [7].

motion required within relatively complicated multi-axes mechanisms, it is rather a matter here of imagining actuator structures which are intrinsically built to manage the different displacement required jointly. As an example, the case of the piezoelectric actuator schematized in Fig. 4 could be quoted. Its working principle lies in the one-phased excitation of a bending standing wave. The adjustment of the supply frequency then enables the control of the trajectory of the mobile on a flat surface. The centimetric demonstrator achieved proves to be relatively interesting from the point of view of performance in so far as it enables the displacement of a 2 kg mass at a speed of 10 mm/s [7].

Seeking original functionalities: distributed actuation for air flow control

The emerging field of the electroactive flow control constitutes a particularly revealing example of the need for electromechanical functions of a distributed nature [8,9].

The field of active flow control raises many relevant questions like internal or external noise control, hybrid laminarity, wave drag control, or friction drag control. A lowering of a few percent of friction drag could provide a non-negligible reduction of fuel consumption.

Several control techniques can be considered in order to decrease friction drag: passive techniques by optimization of wing geometry, and active ones by injecting external energy into the flow. In these cases, two major orientations can be distinguished: on one hand there

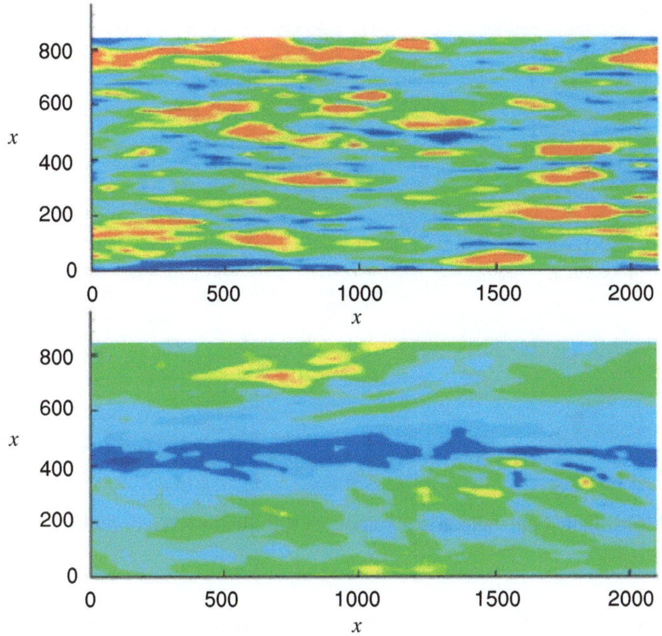

Figure 5. Numerical simulation of streaks agglomeration [10].

is local action at microscopic scale, and on the other hand global action on several structures. Numerical studies have demonstrated the benefit of a transverse traveling wave on the drag force [10].

As illustrated in Fig. 5, the results obtained show a 30% reduction of drag force, by agglomeration of high and low velocity streaks. In the context of a French research program concerning future supersonic aircraft, a specific research project involving ONERA, AIRBUS France, The Institut de Mécanique des Fluides de Toulouse, and the EM3 group of INPT/LEEI (coordinator of the program) has been initiated. The aim of the project is to experimentally reproduce the same effect by using a "smart wall", able to generate transverse traveling wave [11].

Fig. 6 shows the principle of the developed test bench which is based on the use of a multi-bladed structure (PALM concept).

The most popular scenario about turbulence enhancement begins with the formation of high and low speed streaks in the near-wall region, after advection of streamwise swirls. An autonomous cycle of turbulence regeneration appears, due to the creation of new structures on those that already exist. These streaks have great span wise gradient of streamwise velocity, which explains the appearance of vertical vorticity. The vertical and spanwise vorticity is redirected in the streamwise direction creating new turbulent structures. That is why we consider reducing friction drag by inducing progressive waves of spanwise velocity, to vanish artificially created streaks in a Blasius boundary layer [12].

In order to produce transverse traveling waves of at least 1 mm amplitude, a vibrating airfoil, whose surface must be able to move, has to be realized. In this aim, a system of 24 located actuators, distributed on the airfoil, was chosen. The specifications for the required

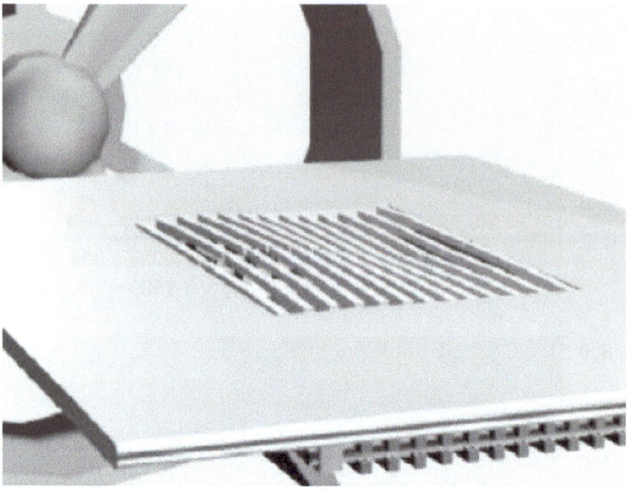

Figure 6. Multi-bladed piezoelectric actuator for turbulence reduction: the PALM concept.

deformation were defined. One other main goal is to produce an adaptive structure in order to study as many different configurations as possible.

Actuator is based on multi-layer piezoceramics stacks from Morgan Matroc Electro ceramics ($5 \times 5 \times 47$ mm^3, 1,000 ppm strain, 770 N max blocked force). In order to obtain large displacement in static use, the ceramic movement is amplified with a lever arm principle, as shown in Fig. 7. The application point of the ceramic is located 1.1 mm above the middle of the flexion blade, so when the ceramic lengthens it produces the flexion of the blade, and makes the upper beam rotate. The rotation angle is so small that the movement of the upper beam can be assimilated to a translation.

To use the ceramics at their maximum power rate, a mechanical device, which produces a 2.5-mm displacement for half of the maximum ceramic displacement, was designed.

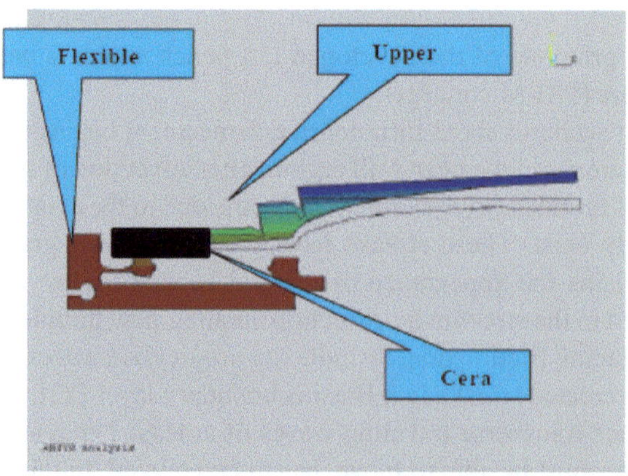

Figure 7. FEM modal analysis: first resonant mode of the actuator.

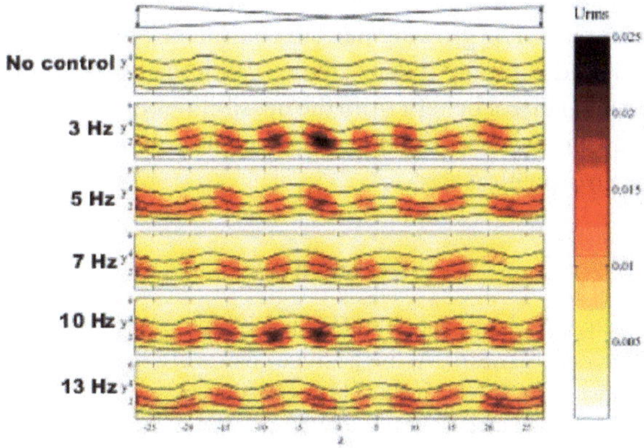

Figure 8. Standing wave on a perturbed laminar boundary layer.

The mechanical structure is made of Duralumin, whose plasticity limit is 75 MPa. The ceramic is linked with the structure through a linear contact allowing its rotation. Modal analysis reveals a first resonance mode at 33 Hz.

When applying 200 V on the ceramic, a 2.52-mm displacement of the upper beam is measured. Tests was made with a laser beam vibrometer insulated from external vibrations which allows us to measure displacements to the order of 1 μm. Dynamic tests revealing that the first resonance mode comes at 40 Hz were also performed.

The first mode frequency is slightly higher than predicted by the FEM study, because the structure is strengthened by the ceramic, which was not simulated in the FEM study.

Actuators are pair-controlled, in order to eliminate every unintentional phase shifting. The active power consumption of a pair of actuators at 20 Hz is about 120 mW, for a reactive power of 2.85 VAR, due to ceramic capacitance.

First wind tunnel tests

After validating the actuator behavior, a prototype was realized for experimentation in a wind tunnel. Several measurement methods have been employed: hot wire velocity measurement and particle image velocimetry (PIV).

The hot wire measurement data describing the effect of different control frequencies of a standing spanwise wave are shown in Fig. 8.

At frequencies of 7 and 13 Hz, certain streaks are reduced. Even though friction drag seems to be enhanced, the produced effect depends on the frequency. The first results obtained are promising, even though deeper investigations are required to precisely quantify the benefit of the produced waves on the flow.

Conclusion

The field of exploitable physical processes is considerably enlarged, notably thanks to the development of electroactive materials of the performances, the working capacities, and the flexibility of implementation pave the way to new designs of converters. Exploited within motors or actuators, this technology makes it possible to envisage substantial gains in terms of force mass ratio. The growing need to control the physical processes of the more and more distributed kind naturally induces the research scientists to imagine devices with distributed drive, drawing direct benefit from the possibilities of structural integration which electroactive materials provide by principle. In this context active flow control corresponds to a very promising investigation area.

References

[1] T. Sashida, T. Kenjo, An Introduction to Ultrasonic Motors, Oxford: Clarendon Press, 1993.

[2] S. Gangbing, K. Brian, N.A. Brij, Active position control of a shape memory alloy wire actuated beam, Smart Mater. Struct., Vol. 9, pp. 711–716, 2000.

[3] B. Nogarede, "Machines électriques: Conversion électromécanique de l'énergie", *Traité de Génie Electrique*, Techniques de l'Ingénieur, D3410, 2000.

[4] C. Henaux, G. Pons, B. Nogarede, "A Novel Type of Permanent Magnet Actuator: the HYPO-MAG Structure", ICEM'2000, Espoo (Finland), August 28–30, 2000.

[5] B. Nogarede, C. Henaux, J.-F. Rouchon, F. Léonard, R. Briot, L. Petit, P. Gonnard, B. Lemaire-Semail, F. Giraud, Ph. Kapsa, "Matériaux électroactifs et génie biomédical: étude d'une prothèse de la main actionnée par une motorisation piézoélectrique", MGE'2000, Lille, décembre 13–14, 2000.

[6] N. Bonnail, D. Tonneau, H. Dallaporta, G.-A. Capolino, "Dynamic Response of a Piezoelectric Actuator at Low Excitation Level in the Nanometer Range", ICEM'2000, Espoo (Finland), August 28–30, 2000.

[7] F. Galiano, B. Nogarede, Un nouveau concept d'actionneur piézoélectrique plan monophasé à onde stationnaire, Revue Internalionale de Génie Electrique, Vol. 2, N°/Ref. 4/1999.

[8] R.G. Loewy, Recent developments in smart structures with aeronautical applications, Smart Mater. Struct., Vol. 6, pp. R11-R42, 1997.

[9] E. Stanewsky, Adaptive wing and flow control technology, Prog. Aerosp. Sci., Vol. 37, pp. 583–667, 2001.

[10] V. Du, G. Karniadakis, Drag reduction in a wall bounded turbulence via a transverse travelling wave, J. Fluid mech., Vol. 457, pp. 1–34, 2002.

[11] B. Nogarede, V. Monturet, D.Harribey, A. Bottaro, H. Boisson, P. Konieckzny, A.Sevrain, J.P. Chretien, A. Sagansan, "Développement et évaluation de nouvelles technologies d'actionneurs répartis pour le supersonique", 1ier Colloque National sur la Recherche Aéronautique sur le Supersonique, Paris, février 6–7, 2002.

[12] P. Konieczny, A. Bottaro, V. Monturet, B. Nogarede, "Active Control of Near-Wall Coherent Structures", FEDSM'2002, Joint US ASME-European Fluids Engineering Summer Conference Montreal, Quebec (Canada), July 14–18, 2002.

III-3.4. ADVANCED MATERIALS FOR HIGH SPEED MOTOR DRIVES

G. Kalokiris, A. Kladas and J. Tegopoulos

Faculty of Electrical and Computer Engineering, National Technical University of Athens,
9, Iroon Polytechneiou Street, 15780 Athens, Greece
tegopoul@power.ece.ntua.g

Abstract. The paper presents electrical machine design considerations introduced by exploiting new magnetic material characteristics. The materials considered are amorphous alloy ribbons as well as neodymium alloy permanent magnets involving very low eddy current losses. Such advance materials enable electric machine operation at higher frequencies compared with the standard iron laminations used in the traditional magnetic circuit construction and provide better efficiently.

Introduction

The impact of innovative materials on the electrical machine design is very important. The advantages involved in machine efficiency and performance are important as mentioned in [1,2]. These materials enable electric machine operation at high frequencies when supplied by inverters, compared to the standard iron laminations used in the traditional magnetic circuit construction. The features and performance characteristics are analyzed by using field calculations and tested by measurements. In this paper, the study of asynchronous and permanent magnet machines based on such materials is undertaken. Low losses and high volumic power associated with high speed and converter machine operation are the main advantages of such applications [3–5].

Design procedure

The proposed machine design procedure involves two steps. In a first step standard design methodology is used for preliminary design. In a second step the method of finite elements is implemented to calculate the machine efficiency and performance. Finally, a prototype is constructed in order to validate and compare the simulated machine characteristics to the corresponding experimental results [6,7].

The method of finite elements, is based on a discretization of the solution domain into small regions. In magnetostatic problems the unknown quantity is usually the magnetic vector potential A, and is approximated by means of polynomial shape functions. In two-dimensional cases triangular elements can easily be adapted to complex configurations and first order elements exhibit advantages in iron saturation representation [8,9]. The size of elements must be small enough to provide sufficient accuracy. In this way the differential

S. Wiak, M. Dems, K. Komęza (eds.), Recent Developments of Electrical Drives, 443–450.

equations of the continuous problem can be transformed into a system of algebraic equations for the discrete problem. The practical problems necessitate usually several tenths of thousands of unknowns. However, appropriate numerical techniques have been developed, enabling to obtain the solution of such systems within reasonable time, even when personal computers are used. It should be mentioned that the 3D problems require considerably higher computational resources than the 2D ones. In the present paper the 2D finite element model adopted, involves vector potential formulation, while the magnetic flux Φ_m per pole can be calculated as follows:

$$\Phi_m = \iint_{S_1} B \cdot dS = \oint_{C_1} A \cdot dl \cong (2A_{gap}) L_0 \tag{1}$$

where L_0 is the length of the magnetic circuit in m, A is the magnetic vector potential, A_{gap} is the vector potential value in the middle of the air-gap, B is the flux density in T, S_1 is the cross-sectional area normal to the direction of flux in m^2, and C_1 is the contour surrounding the surface S_1 in m. The electromotive force at no load can be calculated as follows:

$$E = -\frac{d\Phi_m}{dt} \tag{2}$$

The value of the voltage of the machine operated as generator under load conditions can be calculated by relation (3):

$$V = E - RI - jL_\sigma \omega I \tag{3}$$

where V is the voltage on stator windings in V, E is the electromotive force at no load in V, R is the stator resistance in Ω, L_σ is the stator leakage inductance in H, ω is the rotor angular velocity in rad/s, and I is the stator current in A. Then the magnetic flux and electromotive forces can be derived by using equations (1) and (2). Furthermore, the total resistance of stator winding can be calculated by the following relations:

$$R = \rho \frac{l}{s} \tag{4}$$

where ρ is the electric resistivity of copper, l is the winding total length, and s is the conductor cross-section. The winding length l can be estimated from relation (5):

$$l = 2 \times (l_{ax} + l_p) \times N_w \times P \tag{5}$$

where l_{ax} is the machine's axial length in meters, l_p is the polar pitch in meters, and N_w is the total number of series connected turns.

Results and discussion

The case of a permanent magnet machine has been considered. The machine designed has been checked through a 2.5 kW prototype, which has been connected to an appropriate power electronics converter. The air-gap width has been chosen 1 mm while a multipole "peripheral" machine structure has been adopted. The geometry of the permanent magnet machine is shown in Fig. 1 giving also the mesh employed for the two-dimensional finite element program of the machine involving, approximately 2,100 nodes 4,000 triangular elements.

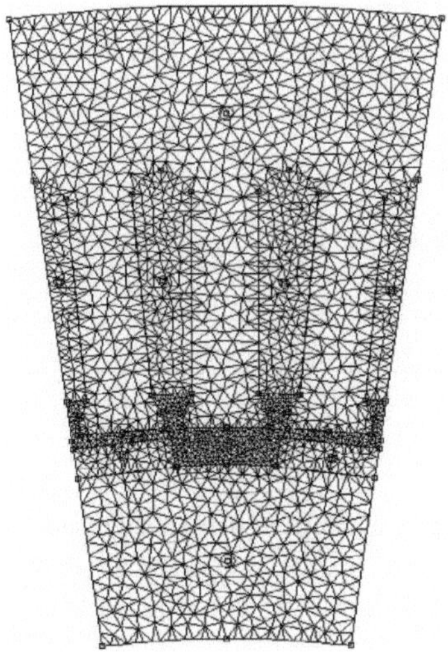

Figure 1. Employed triangular mesh of the one pole part of the permanent magnet machine constructed.

In a first step the no load operating conditions have been examined. The corresponding simulated voltage waveform is shown in Fig. 2 while the measured one is given in Fig. 3, respectively. In these figures a good agreement between the simulated and measured results can be observed. The simulation results concerning full load voltage of synchronous generator are presented in Fig. 4. Fig. 5 gives the measured results under the same operating conditions. A good agreement can be observed in these figures between the simulated and measured results also in the case of full load.

Moreover, measurements were realized for an asynchronous motor, which was supplied by an inverter with variable frequency. The motor is a three phase, four-pole, machine

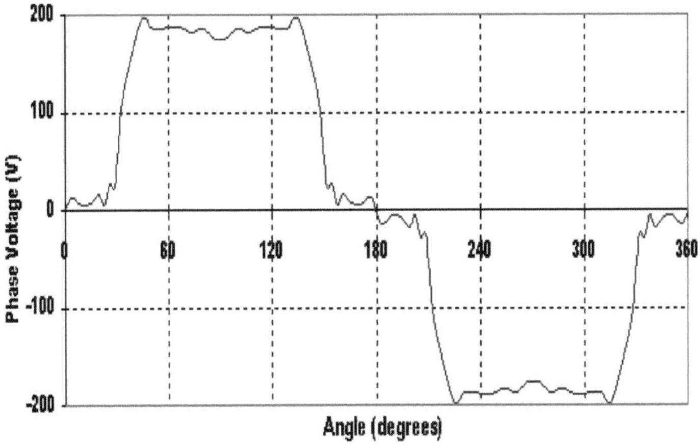

Figure 2. No load voltage waveform of permanent magnet machine (simulation).

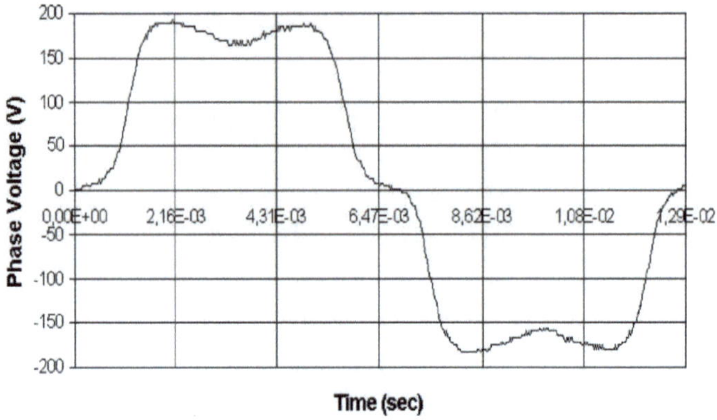

Figure 3. No load voltage waveform of permanent magnet machine (measurement).

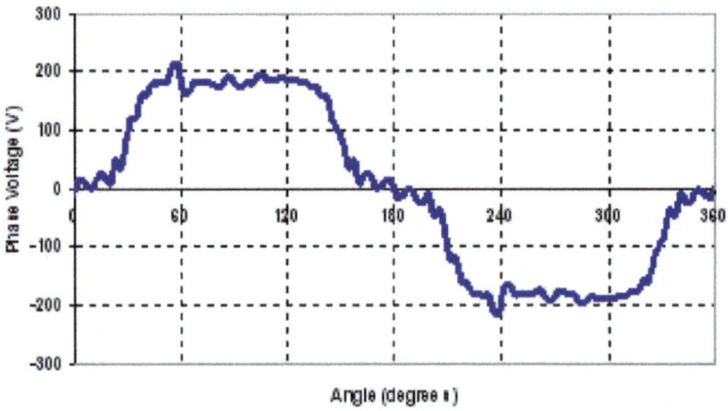

Figure 4. Full load voltage waveform of permanent magnet machine (simulation).

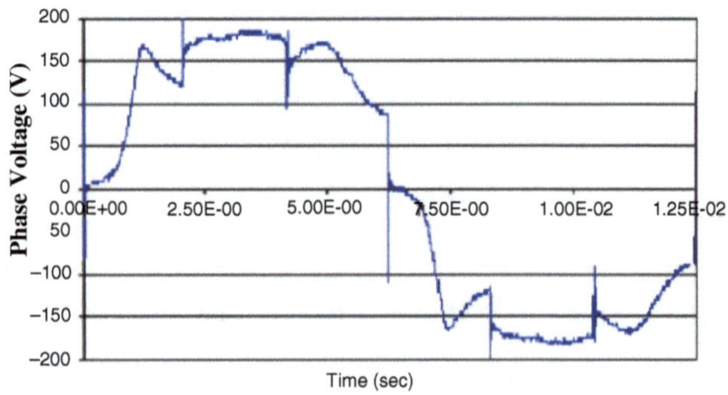

Figure 5. Full load voltage waveform of permanent magnet machine (measurement).

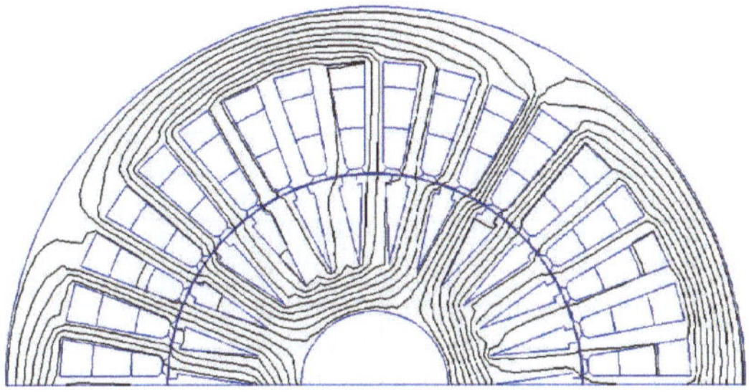

a

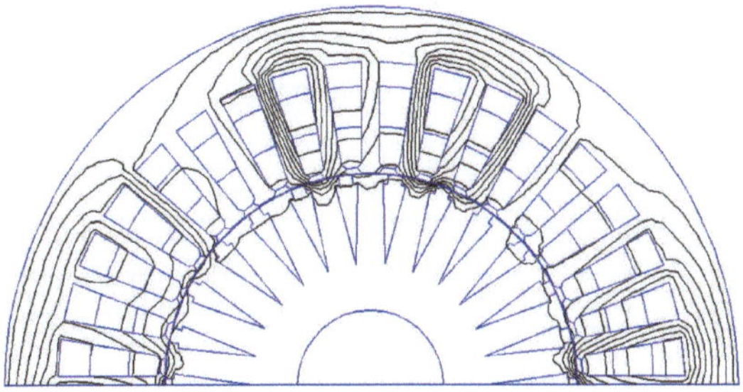

b

Figure 6. Simulated field distribution in the machine under low load conditions. (a) Fundamental supply frequency of 300 Hz, (b) switching frequency of 10 kHz.

supplied at a frequency of 400 Hz, at a voltage of 208 V while the nominal, speed is 10.800 rpm. The motor was tested under no load and low load operating conditions, for various frequencies.

Fig. 6(a) shows the field distribution in the machine supplied at fundamental frequency of 300 Hz, while Fig. 6(b) gives the field distribution at the switching frequency of 10 kHz. Fig. 7 presents the respective measured phase voltage and current time variations.

Fig. 8 shows the field distribution in the machine supplied at fundamental frequency of 100 Hz, while Fig. 9 presents the respective measured phase voltage and current time variations at the switching frequency of 10 kHz.

Table 1 presents the measured and simulation results under no load conditions with a switching frequency of 1 kHz. Table 2 presents the same results under low load conditions. Table 3 presents the results related to a switching frequency of 10 kHz.

The simulated torque T_s is calculated by the relation:

$$T_s = F_t \cdot r_g \tag{6}$$

where F_t is the total circumferential tangential force in N and r_g is the middle air-gap radius in m. The Maxwell's stress tensor is calculated by relation (7):

$$F_t = \frac{1}{\mu_0} \oint_C B_n B_t dl\, L_0 \tag{7}$$

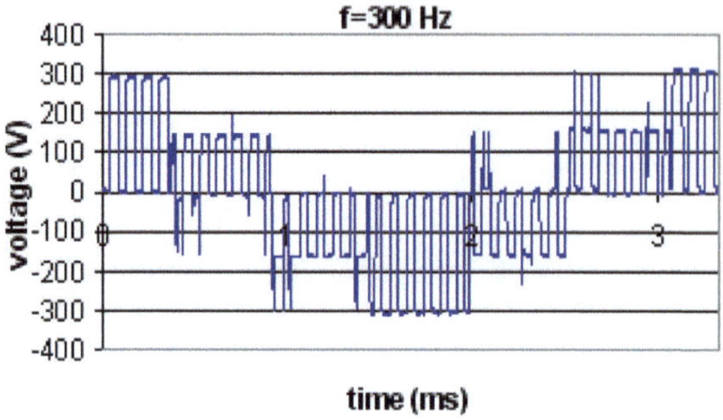

a

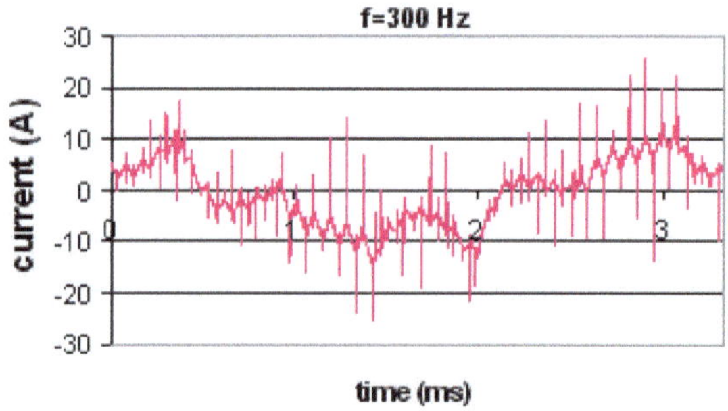

b

Figure 7. Measured supply quantities in the machine for supply frequency of 300 Hz, under low load conditions. (a) Phase voltage time variation, (b) phase current time variation.

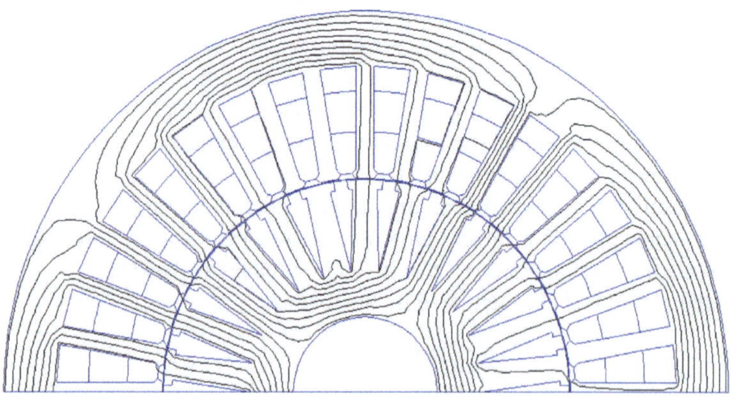

Figure 8. Simulated field distribution in the machine, fundamental supply frequency of 100 Hz under low load conditions.

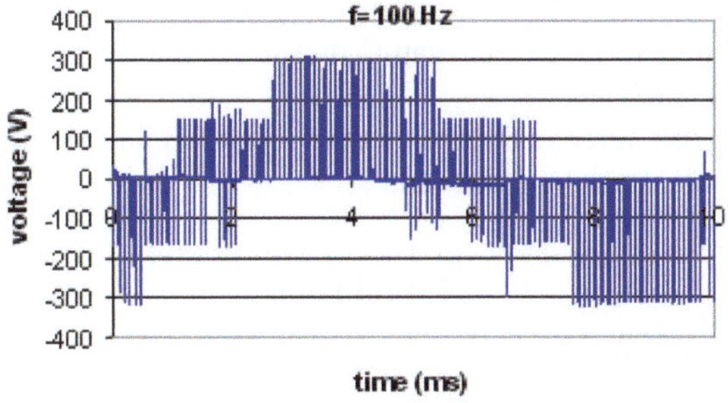

a

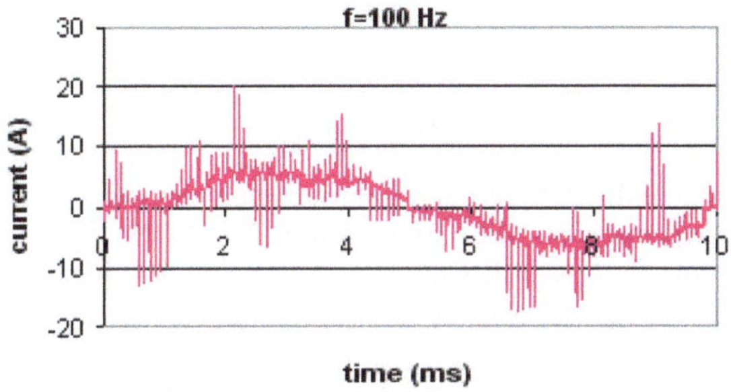

b

Figure 9. Measured supply quantities in the machine for supply frequency of 100 Hz under low load conditions. (a) Phase voltage time variation, (b) phase current time variation.

Table 1. Measured and simulation results under no load conditions and $f_s = 1$ kHz

$f_{1\ fundamental}$ (Hz)	$I_{1\ measured}$ (A)	$V_{1\ measured}$ (V)	$T_{measured}$ (Nm)	$V_{1\ simulated}$ (V)	$T_{simulated}$ (Nm)
20	1.587	4.717	0	2.041	0.055
50	2.267	14.350	0.1	13.279	0.088
75	2.417	20.962	0.15	21.096	0.11

Table 2. Measured and simulation results under low load conditions and $f_s = 1$ kHz

$f_{1\ fundamental}$ (Hz)	$I_{1\ measured}$ (A)	$V_{1\ measured}$ (V)	$T_{measured}$ (Nm)	$V_{1\ simulated}$ (V)	$T_{simulated}$ (Nm)
50	2.754	13.769	0.4	14.540	0.23

Table 3. Measured and simulation results under low load conditions and $f_s = 10$ kHz

$f_{1\ fundamental}$ (Hz)	$I_{1\ measured}$ (A)	$V_{1\ measured}$ (V)	$T_{measured}$ (Nm)	$V_{1\ simulated}$ (V)	$T_{simulated}$ (Nm)
50	2.368	13.584	0.22	13.353	0.18
100	2.874	22.646	0.38	25.790	0.3
200	5.179	53.533	1.05	75.970	1.03

where B_n and B_t are the normal and tangential magnetic flux density components, respectively, to the integration surface of air-gap C in T, μ_0 is the permeability of air, and L_0 is the active part of the machine.

In Tables 1–3 a good agreement between the measured and simulated results for both voltage and torque values can be observed.

Conclusion

The use of innovative materials, in electrical machines supplied by converters, can considerably affect the efficiency and performance in high speed operation. This has been investigated by using the finite element method for the machine analysis and verified by measurements. Moreover, the low cost involved makes such drives attractive rivals of the conventional ones.

References

[1] M.R. Dubois, H. Polinder, J.A. Ferreira, Contribution of permanent-magnet volume elements to no-load voltage in machines, IEEE Trans. Magn., Vol. 39, No. 3, pp. 1784–1792, 2003.

[2] T. Higuchi, J. Oyama, E. Yamada, E. Chiricozzi, F. Parasiliti, M. Villani, Optimization procedure of surface permanent magnet synchronous motors, IEEE Trans. Magn., Vol. 33, No. 2, pp. 1943–1946, 1997.

[3] A. Toba, T. Lipo, Generic torque maximizing design methodology of surface permanent magnet Vernier machine, IEEE Trans. Ind. Appl., Vol. 36, No. 6, pp. 1539–1546, 2000.

[4] G. Tsekouras, S. Kiartzis, A. Kladas, J. Tegopoulos, Neural network approach compared to sensitivity analysis based on finite element technique for optimization of permanent magnet generators, IEEE Trans. Magn., Vol. 37, No. 5/1, pp. 3618–3621, 2001.

[5] D.C. Aliprantis, S.A. Papathanassiou, M.P. Papadopoulos, A.G. Kladas, "Modeling and Control of a Variable-Speed Wind Turbine Equipped with Permanent Magnet Synchronous Generator", International Conference on Electrical Machines, Helsinki, Finland, 2000, pp. 558–562.

[6] N.A. Demerdash, J.F. Bangura, A.A. Arkadan, A time-stepping coupled finite element-state space model for induction motor drives, IEEE Trans. Energy Convers., Vol. 14, No. 4, pp. 1465–1477, 1999.

[7] T.M. Jahns, Motion control with permanent magnet AC machines, IEEE Proc., Vol. 82, No. 8, pp. 1241–1252, 1994.

[8] C. Marchand, Z. Ren, A. Razek, "Torque Optimization of a Buried Permanent Magnet Synchronous Machine by Geometric Modification using FEM", Proc. EMF'94, Leuven, Belgium, 1994, pp. 53–56.

[9] H.C. Lovatt, P.A. Watterson, Energy stored in permanent magnets, IEEE Trans. Magn., Vol. 35, No. 1, pp. 505–507, 1999.

III-3.5. IMPROVED MODELING OF THREE-PHASE TRANSFORMER ANALYSIS BASED ON NONLINEAR *B-H* CURVE AND TAKING INTO ACCOUNT ZERO-SEQUENCE FLUX

B. Kawkabani and J.-J. Simond

Laboratory for Electrical Machines, Swiss Federal Institute of Technology,
EPFL-STI-ISE-LME, ELG Ecublens, CH-1015 Lausanne, Switzerland
basile.kawkabani@epfl.ch

Abstract. The present paper deals with a new approach for the study of the steady-state and transient behavior of three-phase transformers. This approach based on magnetic equivalent circuit-diagrams, takes into account the nonlinear *B-H* curve as well as zero-sequence flux. The nonlinear *B-H* curve is represented by a Fourier series, based on a set of measurement data. For the numerical simulations, two methods have been developed, by considering the total magnetic flux respectively the currents as state variables. Numerical results compared with test results and with FEM computations confirm the validity of the proposed approach.

Introduction

Traditionally in most of power system studies, the modeling of a three-phase transformer is reduced to its short-circuit impedance. The *B-H* curve introduced in some improved models and based on a set of measurement data, is approximated generally by several straight-line segments connecting the points of measurements. But apparently, such *B-H* curve obtained is not smooth at the joints of the segments, and the slopes of the straight lines, representing the permeability, are discontinuous at these joints. Moreover, in the set of differential equations considering the currents as state variables, one needs the expressions of the derivatives of the inductances vs. the currents, which is impossible by using the above mentioned procedure.

For that reason in the present study, the nonlinear *B-H* curve (or *U-I* curve) is represented by a Fourier series technique [1], based on the set of measurement data. An analytical expression of a smooth *B-H* curve connecting the discrete measurement points can be defined. By using a magnetic equivalent circuit-diagram representing the three-phase transformer, all the self and mutual inductances can be expressed analytically in function of the magnetic reluctances of the cores. These inductances (and their derivatives) can be determined precisely using the predetermined series Fourier representation, and adapted at each integration step in the numerical simulations.

S. Wiak, M. Dems, K. Komęza (eds.), Recent Developments of Electrical Drives, 451–460.

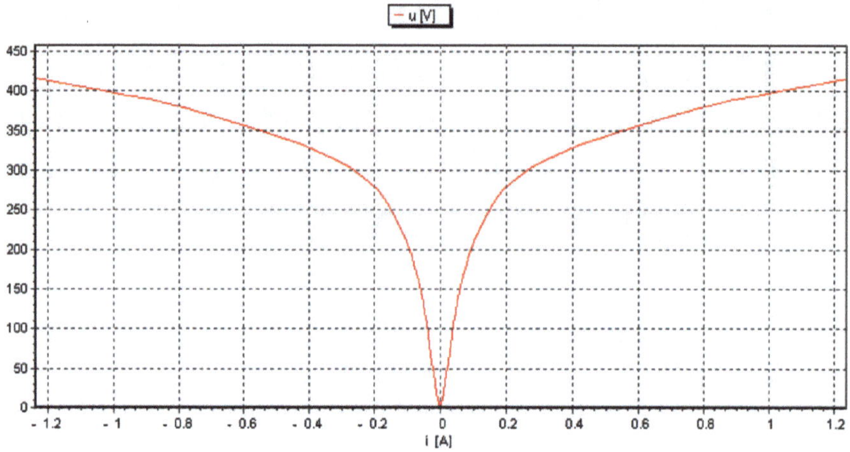

Figure 1. Set of measured data *U-I* and its mirror image.

Fourier series representation of *B-H* curve ($U = f(I)$ curve)

A set of $N + 1$ discrete measurement data U_n and I_n or B_n-H_n of a three-phase transformer ($n = 0, 1, 2, 3, \ldots, N$) is given. For the sake of making use of the Fourier series, a mirror image of this set of data is made about the U or B axis (Fig. 1).

One has:

$$f(H) = a_0 + \sum_{k=1}^{\infty} a_k \cdot \cos(\xi_k \cdot H) \tag{1}$$

with:

$$a_0 = \frac{1}{H_{\max}} \cdot \sum_{n=1}^{N} \left[B_n \cdot (H_n - H_{n-1}) - \frac{1}{2} \cdot \alpha_n \cdot (H_n - H_{n-1})^2 \right] \tag{2}$$

and

$$a_k = \frac{2}{H_{\max}} \cdot \sum_{n=1}^{N} \left[\begin{array}{l} \dfrac{\alpha_n}{\xi_k} \cdot \sin(\xi_k \cdot H_{n-1}) \cdot (H_n - H_{n-1}) + \\ \dfrac{1}{\xi_k^2} \cdot \alpha_n \cdot (\cos(\xi_k \cdot H_n) - \cos(\xi_k \cdot H_{n-1})) + \\ \dfrac{B_n}{\xi_k} \cdot (\sin(\xi_k \cdot H_n) - \sin(\xi_k \cdot H_{n-1})) \end{array} \right] \tag{3}$$

$$H_{\max} = H_N \tag{4}$$

$$\xi_k = \frac{k \cdot \pi}{H_{\max}} \text{ for the } k\text{th term} \tag{5}$$

$$\alpha_n = \frac{B_n - B_{n-1}}{H_n - H_{n-1}} \tag{6}$$

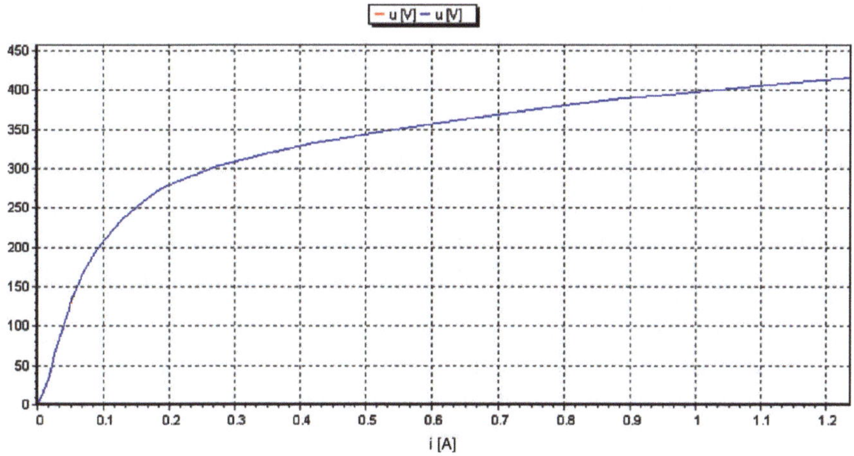

Figure 2. Calculated Fourier series curve and measured curve.

The calculated Fourier series curve and the measured curve are illustrated in Fig. 2. The Fourier coefficients a_k are computed once, then stored for the determination of $B = f(H)$ or $U = f(I)$ for all possible values of H or I. The measured curve considered in this study corresponds to the line-to-line voltage in the primary, and to the current of the phase A of the primary winding.

Numerical approach

Two methods have been developed for this approach. For both methods, the leakage inductances $L_{\sigma 1}$, $L_{\sigma 2}$ of the primary and secondary windings, as well as the zero-sequence inductances L_{01}, L_{02} are considered as constants. The first one considers the total magnetic flux as state variables. The corresponding set of six differential equations is given by:

$$\frac{d[\psi]}{dt} = [B] \tag{7}$$

$$[B] = \begin{bmatrix} u_{\mathrm{ABC}} - R_{\mathrm{ABC}} \cdot i_{\mathrm{ABC}} \\ u_{\mathrm{abc}} - R_{\mathrm{abc}} \cdot i_{\mathrm{abc}} \end{bmatrix} \tag{8}$$

where R_{ABC} is the resistances of the primary windings and R_{abc} is the resistances of the secondary windings.

The total magnetic flux of different windings, including zero-sequence flux are given by:

$$[\psi] = [L] \cdot [i] \tag{9}$$

The self and mutual inductances are expressed in function of the magnetic reluctances R_{1T}, R_{2T}, R_{3T} of the equivalent magnetic circuit-diagrams, with N_1, N_2 the turns of the primary respectively secondary windings. For example, the magnetizing inductance of the primary winding A respectively the mutual inductance between the primary winding A and the secondary one b are given by:

$$L_{h1A} = \frac{N_1^2}{R_{1T} + \dfrac{R_{2T} \cdot R_{3T}}{R_{2T} + R_{3T}}} = \frac{N_1^2 \cdot (R_{2T} + R_{3T})}{R_{1T} \cdot R_{2T} + R_{1T} \cdot R_{3T} + R_{2T} \cdot R_{3T}} \tag{10}$$

$$L_{Ab} = \frac{-N_1 \cdot N_2 \cdot R_{3T}}{R_{1T} \cdot R_{2T} + R_{1T} \cdot R_{3T} + R_{2T} \cdot R_{3T}} \tag{11}$$

Similar expressions are determined for all the inductances. The determination of the magnetic flux at each integration step permits to evaluate and adapt, by the *B-H* curve, the different magnetic reluctances as well as different inductances.

Some numerical software package like SIMSEN [2] (http://simsen.epfl.ch) use essentially the currents as state variables. For this purpose, a second method has been developed. This one considers the following set of differential equations:

$$[A] \cdot \frac{d[X]}{dt} = [B] \tag{12}$$

with:

$$[B] = \begin{bmatrix} u_{ABC} - R_{ABC} \cdot i_{ABC} \\ u_{abc} - R_{abc} \cdot i_{abc} \\ u_{ABC} - R_{ABC} \cdot i_{ABC} \end{bmatrix} \tag{13}$$

$$[X]^T = [i_A \quad i_B \quad i_C \quad i_a \quad i_b \quad i_c \quad \psi_A \quad \psi_B \quad \psi_C]$$

In this case, one needs three supplementary state variables ψ_A, ψ_B, ψ_C and for the matrix [A] the expressions of all the inductances and especially all their derivatives vs. the currents or the total flux (see Appendix). These expressions may be determined analytically by using the Fourier series relations mentioned before and adapted at each integration step.

FEM computations

Based on the detailed knowledge of the geometry and the physical properties of different materials, 2D FEM field computations [3] are performed for symmetrical and unsymmetrical loads in magnetodynamics, on a small transformer of 3 kVA shown in Fig. 3.

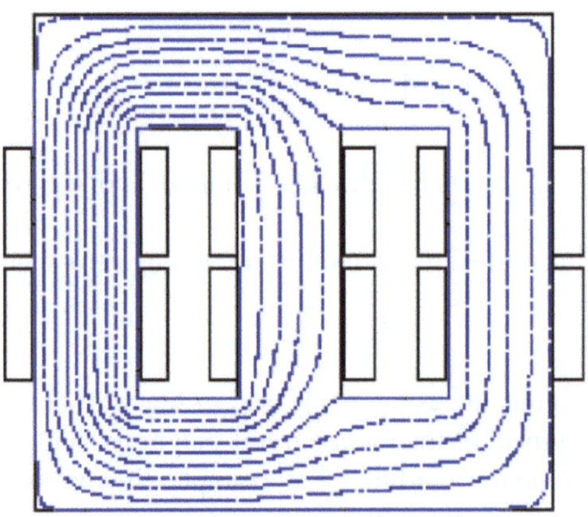

Figure 3. Transformer of 3 kVA: Distribution of the magnetic field in the case of no-load.

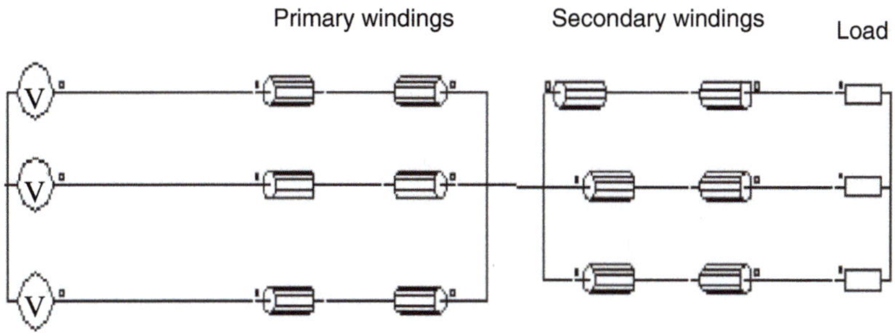

Figure 4. Electrical circuit related to FEM computations.

The electric circuit related to different cases is shown in Fig. 4. One can notice the primary, secondary windings, and the resistances of the load.

Measurements: Comparison of results

Case of no-load for a transformer Yy0 of 3 kVA, 380 V/232 V, 50 Hz, ucc = 3.26%

Fig. 5 shows the computed primary currents given by the two numerical methods and relative to a small transformer of 3 kVA in the case of no-load at rated voltage 380 V.

Figs. 6–8 show respectively the measured respectively computed primary currents i_A, i_B, and i_C relative to this case.

Table 1 shows results coming from different approaches in the case of no-load for a transformer Yy0, without connecting the neutrals in the primary and secondary sides.

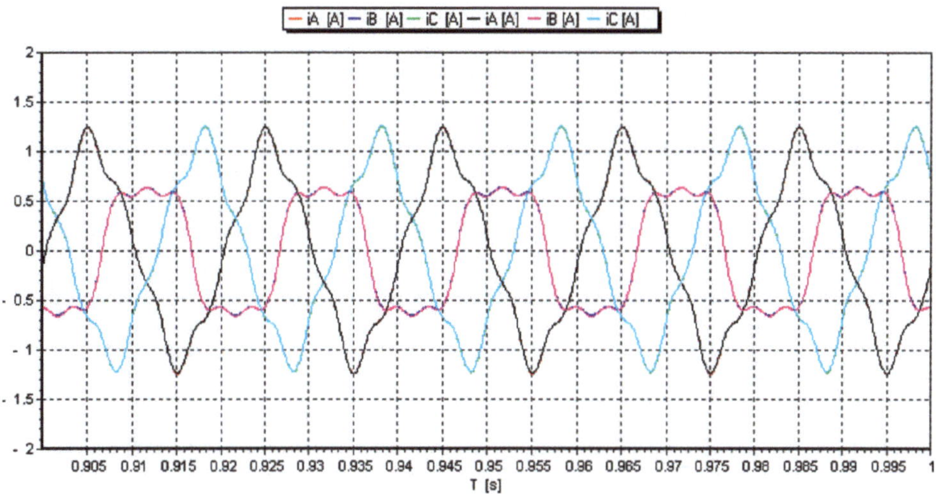

Figure 5. Computed primary currents given by the two numerical methods in the case of no-load.

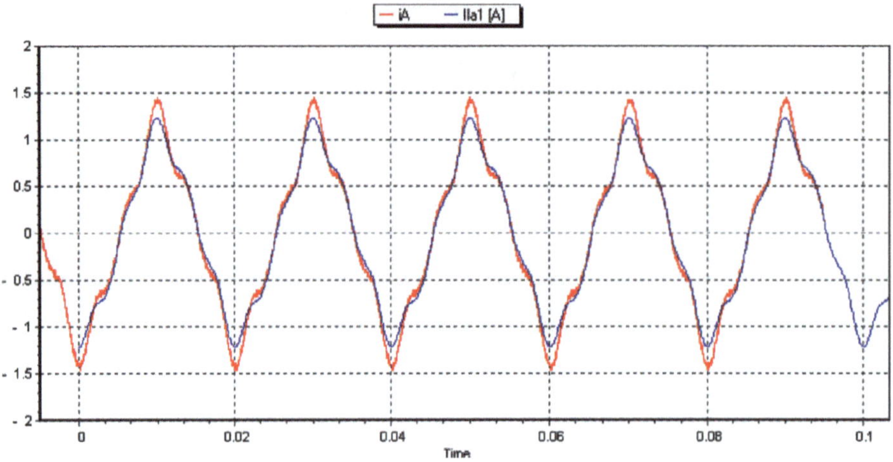

Figure 6. Computed and measured primary current i_A in the case of no-load.

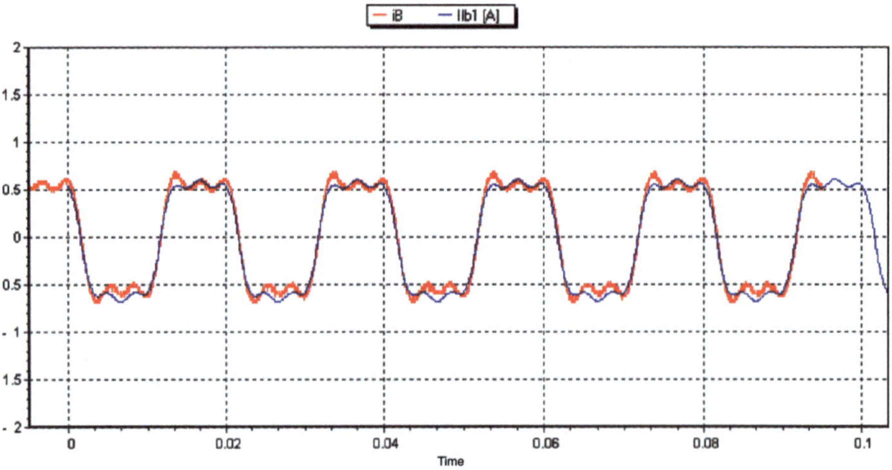

Figure 7. Computed and measured primary current i_B in the case of no-load.

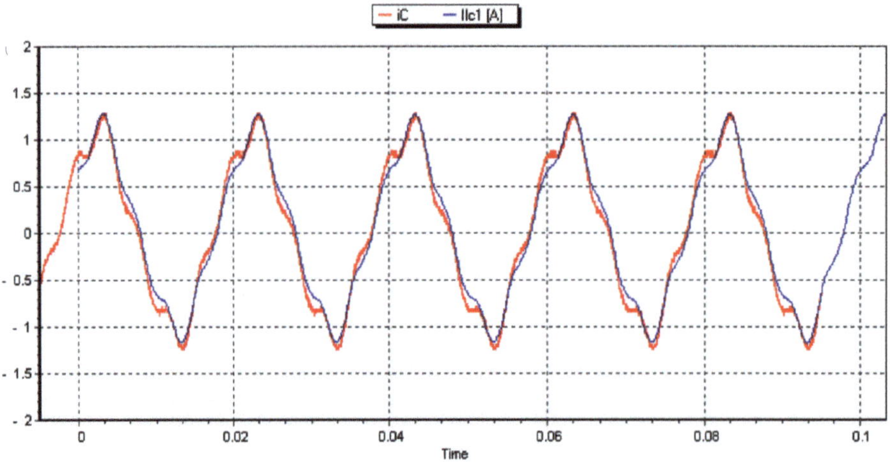

Figure 8. Computed and measured primary current i_C in the case of no-load.

Table 1. Comparison of results case of no-load for a transformer Yy0

	Test results [A]	Numerical approaches [A]	FEM approaches [A]
I_A	0.803	0.745	0.75
I_B	0.52	0.515	0.531
I_C	0.76	0.744	0.749

Case of no-load for a transformer Dy5 of 3 kVA, $U = 230$ V

Figs. 9–11 show the measured respectively computed line primary currents relative to a coupling Dy5 in the case of no-load, under a voltage of 1.045 p.u.

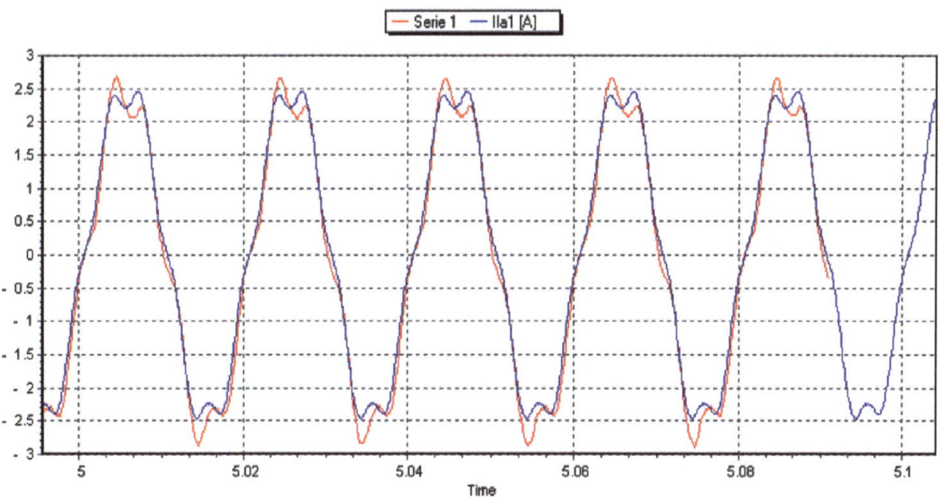

Figure 9. Computed and measured primary current i_{1A} in the case of no-load.

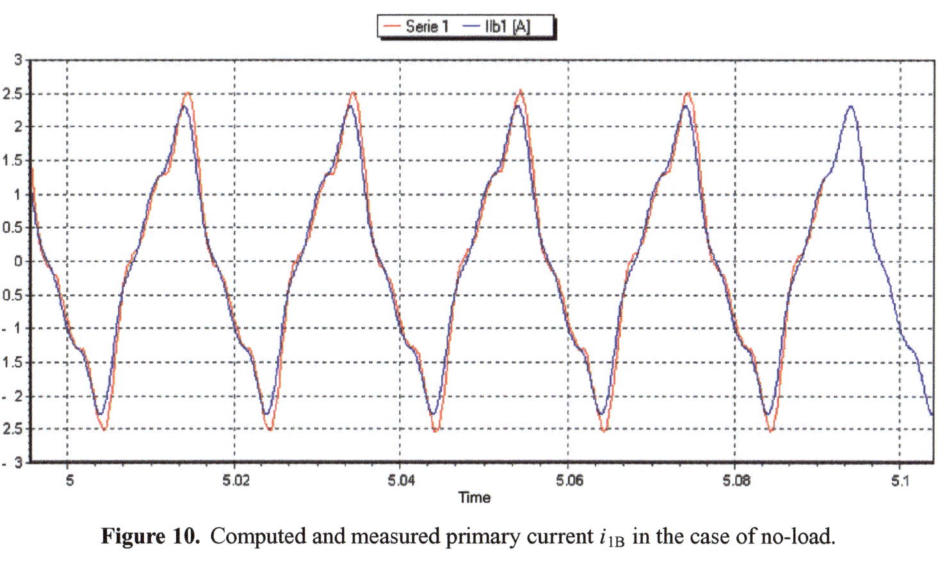

Figure 10. Computed and measured primary current i_{1B} in the case of no-load.

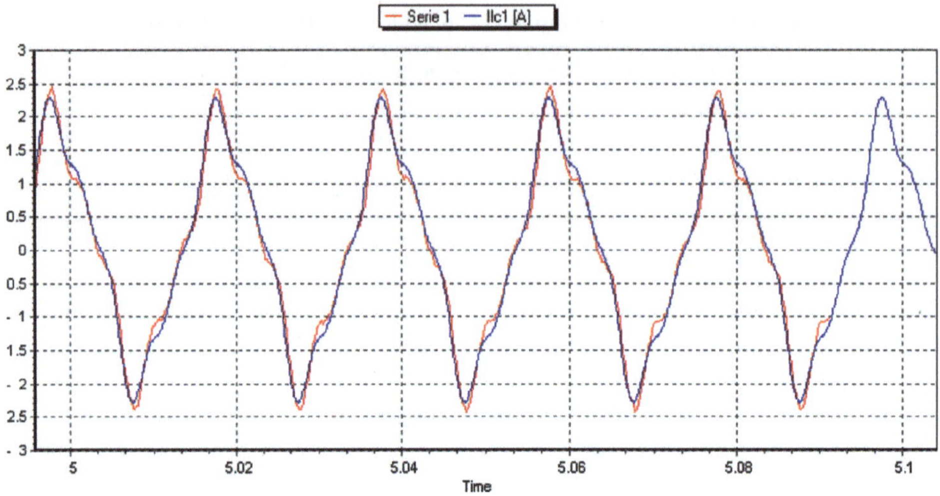

Figure 11. Computed and measured primary current i_{1C} in the case of no-load.

Table 2. Comparison of results case of symmetrical load for a transformer Yy0

	Test results [A]	Numerical approaches [A]	FEM approaches [A]
I_A	3.4	3.18	3.19
I_B	3.34	3.25	3.24
I_C	3.63	3.41	3.37
I_a	5.4	5.26	5.26
I_b	5.38	5.26	5.23
I_c	5.38	5.26	5.19

Case of a symmetrical load in the secondary, the primary supplied at its nominal voltage 380 V, Yy0

Table 2 shows results coming from different approaches in the case of a symmetrical load connected to the secondary of the transformer ($R_a = R_b = R_c = 25\,\Omega$) under nominal voltage $U_N = 380$ V, without connecting neutrals.

Case of an unsymmetrical load in the secondary

Table 3 shows results coming from different approaches in the case of an unsymmetrical load connected to the secondary of the transformer ($R_a = 40.5\,\Omega$; $R_b = 14.6\,\Omega$; $R_c = 39.15\,\Omega$; 63.95% of unsymmetry) under nominal voltage $U_N = 380V$, with the neutral connected only in the secondary side. A measured zero-sequence inductance is taken into account in the secondary side $L_{0s} = 1.9$ mH.

Table 3. Comparison of results case of unsymmetrical load for a transformer Yy0

	Test results [A]	Numerical approaches [A]	FEM approaches [A]
I_A	2.62	2.50	2.55
I_B	4.46	4.33	4.30
I_C	3.41	3.2	3.15
I_a	3.2	3.2	3.22
I_b	9.17	8.89	8.82
I_c	3.42	3.45	3.39

A very good agreement between results coming from different approaches and for different cases can be noticed (relative error less than 8% between different approaches).

The present approach will be applied to a large transformer of distribution (1,000 kVA, Dyn11, 18,300/420 V).

Conclusions

In the present paper, a new approach for the three-phase transformer analysis is described. This one based on equivalent magnetic circuit-diagrams takes into account the nonlinear *B-H* curve and zero-sequence flux. The *B-H* curve is represented by a Fourier series expression, which gives a smooth *B-H* curve, and permits the analytical determination of all the inductances and their derivatives vs. the currents. A very good agreement between results coming from different approaches is obtained.

References

[1] L. Guanghao, Xu Xiao-Bang, Improved modeling of the nonlinear B-H curve and its application in power cable analysis, IEEE Trans. Magn., Vol. 38, No. 4, pp. 1759–1763, 2002.

[2] J.-J. Simond, A. Sapin, B. Kawkabani, D. Schafer, M. Tu Xuan, B. Willy, "Optimized Design of Variable-Speed Drives and Electrical Networks", 7th European Conference on Power Electronics and Applications EPE'97, Trondheim, Norway, September 1997.

[3] FLUX2D, version 7.60/6b, CEDRAT.

Appendix: Numerical approach with the currents as state variables

For example, the voltage equation relative to the primary A is given by:

$$
\begin{aligned}
u_A &= R_A \cdot i_A + \frac{d\psi_A}{dt} + \frac{d\psi_{01}}{dt} \\
&= R_A \cdot i_A + (L_{\sigma A} + L_{h1A} + L_{01}) \cdot \frac{di_A}{dt} + (L_{AB} + L_{01}) \cdot \frac{d}{dt} i_B + (L_{AC} + L_{01}) \cdot \frac{d}{dt} i_C \\
&\quad + L_{Aa} \cdot \frac{d}{dt} i_a + L_{Ab} \cdot \frac{d}{dt} i_b + L_{Ac} \cdot \frac{d}{dt} i_c + \frac{d\psi_A}{dt} \{Val\,1\} + \frac{d\psi_B}{dt} \{Val\,2\} + \frac{d\psi_C}{dt} \{Val\,3\}
\end{aligned}
$$

with:

$$Val\,1 = \frac{dR_{1T}}{d\psi_{\rm A}}\left(i_{\rm A}\cdot\frac{\partial L_{h1\rm A}}{\partial R_{1T}} + i_{\rm a}\cdot\frac{\partial L_{\rm Aa}}{\partial R_{1T}} + i_{\rm B}\cdot\frac{\partial L_{\rm AB}}{\partial R_{1T}} + i_{\rm b}\cdot\frac{\partial L_{\rm Ab}}{\partial R_{1T}} + i_{\rm C}\cdot\frac{\partial L_{\rm AC}}{\partial R_{1T}} + i_{\rm c}\cdot\frac{\partial L_{\rm Ac}}{\partial R_{1T}}\right)$$

$$Val\,2 = \frac{dR_{2T}}{d\psi_{\rm B}}\left(i_{\rm A}\cdot\frac{\partial L_{h1\rm A}}{\partial R_{2T}} + i_{\rm a}\cdot\frac{\partial L_{\rm Aa}}{\partial R_{2T}} + i_{\rm B}\cdot\frac{\partial L_{\rm AB}}{\partial R_{2T}} + i_{\rm b}\cdot\frac{\partial L_{\rm Ab}}{\partial R_{2T}} + i_{\rm C}\cdot\frac{\partial L_{\rm AC}}{\partial R_{2T}} + i_{\rm c}\cdot\frac{\partial L_{\rm Ac}}{\partial R_{2T}}\right)$$

$$Val\,3 = \frac{dR_{3T}}{d\psi_{\rm C}}\left(i_{\rm A}\cdot\frac{\partial L_{h1\rm A}}{\partial R_{3T}} + i_{\rm a}\cdot\frac{\partial L_{\rm Aa}}{\partial R_{3T}} + i_{\rm B}\cdot\frac{\partial L_{\rm AB}}{\partial R_{3T}} + i_{\rm b}\cdot\frac{\partial L_{\rm Ab}}{\partial R_{3T}} + i_{\rm C}\cdot\frac{\partial L_{\rm AC}}{\partial R_{3T}} + i_{\rm c}\cdot\frac{\partial L_{\rm Ac}}{\partial R_{3T}}\right)$$

and

$$\frac{\partial L_{h1\rm A}}{\partial R_{1T}} = \frac{-N_1^2\cdot(R_{2T}+R_{3T})^2}{(R_{1T}\cdot R_{2T}+R_{1T}\cdot R_{3T}+R_{2T}\cdot R_{3T})^2}$$

Similar expressions are established for all the partial derivatives of inductances vs. the magnetic reluctances R_{1T}, R_{2T}, R_{3T} of the cores.

MIX
Papier aus verantwortungsvollen Quellen
Paper from responsible sources
FSC® C105338

If you have any concerns about our products,
you can contact us on
ProductSafety@springernature.com

In case Publisher is established outside the EU,
the EU authorized representative is:
Springer Nature Customer Service Center GmbH
Europaplatz 3, 69115 Heidelberg, Germany

Printed by Libri Plureos GmbH
in Hamburg, Germany